The Sound System Design Primer

The Sound System Design Primer is an introduction to the many topics, technologies, and sub-disciplines that make up contemporary sound system design.

Written in clear, conversational language for those who do not have an engineering background, or who think more in language than in numbers, *The Sound System Design Primer* provides a solid foundation in this expanding discipline for students, early/mid-career system designers, creative and content designers seeking a better grasp on the technical side of things, and non-sound professionals who want or need to be able to speak intelligently with sound system designers.

Josh Loar is a sound, lighting, video, and control system designer who has designed projects all over the world, from theaters, to concerts and festivals, to corporate events and conventions, to theme parks. He has worked as a professional designer for more than 20 years. He has designed and engineered systems for clients including Walt Disney Imagineering, Lotte Group, Busch Gardens, and more. He is co-chair and co-author of the upcoming ESET (Essential Skills for Entertainment Technicians) test in audio. He is chair of the USITT (United States Institute for Theatre Technology) Sound Graphics Working Group, working on issuing new and more comprehensive documentation standards for theatrical sound paperwork. He is an accomplished FOH (Front-of-House) and recording engineer, having worked with artists including Paul Simon, David Byrne, Mos Def, Rickie Lee Jones, Charlie Haden, Philip Glass, and many, many more (he has genuinely lost count). He is an educator who has been teaching at the university level for nearly a decade at Yale School of Drama and now at Michigan Technological University. He is also a multimedia artist, musician, producer, director, and creator of lots of different kinds of weird art. He enjoys long walks in the woods, good whiskey, and the company of intelligent people. Check out his work at www.joshloar.com.

The Sound System Design Primer

Josh Loar

Routledge
Taylor & Francis Group

NEW YORK AND LONDON

First published 2019
by Routledge
52 Vanderbilt Avenue, New York, NY 10017

and by Routledge
2 Park Square, Milton Park, Abingdon, Oxon, OX14 4RN

Routledge is an imprint of the Taylor & Francis Group, an informa business

Library of Congress Cataloging-in-Publication Data
Names: Loar, Josh, author.
Title: The sound system design primer / Josh Loar.
Description: New York, NY : Routledge, 2019. | Includes bibliographical references and index.
Identifiers: LCCN 2018048728 (print) | LCCN 2018049480 (ebook) | ISBN 9781351768184 (pdf) |
    ISBN 9781351768177 (epub) | ISBN 9781351768160 (mobi) | ISBN 9781138716872 (hbk : alk. paper) |
    ISBN 9781138716889 (pbk : alk. paper) | ISBN 9781315196817 (ebk)
Subjects: LCSH: High-fidelity sound systems—Design and construction. | Stereophonic sound systems—
    Design and construction.
Classification: LCC TK7881.7 (ebook) | LCC TK7881.7 .L63 2019 (print) | DDC 621.389/332—dc23
LC record available at https://lccn.loc.gov/2018048728

ISBN: 978-1-138-71687-2 (hbk)
ISBN: 978-1-138-71688-9 (pbk)
ISBN: 978-1-315-19681-7 (ebk)

Typeset in Giovanni
by Apex CoVantage, LLC

Visit the eResources: www.routledge.com/9781138716889

Dedicated to
my teachers, my students, my colleagues in sound,
and, most importantly,
to my wife, Anna, and my daughter, Ada.

CONTENTS

Part I • What Is a Sound System?

Part II • How Do You Set Up, Test, and Calibrate a Basic Sound System?

CONTENTS

A note on images in this book: all images contained in this book were created by the author unless otherwise indicated (public domain, no-attribution-required clip art is used in some images). Technical drawings are generally shown with a proper title block and other labeling information, as any professional technical drawing would sport these features. Other drawings and images do not feature title blocks and legends, but may feature extensive notes. Title blocks may specify scale, but this is scale as drawn, not scale as published (as images may change size in publication).

xi

WHY THIS BOOK? WHO IS THIS BOOK FOR?

Hello and welcome to *The Sound System Design Primer*!

Thank you for picking up this book. There are many sound system design and engineering books available, so why should you choose this one?

As a professional sound system designer and educator, I have spent a great deal of time reading textbooks on sound system design. There are some wonderful books available—if you need an in-depth understanding of the complex acoustic interactions of elaborate speaker systems, you can't do better than Bob McCarthy's *Sound Systems: Design and Optimization*—however, the literature on our discipline consists primarily of advanced, narrowly focused texts, aimed either at readers with a hard engineering and mathematics focus (e.g., Davis and Patronis' *Sound System Engineering*), or at readers who already have a good deal of sound system experience and listening to draw upon as a basis for understanding (e.g., McCarthy). We are sadly lacking a text that lays out, in easy-to-comprehend descriptive language, the fundamentals of sound system design: from functional signal flow and gear selection to the processes and terminology native to different types of system projects.

The Sound System Design Primer is that text. The goal of this book is to lay out in a methodical and logical format the principles of contemporary sound system design. Starting with the basics of signal flow and audio physics, and traveling through gear selection considerations, design process methodology, and delivery formats for final documentation, this book is the first-ever step-by-step guide for those looking to better understand the big picture of contemporary sound system design.

Perhaps you are a sound design student, undergrad or grad, and you need a good foundation in sound systems in order to succeed in your work—*this book is for you*.

Perhaps you are a creative sound designer seeking a better grasp on the technical systems that support your content—*this book is for you*.

Perhaps you are a technical director, a production/project manager, or any other entertainment or architectural professional who needs a better understanding of sound systems in order to communicate with the designers on your team (but who does not need the depth of equations or narrow acoustics/engineering focus of many of the most advanced texts)—*this book is for you*.

Perhaps you are a professional designer working on a different type of systems project than you've experienced, and you are looking for guidance on terminology and process—*this book is for you*.

Perhaps you are **anyone** who is interested in learning the discipline of sound system design and who is looking for a book written in a conversational tone that is as light as possible on scary math (without skipping the critical stuff)—*this book is for you*.

This book is not intended as a replacement for the Davis, Patronis, and McCarthys of the world, but as a bridge to them. Alongside my career as a professional system designer, I have taught sound designers at both the undergraduate and graduate levels, and I have very seldom (at either level) encountered a student who is ready for these advanced texts without having learned a lot of fundamental information first. This book provides that fundamental information. It is my sincere hope that you will find it informative, interesting, and at least a little entertaining.

A properly designed sound system is a magical thing—it can provide critical life-safety information to passengers at an airport, shake us with the backbeat of an amazing song, or surround us with diffuse sound effects until we have been transported to another world. I have often said to my clients and students that my mission in life is to make the world sound better, one room at a time. By writing this book, I encourage you to join me on that mission.

Sincerely,
Josh Loar

I am grateful to many who helped me in my life as a designer, educator, and (now) author.

First, I must thank Krista Elhai, my high school theater teacher, who gave me my first sound design job (for *West Side Story!*), who encouraged me to follow my talent for theater design to UCLA, and who set me off on a path that became my life's work to date. Thank you! I was so shy and confused when I met you, and you helped me immensely in a number of different ways, not the least of which was helping me come out of my shell.

I would like to thank Rob Miller and Jonathan Burke, for whom I worked as a sound engineer when I was a student at UCLA, and who taught me a lot about the nitty-gritty of systems—fixing problems in a flash, keeping cool under high pressure, and enjoying yourself while doing the work.

I would like to thank my teachers at the SAE Institute of Technology New York City, who helped turn my good sound instincts into good sound theory and practice. Particularly, I need to mention A.J. Tissian, John Jansen, Ryan Schimmenti, Mike White, and Rich Tozzoli. Thanks for opening this world to me.

I would like to thank those who have hired me to design things, particularly John Storyk, Joshua Morris, and Bob Chambers. I would like to thank those who have hired me to engineer things, particularly Kevin Curry, David Budries, and Ben Sammler.

I would like to thank my colleagues who graciously agreed to let me grill them about their system design practices for this book: Jamie Anderson, Elizabeth Atkinson, David Budries, Jonathan Deans, Robert Kaplowitz, David Kotch, Bob McCarthy, Vincent Olivieri, and Christopher Plummer. To those who I missed this time—you know who you are. I'll catch you for the second edition!

My thanks to the companies and people who graciously let me use their images or words in this book, and those who helped me along the way: Ahnert-Feistel Media Group, and particularly Pedro Lima; American Heritage Dictionaries, Houghton Mifflin Harcourt, and particularly Leah Petrakis; Anna Ehl; Apple, Inc.; Audient Ltd., and particularly Andy Allen; Audio-Technica, and particularly Jeff Simcox; Auralex, and particularly Josef Milton; Aviom, Inc.; Beacon Capital Partners, ESI Design, Caleb Tkach, and particularly Bruce Odland; Cirrus Logic, and particularly Bill Schnell; Clear-Com, and particularly Brad Kellogg; Countryman Microphones, and particularly Chris Countryman; d&b Audiotechnik, and particularly Roger Keim; Denon Pro, inMusic Brands, and particularly Emily Chrul (even though I didn't end up using the image!); DPA Microphones, and particularly Evan MacKenzie; M.C. Friedrich; Focal Speakers, Audio Plus Services, and particularly Michel Bernatchez; Harman Professional Solutions, and particularly Roslyn Dougherty; Honeywell International, Inc., and particularly Alisha Neubert; L-Acoustics, and particularly Mary Beth Henson and Julie Blore-Bizot; Hosa Technology, and particularly Jose Gonzalez; Lectrosonics, Inc., and particularly Bruce Jones; Line 6, and particularly Barry Cleveland; Los Angeles Convention Center, and particularly Michelle Riehle-Ludtke; Mackie, EAW, Loud Audio, and particularly Shaunna Krebs and Louie King; Meyer Sound, and particularly Bob McCarthy and Caitlin Clausen; Neumann; QSC, LLC, and particularly Megan Strom; Neutrik, and particularly Sarah Tallman and Janet Tufo (even though the images didn't make it into the book!); Radial Engineering, and particularly April Lebedoff; Rational Acoustics, and particularly Jamie Anderson; Sennheiser; Roland/Boss, and particularly Rebecca Eaddy; Shure, Inc., and particularly Davida Rochman and Lou Mora; Sound Lounge NYC, and particularly Liana Rosenberg and Taylor Maggard; SPL Lab, and particularly Kirill Berezin; Joanna Lynne Staub, Ian Dickinson, Maggie Burke, and Autograph Sound; USITT, and particularly David Grindle and Todd Proffitt; Vienna Symphonic Library; Waves Audio Ltd., and particularly Udi Henis; Paul Williams and his manager Nancy Munoz; WSDG, and particularly John Storyk, Josh Morris, and Heinz Zeggl.

Thanks to my partners in crime on the ESET and Sound Graphics Working Group Committees: Michael Backhaus, Brad Berridge, Nicholas Drashner, Sam Kusnetz, Joanna Lynne Staub, and Brad Ward.

xvii

ACKNOWLEDGMENTS

Thanks to my colleagues at Michigan Technological University who supported my desire to write this book. In particular, I must thank Jared Anderson (my department chair and ardent supporter), Christopher Plummer (my faculty mentor and USITT encourager), M.C. Friedrich (who gave me drawings, support, and is adorable with my daughter!), and Kent Cyr (who kept bringing dumb jokes to my office to rightly distract me from impending deadlines).

Thanks to the wonderful editorial and graphic/layout staff at Routledge, Taylor & Francis, Focal Press, and Apex CoVantage, especially Lara Zoble (who signed the book), Hannah Rowe, Claire Margerison, Marie Louise Roberts, and Shannon Neill. Without your hard work and dedication, this book would not have seen the light of day.

Thanks to my wonderful wife, Anna, who transcribed interviews and drew things, who is forever my sounding board to check phrasing or a graphic for clarity, and who put up with me and took care of our lovely daughter while I furiously typed, and typed, and typed some more. I wouldn't have accomplished anything I've done in my life without you, and I am so grateful. I love you so!

And, finally, thanks to all my friends, colleagues, and family members for putting up with me while I wrote this book (and to anyone I've forgotten, to whom I apologize). I won't be (quite) so busy now that it's done! I swear!

—Josh Loar
2019

WHAT THIS BOOK DOES AND HOW IT WORKS

The discipline known as sound system design is a bit hard to define. It incorporates a wide range of skills and knowledge, from the principles of sound transmission, physics, and acoustics, to electrical theory, digital theory, signal processing theory, and an active knowledge and understanding of a host of various disciplines (for example, if you are designing systems for theater, there are things you must know that don't apply to designing for concerts or for theme parks—though there are certainly commonalities). Most books that approach the discipline do so from one or two angles and dive deeply into those topics. This is very helpful if you need that deep dive (which you may at some point), but if you are trying to get a good overall sense of sound systems, how they function, and how we go about designing them, then there isn't a text that accomplishes that.

Or there wasn't, until this book.

The Sound System Design Primer is a guide to the world of contemporary sound systems. It is focused on live and installation sound systems, with some time also spent on recording and production studio facilities. It will attempt to cover in some fashion just about every type of professional sound system, though it admittedly spends little time on cinema systems and no time at all on automobile sound systems.

The Primer takes as its mission two primary goals. The reader of this entire text should be able to walk away with the following:

1. A good, solid understanding of how sound systems work. What components are needed in order to create different types of sound systems, why might we choose one component over another, and how do we hook all these pieces together?
2. A good foundation in how to approach a system design project. What is the design process, how do we consider budget and equipment suitability, and the specific needs native to different disciplines and project types?

The Primer is divided into four parts. Each part asks a fundamental question that a sound system designer must be able to answer and then breaks the answers down into sections that gather related information into logical subdivisions. From there, we break sections down further into chapters that allow the information to be presented in concise, digestible portions.

In Part I, What Is a Sound System?, we begin by examining the basic components of a sound system (with a slight detour through the physics of sound, the electrical conduction of analogous current, and the conversion of that electrical signal into 1s and 0s). How does analog signal flow work? What are the basic stages of sound from acoustic signal to amplified signal? How does digital technology both help us and complicate things? What types of gear are common in sound systems? What do we need to consider when determining what gear to select? We travel piece by piece through the gear used in a typical sound system, starting with microphones, playback systems, and other inputs, and through speakers, headphones, and other outputs. We examine com systems (an oft-overlooked item within the system designer's purview). We talk about power and amplification, mixing and processing, and anything else that is used along the signal flow path. By the end of Part I, the reader should have a good understanding of where sound comes from, where we send it, and what we do to it along the way.

In Part II, How Do You Set Up, Test, and Calibrate a Basic Sound System?, we examine the critical process of turning a sound system from lines on a page to a functional, beautiful-sounding, easy-to-use reality. While the work of installation is not frequently done by professional designers (rather, it is done by their assistants, their systems integrators, etc.), many early career designers will find themselves involved in the hands-on work. Regardless of who is performing the installation, it behooves any designer to have a firm grasp on these processes. A clear understanding of the setup, testing, and tuning of sound systems will influence your system design choices for the better, as you will not make choices that unnecessarily complicate life for the installation team. If there's one thing we all strive to do as designers in today's complex system world, it is to avoid complicating things for little or no apparent benefit (something your clients will appreciate as well). By the end of Part II, the reader should be prepared to understand the installation process and how it impacts the overall design process.

In Part III, What Type of Sound System Do You Need?, we begin by establishing a basic design process that applies to any sound system design project. Regardless of discipline, there are questions we need to answer before we can design the sound system (what needs to be amplified, who is the audience, and what is the venue?). We discuss in-depth the thought processes that start any system design project and establish an orderly method of building up your system's specifications. From there, we move into an examination, chapter by chapter, of many specific disciplines—from theater, concerts, and conventions to theme parks, retail spaces, and production studios—and discuss what questions a designer needs to ask (and answer) when working in each field. Discipline by discipline, we detail terminology and processes specific to sound system design for each of these fields and provide a guide to the project hierarchies we commonly find on these jobs. By the end of Part III, the reader should be ready to take on a new design project in a new disciplinary area—knowing what questions they typically need to ask and answer based on the particulars of each field—without being left in the dark when the acronyms start to fly!

In Part IV, How Do You Document and Present Your Design?, we examine document formats, drafting standards, and other such details involved in preparing a professional design documentation package. Technical documentation is often not the most thrilling part of the system designer's work (though I know a few designers who love to draft and would beg to differ!), but clear technical documentation is absolutely essential for communicating your ideas properly to clients, integrators, and the like. Without strong documentation, you cannot work professionally as a system designer, so in this part, we cover all the standard document forms in today's production world and what formats are expected for their delivery. We also discuss some tactics for working with production staff, handling difficult situations, and leaving your clients satisfied with your work. By the end of Part IV, the reader should be ready to present their ideas clearly and concisely to any project manager, director, producer, or whoever is signing your paycheck!

Seeded throughout the book are two recurring segments of stories and advice from designers who work everywhere from Broadway and Cirque du Soleil to the largest concert tours and theme parks. "Tales From the Field" is a segment featuring stories from the front lines of the sound system world, lessons learned, and the like. "Message in a Bottle" segments include these professionals' approach to design, advice to young designers, and other thoughts on the art and science of sound systems. The contributors to this section are as follows:

Jamie Anderson—Owner, Rational Acoustics; SysTweak
Elizabeth Atkinson—Independent Sound Composer/Designer
David Budries—Chair, Sound Design, Yale School of Drama; Independent Sound Designer
Jonathan Deans—Independent Sound Designer: Broadway, West End, Cirque du Soleil, etc.
Robert Kaplowitz—Independent Sound Designer (Tony winner for *Fela*)
David Kotch—President, Criterion Acoustics; Sound System Designer
Bob McCarthy—Director of System Optimization, Meyer Sound; Sound System Designer
Vincent Olivieri—Professor, Associate Dean for Graduate Affairs, UC Irvine Claire Trevor School of the Arts; Independent Sound Designer
Christopher Plummer—Professor, Program Director, Sound Design/Audio Production, Michigan Technological University; Independent Sound Designer
. . . and myself

The Primer is structured so that you can read it in either linear or non-linear fashion: it is primarily ordered as an instructional guide that builds concepts from one chapter to the next; it also functions as a reference manual that can be understood reading chapters à la carte (though in reading this way, you will miss some terminology and topics that have been introduced in earlier chapters).

The Primer does not skimp on the mathematics required for our discipline, but it is designed to be as light on equations as it can be while not skipping core concepts. Each time a math equation is introduced, we present samples of how to use that equation in real-world situations, so that the theoretical is connected clearly to the practical. While many sound system designers approach the process from an engineering or math background, many do not, and this guide is the first to approach sound system design from a linguistic position—in addition to the included illustrations and equations—to allow the greatest number of readers, and types of thinkers, to understand the concepts at play.

The Sound System Design Primer is informed by my many years of working as a professional sound system designer and as a university educator in the field of sound and sound system design. Every effort has been made to reflect current industry standards and terms as of this publication, but inevitably, the industry moves fast, and by the time this book is published, one or more bits of information may be out of date. While no book can substitute for listening to gear, working on systems, and the trial and error of real-world design projects, this book aims to be the next best thing.

Happy reading!

No book is a substitute for real-world training and experience. Reading this book does not mean you are a certified rigger, even if, after you've read it, you have a good idea of how to rig your loudspeakers. Safety should always come first in any production process. When working in audio, you will often be working with heavy equipment, loud noises, overhead rigging, heights in general, and electricity, all of which can pose risks. Use caution, and if you are unclear about whether or not your plan of action is safe, check with trained supervisory personnel (your theater's technical director, for example). If you have no such personnel with whom to check, stick to the safest possible course of action in executing your design. No sound system is worth injury or death.

The author and publishers of this book assume no liability or responsibility for any loss or damage suffered by any person as a result of the use or misuse of any information in this book.

PART 1

What Is a Sound System?

What is a sound system? How does sound itself work? What are the components we need in order to get signal to our audience? Part I examines the fundamentals of sound and all the standard components of a contemporary sound system.

CHAPTER 1

The Four Standard Elements of the Analog Sound System

WHAT IS A SOUND SYSTEM?

This seems like a straightforward question—a sound system is that thing your neighbor turns up way too loud when you're trying to sleep, right?

As is the case with most seemingly straightforward questions, the answer proves to be more complex than we might think at first glance. Do you think of your cell phone as a sound system? How about a piece of cardboard rolled up in a cone shape, used as a megaphone? How about a greeting card that allows you to record your voice saying "happy birthday!" to your Grandma? All of these are, in some fashion, sound systems.

In the contemporary world, sound systems surround most of us all day, every day, and yet until we are standing in front of a giant subwoofer stack at a concert, most of us never give sound systems a second thought (even then we may not give them a second thought unless the mix engineer allows the system to feedback). In Part I of this book, we will be examining sound systems, breaking them down to their component pieces, and discussing when, why, and to what ends we use them.

This book is focused on professional sound systems for live entertainment and installation sound systems for themed entertainment, retail, and corporate environments, with a little focus on production spaces such as studios. While we may not spend dedicated time on that system inside your Nana's greeting card, the principles we define in this chapter apply not only to that greeting card but also in one way or another to any sound system anywhere.

To begin, we must first define the term sound system.

Sound System

A collection of components, assembled for the purpose of taking an audio signal (acoustic, electrical, or digital, live, or pre-recorded), processing it, amplifying it, and reproducing it.[1]

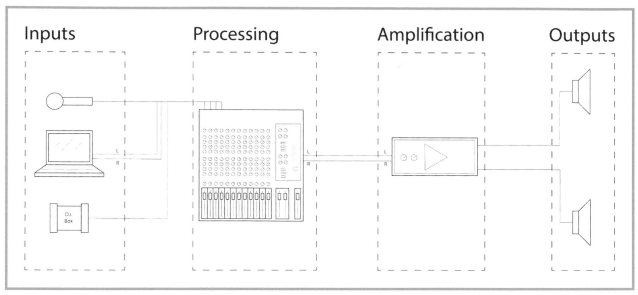

FIGURE 1.1
Signal Flow of the Four Standard Elements

Every sound system known to humans (even the rolled-up paper cone megaphone) consists, at root, of four standard elements:

1. Inputs
2. Processing
3. Amplification
4. Outputs

In an analog sound system, the above elements (which we will discuss in detail in a moment) may be the only elements of the system. In the digital world, there are often added elements that allow us a range of functionality and flexibility we don't possess in the analog world—those will be addressed beginning in Chapter 2. Figure 1.1 shows the signal flow of the standard four analog elements.

In any given sound system, audio content enters the system via the input stage. That content is processed (changed) in some fashion, amplified (made louder), and then made audible again in its final form at the output stage. To consider the example of perhaps the most rudimentary sound system—the cone-shaped cardboard megaphone—the four elements break down like this:

The INPUT stage is your voice, sending sound into the system via an acoustic path.
The PROCESSING stage is the narrow throat of the cone, which imparts specific acoustic changes to the source acoustic signal.
The AMPLIFICATION stage is the throw from the narrow throat of the cone to the wide mouth of the cone. The cone works as an acoustic resonator (among other things), which helps add constructively to the loudness of your voice.
The OUTPUT stage is the wide mouth of the cone, which spreads the sound out into the space, acting as a *waveguide*.

Waveguide

A structure that shapes the path of oscillating energy—most commonly used in sound system design as a term to describe acoustic energy shapers, such as loudspeaker horns,[2] but also used to describe any shaped path through which signal can be forced to move—*e.g., optical fiber is a WAVEGUIDE for optical data transmission.*

While that describes (albeit in a simplified fashion) how a cone megaphone serves as a rudimentary sound system, most sound systems are at least a *bit* more complex than this.

We will begin by examining analog sound systems, because analog systems in their elemental form represent the most straightforward possible signal path.

Analog

When used to describe a sound system, this is a term referring to a signal path in which acoustic energy (voice, music, sound effects, anything else worth hearing) is transduced into ANALOGOUS electrical energy, that electrical energy (now an electronic signal) is manipulated and amplified, and then the amplified signal is transduced back into acoustic energy and played via loudspeakers, with no computer/digital signal path in between.

Transducer

A device that changes one form of energy to another. Transduction is a constant function of sound systems: a MICROPHONE transduces acoustic energy to electrical energy; a LOUDSPEAKER does the same in reverse.[3]

Digital

When used to describe a sound system (or part thereof), this is a term referring to a signal path in which electrical energy (the electronic audio signal) is sensed and then converted into a data stream in order that it may be stored in a computer's memory and/or processed by software.[4]

A note on the terms "Electrical Energy" and "Electronic Signal":

Electrical Energy is an imprecise common-language term that will be used in this text to mean "current flowing along conductors." "Electronic Signal" is another imprecise common term, that will be used in this text to mean "audio flowing through electronic equipment, e.g., microphones, mixing consoles, etc."[5]

Let us imagine a very simple entertainment event, with a very simple sound system.

We are in a public park. There is a small stage, and on that stage is a solo performer—a singer with an acoustic guitar. The park is so small that there is no need for onstage monitoring (speakers aimed back at the artist so that they can hear themselves or their bandmates).[6] Before the musician's set, quiet music is playing from a cassette tape deck that is jacked into the system. The musician's set will be amplified. This concert is a typical small event, like those held in parks all across the world during the warm months. The simple needs of such an event predicate a simple sound system—often a basic analog sound system.

Our stage features two microphones, both wired. One is for announcements and for the singer's voice. The other is on

Mixing Console

An electronic device that combines a variety of input signals and sends them to the amplification stage in that combined form. Mixing consoles are also used to adjust individual source loudness, to bring microphone-level signals to line level,[7] and to implement other processing options. In the simplest systems, multiple audio inputs are combined to a single output pathway. In the most complex systems, the mixing console may route a signal to a wide variety of output locations, via different types of output paths.

Tales From the Field

In some audio production circles, it is fashionable to adore analog gear and claim its superiority to digital. While I love the sound of warm tape compression in the studio as much as the next engineer, in the live realm, digital gear has become standard for a reason. In addition to working as a designer, I have worked as a live sound engineer on thousands of events, and since most of those took place in the 2000s, a majority of the systems I have operated have been digital—but that certainly wasn't always the case. Back around the turn of the 21st century, I was working a benefit show at the Geffen Playhouse in Los Angeles, as Front-of-House (FOH) engineer. One of the acts was a venerable comedy duo who shall remain nameless. I was told they had sound effects for their act, and so I sent a stage assistant down to retrieve what I was expecting to be a CD of their cues. The stage assistant returned with the *very* aged cassette tape on which their cues were recorded. This sent me into a panic, as I wasn't even sure the Geffen still possessed a working tape deck. After much searching, I located one, hooked it into the system, inserted their cassette, hit play, and immediately saw the tape spool out onto the floor and heard it snap in more than one spot. I sent the assistant back down to the dressing room to ask after another copy of the cues. As it happened, not only did the brothers not have another copy of the cues with them, they didn't have another copy . . . *period*. They had been using this one cassette tape, that someone had prepared for them 20 years prior, at every single show for those last two decades—and this was the day it decided to break. We didn't have any proper splicing tape at the theater, so for the next 10 min, I feverishly reassembled the magnetic tape with Scotch tape, wound it back into the cassette with a pencil, and crossed my fingers that these cues (which were critical to the act) would be audible and would not contain massive glitches in the middle. To my great delight, the repairs held up, both during the rehearsal and during the show. Whew! After the show, I asked the brothers if they would like for me to burn those sounds onto a CD for them before they left, but they had another function to attend and couldn't wait around, so they left with their hodgepodge cassette in hand and, presumably, continued using it at subsequent shows. While I was glad in that moment that I knew how to splice tape, always remember, when designing a system, reliability is key. *Digital is our friend.*—J.L.

a lower stand, aiming at the guitar. The microphones are connected to our mixing console via XLR cables, and the tape deck that plays our entry music is connected via RCA cables.[8] These cables carry electronic signals over copper wires into the mixing console.[9]

Our mixing console may or may not have effects built into it, or there may be some external effects devices connected to it.[10] From the console, the signals, now combined, are sent to an amplifier, and then to loudspeakers. (The amplifier may be built into the loudspeakers or may be external to them.)[11] We hear the sounds, from the tape deck, and from the performer, louder than we would hear them without the system. Our goal is achieved!

In this example,

the INPUT stage consists of our microphones and our tape deck playing music;
the PROCESSING stage consists of our mixing console and any effects;
the AMPLIFICATION stage consists of our amplifiers (wacky, right?); and
the OUTPUT stage consists of our speakers.

In this small system, the sound flows from input to output stages in a very orderly fashion. We are not routing audio over a computer network[12] to three destinations at once. We do not have a digital signal processor (DSP)[13] to calibrate the output system. We are not recording this event. It is a basic live event sound system, analog all the way. As we shall see, the four standard elements of this system are found in one form or another in every single sound system in the entire world.

THE INPUT STAGE

The input stage of a sound system begins, always, with whatever sound we are trying to amplify. While we won't claim that the singer herself is a part of the sound system, without her, we probably wouldn't have assembled all this gear in the first place—so whenever we are planning a sound system, the very first thing to consider is: what sounds do we need to amplify? While we discuss the input stage in greater detail in Section 2 of this book; for now, it will suffice to identify the fundamental considerations.

> **DI Box**
>
> A DI (an acronym standing for direct injection, and often "backronymed" to stand for "direct input") is an electrical impedance transformer.[14] A DI is designed to take "instrument" level signal and convert it to "mic" level signal, which allows us to send a signal directly from an electric guitar or bass, a keyboard instrument, or something similar, into our mixing console's mic inputs, with no actual microphone involved.[15]

1. Live acoustic sounds—if your event, production, or installation will need to amplify some form of live acoustic sounds, we know that the first type of equipment we need will be microphones. Whether the event consists of a concert, a lecture, a corporate board meeting, or anything else, if we need to take acoustic energy and amplify it, then we must transduce it into an electronic signal (and potentially digitize it) that can then be manipulated in real time and amplified.[16]

   Note: We can directly amplify live sound without transduction, of course, by use of band shells, reflecting walls, megaphones, and the like, but those solutions fall more into the realm of acoustic design than sound system design, so we are going to presume that with the exception of certain unique situations,[17] we are not making acoustic adjustments to the performance, only adjustments predicated on sound equipment.

2. Live electric sounds—most nonclassical musical acts will include some kind of electrified instrument. In many cases, those electrified instruments will be played through speaker cabinets onstage (e.g., guitar combo amplifiers, which are themselves self-contained sound systems, featuring the input stage, processing, amplification, and the output stage). When this is the case, we will often place a microphone directly in front of that dedicated speaker cabinet and route it into our system that way. However, there are times when, for whatever reason,[18] it suits us to take a direct feed from the electric or electronic instrument into our system, rather than mic-ing a dedicated speaker cabinet. In this case, we use a DI box.

3. Playback sounds—many events, not just our little concert in the park, require pre-recorded sounds to be played back. Whether we are just streaming Spotify from our phone as a little audience-entrance music or playing a list of 600 complex sound cues for a 3-hr stage play, the principle remains the same. We are taking audio content that has been stored, either as an analog signal on a magnetic tape or vinyl record, or as digital data on a hard drive, CD, or the like, and running it at line level[19] through our sound system. Playback devices vary wildly, from the simplest CD player to dedicated playback servers with highly customizable software/hardware configurations.[20]

THE PROCESSING STAGE

The processing stage of a sound system can take on an alarming range of forms. In our little concert in the park system we imagined earlier, the processing stage consists primarily of our mixing console.

When we approach the selection of processing stage equipment, we must ask: how many signals do we need to process? What types of signals do we need to process? To how many destinations does signal need be routed, and of what type are those destinations (audience speakers, monitor speakers, etc.)? These questions will be examined in-depth in Section 3.

In addition to mixing consoles, the processing stage also consists of all the effects we use in order to alter the sounds we are amplifying. These may consist of creative mixing effects, like reverberation units, delays, dynamic effects, equalizers, and the like,[21] or they may consist of system calibration effects that we use in order to "tune" our system—to ensure that it produces the proper range of frequencies, in the proper time-alignment, and at the proper intensity—such as all-pass filters, delays, limiters, equalizers, etc.[22] You will notice that some of our effects (equalizers, delays, and limiters, particularly) may be used both in the creative mixing stage and in the system calibration stage. While these effects have the same names, and often the same basic mechanisms, they are used in very distinct fashions for mixing versus system tuning.[23]

The processing stage also includes any device housing those effects, which may be your mixing console, or may be dedicated outboard effects units, or dedicated digital signal processor units.

THE AMPLIFICATION STAGE

The amplification stage takes all of our processed electronic signals and makes them . . . well, *bigger*, which equates to more loudness.

> ### Amplifier
>
> A device used to increase the amplitude of electrical signals, especially in the context of sound reproduction.

Amplification, in one sense, is very straightforward. We take a signal, we make that signal larger/louder, and we send it on its merry way. However, the details of how that works are important for a systems designer to understand. We discuss amplification in detail in Section 4. Amplifier choice depends primarily upon speaker choice, so we ideally will select our speakers first, and then select amplifiers that will ensure that we can power those speakers properly.

THE OUTPUT STAGE

The output stage consists of the devices that actually produce the new, louder sound that is the intended result of the sound system. Most of the time, your output device(s) will be some form of loudspeaker. In order to determine what kind of output devices we need, we must answer the following basic questions: who needs to hear the sound? How loud does the sound need to be? Where can we reasonably place a device to produce that sound (and what are the spatial limitations of that location)? There are three basic destinations for sound at the output stage:

1. Audience—Who is the audience, where are they located, and what do they need?[24]
2. Performer—Whether the performers are musicians and actors onstage or board members delivering a quarterly report around a conference table, there are times when performers need to hear themselves (or their fellow performers) in a manner that is isolated from the audience. These systems are referred to as *monitor* systems.[25]
3. Other—Destinations could include an audio feed into a sensor that triggers an action based on the amplitude of the input signal, or the like;[26] a program feed, sending onstage audio backstage to help performers keep track of time (this is not the same as a stage monitor system);[27] or a feed to a recording system that is not re-amplified.[28]

Inputs feed to processors. Processors feed to amplifiers. Amplifiers feed to outputs. Outputs make sound. Every sound system you will ever encounter, or will ever design, can be broken down in very broad strokes to these four basic elements. We will see this basic signal path recur again and again throughout our work (and throughout this book). We will complicate it, add to it, twist it, manipulate it, and adorn it with a host of modern technology, but these same four elements will remain.

So, that said—we're done here, right? You know everything there is to know about sound systems?

Nope! As the song says, "*We've only just begun.*"[29]

Notes

1 The term "sound system" may include recording systems as well, but recording systems all generally incorporate some form of reproduction (even if you are working with a field recording rig, you are still listening through headphones); thus, they will be referred to as sound systems in this text.
2 Discussed in Chapter 16.
3 Discussed in Chapter 4.
4 Discussed in Chapter 5.
5 Generally, the distinction between electrical and electronic is that an electrical device is a simple manipulation of electricity (like a lightbulb with only an on/off switch), where an electronic device features circuitry that

alters the function of the electricity (like light used as a data transmitter in fiber optics). For more on electricity in audio, see Chapter 4.

6 Discussed in Chapter 17.

7 Discussed in Chapter 4.

8 For more on cables and connectors see Chapter 24.

9 For more on consoles, see Chapter 10.

10 Discussed in Chapter 11.

11 We discuss the difference between these two types of configuration in Chapter 15.

12 Discussed in Chapter 18.

13 Discussed in Chapter 12.

14 Electrical levels, impedance, and transformers will be discussed Chapter 4.

15 DIs are discussed in detail in Chapter 8.

16 Microphones are discussed in detail in Chapters 6 and 7 (wired and wireless, respectively)

17 Discussed in Chapter 29.

18 Reasons we will discuss in greater detail in Chapter 8.

19 For more on line level, see Chapter 4.

20 Playback devices are discussed in detail in Chapter 9.

21 Discussed in detail in Chapter 11.

22 Discussed in detail in Chapter 13.

23 We will discuss those distinct uses in Section 3

24 We examine the principles of audience speaker selection in detail in Chapter 16, and we examine discipline-specific speaker choice considerations in Section 12 (Chapters 37–47).

25 These systems, sometimes referred to as "foldback" systems (particularly in the United Kingdom), will be discussed in detail in Chapter 17.

26 Triggered systems are discussed in Chapter 20.

27 Program feeds are discussed in Section 7.

28 Discussed in Chapter 45.

29 Lyrics by Paul Williams. Irving Music, Inc. (ASCAP), used with permission.

CHAPTER 2

The Four Standard Elements of the Digital Sound System

Now that we have a working definition of a sound system, let's complicate things a bit.

In the 21st century, a vast majority of the sound systems you will encounter in the professional world will have some digital components, and these components will add functionality to our system in a variety of ways. At the most complex end, computers and digital technology allow sound systems to achieve seemingly magical feats, but we'll get to those complex configurations much later in the book.[1]

However, we need to start by addressing the fundamental ways in which digital technology invades our four standard analog elements—to expand our periodic table, if you will. There are four standard elements of the contemporary digital sound system. Every digital sound system will contain not only the four standard analog elements but also will feature (in one form or another) the four standard digital elements:

1. Converters
2. Advanced processing
3. Control
4. Clocking

If a sound system is called a "digital" sound system, at its simplest this means that at some stage in the audio signal path, the audio is made up of 1s and 0s instead of voltage or acoustic pressure waves. If we return to our concert in the park from Chapter 1, then there is audio that starts as music performed by our guitarist/singer onstage. It travels acoustically into the microphones and is transduced into analog electronic signals. In a digital system, the next place the signal arrives is at a device called a converter.

Once the analog audio from the microphones is converted to digital data (the input stage), it can be processed by all manner of digital toys, such as a digital mixing console (the processing stage). In our simplest digital system, it will then again encounter a converter on its way out of the mixing console, so it can send an analog signal to

Converter

A device that either changes analog audio signal to digital audio signal, or vice versa. In an A/D (analog-to-digital) converter, the device accepts the incoming current flow, senses the oscillation rate (frequency) and voltage intensity (amplitude), and assigns numeric values to them, creating a digital (numeric) stream of data that represents the analog audio signal. In a D/A (digital-to-analog) converter, the device takes the data stream and generates an electronic signal with equivalent oscillations and amplitude changes, creating analog audio signal based on the digital data.[2]

the power amps (the amplification stage) and on out to the loudspeakers (output stage). Figure 2.1 shows the signal flow of a sound system with a digital console elements added.

Now, for simplicity's sake, Figure 2.1 is shown with the converters as separate steps from the digital mixing console, but in reality, the converters are very often built into the console itself, so you would not wire them up separately (though there are systems in which external converters are used). In advanced sound systems, other digital components might be in place as well: a playback system sending a digital signal directly to the console,[3] a digital signal processor after the console but before the amplifiers (for use in system calibration),[4] and even sometimes amplifiers with digital inputs.[5]

You might be wondering: why add this step? If it's just adding complexity to our system, why not just keep the system analog and save the hassle? In addition, any A/D or D/A process adds a tiny delay to the audio signal traveling through the system (called *latency*), which means that the signal isn't 100% in real time as it would be in an analog system. So what's the point?

I'm glad you asked!

Digital audio systems provide a number of extremely useful advantages, in terms of functionality and ease of use, that are simply not possible with analog systems. The core of these advantages makes up standard digital elements #2 and #3.

As we already know, all audio systems contain a measure of processing. Whether the simple cone megaphone from Chapter 1 or the sound system powering your favorite Disney attraction, taking an input sound and altering it is one of the core functions of a sound system. In a digital system, our ability to perform all the functions typically associated with the processing stage of a system multiplies exponentially. This is digital system element #2: advanced processing.

In an analog mixing console, for example, a typical input channel strip (the path that a signal from, say, one of our aforementioned microphones, takes through the console) may feature a few bands of equalizers, possibly

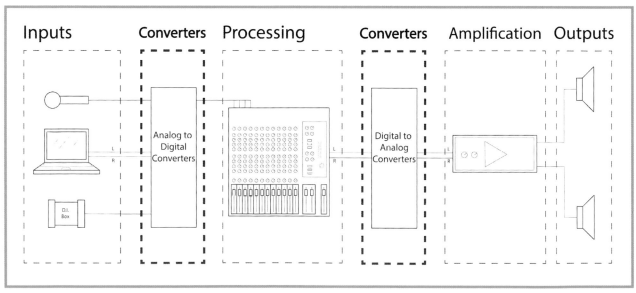

FIGURE 2.1
Signal Flow of a Basic Sound System With Digital Console

a compressor, and often the ability to route signal out of a couple of auxiliary sends.[6] In order to send that microphone's signal to a different compressor, a noise gate, or a reverb unit, you need outboard (external, rack-mounted) devices for each of those effects.

In a typical digital mixing console, however, those effects and numerous routing paths will be built in. Sending your microphone's signal to them is as simple as a few clicks of a mouse (or a touchscreen, depending on the console). Digital mixing consoles, processors, and other such equipment contain vastly more options for sonic alteration than your typical piece of analog equipment. Instead of taking an equipment rack that contains 12 analog noise gates for your drum microphones (that easily weighs 100 lbs.), why not just pack the digital console that has all of them built in? In Figure 2.2, we see how many cables we need to connect just four outboard devices to an analog console versus the total lack of cables in a digital console to use the exact same effects.

In addition, sending signals to a variety of locations is much simpler and easier with digital gear than with analog gear. If I have an analog system, and I want to send my playback pre-show music to the audience, the backstage, the balcony, the owners' private box, and the lobby, then I will need a playback device (like a cassette deck, CD player, or laptop), a console with many physical output connections (or a hardware interface for the laptop with many physical outputs), and copper wire from each of these connection points to the amplifiers. If I want to do the same with a digital system, then I can just play the audio file on my laptop, and use a digital routing software (such as *Dante Controller*)[7] to route the signal to my digitally capable amplifiers, over a standard network switch,[8] and boom, I've accomplished the same thing with much less wire and (in this case) no console necessary! See Figure 2.3 for this comparison.

Beyond the mixing stage, we introduce advanced processor devices known as DSPs (digital signal processors), which take all the output signals from a console/playback system, and route them to the amplifiers and then to speakers. Inside the DSP, we have a range of customizable options[9] that allow us to "tune" or calibrate our speaker system for the finest audio quality possible. We see this device (and our other standard elements) in Figure 2.4.

Now, even if you don't understand what half the above words mean (we will explain them in more detail in subsequent chapters, I promise!), understand that one of the key reasons we embrace digital systems is because they allow for advanced processing (and routing) options that are very desirable, with a smaller equipment footprint and more control than their analog equivalents.

This brings us to the third digital element: control!

13

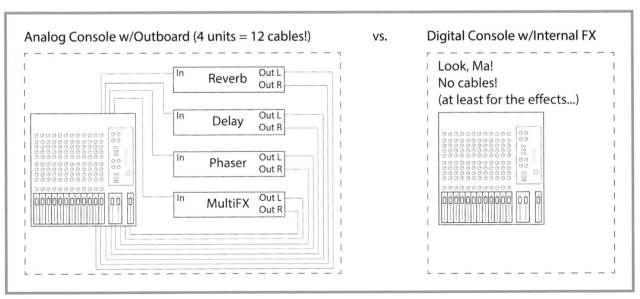

FIGURE 2.2
Analog Console With Outboard vs. Digital Console with Built-In Effects

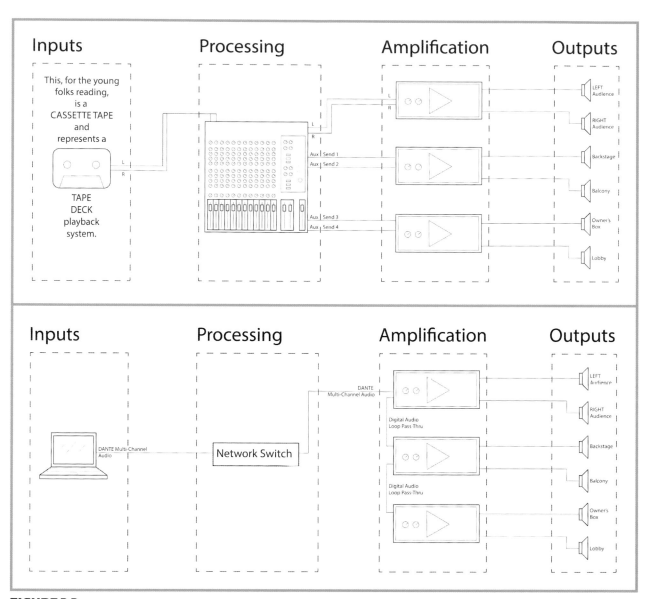

FIGURE 2.3
Analog Playback Routed to Multiple Locations vs. Digital Playback Routed to Multiple Locations

Let's imagine our concert in the park. Instead of just the one performer, however, this time, we have five bands who are set to play. Each band gets a 20-min soundcheck earlier in the day, where we take the time to set their microphone positions and levels, and make sure they sound just right. In an analog system, once we've set the levels of all the console controls, all the effects in our outboard rack, and so on, we have to figure out how to record this information, so we can reset for the next band. In most analog systems, this means you will be taking out a pen and paper and taking copious notes. Sure, you could type these notes into a laptop, but you still have to write down the settings for every channel, every knob and button, every routing patch, everything, to make sure that once the band gets onstage during the show later, it sounds just like it did during soundcheck.

This, suffice it to say, kind of sucks.

In a digital system, we have an enormous advantage, one shared by just about any computerized system, which is to say, we can hit the *save* button!

Now, it might not be actually labeled "save," every piece of equipment is different, but in any digital console, we can set up Band #1 perfectly, save the settings, wipe them all clean to set up Band #2, save those, and so on through the

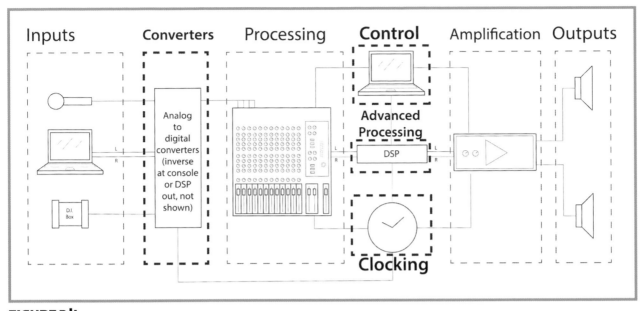

FIGURE 2.4
A Simple Digital System Showing All the Standard Analog and Digital Elements

rest of the bands. Once soundcheck is done, we can hit "recall/load" on our Band #1 settings, and they are instantly recalled without having had to take any notes at all! This not only saves enormous amounts of time (because instead of waiting for you to write down all of Band #1's settings, we can just get Band #2 onstage and get into the next soundcheck), but it saves the worry of bad handwriting, of writing something down incorrectly, and of spilling coffee on our notepad.[10]

This is but one piece of what control can do for us in the digital realm. Imagine now that you are mixing our concert in the park, but because the park is small, and the audience is large, instead of putting the mixing console out in the audience in front of the speakers where it's easy to hear, the console is back in some weird corner. Your view is blocked by trees, and the sound traveling to you through those trees is different than what anyone in the audience hears. How are you going to make sure that this show sounds good?

Thankfully, most contemporary digital systems have an answer. The iPad.

If you're in a bad mix position, many digital consoles allow you to connect an iPad to the console wirelessly and take that iPad and mix from anywhere in the venue where the Wi-Fi signal will reach. Most of these apps allow at least fader, mute, and pan control, though many also allow you to change equalizers, routing, and other features. Now, you can stand out in the middle of the audience, taking up no more room than you would as an ordinary audience member, and mix the show so it sounds good for the majority of the attendees—not just for your weird, arboreal mix position.

In our multi-destination playback scenario earlier, if you need to play the CD (or cassette, or laptop, or whatever) and have it sent to all the destinations throughout the venue, you pretty much have to be standing at that CD player to hit the play button. Sure, some CD players have remote controls, but at most that gets you 3' away or so.

However, if you are playing from your laptop, over a distributed digital system, your laptop will presumably be connected to a show system control network,[11] and if you set it up properly, you can dial in from your smartphone and hit play from anywhere. This technique knows no bounds: if you configure your network properly, you can be in California and hit play on a file in Florida, and it will play on time and through all your desired speakers.

Finally, the fourth standard element of any digital system is clocking. This is the least glamorous element but is vital to any digital system. Since digital audio is a set of numbers[12] sent over a period of time, we need to ensure that the numbers sent from our transmitting device time out properly with the counters in the receiving device so that when they are re-converted to audible sound, there are no errors.

In an oversimplified analogy, if I send Values 1–5 from Device A to Device B, and each of those values corresponds with a second on the clock, this means that Device B needs to read Value #3 as Second #3, not as Second #4, or Second #2. If Device A's Value #3 arrives at Second #2 on Device B, we will encounter a problem called *jitter*, and the sound we hear at Device B will be broken up, distorted, and just generally no good. Device clocking sends a calibrated timing signal that allows all these devices in your system to line up all our signals exactly the same so that we hear them clearly and cleanly.[13]

Clock

In digital audio, a clock is a device that sends out a square wave signal (regular oscillations, timed via an internal crystal, just like a quartz watch, but more accurate), to each device in a digital audio system to ensure they all stay aligned. In a digital audio system, the device that sends the clock signal is called the MASTER, and all other devices are the SLAVES. A clock may be a dedicated master clock, whose sole system purpose is sending out clock signals to other devices, or a clock may be internal to a digital mixing console, DSP, or another such device. Clocking signals do not necessarily proceed in signal flow order, even though certain signal types can contain clock information (like Dante digital audio).

So, from the simple analog model of input to processing to amplification to output, we have complicated things a bit:

- Analog inputs feed to converters, digital inputs send a direct signal
- Converters feed to processors (of increasingly advanced design)
- Processors are controlled by computers
- Signal paths are calibrated in time with clocks
- Processors feed to amplifiers (usually converting back to analog along the way)
- Amplifiers feed to outputs

A little more complex than the standard analog model, but still not exactly rocket surgery.

Now we're ready to design a system, right?

Not quite. First, let's take a detour through the fundamental theory behind audio, in the acoustic world, the analog electronic world, and the digital world. Then we need to learn about gear—a *lot* about gear.

Onward!

Notes

1   See Section 12 for more on advanced systems.
2   Details of the conversion process, and what this means for audio fidelity, are discussed in Chapter 5.
3   Discussed in Chapter 9.
4   Discussed in Chapter 12.
5   Discussed in Chapter 15.
6   Discussed in Chapters 10 and 11.
7   Discussed in Chapter 5.
8   Discussed in Chapter 18.
9   See Chapter 12.

10 Now, if you spill coffee on the console, digital still won't be much help. Don't spill coffee into sound equipment!

11 Discussed in Chapter 18.

12 Discussed in Chapter 5.

13 For more on clocking, see Chapter 5.

CHAPTER 3

What Is Sound?—Part 1

Acoustic Version

There are dozens of textbooks that go into great detail describing and explaining the physics of sound. This is not one of those books.

This chapter will serve as an introduction (or a refresher course) on the basics of sound propagation, terminology used to describe sound, and a few essential equations for the system designer. If you already have a good grasp on the physics and terminology of acoustic sound, feel free to skip this chapter.

First, we must once again ask a seemingly basic question: what is sound? Easy, right? It's what we hear! Well, yes, but also the following:

> **Sound**
>
> 1) Vibrations transmitted through an elastic solid or a liquid or gas, with frequencies in the approximate range of 20 to 20,000 Hertz, capable of being detected by human organs of hearing. 2) The sensation stimulated in the organs of hearing by such vibrations in the air or other medium.[1]

There are a few things to unpack here: what is an elastic medium? What is a frequency? What are Hertz? We'll get to all of those things in just a moment, but first, let's try a little thought experiment.

Imagine an inflated rubber/plastic birthday balloon (not the Mylar kind). It is floating motionless in free space (which we define as being a space with no obstacles within 3' of it on any side). A pin pricks the balloon and it pops. In what direction does the sound of the pop travel? In every direction at once! This is what we think of as an "omnidirectional radiator," which is to say a sound source that sends its sounds in every direction equally.

What is the sound of the balloon popping? In the English language, we define it with an onomatopoeia, and we just call it a "pop," but this doesn't help us here. What is actually creating the sound?

Inside the balloon, the air is held in a high-pressure state. That pressure can be indirectly felt by pressing on the sides of the balloon (before it is popped). The tension in the balloon's walls are a reflection of the internal pressure, and the higher the pressure inside the balloon (the more air inside), the tighter the skin of the balloon will feel. So, when we pop a balloon, the sound we hear is a combination of two actions. The first is the rubbery sound of the balloon skin itself rupturing. This is actually a relatively small sound, as you can find if you take an uninflated balloon and puncture it with a pin. You will hear it, but just barely. The primary source of our sound is the high-pressure air in the balloon suddenly expanding and creating a pressure wave that is audible to our ears. Pressure (like electric charge)[2] always seeks to level itself out, so the air that is pressurized inside the balloon immediately dissipates to the standard atmospheric pressure when the balloon is popped. (To put this another way, think of a garden hose. If you turn it on and hold your thumb over the end, the pressure behind your thumb builds up until you will be unable to hold it. When you release it, the rush of water is stronger and faster than it would be at normal running level. It's louder for an instant, too!)

Going back to our definition, the change in pressure makes vibrations in the "elastic medium" of air. This means that a pressure wave moves *through* the air toward us, and because the vibrations happen a certain number of times per second, we can hear them. In Figure 3.1, we see a representation of a balloon popping, sending sound in all directions.

Now, if a pressure wave moves through the air, does that mean the air moves with the sound? Well . . . sort of. Air molecules are compressed and rarefacted by the pressure waves, which is to say that they are squashed together and then spread apart. An easy way to understand this is by imagining the front of a large speaker.

When a speaker makes a loud sound, we often see the speaker cone itself vibrating back and forth in the air. At the first edge of a sound, the speaker cone will press forward in space. This compresses the air molecules directly in front of the cone. Immediately after it compresses, the speaker cone pulls back farther than its original resting state, which creates a negative pressure that rarefacts (spaces out) the molecules in front of it. In Figure 3.2, we see a speaker compressing and rarefacting.

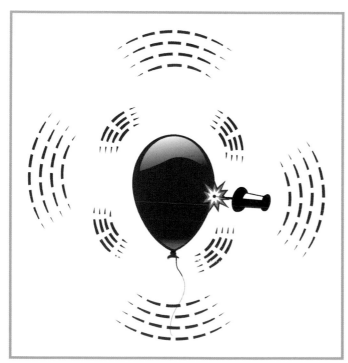

FIGURE 3.1
A Balloon Pops and Sound Travels Away in a 360° Sphere

Now imagine what happens as the speaker continues to vibrate. It pushes forward again and back again, very quickly and in repetition. This creates numerous shocks of compression and rarefaction in front of it in very quick succession. Those shocks form a pressure wave that travels through air, and to our ears, which we perceive as sound. It is important to note that the air molecules, while bunching and spreading, otherwise remain in place. Otherwise, a heavy metal drum solo would create a tornado out in front of it! While that might be awesome to see, that's just not how physics works.

Now, a single stroke forward and backward (compression and rarefaction) is known as a single cycle. This is true of any vibrating object. A guitar string, when plucked, vibrates back and forth along the axis at which it was plucked. One full vibration (in and out of the position where it was released) is a cycle. The number of cycles of vibration completed in a single second is what we call the frequency of the vibration. Frequency is measured in Hertz.[3] The range of human hearing—the vibrations we perceive as sound instead of as radio waves or other phenomena—is clearly defined.

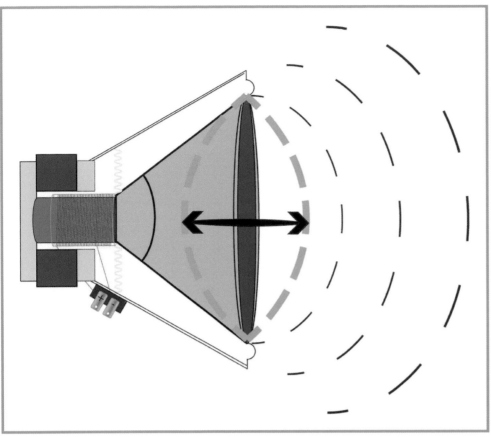

FIGURE 3.2
A Speaker Compressing and Rarefacting Air Molecules (Compressed Bands in Red, Rarefacted as Open Space)

Human Hearing Range

Vibrations that oscillate in complete cycles at a rate of between 20 and 20,000 per second. 20 Hertz (abbreviated Hz) is the lowest note most humans can perceive as sound, and 20 kilohertz (kHz) is the highest.

Frequency is closely related to pitch in musical content. The standard "A" note around which contemporary orchestras tune their instruments is 440 Hz (though orchestras throughout history often tuned to different notes).

Now, when we're talking frequency, there are a few other basic terms and concepts we need.

Octave

A doubling or halving of a given frequency. For example, the A as 440 Hz is doubled to 880 Hz for the next A note higher on the keyboard. The A note below 440 Hz is 220 Hz, etc.

Harmonics

A whole number multiple of a given frequency. The root frequency under discussion is referred to as either the "fundamental frequency" or the "first harmonic." The next multiple up (two times the original) is referred to as the "second harmonic," three times the fundamental is the "third harmonic," etc.

These terms allow us to identify fundamental and related frequencies easily. Most instruments that produce musical notes do not produce fundamental frequencies alone (the exception being a sine wave synthesizer, which generates a pure fundamental and nothing else). If you pluck a guitar string, there is a combination of the fundamental with harmonics (vibrations shaped by the material of the string and guitar, the shape of the guitar, etc.) that creates the signature tone (known as timbre).

Now, if you've ever been at a giant concert and sat very far from the stage, you might have noticed something funny. You might have, for example, seen the drummer hit a cymbal, but not heard the sound of that cymbal till a half a second or more after you saw the hit. Why does this happen? Because sound is slow. In comparison with light, which travels at 186,000 miles per second, sound is a laggard.

Speed of Sound

At 68° F, at 50% barometric pressure and humidity, sound travels at approximately 1,125'/sec through air. Speed increases as heat increases and slows as heat decreases. Barometric pressure and humidity can also impact the speed of sound, but only in such small amounts as to not impact practical considerations.

This means that if you were 1,000' from stage, it would take sound almost a full second to reach you from the cymbal crash, by which time the drummer is visibly onto something else. Now, when we get to large system considerations for outdoor work,[1] we'll return to the topic of temperature impacting speed of sound, but for now, we can consider that average speed of 1,125'/sec (343 m/sec) as close enough for our work.

To get back to our balloon, because there are no obstacles around it for the sound of the pop to interact with, sound travels omnidirectionally, with no interaction with other objects. How do we measure how loud the pop is? And how do we know how loud it will be to an observer a given distance away?

The basic term of loudness that is used conversationally is the decibel. You will often see this written in articles about sound. Even the *New York Times* will say that something is "94 decibels,"[5] but using this term without a qualifier at the end is actually surprisingly imprecise.

Decibel

A logarithmic unit of measurement, expressing the ratio of two physical values.

As you can see, this does not mean that a decibel always measures sound. We definitely use it for sound, but we often use it for ratios of voltage or wattage quite frequently. In a vacuum, saying something is "X decibels" technically means nothing. However, pedantry aside, when someone says something is "94 decibels," what they *almost always* mean is dB SPL,[6] or decibels of sound pressure level.

Sound Pressure

The localized pressure variation, within the audible frequency range (20 Hz–20 kHz), from the atmospheric pressure baseline.

The air around us has a given standard pressure (it varies a little from place to place but is generally stable enough for our purposes). Pressure, in scientific terms, is measured using the unit of the pascal (Pa).[7] In sound, 1 Pa≈94 dB SPL. This number will come back when we examine microphones, as 1 Pa is always the standard measure used to explain how electrically sensitive is (how much voltage is produced by) a given mic. The amount of variation in that standard pressure level is what is measured as sound pressure, which is the technical term we use to define how loud a given sound is in acoustic space.

The sound pressure scale is reasonably straightforward. It begins at 0 dB SPL, which we can think of as the threshold of silence. This is the quietest sound that the average human being can hear. Though there is no theoretical maximum level to the sound pressure scale, the loudest sounds we hope to encounter without protection are about 130 dB SPL, which we colloquially refer to as the "threshold of pain." This, quite simply, is the level at which sound begins to physically hurt the ears. Figure 3.3 shows SPL levels of some common sounds.

So, how loud is our balloon pop? According to researchers from the University of Alberta, who measured as close to the balloon as possible, a balloon pop can be louder than a shotgun blast: almost 168 dB SPL.[8] That's way beyond the threshold of pain, and very dangerous to our ears. But, unless someone is being mean to you at a birthday party, chances are that you won't have your ear

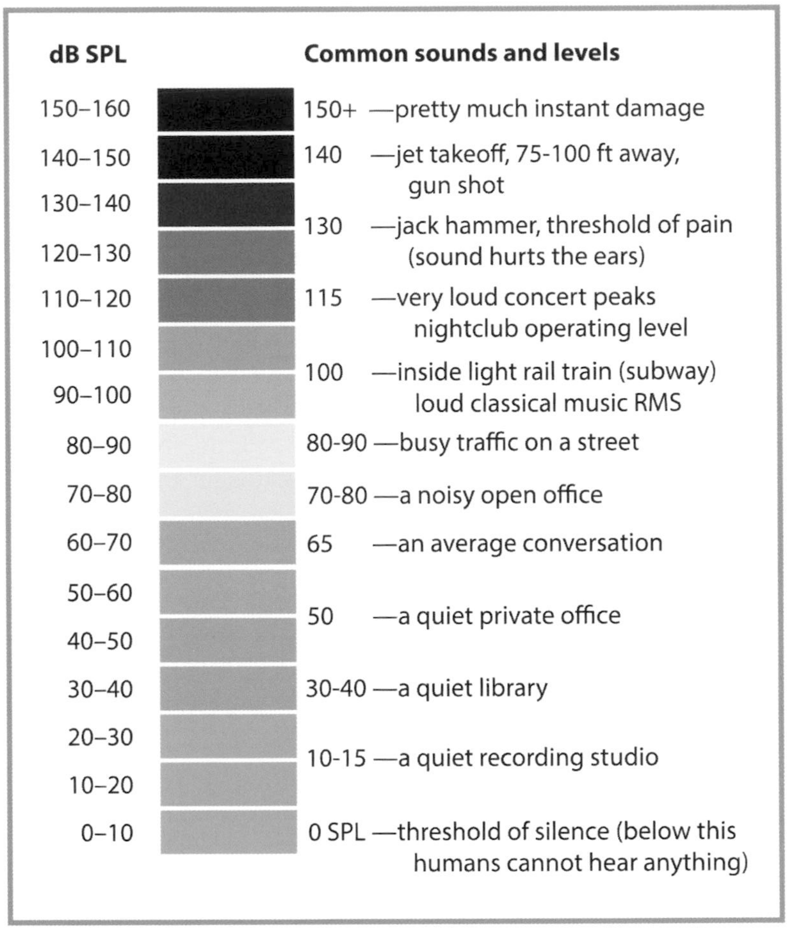

dB SPL	Common sounds and levels
150–160	150+ —pretty much instant damage
140–150	140 —jet takeoff, 75-100 ft away, gun shot
130–140	
120–130	130 —jack hammer, threshold of pain (sound hurts the ears)
110–120	115 —very loud concert peaks nightclub operating level
100–110	
90–100	100 —inside light rail train (subway) loud classical music RMS
80–90	80-90 —busy traffic on a street
70–80	70-80 —a noisy open office
60–70	65 —an average conversation
50–60	
40–50	50 —a quiet private office
30–40	30-40 —a quiet library
20–30	
10–20	10-15 —a quiet recording studio
0–10	0 SPL —threshold of silence (below this humans cannot hear anything)

FIGURE 3.3

Sound Pressure Levels of Common Sounds

Dynamic Range

The difference between the softest and loudest parts of either an audio program, or a piece of equipment. The dynamic range of the human hearing system is often stated to be about 130 dB, since it ranges from 0 dB SPL at the threshold of silence to about 130 dB SPL at the threshold of pain.

Message in a Bottle

The biggest pieces of advice that I would have: number 1, never forget that it's a business of people, that you have to keep really good relationships, and to value your relationships with people and work as part of a team. I always really liked being a part of the team, it really is a team sport. Number 2, learn your physics and your fundamentals so that you're based on reality rather than on mythology. Reality is going to persevere, and mythologies will blow around with the wind. The old, "I do it this way because this guy who mixed this really famous band did it this way" is not a really good approach, but "I did it this way because the physics line up in the following reality-based ways"—that is going to adapt toward the future whereas copying the way the guy did it that did this famous band really has no future, it's an idea that's waiting to become obsolete.—Bob McCarthy

directly next to the balloon when it pops. So, how loud is it by the time you hear it? Well, it depends on how far away from it you are. Logic tells us that the farther from a sound source the listener is, the quieter the sound will be, but *how much quieter*?

It's time to meet our friend (and first equation of note), the Inverse Square Law.

> **Inverse Square Law**
> In a free field, for every doubling of distance from a source, sound pressure decreases approximately 6 dB.

Now, we can't easily double the distance of *zero*, so let's turn to the next distance measurement in the study: at 0.5 m, the pop was measured at 157.03 dB SPL. So, following the shorthand rule for the Inverse Square Law described earlier, we would guess that, measured at 1 m, the SPL would be something like 151. What was it in the study? 151.26. So the generic rule holds!

However, what happens when we need to estimate sound level at a distance that isn't an even doubling? Most loudspeaker specifications list the loudness they produce at a distance of 1 m from the speaker. What if we need to know how loud the sound will be at 6 m? This is when we need to know the Inverse Square Law Equation.

$$\left[\text{INVERSE SQUARE LAW EQUATION}: x\,dB\,\Delta = 20log_{10} \frac{D_1}{D_0} \right]$$

In English, this reads, "A given amount (x) of change (Δ) in decibels (dB) is the result of calculating 20 times the base-10 logarithm (log_{10}) of distance 1 (D_1_our desired measurement distance) divided by distance 0 (D_0_our reference distance)."

In practice, this is much easier than it sounds. We're not going to spend pages explaining how logarithms work, that's for books with more math. For now, suffice it to say that a logarithm is the inverse of an exponent, and that when we are calculating a log in "base 10," what that means is that you can hit the *log* button on your calculator at that step (as calculators default to log base ten). Let's put this to use.

If our balloon, when popped, is measured at approximately 151 dB SPL at 1 m, how loud will it be when we measure it at 6 m? Since we are solving for "x," we need to fill in the variables on the other side of the equation. In this case, 6 m is our desired measurement distance, and 1 m is our reference distance. Thus, the equation looks like this:

$$[x\,dB\,\Delta = 20log_{10} \frac{6}{1}] \,.$$

In order to solve this, we will first divide 6 by 1. The answer (for those who skipped fourth grade) is 6. Then we need to get the log_{10} of 6. So, I enter 6 into my MacBook calculator, hit the log_{10} button, and the result is 0.778. Then we multiply this by 20. The answer is 15.56. . . *what?* Looking back to the left of the equation, we learn that this is 15.56 dB of change. Now there's one more step.

If our original measurement at 1 m is 151 dB SPL, we must subtract 15.56 from that value. The answer, then, is that at 6 m, the sound of the balloon popping is approximately 135.24 dB SPL. That's still pretty loud!

This is the kind of calculation sound system designers do every single day in order to determine how loud their speakers will be for the audience at which they are targeted. While, in reality, sound performs a bit differently coming out of a speaker (it isn't truly omnidirectional, and in fact in the case of speakers with horns, is very directional, which changes the falloff a bit), and it also varies a bit by frequency, this general rule of the Inverse Square Law is valuable to understand, and we return to it over and over again throughout our work.

Now, once the balloon has popped, what happens to the sound? Again, in a free field, it spreads omnidirectionally, and since it is a brief loud sound (known as a transient), it falls off to silence fairly quickly. However, we are almost never designing systems for a free field, and if we're working, say, in an auditorium, there are tons of objects around us that will interact with the sound and change how we perceive it. When sound travels through a space with obstacles, there are basically five different things that can happen to it when it encounters those obstacles.

1. Reflection
2. Diffusion
3. Diffraction
4. Refraction
5. Absorption

We will define these terms in a moment, but in order to understand them, we have to understand another key concept of how audio works, which is *wavelength*. When we draw a sound wave unfolded in linear time (as we see in software displays), a pure sine wave fundamentally looks like this (Figure 3.4):

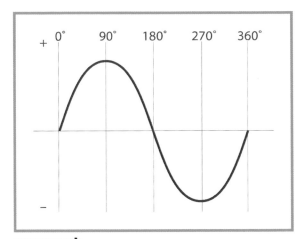

FIGURE 3.4
A Single Cycle of a Sine Wave Shown With Degrees of Phase

The numbers shown around the curve of the sine wave are *degrees of phase*, which we will return to in a moment. This drawing shows one complete cycle, from compression through rarefaction, and back to rest.

Wavelength

The amount of physical space it takes for one complete cycle of a given frequency.

Wavelength varies by frequency. The higher the frequency, the smaller the wavelength.

$$\left[\text{WAVELENGTH EQUATION: } \lambda = \frac{S.O.S.}{C/S} \right]$$

In English, this reads: wavelength (λ) equals speed of sound (SOS), divided by frequency (cycles per second). So, let's imagine a sound in the middle of the audible range, at 3 kHz. How large is the physical space in which a single cycle of that frequency can complete?

$$\left[\lambda = \frac{1125 \frac{ft}{sec}}{3000/sec} \right]$$

Since we are dividing, the units of "seconds" cancel one another out. Then we have 1,125/3,000, which equals 0.375', or 4.5". So, it takes only 4.5" of physical space for 3 kHz to complete its cycle.

Let's try one more. What is the wavelength of 50 Hz?

$$\left[\lambda = \frac{1125 \frac{ft}{sec}}{50/sec} \right]$$

So 1,125 divided by 50 is 22.5'. It takes 22.5' for 50 Hz to complete its cycle.

So why is this important? Two reasons:

1. It takes physical space for sound to be heard. High frequencies can be heard at very short distances, but low frequencies take space to be heard.[9]
2. All of the five earlier listed interactions are defined by the wavelength of the sound versus the size (or in the case of absorption, density) of the obstacle in question.

Let's define these interactions then!

Reflection

When a sound hits a surface and bounces off with the angle of incidence equal to the angle of return. Occurs when a surface is at least four times the wavelength of the sound (see Figure 3.5).

This means that if a sound hits a wall at 45°, it bounces off also at 45°. For our 3 kHz sound, the obstacle would have to be at least 18" diameter in order to properly reflect.

Diffusion

When a sound hits a surface and bounces off, but the angle of incidence is different from the angle of return, scattering the sound. Occurs when a surface is between one and four times the wavelength of the sound (see Figure 3.6).

For our 3 kHz sound, any obstacle with diameter of between 4.5" and 18" is likely to diffuse the sound.

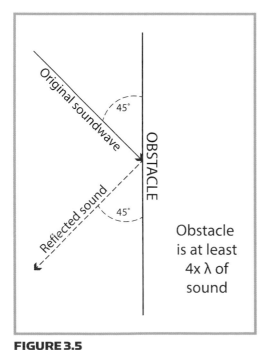

FIGURE 3.5
Acoustic Reflection

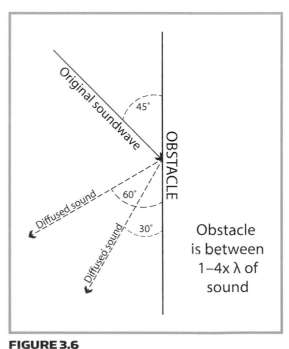

FIGURE 3.6
Acoustic Diffusion

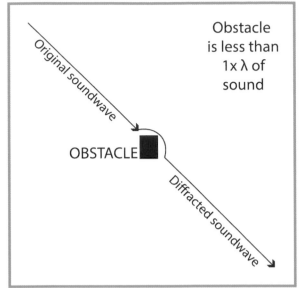

FIGURE 3.7
Acoustic Diffraction

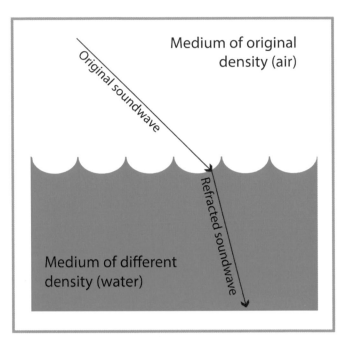

FIGURE 3.8
Acoustic Refraction

Diffraction

When a sound approaches a surface but does not bounce off and instead simply bends around it and continues in its original direction. Occurs when a surface is less than one time the wavelength of the sound (see Figure 3.7).

Our 3 kHz sound will diffract around a 2" pole. Low-frequency sound, which has large wavelengths, will diffract around a lot of objects (imagine listening to loud rap music: you are sitting in front of the speaker; someone comes to stand right in front of you; the high frequency material—voices, cymbals—gets quieter, but the bass is just as loud . . . diffraction!).

Refraction

When a sound passes from a medium of one density (e.g., air) into a medium of a different density (e.g., water), its angle of travel is changed (see Figure. 3.8).

Absorption

When a sound hits an obstacle and sound is passed into it, some transducing to heat energy and dissipating, and some passing through to the other side of the obstacle (see Figure 3.9).

The calculations for how much sound of a given frequency an obstacle will absorb are beyond the scope of this book, but in practical terms, dense, soft, porous materials (like foam) are usually good absorbers, where flat, hard, nonporous ones tend to act as reflectors or diffusers.

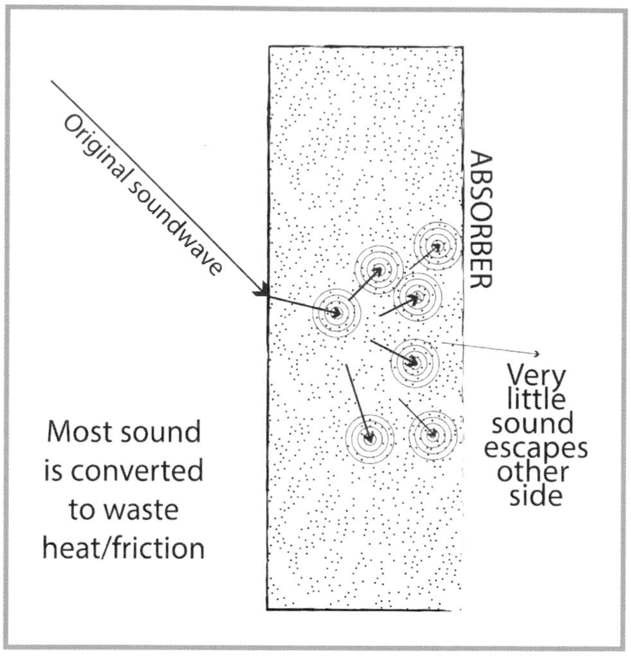

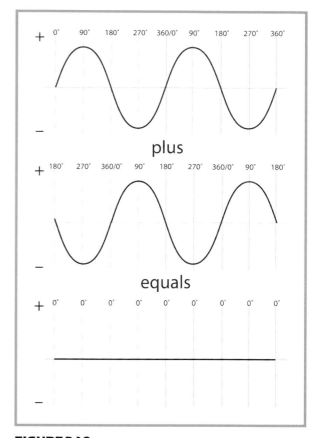

FIGURE 3.9
Acoustic Absorption

FIGURE 3.10
Inverting the Polarity of Two Identical Sine Waves Summing to Zero

When a sound hits a surface, one of these five things (or, more likely, a combination of them) happens. It is important to understand this—as we design systems, we need to anticipate how sound will interact with our venue. We will return to these concepts many times throughout this book.[10]

There are two other salient principles we need to understand before we are ready to move forward:

1. Phase and cancellation
2. Localization

As mentioned earlier (see Figure 3.4), sound cycles are measured in degrees of phase. A compression is measured from 0° at the start, to 90° at the peak of compression, to 180° at the resting point between compression and rarefaction, to 270° at the trough of rarefaction, and to 360° at the resting point before the next compression. If we have two copies of the same sound, for example, the sound as played through a speaker, and the reflection from a nearby wall, they will interact, summing in space to form a signature tone that we think of as *reverberance*.

When a signal and a copy of that signal are exactly 180° out of phase with one another (and the same amplitude), they will cancel each other out completely when summed. We can see this in digital software, when we take a signal, make a copy, invert the polarity, and sum them together . . . we will hear nothing (see Figure 3.10).

In acoustic space, it is nearly impossible to completely cancel a signal out, especially when you are working with a complex signal like a voice, rather than a simple signal like a sine wave. However, when two copies of a complex signal interact, they will still sum together, and if they sum together in certain time and space relationships, then they will create sharp peaks and deep valleys of compression and rarefaction, creating a sort of swirling sound that we call *comb filtering* (because the graphic representation of the sound looks like the teeth of a comb—see Figure 3.11).

This is an undesirable sound, and as we will find repeatedly in sound system design, when we have more than one speaker playing the same sound (which we very often will), comb filtering is challenging to avoid or minimize.[11]

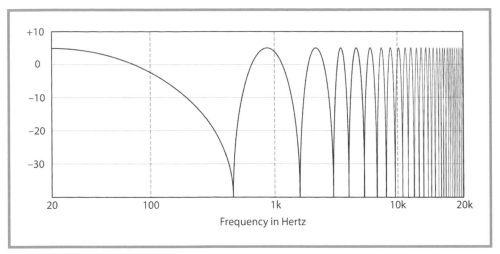

FIGURE 3.11
Comb Filtering

Finally, we need to talk about a principle called *localization*. To return to our popping balloon, if we are standing with the balloon to our left (and our eyes are closed), and it pops, how can we tell what direction it comes from?

Our brain uses our two ears to interpret time, level, and frequency differences in the signals it receives, and these differences allow us to localize (determine the origin of) sounds. If the balloon pops on our left, the high frequencies (with the tiny wavelengths!) will be absorbed by our head, so the sound that reaches our left ear will be higher in frequency than the sound that reaches our right ear. The right ear is farther from the balloon than the left ear, so the sound that reaches it will be a bit quieter. Finally, the part of the sound that reaches the left ear will arrive slightly sooner than the sound that reaches the right ear. These three differences add up, and we can easily tell that the balloon is on our left. This principle guides everything we hear.

It also leads to a problem we deal with over and over again in sound system design, which is embodied by the Haas (or Precedence) effect. The Haas effect[12] states the following:

- If we hear two copies of a sound (e.g., a direct sound and its reflection) within quick succession (1–5 milliseconds {ms} of each other for short click sounds, up to 40 ms for longer sounds, they will merge into one sound in our minds).
- If we hear two copies of a sound spaced beyond the above stated range, whichever copy arrives first at our ears will be perceived as the source to which we localize.

This means we can create a "phantom image" in sound design, convincing a listener that a sound is coming from a place other than its source by manipulating their distance and a speaker's distance to a reflector (see Figure 3.12).

29

Tales From the Field

I worked at Covent Garden, the opera house, as a sound person. There we were—not the *taboo* department—but certainly the department that was not allowed to be seen, only heard. So we did lots of things that the opera audiences didn't know that we were doing. So we had to do it very well, and that's when I started learning about . . . the head of sound at the time, he taught me so much about the art of mic-ing instruments and singers, and putting sound in different locations. Because it was written, "Oh the singer's got to be over here on stage left," but there's no room on stage left, so what we'll do is we'll mic him and put speakers over there, so it'll sound like he's stage left. So the audience still felt like it was coming from the correct place as per the score, and as per the intention. But, in fact, sometimes we'd put musicians across the rows in different places and amplify them into the opera house and nobody would know.—Jonathan Deans

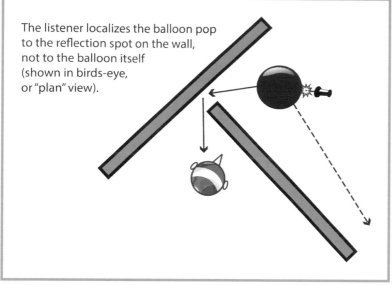

The listener localizes the balloon pop to the reflection spot on the wall, not to the balloon itself (shown in birds-eye, or "plan" view).

FIGURE 3.12
A Balloon Pops Near a Wall, and the Listener Hears the Reflection Before the Original Sound Based on Position

These Haas principles will come up again and again when choosing speaker positions, acoustic treatments, and more.

Well, that was a lot in one chapter, right? I know, I know, I said this was going to be easy! Believe me, it could've been more difficult, there are whole books written on the material in this chapter alone. However, if you can manage to understand the principles outlined in this chapter, everything else we do in this book will make a lot more sense.

So . . . now that we understand how sound travels through acoustic space, we need to get that acoustic sound into our sound system. That means we have to wrestle with that mysterious (and ill-defined) force: *electricity*.

Notes

1 Definition from *The American Heritage® Dictionary of the English Language*, Fifth Edition, by Editors of the American Heritage Dictionaries. Reprinted by permission of Houghton Mifflin Harcourt Publishing Company. All rights reserved.

2 As we will see in the next chapter.

3 Named for Heinrich Hertz, who figured this out.

4 Discussed in Chapter 38.

5 "Working or Playing Indoors, New Yorkers Face an Unabated Roar," by Cara Buckley, *New York Times*, July 9, 2012. Accessed online July 11, 2017: www.nytimes.com/2012/07/20/nyregion/in-new-york-city-indoor-noise-goes-unabated.html

6 Note that decibels are always abbreviated dB, with the B capitalized. Decibel is a unit derived from an earlier, blunter measurement called the bel, which itself was named after Alexander Graham Bell, who did early work that led to the measurement. A deci-bel is a tenth of a bel and came into use because the bel itself is too large an interval to be useful. Any abbreviated symbol based on the name of a scientist is capitalized, hence kHz for kilohertz capitalizes the H after Heinrich Hertz.

7 Named for Blaise Pascal.

8 "Did You Know How Loud Balloons Can Be?" by Bill Hodgetts and Dylan Scott, University of Alberta, printed in *Canadian Audiologist*, Vol. 3, Issue 6, 2016. Accessed online 7/11/2017: www.canadianaudiologist.ca/issue/volume-3-issue-6-2016/column/science-matters/

9 Now, there are some tricks at work here . . . we can often hear a frequency in a space that is half the size of its wavelength, if the sound can reflect at one end and foldback to complete at the other (this works at one-fourth length and other fractions as well—a phenomenon used in speaker design—which is part of why we can hear low frequencies in small spaces at all . . . but also why they are a bit blurred, because by the time they have folded back to complete, they are out of phase with the original broadband signal impulse). Also, headphones make use of a phenomenon called "proximity effect" (which we will revisit when examining microphones), which makes low frequencies appear louder when a source is very close to a receiver. More on this in Chapters 6 and 16.

10 Particularly in Chapter 29.

11 We will return to this subject frequently, in the book, and in our work.

12 After Helmut Haas.

CHAPTER 4

What Is Sound?—Part 2

Electric Version

Unless you live in a dark cave somewhere (in which case, why are you reading a book about sound systems?), you probably use electricity every single day of your life. In the 21st century, our lives depend constantly on electricity—when a big storm knocks electricity out in our neighborhood, everything around us grinds to a halt. So, given that electricity is so important a part of our lives, it should be pretty easy to define, right?

So let's ask: *what is electricity?*

This question isn't as easily answered as it is asked. Electricity is sometimes defined as potential energy stored as a "charge." It is sometimes defined as electrical current flowing through a circuit. It is sometimes defined as the magic when two people kiss!

Aside from that last one, we're on the right track:

Electricity

1a) The physical phenomena arising from the behavior of electrons and protons that is caused by the attraction of particles with opposite charges and the repulsion of particles of the same charge. 1b) The physical science of such phenomena. 2) Electric current used or regarded as a source of power.[1]

OK, so there are charged particles called electrons. We know from observation that particles of "negative" charge (electrons) flow toward particles of "positive" charge (protons). Why are these particles charged? This is one of those questions that science tries repeatedly to answer but hasn't quite solved. Atoms are made up of protons, electrons, and neutrons (neutral particles), and the fact that the protons and electrons are charged is simply a part of our physical universe. When we find items in the natural world that innately have a charge, we call them *magnets*.[2] Most of the magnets we use (and we use a lot, in sound equipment) are not naturally occurring magnets, but are metals that because of their properties are able to hold a relatively stable polarized charge for a long time. We apply electric current to them in order to set them to a particular polarity (positive or negative), and then we can use them in our equipment. As we will soon see, both dynamic microphones and most loudspeakers rely on magnets to do their work.

In the sage words of none other than Insane Clown Posse: "Fucking magnets, how do they work?"[3]

The truth is *we don't really know.*

We know a lot about how to manipulate magnetism and electricity; we can generate electricity (usually using magnets in some configuration); we can use magnets to pick up other metals or drive our speakers, but why all that works is a question we're still trying to understand. However, this isn't an electrical engineering textbook, nor is it a high-level physics textbook, so for the time being, we have to accept two facts:

1. Magnets can and do hold a charge.
2. Electricity, which can be defined as the flow of charged particles from one place to another, is real.

So, we know that in the acoustic world, sound is a series of vibrations passing through a medium, yes? (If you don't know this, re-read Chapter 3!) When sound hits a microphone, we transduce (turn energy from one form to another) that acoustic energy into electrical energy. How does this work?

Before we can explain how the microphone magically captures our sound (and no matter how long I study this subject, it still feels like magic), we need to understand some basic things about the workings of electricity, and a few terms that we will use again and again in our work (particularly when we work with power amplifiers).

> ### Circuit
>
> A path in which electrons from a voltage/current source flow. The path is circular, in that the charged particles flow through the circuit, through devices, and back to the origin point, where they are drained out of the device via the "ground" or "earth" terminal.

So, if we have charged particles, they can be drawn out along a circuit path, and in the middle of that path are our electrical devices, which make use of the energy in the charge to do work. Electricity always flows from negative to positive (electrons seeking protons). Charged particles, like unequal air pressures in the previous chapter, want to level themselves out. Why? It is their essential nature.[4] Negatively charged particles always seek positively charged particles, and when they meet, they neutralize one another, dissipating the electric charge. Because of this property, we can attach metal wires to a source of charge, and if the wires connect from the negatively charged source to the positively charged destination, the particles will flow along the wire from − to +.[5]

What is the source of our charged particles in a circuit? It's a form of stored potential energy:

> ### Voltage
>
> A standard measure (volts,[6] referred to by the symbol V) of the electrical potential of a power source. The potential is measured by virtue of examining two points in physical space—the difference in charge between the two points is expressed as voltage. In a battery cell, voltage is dependent on two separate chambers filled with electrically polarized materials; in a home electrical circuit, the difference is between the source poles feeding power from the local power station.

In short, voltage is the energy stored in a usable form. In the United States, for example, the standard voltage is listed as 120 V (in practice, the usable amount is slightly lower). A single AA battery provides 1.5 V.

Now, if we have a voltage source, but it is not connected to anything, we can't do much with it. A battery sitting alone on a table isn't providing its power to do any work.[7] We have to connect it to something. What do we use to carry electricity from one place to another? Wires. The technical term for these wires is a *conductor*.

> **Conductor**
> A substance in which charged particles easily move from atom to atom with the application of voltage.

These conductors (our wires) are typically made of metals that naturally conduct well, often copper. Now, if you take a copper wire, and connect one end to the negative terminal of a battery, and the other end to the positive terminal,[8] the electrical charge will race around the wire so fast that the wire might actually melt, and at the very least, anyone touching the wire would get a very unpleasant shock. The electricity races around, but because the *resistance* is so close to zero, it connects from one side to the other almost instantly, and we can't do much with that voltage we've just neutralized.

> **Resistance**
> The measure of the difficulty with which charged particles move through a conductor (referred to by the symbol Ω, and the term ohms).[9]

If we want to make some use of the electricity flying around that wire (and we do), we need to put a resistor in place. In practice, every single electrical device of any type is a resistor (or, more accurately, has *resistance*, as complex devices often incorporate a number of resistors at different points of their internal circuit paths, but we're getting ahead of ourselves).

Imagine a lightbulb. A simpler electrical device is hard to find. In its basic form, a lightbulb has two terminals (for the − and + parts of the circuit), and a material in the middle that glows when electricity is applied to it. If we connect the terminals to our battery, using wires, the light glows. The lightbulb, in this case, is the resistor. In order to make use of any electricity, we need to have a resistor in place. If the lightbulb were made of materials with no resistance, the bulb would burn out instantly upon being connected to the voltage source. Figure 4.1 shows a simple circuit with a lightbulb.

Now, let's move our lightbulb from our battery to a wall socket. A standard electrical socket provides (in the United States) 120 V of potential but is usually listed with another figure as well: "amps." Standard electrical circuits in the United States provide 15 amps of current, but what are amps? I don't mean the thing you plug your guitar into onstage.

> **Current**
> A measure of the rate of flow of electricity through a circuit, measured in amperes[10] (shortened to amps, represented by the symbol A). A given electrical outlet is connected to a circuit, and that circuit is rated by how many amperes of current it can deliver before it overheats/melts. A "circuit breaker" is a switch put in place that interrupts the circuit path if it senses too much current being drawn through the connection.

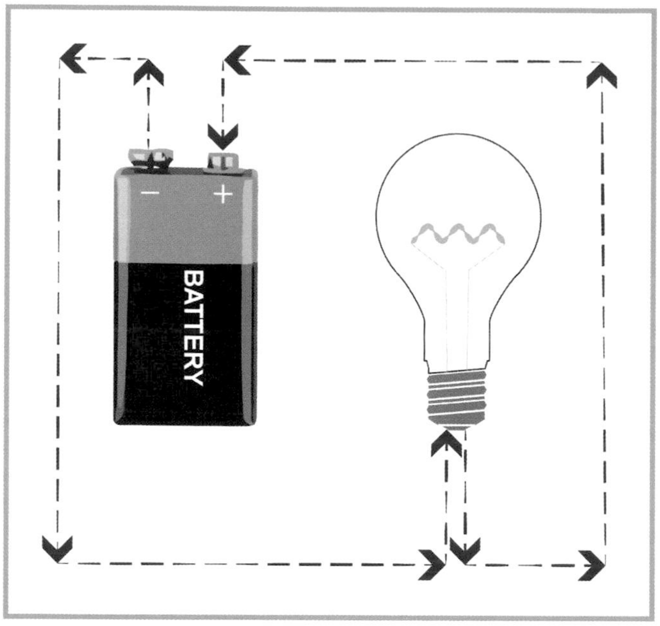

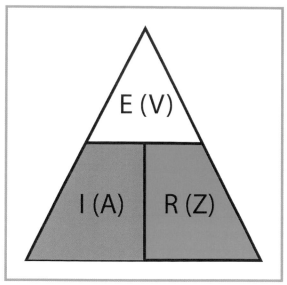

FIGURE 4.2

Ohm's Triangle: E (V) = Voltage, I (A) = Current, R (Z) = Resistance (Impedance)

FIGURE 4.1

A Lightbulb Connected to a Battery in a Simple Circuit

34

These three basic elements of electricity, voltage, resistance, and current, interrelate in a simple equation that is known as Ohm's Law, and that can be easily demonstrated by Ohm's triangle (Figure 4.2):

The elements of the triangle are voltage (represented by either the letter V, or the letter E, which stands for Electromotive Force, the potential energy available for the accomplishment of work), Amperage (represented by either the letter A, or the letter I, which stands for Intensity of flow, an older way of expressing the term "current"), and Resistance/Impedance (represented by the letter R for Resistance, or Z for Impedance, to avoid confusion with other terms).

Impedance

Resistance + reactance, which is the variable response of a conductor to input signal of varying frequency (represented by the symbol Ω, and the letter Z). In practice, most sound equipment has measured Impedance, not Resistance, because sound signals vary in frequency.[11]

The triangle is laid out the way it is in order to simply express the mathematical relationships between the three elements.

$$[E = I * Z]$$
$$\left[I = \frac{E}{Z}\right]$$
$$\left[Z = \frac{E}{I}\right]$$

This simple triangle thus allows us to solve for any one of the three variables, so long as we know the other two. To return to our lightbulb now connected to a home outlet, we know that the home outlet is rated as delivering 120 V.

The typical 100-W incandescent lightbulb has a fixed resistance of about 144 Ω once it is lit/heated.[12] How much current will it draw?

$$\left[I = \frac{120}{144} \right]$$
$$[I \approx .83A]$$

This lightbulb thus draws approximately 0.83 Amperes of current. But there was a critical power term in there that we haven't yet addressed: watt. A 100 W[13] lightbulb will perform roughly as noted earlier, but what is a watt?

> **Watt**
>
> The standard unit of measure for power (represented by the letter W, and sometimes the letter P for "power"). Power is defined as the rate of doing work.

Note that this is distinct from current, which is the rate of flow of energy, whereas watt is the measure of the rate at which that energy does work in a specific device. Power amplifiers are often listed by wattage (e.g., "a 500-W amp channel").[14] How does wattage relate to our basic Ohm's triangle? Well, we must expand our Ohm's Law model from the triangle to the wheel (Figure 4.3)!

This can be a little overwhelming, but what you will quickly see is that there is a tight relationship among the four power elements. We won't go into the wheel in great detail here, for the time being, the most important part of this setup is what we refer to as the "West Virginia" equation:

$$[W = V * A]$$

(also shown on the wheel as $[P = E * I]$).

This means that if we know the voltage of our circuit, and the amperage drawn by a device, we can determine how many watts it will use (at what rate it does work). To once again return to our lightbulb, since we know that it is connected to a 120-V circuit, and it will draw approximately 0.83A of current, this means

$$[W = 120 * .83]$$
$$[W = 99.6]$$

Our lightbulb uses electricity at a rate of 99.6 W, so the 100-W designation under which that bulb is sold is essentially correct.

What happens in our lightbulb circuit if we increase the resistance? If the resistance were, for example, 1440 Ω instead of 144 Ω?

$$\left[I = \frac{120}{1440} \right]$$
$$[I \approx .08A]$$

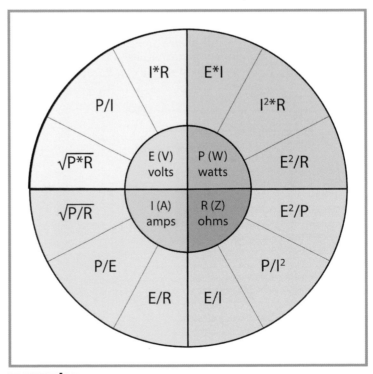

FIGURE 4.3
Ohm's Wheel: V = Voltage, A = Current, Z = Impedance, W = Wattage

This is practically no current flow at all. In that case, the wattage used would be

$$[W = 120 * .08]$$
$$[W = 9.6]$$

As anyone who has used conventional lightbulbs knows, a 9.6-W lightbulb would be much dimmer than a 100-W lightbulb. Now, imagine a lightbulb with infinite resistance. What happens to the current?

It stops. No current flows.

This is what happens when a loudspeaker melts its voice coil (one of two basic ways you can "blow" a speaker).[15] When this happens, internal parts overheat, preventing them from moving when power is applied, and thus creating a "perfect resistor" which is to say, one that has infinite resistance and effectively allows no current to pass (and therefore makes no sound).

What happens if we go the other direction, and resistance is 0? Remember our wire directly connected to the battery terminals? Our lightbulb would effectively become a piece of wire and melt with the instantaneous power surge.

Now, when we plug a device into the wall, there are typically three prongs that connect. In the United States, we use the so-called Edison power connector, which is properly called a NEMA 5 receptacle (Figure 4.4).

These receptacles are sometimes referred to as NEMA 5–15, where NEMA 5 refers to the form factor, and the 15 refers to how many amps the connector is rated to deliver.

In a typical three-pronged connection, each prong has its own function. The two vertical slots, known as the neutral and hot slots, are where the actual current flows. In older residences, you will sometimes still find receptacles that only have two slots—these are simply the neutral and hot wires, without the third ground wire. But, what do all these do? Before we can explain their functions, we have to talk a bit about AC/DC, and no, we don't mean the hard rock band.

Home receptacles deliver what is known as alternating current (AC). This is opposed to direct current (DC), which is what most electronic devices require (see Figure 4.5 for AC and DC waveforms). To oversimplify a bit, alternating current is current in which the polarity of the signal switches back and forth constantly, where direct current is one in which the polarity is static. This means that the poles at the transmitting end of an AC system reverse themselves, so at one moment, Pole A is negative, and Pole B is positive, and at the next moment, Pole A is positive, and Pole B is negative; thus, the direction current flows is switching as the poles switch. In US AC circuits, the polarity switches 60 times per second (at a rate of 60 Hz), in Europe and other parts of the world, it typically switches at 50 Hz.

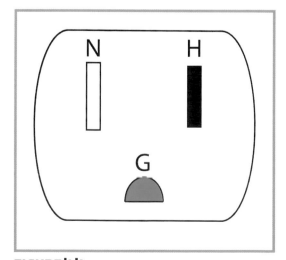

FIGURE 4.4
NEMA 5–15 ("Edison") Receptacle (H = Hot {Black Wire},
N = Neutral {White Wire}, G = Ground {Green Wire})

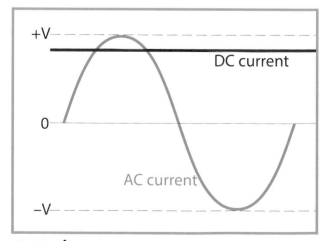

FIGURE 4.5
AC vs. DC Waveforms

The hot wire is the source of power that a device plugged into our receptacle uses. The neutral wire is the wire that returns current from the device. Why does this need to exist? Aren't we using the electricity in our device?

Well, sure, but no device is 100% electrically efficient. If a device is, say 85% efficient, that means 15% of the power sent to it is either burned off as waste heat or is simply returned out of the device essentially unused. Different devices have different rates of efficiency, but even the most exactingly engineered devices still waste or return a little bit of current.

So, in an old-fashioned, two-prong circuit, current arrives at the device via the hot, and leaves the device via the neutral. Where does it go once it is sent over the neutral? The neutral wire eventually connects through the power system to what we call the "ground" or "earth" connection, which is usually a literal metal pipe sunk into the earth. Why is this necessary?

Remember when we said that charged particles will always seek their opposite in order to neutralize themselves? Well, the ultimate neutralizer is the Earth itself. It is a more powerful attractor of electricity than just about anything else, and so by returning the current to earth, we dissipate it and thus ensure that it does not harm us.[16] When we get an electric shock from a device, it is generally because we are physically connected to earth, and the electricity will take the shortest possible path to earth, even if that means traveling through us. What do we mean by "shortest"? Generally, we mean the path with the lowest electrical resistance.

Short Circuit

A malfunction in an electrical device whereby electricity takes a path other than the one intended because a lower-resistance path has become available to it.

A short circuit is sometimes related to physical distance—e.g., if a circuit usually takes 100 m to complete, but wires touch across the path at a distance of 5 m, the electricity will take the physically shorter path. This isn't strictly because it is physically shorter, however, it is because—all paths being made of the same materials—the resistance will be lower for the physically shorter path. Any wire has a nominal resistance, and the longer the wire, the greater the resistance overall. The physically shorter path is taken because (assuming the same wire in both paths), it presents less resistance. If the physically shorter path had a large resistor wired into it, the electricity would still take the physically longer (lower-resistance) path.

When we get a shock, it is because we are a lower-resistance path to earth—we are the "short circuit." This is why electricians wear heavy rubber-soled shoes, as the rubber is an insulator (opposite of a conductor), and disconnects them from the ground. It is also why our wires are always enclosed in some kind of plastic or rubber insulator, to avoid having electricity leaping off of them and into us!

Now, if the neutral path takes electricity back to the ground, why do we have the third pin, which we actually call "ground"? Well, in two-wire systems, static electricity would build up between the two wires, creating a fire hazard. Also, when lightning struck power lines, the surge over the hotline didn't have an escape path without going through the devices and receptacles, so devices were easily damaged by the resultant extra flow of current. Modern electrical systems give us the ground pin so that there is a dedicated escape path for surges, static build up, and the like. While the hot and neutral both have electrical potential (voltage) on them, the ground pin has no such potential and is just wired to that aforementioned metal pipe.

You might notice that the ground pin on a NEMA plug is slightly longer than the other two. This is designed to allow the escape path to be established before current races through the device you are plugging in. That way, when we complete the circuit by plugging a device in, the return path is already available, and is a shorter/lower-resistance path to earth than *through you!*

We do still have two-prong devices, though, so are these unsafe? Not generally. If they are contemporary devices, they are devices that use very little power and that have a "double insulated" configuration on all of their power connections, so there is very little risk of shock. If they are old devices (prior to 1961, when the NEMA 5 standard was made official), there might be a risk of shock, especially if any of the old wiring is internally frayed. Caution should be used with such devices.

Every once in a while, you may find a guitar or bass amplifier that was originally a three-prong device that now has its ground pin cut off. The people who cut this off usually say something like: "It hummed too much, and when I cut the ground pin off, it stopped humming." Yes, sometimes an amp can hum, because the building where the amp is being used has a poor grounding system (or occasionally because of poor wiring inside the amp). If the ground wire connects to either of the signal leads in the power system, this can generate a low hum (which in the United States we call "60-cycle hum"). However, while sawing off the ground pin may solve the hum problem, it creates a much worse problem, which is that now the escape path designed into the amp is missing, and while the neutral wire will still allow some current to escape, the current that was designed to escape via the ground wire builds up in the amp itself, creating—at best—way more overheating and—at worst—risk of intense shock. Don't saw the ground pin off of three-prong devices, ever.

To return to our AC/DC dichotomy, I did say that most electronic devices require DC, so why are we sending AC through the walls, and how does it turn to DC? The short answers here are

- AC is sometimes thought to be easier to send over long distances than DC. This is false, but what *is* easier is to transform AC from the extremely high voltages that are sent from power stations down to the 120 V we get in our homes. Thus, AC is what we use for transmission in the United States.
- To turn AC to DC, we use a device called a *rectifier*. To oversimplify, a rectifier generally takes the part of AC that is in one polarity as is and flips the polarity of the rest of it to match, and then sends the new, coherent, DC signal onward.[17]

Why do we require DC? Because the heart of complex electronic devices is the *transistor*, and transistors require a "DC bias," or steady state of electricity, in order to perform their operations. Why *that* is the case is an electrical engineering topic beyond the scope of this book.

Tales From the Field

I was working at an outdoor amphitheater in Los Angeles as part of an audio strike (uninstallation) crew. We were working with what is known as a "three-phase" power system. This is a system wherein a very high-voltage signal is sent broken up over three identical hot legs, each offset in time by 120° of phase, and each sharing a common neutral and ground connection. The leads of such a system in entertainment are usually very large locking connectors known as camlocks. Since each lead is a separate cable in such a system, there is no "longer ground pin" for safety, you simply have to attach and detach the cables in the proper order. When attaching, you always attach ground first, neutral second, then hot leads, and you always detach in reverse of that order. This is done to ensure that the ground path is always available as the shortest path to neutralizing the charge, to prevent you from getting shocked. Well, a fellow crewmember who shall remain nameless was asked to disconnect the three-phase camlocks. He said OK. The crew chief asked him if he knew how to do it. He said yes. He promptly walked over and disconnected the ground connector first. He dropped that connector and then disconnected one of the hot leads, and residual charge in the system arced visibly between his camlocks and the wall panel, like a giant lightning spark out of an old Frankenstein movie, and the force of that shock knocked him clear across the room, a good 8' with his back against a wall. He was extremely fortunate that he didn't touch any of the metal parts, or he would have been fried by 240 V of electricity and could have died. My crew chief saw this happen and raced to reconnect the ground lead, and we checked to make sure our colleague could see how many fingers we were holding up. He was lucky and wasn't truly injured, but let this be a warning: electricity is dangerous! Ground needs to be first connected and last disconnected, whether in an Edison system with a longer ground pin, or a large-scale industrial power system like three-phase.—J.L

So!

Let's return to our original question, which is *how does our microphone magically turn acoustic energy to electrical current?*

While there are several types of mics, and each works differently,[18] the basic operation is that in a microphone, there is a stable charged field. At the front of that field is the microphone's diaphragm—the very thin round membrane into which we sing or play instruments. When sound waves hit that membrane, the pressure changes make the membrane fluctuate in space back and forth (over a very short distance). That motion, in turn, creates variance in the stable charged field. That variance changes the stable charged field into a fluctuating current, which then flows up our conductor (mic cable) to the mixing console, preamplifier, or whatever device the cable is plugged into. The current that flows fluctuates at a rate *analogous* to the rate of pressure changes in the acoustic signal (see Figure 4.6).

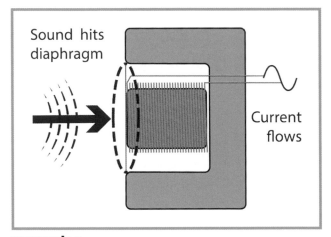

FIGURE 4.6
A Dynamic Microphone Diaphragm Oscillating to Induce Analogous Current

Thus, the *frequency* of the acoustic sound is captured as the *analog electronic* audio signal that we manipulate in our equipment.[19] The exact opposite process happens in a loudspeaker, where the large magnet that creates a stable charged field in the speaker driver is induced to fluctuate by the electrical current sent from the power amplifier, which thus induces the speaker cone to vibrate and create the acoustic sound we hear.

Why does destabilizing the stable charged field make current flow? It's kind of like our pressure escaping the balloon in the last chapter. In its oscillations, the charged field will vacillate between positive and negative charge. As we know from earlier in this chapter, negative charge always rushes in to seek positive charge, and so the oscillations in the charged field in the microphone create a situation in which the charge goes negative/positive/negative/positive . . . and on down the line, chasing itself until it arrives at our system inputs.

Now, the microphone will have a rated output impedance, a number listed in Ohms. For the world-famous Shure SM58 (Figure 4.7), the output impedance is listed as 300 Ω.

If you plug that into a little Mackie 1604 mixing console, the input impedance of the mic input is rated at 2500 Ω. This means that when the mic gain is turned all the way down, the impedance of the input is much higher than the output impedance of the microphone, and thus the channel will resist the signal—meaning that turning the gain all the way down turns off the mic signal. By turning up the gain knob, we reduce the input impedance (and feed into an amplifier circuit), which induces the current to exit the microphone and enter the mixer channel, where we make use of it.

FIGURE 4.7
Shure SM58
Source: Courtesy Shure, Inc.

Whew!

Electricity is a complex topic, and indeed many books have been written on the material we've grazed over in this chapter, but you can rest assured that

A. When we encounter these topics again in later chapters, we will reinforce the important parts with examples, so we will expand, reaffirm, and further clarify what we've just learned.
B. A lot of system design relies on the manufacturers to have created gear that naturally works together electrically without too much fuss. It's not until you get to super-advanced systems with tons of triggers and interactivity that hardcore electrical system knowledge comes back into play.

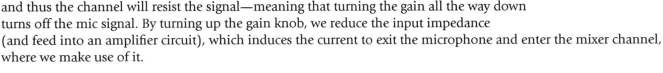

39

OK, so now we understand everything about how audio works, right? We understand acoustic pressure, and how it turns to analog electronic signals, and back, so we can move on to the fun gear stuff, right?

I know . . . that joke is getting old. But no, we have to first understand some digital theory, so we understand the terms and processes by which that analog electronic signal turns into the 1s and 0s racing around contemporary sound systems.

Then we get to the gear. I promise!

Notes

1. Definition from *The American Heritage® Dictionary of the English Language*, Fifth Edition, by the Editors of the American Heritage Dictionaries. Reprinted by permission of Houghton Mifflin Harcourt Publishing Company. All rights reserved.
2. Well, to be fair, naturally occurring magnets are actually called *lodestones*, and human beings first discovered magnetism by finding lodestones to which other metals stuck. We are not sure how lodestones are created, but the leading theory is that it is a result of lightning striking certain types of minerals on the earth.
3. Insane Clown Posse, "Miracles" from the album *Bang! Pow! Boom!* 2009
4. And this isn't a book about entropy. At least not directly.
5. You might ask why negative flows to positive and not the other way around. The truth is that the particle polarities were named before we knew which direction they flowed, and by the time we discovered the direction of flow, the names had stuck. Science is filled with terms like this that are just the accidents of history. Why do Americans measure distance in "feet"? Same basic answer.
6. Named for Alessandro Volta.
7. The physics definition of work involves any situation in which a force is applied to create a physical change in an object.
8. Note: DO NOT DO THIS.
9. Named for Georg Simon Ohm.
10. Named for Andrè-Marie Ampere.
11. Different electronic analog signals have different general impedances and strengths/levels. You may have heard of "line level" or "mic level" signals. See Chapter 8 (on direct inputs, which are impedance/level transformers) for a clear explanation of these signal levels.
12. "Say Goodbye to the Incandescent Lightbulb (As We Know It)" by Chuck Newcombe, *Fluke.com*, accessed online 7/15/2017 www.fluke.com/fluke/sgen/community/fluke-news-plus/articlecategories/energy/say-good-bye-to-the-incandescent-lightbulb
13. Named for James Watt.
14. This will be discussed in detail in Chapters 14 and 15.
15. Voice coils will be explained in Chapter 16.
16. You might wonder why we don't re-use this current as it exits devices. The general answer without going into too much detail is that upon exiting the device, the current is no longer regulated in its timing and strength, and so it can actually damage devices it encounters, it is easier to simply dissipate it back to earth than to try to re-shape it into a usable form.
17. This is actually only one type of rectifier, called a "full-wave" rectifier. There are other types, but since audio devices require very steady DC, many if not most of them rely on this type of rectifier in their power supplies.
18. Examined in detail in Chapter 6.
19. You might have noticed that microphones have three pins connecting to them (via an XLR connector), whereas something like a vinyl record player typically has two pins (via RCA connectors, which visually look like one pin, but are actually two including the metal ring around the connection point). Why is this? The three-pin connection is a *balanced* connector, where the two-pin is an *unbalanced* connector. We will explain the difference between these, and how they work, in Chapter 24, but for now, suffice it to say that balanced connections can travel much longer distances (up to 1,000') without loss of fidelity or added noise, where unbalanced connections are for much shorter distances (less than 25') and are much more susceptible to noise.

CHAPTER 5

What Is Sound?—Part 3

Digital Version

Sound is made in acoustic space. Microphones pick it up and transduce it to an analogous electronic signal. Then it magically transforms into data, and we can manipulate it endlessly with our computers.

End of chapter!

No, not quite. How does digital audio work? What are the terms we need to know in order to comprehend its operation and properly configure digital audio equipment?

In our simplified system model in Chapter 2, we showed analog inputs feeding into analog-to-digital (A/D) converters. What is a converter, and how does it turn analog signal into digital signal?

Any sound source is essentially made up of two components:

- Frequency
- Amplitude

Frequency defines what pitch or series of pitches we hear. In the case of a simple sine wave, we hear one pure frequency with no complex added material. Once we play that sine wave through a keyboard amplifier onstage, we already have a less-than-pure signal, as the amplifier electronics, speaker materials, and room in which they sit, all contribute sonic artifacts to the tone. Any sound that is *not* a pure sine wave, is—by definition—more complex. In the case of something like a crash cymbal (see Figure 5.1), there is a whole range of frequency information that adds up to make the complex tone of that instrument.

Regardless of complexity, frequency is a primary component of any audible material.

> **Amplitude**
>
> The objective measurement of the degree of change in pressure a given sound causes, when measured versus the atmospheric baseline.

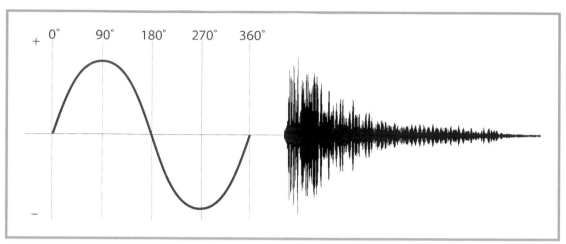

FIGURE 5.1
(Left) Sine Wave; (Right) Cymbal Crash (w/Kick Drum) Waveform

Amplitude is essentially *loudness* quantified, a measure of its relative dynamic range.

The two essential components for recreating a sound, therefore, are the frequency/pitch information, and the amplitude/loudness information. With those two, we have enough information to recreate a sound—an important concept for digital audio, as we shall see.

In the acoustic and analog electronic worlds, audio signal is a continuous stream. Even when things fall "silent" they are never truly silent, as rooms and devices have a level of (usually quiet) self-noise that sensitive equipment can still read. If you are recording to analog tape—a medium that takes the electronic signal, runs it through magnetic tape heads, and uses the fluctuating intensity of the signal to polarize minuscule magnetic particles bound to a plastic tape, charging them in an analogous polarity to the audio signals seen at the inputs of the device—the difference between loud and quiet is a continuous slope. Everything, in fact, is a continuous stream of information. When a band is playing together, the complex combined frequencies of their blended sound are moving smoothly through time, without clear breaks.

But that's not how computers work. In order to function, computers need everything to take the form of individual points of data, points that can be stored numerically and retrieved later. So, in order to save what is in nature a continuous stream of data, we must undergo a process of *discretizing* the information, which is to say we have to break up the continuous audio into tiny manageable slices and store exact numeric information about those slices.

In order to properly understand what is happening when we digitize analog audio signal, we should take a brief tour through history. As early as 1928, a scientist named Harry Nyquist wrote a theorem[1] explaining the basic concept of how to digitize audio (he was writing about telephone and communications data). His basic concept was this (*paraphrased*):

> *If a system were able to capture twice the number of samples (digital snapshots of position) as the frequency rate of an audio signal, it should be possible to perfectly recreate that frequency based on the sampled data alone.*

Think of our sine wave again. A sine wave oscillates at a completely predictable frequency, and each compression and rarefaction is identical to the previous compression and rarefaction. Therefore, it was thought that if we could capture one piece of data measuring the compression, and one measuring the rarefaction, we would have enough information with which to recreate said sine wave (see an example in Figure 5.2).

The essential idea here is that for any given waveform, so long as we have enough computing power to capture the compression and the rarefaction, we'd capture the wave perfectly. For the human hearing range, which tops out at 20,000 Hz, this implies that if we have a sampling frequency of 40,000 Hz (two samples per cycle, one each for compression and rarefaction), we would have a perfect representation of the sound.

Sample Rate

The number of "snapshots" (samples) per second captured or reproduced by a digital audio system. The sample rate must be *at least* twice the highest frequency to be captured/reproduced. Sample rates of digital audio devices in the same signal chain must match.

Other scientists and engineers (including Claude Shannon, for whom the eventual Nyquist-Shannon Sampling Theorem would be named) took these concepts and ran with them. By the mid-1940s, the first practical digital audio system was in use by the military for communication transmission.[2]

The Nyquist-Shannon theorem is focused on sine wave reproduction, and so does not address the questions of fidelity of capture of complex audio signals; however, the work did identify the central concept that remains with us today: in order to digitize audio, we must turn a continuously flowing analog signal into discrete steps of information (see Figure 5.3, where a sine wave has been "discretized," albeit with an extremely reduced sample count for visual clarity's sake).

Now, this set of theorems works very well for sine waves, as the periodicity (variance in time of the compressions and rarefactions) of a sine wave is completely stable. A 440 Hz sine wave is always producing oscillations of 440 cycles/second, spaced evenly across that second, every single time it sounds. However, a sound that has a 440 Hz fundamental and a bunch of overtones as well as some wobble in pitch won't be as stable in periodicity as the sine wave (see Figure 5.4).

Since our fundamental note here is still 440 cycles/second, the 40,000 samples/second system needed for basic full-range audio reproduction provides plenty of data points to capture the variations in timbre/overtones produced by the guitar.

In Figure 5.5, we see a complex signal. The sample rate shown here (which, again, is vastly reduced from a real-world sample rate) is higher than the very broad steps we saw in Figure 5.3, but while more detail is captured, there is still some lost. Now, if this sound were a crash cymbal, or another sound with complex information in the upper reaches of human hearing (10–20 kHz) and beyond (some crashes will create tones up to 30k), we can see a problem.

First, at the top end of the human hearing range (20 kHz), a 40,000 samples/second rate *only* allows us to capture and reproduce a sine wave (one sample for compression, one for rarefaction, in even periodicity). Anything with

43

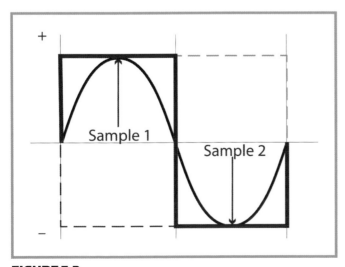

FIGURE 5.2
A Sine Wave With Two Data Samples Drawn Around the Compression and Rarefaction

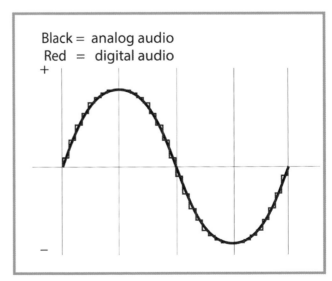

FIGURE 5.3
Analog Signal vs. Digitized ("Discrete") Signal

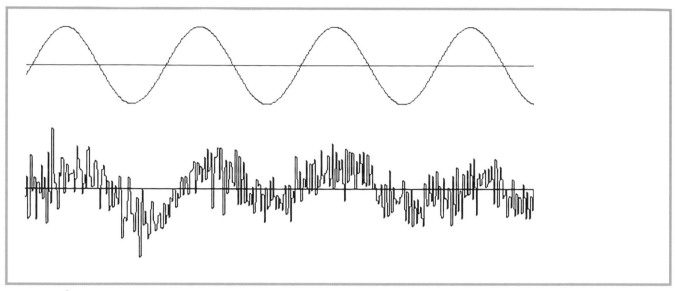

FIGURE 5.4
A 440-Hz Sine Wave (Top) vs. A440 on an Acoustic Guitar (Bottom)

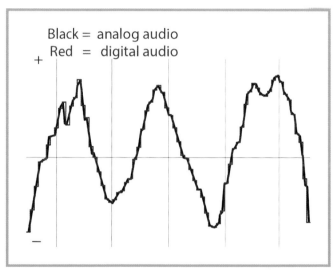

FIGURE 5.5
A High Sample Rate Capturing a Complex High-Frequency Sound
Source: Drawing by Anna Ehl, Courtesy Anna Ehl

more complexity will be lost, because there are simply not enough data points to represent the waveform. For many years, the standard of consumer digital audio quality was the CD. By definition, the sample rate of CD media is 44.1 kHz. Leave aside that 4.1 kHz (it has a different function, which we will address in a moment), this means that CDs are sampling effectively at 40,000 samples/second, which means that low-frequency information will be faithfully represented (because 40,000 data points applied to 40 cycles/second means that each cycle can be recorded with 1,000 data points), whereas any complex high-frequency sound will be recorded with limited fidelity. This is why many studios record at much higher sample rates (up to 192 kHz), so that high-frequency content is captured faithfully, and any loss of quality when converting to CD sample rate can be managed intelligently rather than done automatically on capture.

The second problem with our cymbal crash, however, comes from the frequencies it produces that are above the human hearing range. If we simply sample at a rate of 40 kHz, we can capture some version of the sound up to 20 kHz, but once the vibrations exceed this range, the gear will still try to capture those vibrations, but will not be able to. As a result of its attempt, the equipment will generate a phantom tone within the audible range, which we know as an *alias frequency*.

Imagine a pure sine wave at 30 kHz. If our sample rate is 40 kHz (meaning a top *captured* frequency of 20 kHz), then the induced alias frequency will be a 10 kHz tone in our recording that was not produced acoustically but is simply an artifact of the digitizing process. The equation for determining alias frequencies runs as follows:

$$f_s - f = alias,$$

where f_s is the sampling rate and f is the audio being captured. Thus, if our sample rate is 40 kHz, and we are recording 30 kHz, we get a signal that only captures a portion of the 30 kHz compressions and rarefactions that is represented as 10 kHz. If our sample frequency stays solid at 40 kHz, this aliasing has a limit. If there were a 40 kHz tone in the acoustic recording, a 40 kHz sample frequency produces the equation $40 - 40 = 0$. There is thus no aliasing from any frequencies 40 kHz and above, when recording with a 40 kHz sample rate, aka 44.1 kHz. Up to 39,980 Hz signals can produce alias frequencies at 40 kHz sample rate—a 20 Hz alias frequency. This assumes your mics, interfaces, and other hardware pass on that high frequency to the sampling engine—which they may not.

When capturing audio sources with frequency content up above the hearing range, we run the risk of introducing alias frequencies down in the hearing range. This, perhaps obviously, is not cool. We don't want ghosts hanging around in our sound, so how do we fix this?

Digital audio systems have a built-in solution for this, called *anti-alias filters*. Now, in a perfect world, a digital audio system would perfectly reproduce audio up to 20 kHz, and then have a brick-wall cutoff so that nothing beyond 20 kHz makes it into our signal. However, in practical reality, a brick-wall filter is something that, despite decades of attempts, engineers have been unable to create. Therefore, we need some bandwidth (space in the frequency spectrum) for the slope of that filter. Thus, CD audio is sampled at 44.1 kHz (instead of 40 kHz) to allow for 4.1 kHz of space for the filter to slope down and out of audible range. The wider the filter space, the smoother its performance (creating fewer audible artifacts at the top end of the frequency spectrum), so DVD video (which set its standards later in history than CDs) has a sample rate of 48 kHz, for even smoother filtering. Every professional sample rate above either 44.1 or 48 is a whole number multiple of those figures (88.2 kHz, 96 kHz, 176.4 kHz, 192 kHz). The added samples in those higher rates provide more samples/second, primarily to the benefit of high-frequency content, and less phase delay at the filter (the smoother and longer the filter, the more reduced the natural filter artifacts).[3]

Now, while we should have a pretty good understanding of how we capture frequency data (with a number of samples, or "snapshots," per second), we still haven't addressed how to capture the amplitude data. For any given input signal, the sample rate determines how the frequency data itself is captured, but there is another figure that corresponds to amplitude:

> ### Bit Depth
>
> The number of bits of information in each individual sample captured. The bit depth of a medium relates closely to the signal to noise ratio, or available dynamic range, of a digital medium.

What is a bit? A bit is the fundamental piece of digital information in a computing system. The root language that all computers speak (albeit usually encoded in a higher-level language that shortens the data forms) is binary. In a binary system, each piece of information is represented as either an ON or an OFF (a 1 or a 0). Thus, digital data is so-called because it is stored as one of two digits, 1 or 0. Now, a single slot of information, for example, a "1" is one bit of information. Bits laid end to end form larger "words" of data. The standard "word" of basic computing systems is an eight-bit word known as a byte. An eight-bit byte looks like this:

01101011.

We use bytes in order to compute larger numbers than just 1 or 0. In binary systems, we count starting from the right side, and each value to the right represents a higher power of two. In the example noted earlier, the byte represents the numeric decimal value 107. Figure 5.6 shows how this is counted.

The byte is the fundamental counting number for data storage, which we expand into the familiar kilobytes, megabytes, and gigabytes, which we will explore further later in this chapter when discussing data storage.

Returning to audio, the bit depth of an audio medium is a number related to the fidelity with which an audio system or device can represent the amplitude of a given audio signal. To understand this, first, let us once again consider our sine wave. In order to capture the amplitude of a sine wave of stable amplitude, we only need a system capable of capturing two values, the peak of the compression and the trough of the rarefaction.

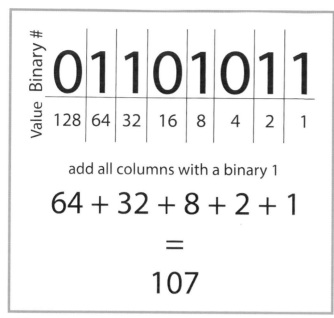

FIGURE 5.6
Counting the Decimal 107 in Binary

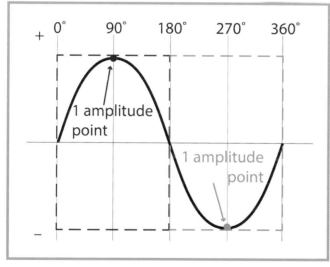

FIGURE 5.7
One-Bit Capture of Sine Wave

46

This would be a one-bit system, recording either a 1 or a 0 to represent peak or trough. We assume that the slope and timing between the two is fixed and repeating, because that is the nature of a sine wave. See Figure 5.7.

However, as we already established with regard to frequency, the real-world amplitude of sounds is seldom mathematically perfect, and we need more data points in order to properly capture the information.

By increasing to three-bit, we can now represent eight data points. And so on, growing by multiples of two. An eight-bit word can represent values from 0 to 255 (256 data points).[4] A 16-bit word (the standard bit depth for CD-quality audio) can represent values from 0 to 65,535 (65,536 points of data).

This means that in CD-quality audio, we have over 65,000 points of data with which to discretize our amplitude values. This sounds like a lot! CD-quality audio must be able to perfectly capture and reproduce real-world amplitude, right?

As you have no doubt guessed by now, whenever I ask one of these leading questions, the answer is generally no.[5]

In the real world, as we have established, our human hearing system has an effective range of about 130 dB. Without going into a ton of unnecessary math, in practical application, each bit of depth corresponds roughly to about 6 dB of dynamic range in a medium (whether a CD, a digital audio channel in a live console, or whatever). Therefore, CDs, which have a bit depth of 16, have a maximum dynamic range of approximately 96 dB. While that is still superior to the dynamic range available on a vinyl record (which is generally between 60 and 80 dB, depending on the manufacturing technique and density of content on the disc), this still doesn't live up to our own ears. DVD and cinema-quality audio have a bit depth of 24, which in theory provides 144 dB of dynamic range. This is definitely enough to accurately reproduce human hearing, but in practice, that dynamic range is limited by the electronics doing the encoding/decoding. If the converter being used has a maximum dynamic range of 127 dB, it doesn't matter that it is set to theoretically capture 24-bit audio, your recording will still be limited to 127 dB.[6]

At any rate (pun definitely intended), once we have assigned one of 65,000 + available data values to each of at least 40,000 samples per second, we have captured digital audio. Awesome! However, there are still two major problems we can encounter:

- Jitter
- Quantize error

First, jitter. In a digital audio converter, how do we ensure that each of the 40,000 audio samples each second are representative of *exactly* 1/40,000th of a second? If a system isn't calibrated properly, maybe sample 1 is 1/40,000th of a second, but sample 2 is more like 1/25,000th of a second, and so on. To remedy this, every digital audio system needs a clock. In digital audio equipment, the clock source is usually provided by a small crystal, which (when current is applied to it) will oscillate very regularly, emitting a square wave signal (see Figure 5.8) that tells the device exactly when each sample is supposed to start and end.[7]

If this clock is inside your digital device, and it is the only digital device in your system, the clock is said to be *internal*. However, once we send digital audio from one machine to another, we must make sure they are clocked together, such that sample #1,244 on the first device is exactly the same as sample #1,244 on the second device. Thus, we have to send clock signals from one device to another.

Clock signal (or word clock, as it is often called, after the fact that one digital word represents each sample) is sent from one device only (the clock master) to all other digital audio devices in the system (the slaves). Digital audio devices working together must all be set to the same sample rates and bit depths, and must only have one master clock. If you try to configure a system with more than one master clock, they compete for dominance and cause system errors and crashes.

Now, what happens if sample #1,244 from the first device arrives instead at sample #1,246 on the second device? This is what we call *jitter* (see Figure 5.9 for an example of properly clocked signals versus jittery signals). Jitter is a kind of clocking error that can result from improperly configured systems, mismatched sample rates, or a poorly manufactured master clock. Jitter sounds like digital equipment having a stroke. Sounds glitch and appear to stutter and crunch in an unmistakably digital way. We want to avoid jitter at all costs. Thus, we must make sure to use the most stable clock device available to us for our master. How do we know which device is most stable? Read reviews, listen to word of mouth, and learn from experience (unless, that is, you have sample-accurate testing devices available to you, which, if you are reading this book, I am guessing you don't).[8]

The process of capturing amplitude and assigning it to discrete numeric values is called *quantizing*.[9] The practice of quantizing can run into its own native problems. What happens if you have a complex waveform whose amplitude in real terms falls at a value between two available bits of data? Or if a device just records the wrong value (as sometimes happens)?

In Figure 5.10, Point A represents a piece of amplitude information that falls squarely between two available digital data points (horizontal lines, representing bits). How does the computer decide where to record that amplitude? If the analog value were, say, two-thirds of the way to one or the other value, it would record the amplitude as the nearest available value, but when the analog amplitude falls exactly between the two data points, the converter has to make a judgment call.

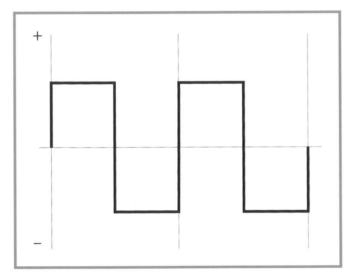

FIGURE 5.8
Square Wave Clocking Signal

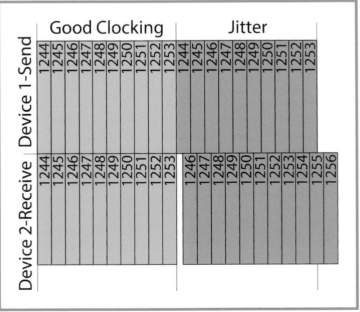

FIGURE 5.9
Properly Clocked Signals vs. Jittery Signals

47

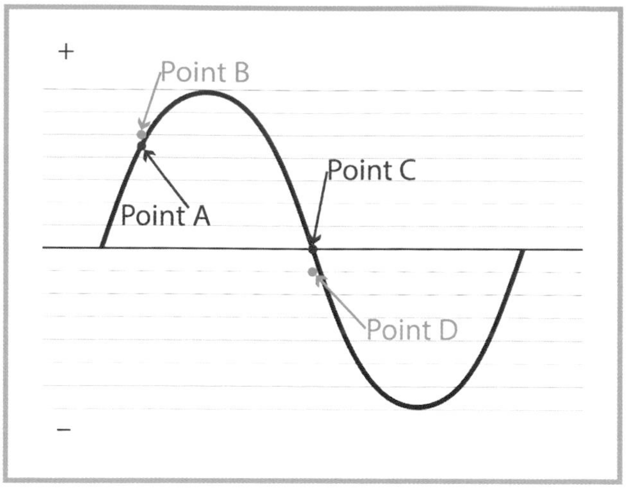

FIGURE 5.10
Quantize Error (Amplitude)

Different converters have different algorithms by which they make this decision—some randomize the choice, some insure a Gaussian distribution over the length of a given period of time, some operate in another proprietary hybrid mode. Point B in Figure 5.10 represents the choice the computer has made for Point A's amplitude. This is a reasonable estimate.

However, every now and then, the converter will make a mistake, which is called a *quantize error*. In Figure 5.10, Point C is a piece of amplitude information that is misinterpreted by the computer, at Point D—a quantize error.

There is no clock that can fix quantize errors, however, the greater the bit depth, the less likely these errors will be noticeable (this is the theoretical advantage of 32-bit floating point systems). Quantize errors are not nearly as audible as jitter, but they sound simply like less-faithful dynamics in the digital audio when compared to the analog.

Audio people love to talk about "better converters" in audio gear. When they know what they are talking about (which isn't all the time), what they generally mean is that sensors that read the frequency and amplitude of the incoming analog signal perform with great fidelity, accurate transient response (responding quickly to notes that start sharp and end fast, like a snare drum hit), and deliver results that are pleasing and musical to the ear. Good clocking, high sample rates, and high bit depth are key to "good converters."

Finally, before we leave digital theory, a quick word on storage of digital recordings. While this book is primarily focused on live systems, we are constantly recording live shows these days, and one of the persistent questions we have to answer is *how much hard drive space do we need for our recording?*

This is a really simple thing to calculate. For any given audio recording, we use the following equation:

$$[data\,in\,bits\,of\,recording = sample\,rate \times bit\,depth \times 60\,seconds\,in\,a\,minute \times$$
$$\#\,of\,minutes\,you\,plan\,to\,record \times \#\,of\,tracks\,you\,are\,recording].$$

Of course, we don't see hard drives listed in the number of bits, so we then divide the number from the earlier equation by 8 (for bytes of data). Then by 1,024 (for kilobytes . . . I know kilo usually means 1,000, but because computers work in binary, the value is 1,024). Then by 1,024 again (for megabytes). Then by 1,024 again (for gigabytes).

Tales From the Field

When the MBOX (Pro Tools' original home recording interface for computers) was released, the original version featured a word clock input jack, as well as an internal clock. When I was in engineering school, one of my teachers did a demo—they recorded a mic using only the internal clock, then recorded a mic using an external (Apogee Big Ben) clock. The version recorded with the Big Ben was *so much clearer and more lucid* that my mind was blown. I had no idea that clocking could make such a huge difference to sound, but the MBOX internal clock was not great, and the Big Ben is known as one of the best master clocks in the biz. Every piece of gear (even the mundane things like a clock source) can make a huge difference to your sound, so don't write anything off when designing systems!—J.L.

Thus, if you are recording 10 tracks of audio, for 5 min, at 24-bit, 48 kHz sample rate, the equation looks like this:

$$[x\,bits = 24 \times 48000 \times 60 \times 5 \times 10]$$

which equals 3,456,000,000 bits of data, which equals 432,000,000 bytes of data, which equals 421,875 kilobytes of data, which equals 411.98. . . megabytes of data, which equals approximately 0.4 gigabytes of drive space needed for this recording. Of course, you always want to leave extra space just in case you run long (and drives should always leave at least 10% of their storage room free to have space for temp files and other moving of data that programs need to do).

So!

We've now been through the acoustic, the electric/analog, and the digital worlds of sound and how they work. I won't ask if that's all we need to know in order to design our system because you hopefully already know that it isn't. In the next section, we take a look at the input stage of our system, with a deep dive into microphones, playback systems, and the like.

Without further ado. . . *let's talk about gear!*

Notes

1 "Certain Topics in Telegraph Transmission Theory" by Harry Nyquist, 1928.
2 "Sigsaly—The Start of the Digital Revolution" by J.V. Boone and R.R Peterson, *NSA.gov*, accessed 7/17/2017 www.nsa.gov/about/cryptologic-heritage/historical-figures-publications/publications/wwii/sigsaly-start-digital.shtml
3 See Chapter 11 for info on filters and phase changes.
4 Eight-bit words are the standard data format for MIDI (Musical Instrument Digital Interface) instruments and devices.
5 That does not mean I will stop asking them.
6 This doesn't address so-called 32-bit float systems. In these systems, the added bits of information beyond the 24 bits for dynamic range are in theory distributed to the areas of sound in which the system senses the most amplitude variance, and thus produces a finer-grained representation of those amplitude ranges. In practice, some of these systems have no audible advantage, and the use of "32-bit float" in marketing is usually just that: a marketing device.
7 My lovely wife, also an engineer, used to run a studio that had a ton of MOTU audio devices. MOTU originally stood for "Mark of the Unicorn." Since a MOTU device provided the master clock in the studio, she would joke with her employees that her studio "ran on the magic of crystals and unicorns," and she was not wrong!
8 Neither do I. They are unnecessary for the work of a system designer.
9 This is not to be confused with the MIDI programming version of quantizing, which is generally used to describe the aligning of a bunch of programmed notes to a fixed tempo grid for musical production. Why the industry decided to use this same word for these two different processes is beyond me, but that's what they are called, so we're stuck remembering both.

CHAPTER 6

Microphones (Wired)

In the following chapters (6–24), we will examine the component parts of a contemporary sound system one by one. Each of these chapters[1] will follow the same format:

- DEFINITION: What is the component in question, and what is its function in a sound system?
- CLASSIFICATION: What are the basic types of this component, and how do they work?
- SPECIFICATION: What are the important specifications to note when assessing a given component's application, and how do we read them?
- CONSIDERATION: What are the essential questions a designer must ask (and answer!) before specifying these components in a system design?

We examine the four standard elements first, in signal order, and then expand to cover computers, communication technology, and cables/connectors. By the time you finish these chapters, you should have a pretty good idea of what pieces you will need to include in your sound system.

We begin at the beginning, with wired microphones.

DEFINITION

Microphone

A transducer that converts acoustic pressure variations within the audible frequency range to variations in electrical energy, which then flow as current into processing devices.

FIGURE 6.1
(Left) Shure SM58 on a Stand; (Right) Cirrus Logic WM7331 MEMS Circuit Board Microphone
Source: (Left) Courtesy Shure, Inc., © Lou Mora; (Right) Courtesy Cirrus Logic

A microphone is a device we point at a sound source, hoping to amplify or record that sound (or both). Microphones come in a very wide variety of shapes and sizes (taken together, shape and size are what we will often refer to as *form factor*), and they are designed with different purposes in mind (see Figure 6.1 for some examples). From the most common handheld microphone on a stand in front of a singer to minuscule microphones mounted to circuit boards and used to monitor industrial processes, every microphone is essentially performing the same function: take sound from one place and help us transmit it to another place.

In the simplest possible terms, a sound system needs microphones when it needs to amplify the sound of performance. Whether this is the amplification of singers during a concert or actors during a play, we often work on productions where some element (if not all elements) of the performance need to be louder than they would naturally be, so we mic those elements. Pretty straightforward.

So, why are there so many different types of microphones? Because each type has signature strengths and weaknesses that make them more or less ideal for a given application. How do we know which microphone will be the correct choice for a given application?

- First, we must understand the types and how they work, which will tell us a great deal about what tasks they are suited to.
- Second, we must understand how to read the specifications listed for each device in order to estimate how they will perform within their general class.
- Third, we need to listen to a wide variety of microphones applied to a wide variety of sources and assess how they perform.

While I always prefer to select a microphone I have heard and know intimately for a task, there are plenty of circumstances in which hearing a given microphone won't be possible before you need to use it on a project, so a solid understanding of the first and second items noted above is essential.

CLASSIFICATION

Microphones are typically classified by three main categories:

- Engine type: what is the source of the charge in which sound pressure induces variance, and what is the physical object that responds to acoustic vibration? Some mic types use permanently charged magnets to store a charge that we manipulate (dynamic and ribbon mics), others use capacitors or crystals (condenser mics, piezoelectric mics). I refer to the mechanism of generating charge and fluctuation as the *engine* of the microphone (because it drives the current), and which engine your microphone uses will change the performance quite a bit.
- Polar pattern: what is the shape of the coverage pattern of your microphone? Does it pick up everything around it equally (omnidirectional)? Does it pick up only in front on one side (unidirectional, like cardioid)? Does it have the option to change the shape of coverage with a switch (multi-pattern)? Understanding polar patterns is key to selecting the right mic for your application.

52

■ Form factor: how large is the mic? Is it designed to be held in the hand of a performer without substantial handling noise? Does it require its own dedicated power supply? In any production (and especially for live events), space is a major consideration, and selecting the right mic for the task involves asking whether the shape and size will suit the available area in which it will be mounted.

ENGINE TYPE

All microphones essentially do the same thing: they possess a static charged field, and when acoustic sound vibrations hit the microphone, the diaphragm vibrates, causing the charge to oscillate in response, which causes current to flow down the microphone cable and into our system. However, the source of that initial static charge varies by microphone type and has a lot to do with the quality[2] of sound the microphone will transmit.

There are four basic engine types found in commercial microphone design (with some variation among subtypes):

1. Dynamic
2. Ribbon
3. Condenser
4. Piezoelectric

We will identify how each works and how this will impact our sound, but first we need to know some fundamental terms for parts that all microphones share:

> **Diaphragm**
>
> A thin membrane inside a microphone that vibrates when hit by sound waves. Usually made of very lightweight plastic or thin metal like Mylar.

> **Capsule**
>
> The housing that contains the diaphragm and other mechanical pieces that influence the sound of a mic. Some microphone bodies have changeable capsules, which means that the internal electronics of the mic remain the same, but the diaphragm and associated housing are swappable (typically to allow one mic body to be configured in different polar patterns).

> **Preamplifier**
>
> A circuit that raises a very low-level electronic signal to a higher level. All microphones must connect to a preamplifier in order to raise their low "mic" electrical level to a higher "line" level that allows it to be easily manipulated within a system.

DYNAMIC MICROPHONES

The dynamic microphone is the most commonly seen onstage microphone in concert production. If you've ever seen music performed live, or live on video, you have likely seen dynamic microphones. The world-famous Shure SM58 (which we have already shown twice in this book!) is a dynamic microphone (Figure 6.2). So, what is a dynamic microphone and how does it work?

In a dynamic microphone, our power source is internal to the mic itself. The power is stored as a permanent magnetic field (or *flux*). This means we have a magnet with two poles that are magnetically charged (one positive,

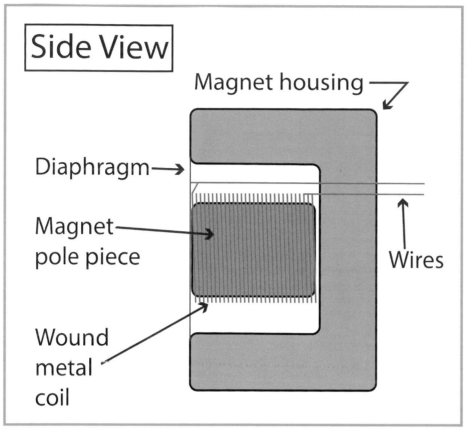

Side View

Magnet housing

Diaphragm→

Magnet
pole piece

Wires

Wound
metal
coil

FIGURE 6.2
Dynamic Microphone Capsule Diagram

one negative, or *north and south*). If you've ever played with magnets for fun on your kitchen table, you may have noticed that when you point the north pole of one magnet at the south pole of another, they want to come together and close the gap between them. If you hold a magnet in your left hand with positive pole facing your right hand, and a magnet in your right hand with negative pole facing your left hand—close enough so that they exert force on one another, but far enough to prevent them from closing the gap—you can feel the potential energy between the two pieces. A dynamic microphone's magnetic poles are held in this state permanently—close enough to evince force but prevented from neutralizing by preserving the gap between them.

In Figure 6.2, we see a cross-section of the capsule of a dynamic microphone. The magnet is shaped such that there is a central pole piece (the north polar part of the magnet), surrounded by a ring magnet (the south polar part). Around the pole piece, there is a wound coil of metal, known as the *voice coil*. At rest, this coil is perfectly suspended in the magnetic field, touching neither north nor south poles. The voice coil is physically coupled to the diaphragm, and when the diaphragm oscillates in response to sound, the voice coil moves along with it. The motion of the voice coil cuts across the permanent lines of flux stored by the magnet, which causes the stored charge to vary, thus driving current down our signal wires and into our system. In a dynamic microphone, the capsule contains the diaphragm, voice coil, and magnet, as well as such physical features as "delay ports," which shape the polar pattern of the pickup (about which more later).

ADVANTAGES

Dynamic microphones are incredibly robust and can stand up to a lot of abuse.[3] As a result, dynamic mics are often specified for tours in which equipment might be susceptible to a lot of jostling in transit and use.

Because of the weight of the voice coil, the diaphragm does not move as quickly as it does in mics without the voice coil (like condensers). As a result, dynamics can typically handle very high SPL without distortion (for a kick drum, for example) and can smooth out tones that might otherwise be jagged (they are very common on brass instruments, for this reason).

DISADVANTAGES

Because of the weight of the voice coil, and the diaphragm's lack of quick motion, dynamic microphones do not have the fastest transient response. For some instruments (particularly mallet percussion), fast transient response is desirable so that the true character of the instrument can be transmitted, and a dynamic microphone might blur the attack of instruments of these types in an undesirable manner.

> ### Transient
> A high amplitude, short duration sound—a snare drum hit, or fast notes on a xylophone, are good examples.

With rare exception, dynamic microphones produce a sound that is more "colored" (altered) by the internal electronics than condenser mics (generally), and so for high-resolution transmissions where no such alteration is desired (like recording a classical orchestra, where fidelity to the live tone is imperative), dynamic mics will typically not perform well enough for the task.

RIBBON MICROPHONES

Ribbon mics are essentially a variation on the dynamic microphone type (though they pre-date conventional dynamics), in that they too derive their charge from a permanent magnet. How they differ from dynamic mics is in the diaphragm itself. Where a dynamic mic has a round diaphragm coupled to a voice coil, a ribbon mic instead has a very thin, corrugated metal ribbon suspended between the poles, which effectively acts as both diaphragm and coil (Figure 6.3).

The ribbon itself vibrates and cuts against the static magnetic flux, inducing current to flow.

ADVANTAGES

Ribbon mics often evince an even smoother transient response than dynamic mics. While this does not recommend them for a covering a xylophone, it does mean that they can give warm and buttery tones for horns, jazz vocals, and the like. Many (if not most) classic jazz vocal recordings were done with ribbon mics.

DISADVANTAGES

With rare exception, ribbon mics are more fragile than typical dynamics. Vintage ribbon mics also can't withstand the external power source (phantom power) that is necessary for condenser mics, and if an old ribbon mic is connected to phantom power, it can literally catch fire.[4]

CONDENSER MICROPHONES

Condenser mics,[5] so named after their central electrical component, the capacitor (which historically was known as a condenser), have a very different engine from dynamics and ribbons. In a condenser mic, energy is stored in a capacitor. A capacitor is an electrical component that consists of two plates, separated by an insulator, that can be energized when attached to a power source and that stores electric

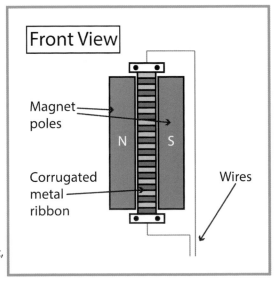

FIGURE 6.3
Ribbon Microphone Capsule Diagram

charge. In a condenser mic (see Figure 6.4), the capacitor itself consists of the diaphragm (which is typically a very thin metal and acts as the front plate) and the back plate (which is a very thin coating of metal applied to a ceramic insulator). The mic itself has no charge at rest, and when it is connected to a system, it needs an external power source applied in order for the capacitor to charge. This can take the form of either a battery or a form of direct current sent to the microphone over its cable, which we call *phantom power*.

> **Phantom Power**
> Direct current, typically 48 V, sent to a condenser microphone to power its operation.

When sound waves strike the diaphragm of a condenser mic, the diaphragm oscillates, and thereby the static charge stored in the capacitor varies, inducing current to flow into our system. Condenser mics are often described by their form factor (large diaphragm, small diaphragm, or handheld).[6] Condenser mics generate a very low electrical signal, so there is the need of a transistor in the design to amplify the signal to be usable (this amplification is a *pre*-preamplifier process that takes place in the microphone itself). Condensers will typically have one of two types of transistor: FET (field-effect transistor, known as *solid state*) or valve (also known as vacuum tube). Tube microphones, like most tube electronics (guitar amps, for example), are known for emphasizing even-order harmonics of the input signal, which our ears tend to associate with a quality known colloquially as "warmth." FET microphones tend to be more neutral in character and thus more accurate to the original sound source. Each has their uses.[7]

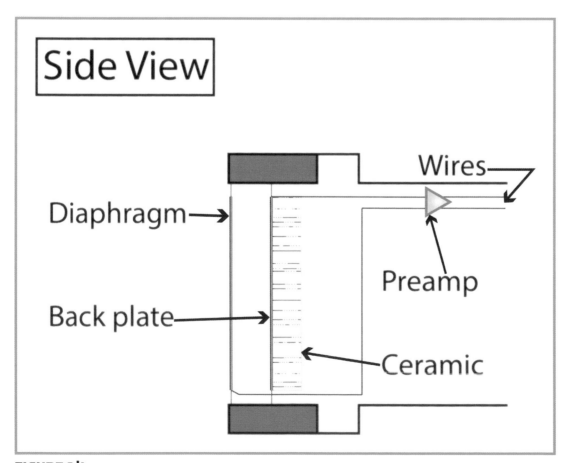

FIGURE 6.4
Condenser Microphone Capsule Diagram

ADVANTAGES

Because there is no voice coil, the diaphragm of the typical condenser microphone is very lightweight and moves very quickly. Thus, condenser mics tend to have the fastest transient response of any of our basic engine types. This makes them ideal for use on mallet percussion, pizzicato strings, or any other material that depends on crisp and clear transient response to preserve the musicality of the content.[8]

It is generally easier to get a very neutral frequency response (without peaks and valleys in certain ranges) from a condenser microphone.

Many condenser microphones have more than one diaphragm and make use of advanced switching electronics to allow one microphone to operate in multiple polar patterns, making them very flexible tools for multiple setups.

DISADVANTAGES

Due to the back plate being applied to a ceramic insulator, condenser mics are very fragile. The ceramic can easily shatter if dropped, rendering the mic useless.

Due to the ease of the diaphragm's motion, many condenser mics are not suitable to being held in a performer's hand,[9] and they are very susceptible to environmental rumbling (footsteps onstage making vibrations that transmit up the mic stand and into the mic, for example). Many condensers are sold with *shock mounts*, which are clips that make use of elaborate rubber-band baskets to decouple the mic from the stand, thereby reducing (if not always eliminating) this kind of rumble, but those shock mounts (while often quite effective) take up space that may not be available in a given setup.

Sometimes a condenser's tone can be deemed "too bright" for an instrument, leading, for example, to harsh tones on some brass instruments that may not be desirable.

PIEZOELECTRIC MICROPHONES

Piezoelectric mics (colloquially shortened to "piezos"), make use of a crystalline engine. In between the diaphragm and a back plate, there is a crystalline medium that responds to pressure changes by generating a voltage (Figure 6.5).

When the diaphragm oscillates, it applies and removes pressure to the crystalline filler, and that pressure induces current to flow. This is a known property of some types of crystalline matter. Why does it work that way? We aren't entirely sure; we just know that it does.

ADVANTAGES

Piezo mics are usually difficult to damage. Thus, they are often used as *contact microphones*, which are applied directly to instruments (like a drum head) to transmit audio through physical vibration.

They are often inexpensive.

DISADVANTAGES

Piezo mics usually produce lower-resolution audio. Contact mics are not known for their clarity of tone, and for most high-fidelity applications, they are therefore an undesirable choice.

POLAR PATTERN

The polar pattern of a microphone is a description of the directionality of its pickup. In specifications and manuals, polar patterns are represented as shapes on a polar chart, with 0° representing the "on-axis" point (the position directly aimed at

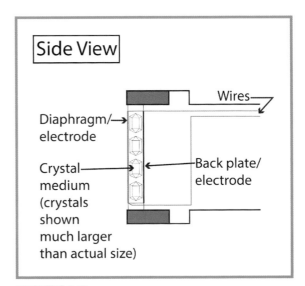

FIGURE 6.5
Piezoelectric Microphone Diagram

the front of the diaphragm) and 180° representing the "off-axis" point (the point directly behind the diaphragm). A microphone's polar pattern will define a lot about its use. While a mic will often be defined by its polar pattern, note that mics have different polar responses at different frequencies.

There are six basic polar patterns (Figure 6.6):

- Cardioid ("heart shaped")—a unidirectional polar pattern with a wide lobe of coverage at the on-axis point and good rejection at the off-axis point
- Supercardioid—a unidirectional polar pattern with slightly narrower coverage on axis and a small lobe of coverage at the rear
- Hypercardioid—a unidirectional polar pattern with even narrower on-axis coverage than the previous two, but also with an even larger lobe of coverage at the rear and sometimes slight lobes at the sides
- Lobar—a highly unidirectional polar pattern with the narrowest possible forward coverage but with pronounced lobes of coverage at the rear and sides.[10]
- Figure 8—a bidirectional polar pattern, with lobes of coverage at both the on- and off-axis points, and good rejection at the sides
- Omnidirectional—a polar pattern that captures sound more or less equally from all directions

Polar patterns are of great importance in mic choice. For one thing, onstage, the difference between a cardioid mic and an omni mic might be the difference between controlled sound or massive feedback. Consider a lead vocalist. They are situated at the downstage edge (closest to the audience) with monitor speakers on the floor in front of them aimed at their ears so they can hear their bandmates and their own vocal. With a cardioid mic, the off-axis

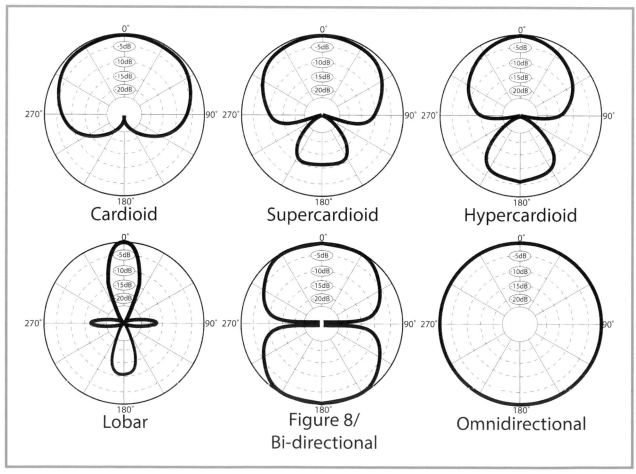

FIGURE 6.6
Microphone Polar Patterns

rejection means that the vocal performance can be sent through the monitors at a very high level before the mic will pick up that amplified signal. With an omni mic, the slightest bit of amplified vocal sent through the speakers will be picked up by the mic, re-amplified, sent through the speaker again, again picked up by the mic, again re-amplified, and so on until you get that horrible whining feedback sound that everyone hates (see Figure 6.7).

Beyond that, the mechanism whereby microphones become selectively directional (as opposed to omnidirectional) creates a phenomenon known as *proximity effect*. Proximity effect is the tendency of directional mics to increase the bass frequency response when placed in close proximity to a sound source. Radio broadcasters often make use of

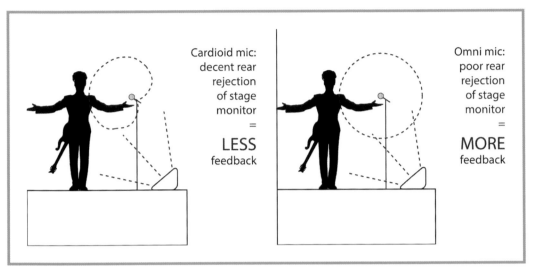

FIGURE 6.7
Cardioid vs. Omni Mics for Stage Vocals

FIGURE 6.8
Front-Address Microphones: (Left) Shure SM57; (Right) Neumann KM185
Source: (Left) Courtesy Shure, Inc.; (Right) Courtesy Neumann

this effect by getting very close to a mic and thereby making their voices sound lower and deeper than they do in real life. Proximity effect must be considered in selecting a mic for a task. In live production, mics are often placed as close to the sound source as possible to avoid "bleed" from surrounding sound sources (thereby giving us the most individual control of that mic-ed source in our signal path), but whether or not proximity effect is desirable is up to the creative decisions of the system designer and engineer.[11]

When working in a professional recording studio, on the other hand, the sound of the room itself is often a desirable part of the tone intended for capture, in which case mics may be placed farther from their sound sources and/or omni or figure-eight mics may be desirable, as they capture more of the "room tone."

FORM FACTOR

Wired microphones come in three basic form factors (though infinite varieties are found within each category):

- Front-address
  - Handheld (a subtype of front-address)
- Side-address
- Miniature/sub-miniature/lavalier

Front-address microphones (see Figure 6.8) typically have a cylindrical body, with the diaphragm's on-axis point being one end of the cylinder, and the diaphragm itself situated perpendicular to the body of the mic. Most (but not all) small-diaphragm condensers are front-address mics, and all handheld vocal mics are front-address.

Side-address microphone bodies are often (but not always) larger than front-address mic bodies, with the diaphragm situated parallel to the body itself. Most large-diaphragm condensers are side-address (see Figure 6.9).[12]

Miniature, sub-miniature, or lavalier type microphones are small microphone capsules with built-in wires (which as a unit are usually dubbed "mic elements") used because they are small enough to be hidden (see Figure 6.10). These mics can be wired direct to a system, or to a wireless transmitter (see the next chapter). Lavalier elements are so-called because they originally were designed to be worn as a lavalier, which is to say, a pendant at the end of a lanyard or necklace. Today, most of these elements are worn hidden in performers' hair, attached to lapels of clothing, in headset-style mounts, or attached directly to instruments (often to the bridge of string instruments).

Form factor is an important part of selecting a mic. You must know that the mic you select will fit in the space allocated and that it can be positioned optimally to aim at the source of your choosing.

SPECIFICATION

When selecting any component for inclusion in a sound system, we would be well served to understand how to interpret one of the standard documents provided for any piece of equipment, and that's the list of specifications, known as the *spec sheet*. Microphone spec sheets, as we will find in all classes of components, vary in level of detail, but there are certain fundamental pieces of information we expect to be able to glean from any spec sheet. While there are other categories that we will leave

FIGURE 6.9

Side-Address Microphones: (Left) Neumann U87-Ai; (Right) Audio-Technica AT4050
Source: (Left) Courtesy Neumann; (Right) Courtesy of Audio-Technica

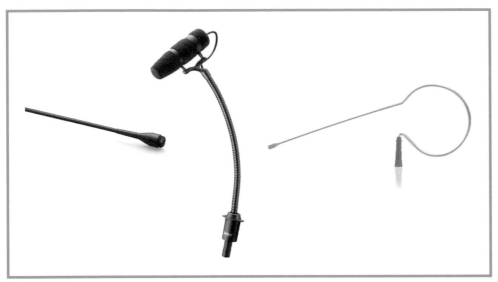

FIGURE 6.10
Miniature and Sub-miniature Microphones: (Left) DPA 4060 Lavalier Mic; (Center) DPA d:vote 4099 Cello
Mic; (Right) Countryman E6 Headset Mic
Source: (Left and Center) Courtesy DPA; (Right) Courtesy Countryman

unexamined at present, the following are the key data points we need to understand in order to properly select a
mic for our application:

- Mic type: dynamic, condenser, ribbon, piezo (discussed earlier)
- Frequency response: this specification can be listed in a number of different ways. At minimum, you will find
  a numeric range—for example, "50 Hz–15,000 Hz" (the stated frequency response of the SM58). However,
  all this tells us is the outside edges of the mic's frequency performance. It tells us nothing about how flat the
  response is within that range. Does it have a peak in the vocal register (between 1 Hz and 6 kHz)? Does it have
  a weird dip around 7.5 kHz? This stated range doesn't tell us any of this. Both of the earlier "quirks" are also
  found in the SM58 (and
  both are quite intentional,
  since it is primarily a vocal
  microphone, giving a
  boost in the vocal range
  and a dip in the area often
  associated with *sibilance*,
  the sometimes-harsh tones
  associated with the sound of
  the letter S), but we wouldn't
  know it from this stated
  range alone. Thankfully,
  Shure, like any professional
  manufacturer worth their
  salt, gives us a frequency
  response chart (Figure 6.11)
  to better help us understand
  the frequency performance
  of the mic.
- Polar pattern: here is another
  spec that is usually listed by
  a single term or set of terms

61

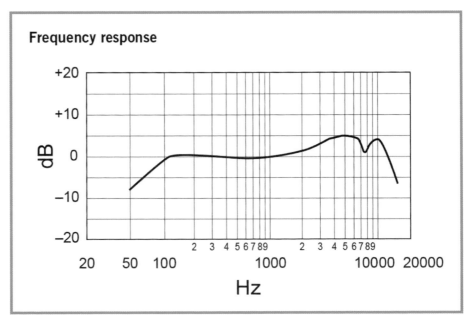

FIGURE 6.11
Frequency Response Plot for SM58
Source: Courtesy Shure, Inc.

(in the case of our beloved SM58, it is listed simply as cardioid). However, as mentioned earlier in the chapter, polar patterns are not consistent over the whole frequency range.[13] Thankfully, most reputable manufacturers will again give us polar pattern charts illustrating the different pickup shapes at different center frequencies. Note that in Figure 6.12 (also from the SM58), the pickup basically gets narrower as frequency increases.

- Sensitivity: how much voltage is produced by the microphone when a given amount of acoustic pressure is applied to the diaphragm. This spec is a good indication of the transient response (or lack thereof) of a given mic. The terms in which this spec is expressed are "x mV/Pa." A Pa (Pascal) of pressure is equivalent to roughly 94 dB SPL. Dynamic and ribbon mics are typically low sensitivity, while condensers are typically high. The SM58, our constant companion, has a stated sensitivity of 1.85 mV/Pa. By comparison, a Neumann U87—a world-class, large-diaphragm condenser mic—has sensitivity ranging from 20 mV/PA to 28 mV/Pa (depending on the polar pattern you select).

Beyond these specs, you will find standard info like polarity (any mic made for the US market produces voltage on pin 2 with positive sound pressure, where mics made for international markets are generally wired to produce voltage on pin 3 with positive sound pressure), connector type (mics generally use 3-pin XLR connections), weight, dimensions, etc. Some mics will give further details like dynamic range, signal to noise ratio, and impedance. These terms will be important when we get to mixing consoles, and we'll talk more about them at that time. There is, however, one further spec that is critical to observe if and when it is provided:

- Max SPL: some mics specs will state a maximum SPL that the mic can handle. If this spec is absent, typically it means that in practical terms, the mic can handle any sound that we are likely to throw at it. The SM58, for example, has no stated max SPL, but in research published by Shure,[14] the practical SPL ceiling for the SM58 varied by frequency and is somewhere between 150 dB and 180 dB SPL. Condenser mics often have lower max SPL, for example, the Neumann U87 has a stated max SPL of 117 dB SPL for THD (total harmonic distortion)[15] of 0.5% or less. This means that when applied to sounds *over* 117 dB SPL (think heavy metal kick drum), distortion will increase. Observe this spec carefully—while distortion is great when intentional (like in a guitar effect pedal), undesirable distortion is unacceptable in a professional setting.

If we can make sense of the specs, we can select the right mic for the job!

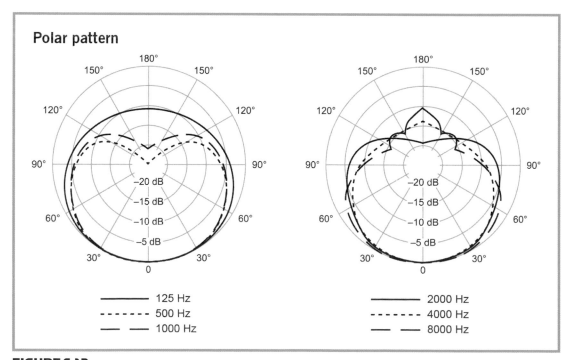

FIGURE 6.12
Polar Pattern Plots for SM58
Source: Courtesy Shure, Inc.

CONSIDERATION

So, what questions must a designer ask (and properly answer) in order to properly select mics for the system?

1. Do I need to amplify any sounds? System designers often rush to select mics before asking this basic question. If you have well-trained actors in a small theater space, there is no need to mic them unless you are doing so in order to apply special audio effects. If you have an acoustic folk group performing for an audience of 20, you might want to mic any very soft singers to make sure the lyrics are understood, but whether you mic the instruments is an open question and one you have to weigh based on your situation.

2. What sounds need to be amplified? Am I picking up singers in a concert? Guitar amps and drums? Actor voices? The breaking of a glass prop at a key dramatic moment in a play? The answers to this question will help you understand first what form factor of mic you need (can it be visible or does it need to be hidden?), how sensitive the mic should be (is the kick drum a subtle jazz kick with low SPL, or a heavy metal double-kick producing thunder?), and what polar pattern you need (is the trumpet by itself onstage with no one around, or packed in a line of ten brass players?).

3. What mics do I have access to? In a perfect world, we all have infinite budgets and can choose whatever mics we want. In the real world, budget is always a consideration. Are you designing a system for a repertory theater that has a fixed stock of gear you must select from? Then research the mics in your stock and choose the best possible options. Are you designing a system for a touring show, where all the gear will be rented? Choose ideal mics but be prepared to make substitutions when you find out that the rental vendor can't find that vintage Coles 4038 ribbon you wanted. Are you designing a system for a new nightclub and specifying the stock they will use day in and day out? You will likely have a financial target limit in advance, so choose the best options within your range.[16]

4. What do I want my sources to sound like? In theory, a "perfect" mic would reproduce the exact sound we aim it at, without changing a bit, right? Those mics exist. They are called measurement mics, and we use them extensively in system calibration,[17] but while we *can* use them for other applications, we seldom do. We choose mics *because* they provide a certain character, coloration, or tonal imbalance. We like the slow transient response of ribbons for some instruments, and we like the brittle response of a Shure SM87 for certain voices. What's more, what *I* like as a designer for a certain application may not be at all what *you* like for the same application. When you are learning, it is good to talk to other practitioners, read reviews and specs of equipment, and most importantly, whenever you have access to mics—*listen* to them.

There's a world of difference between specifying mics for a single theater production system, where each mic only needs to be suited to the task at hand, and specifying mics for a new venue, where the collection of mics may need to accommodate a range of performances and applications.

At the end of the day, how important is mic selection? After all, why would anyone pay thousands of dollars on fancy condenser mics to record vocals if mic choice isn't *super important*, right?

I don't want to discount the importance of mic choice. If you need a hidden mic in tight quarters in a theatrical set piece, the Blue Mouse is probably not your best option, so mic choice is definitely important. That said, its importance can be overstated. When recording studio vocals, for example, do we always need the most expensive large-diaphragm condenser? No. For example, U2's Bono has recorded a ton of his studio vocals into a mic that costs about $100. Which mic? Our recurring friend, the SM58. Of course, what this leads us to is that the mic isn't the only thing that colors our sound—a very important part of the sonic coloration is what the mic connects to: the preamplifier. While Bono might have recorded into a $100 mic, it is highly unlikely that his preamp cost $100.[18]

Wired mics are found in most (though not all) sound systems you will encounter or design, and understanding how they work and how they might be used is critical to your success as a system designer. But what happens when we need to mic actors who are moving around a stage, and the mics need to be hidden? We need to venture into the realm of Chapter 7—wireless mics!

Notes

1 Except for Chapter 14 on power math.
2 I refer to "quality" here in the subjective sense (as in "what are the qualities the item possesses") not in the sca-

lar sense (as in "this is the highest quality item"). Mics have varying qualities, some of which will be desirable for one application and undesirable for another.

3 I once saw an SM58 dropped from a height of about 30', onto a concrete floor. The windscreen was a bit dented, but the mic otherwise suffered no damage and sounded just fine!

4 Making things even more complex, some contemporary ribbon mics are designed to *require* phantom power for operation. Always read the manual/spec sheets of any mic you use, especially if it is a ribbon! You don't want to light a $4,000 Vintage RCA 77 on fire, but you also don't want to fail to power a mic that needs the external source.

5 Known as capacitor mics in the United Kingdom, after the capacitor component that stores the electric charge.

6 Large-diaphragm condenser mics are very commonly used in studio recording, for a variety of instruments, including vocals. Small-diaphragm condenser mics are often used on instruments, but (with the exception of handheld vocal condensers) are not used on vocals, as the tiny diaphragm tends to judder with the application of breath and thereby create unwanted artifacts in the sound. Small, thin, front-addressed condenser mics are often referred to as "pencil mics."

7 A subtype of condenser microphones, the *electret* condenser, is very common. In an electret mic, the external power source is not derived from a steady direct current (phantom power) but from a permanently charged source. Due to their ease of manufacture, the vast majority of microphones for non-production uses (e.g., the mic in your cell phone, laptop, or the like) are electret condensers. Electrets are sometimes used in production work, but in those cases, the charge source is often a battery, or is *optionally* energized itself by the application of phantom power—which seems contradictory, I know, but some mics feature the ability to be powered by either battery or phantom and are still usually classified as electrets in that case. Suffice it to say, read the manual, and you will know if you need a battery or phantom power to be applied to your given microphone.

8 The snare drum is an instrument that sees wide use of both dynamic mics and condenser mics. Some engineers tend to prefer dynamics due to the fat and thick tone that the slower voice coil gives to the smack of stick on head, where others prefer the quick and crisp transient response of a condenser that gives the snare a bright and cutting tone. Some engineers will use either or both types, depending on the type of music being captured, the style of the performer, and the tone of the particular drum and head. Either is a valid choice, if made intelligently.

9 Though of course there are condensers designed specifically for handheld use.

10 Typically, only found in so-called shotgun mics, used for tv/film production to capture somewhat isolated sound from a distance.

11 For a good description of how proximity effect works (and what delay ports have to do with it), see www. creativeedgemusic.com/2014/09/science-behind-the-proximity-effect-microphones.html.

12 There are some microphones that defy easy form factor categorization. For example, the AKG D112, commonly used on kick drums, might be classified as either front- or side-address, and given its strange right-angle shape, an argument could be made that it is neither. The Blue Mouse microphone has a rotatable capsule, which turns what would ordinarily be a side-address mic into one with flexible aim. As with all rules, the rules of form factor classification are generalizations.

13 Because low frequencies have larger wavelengths and high frequencies have smaller wavelengths, it is physically easier to make a microphone act directionally at high frequencies than at low frequencies. All frequencies approach the mic from multiple angles. High frequencies get canceled out by being delayed by physical ports around the sides of the capsule, whereas low frequencies are not able to be delayed by these tiny openings (this is the basic mechanism behind proximity effect as well). Therefore, low frequencies approaching from all sides sum together producing both the bass proximity in directional mics and also an essentially omnidirectional performance, whereas high frequencies are diverted and thus null some on the sides and substantially toward the rear (via *phase cancellation*) creating the directional performance we desire.

14 "Realistic Maximum Sound Pressures for Dynamic Microphones" http://shure.custhelp.com/app/answers/detail/a_id/75, accessed July 25, 2017.

15 Harmonic distortion is the tendency of a circuit or piece of equipment to produce artifact frequencies in a harmonic sequence when overloaded. For example, a circuit driven with a 1 kHz signal to overload would produce not only a distortion of the 1 kHz signal, but also induce distortion at 2 kHz, 3 kHz, and other harmonic multiples of the input signal.

16 Note: this question of budget and available resources will apply to every single item in your system design. I will not repeat this question in subsequent chapters, but you should assume that you should *always* ask this question for any item or choice.

17 Discussed in Chapters 25 and 26.

18 Don't let my cheeky description lead you to believe that cost is a true indicator of quality in preamps (or for any piece of gear). I own a $200 preamp (an ART TPS II variable impedance tube pre) that I like a lot more on drums than some preamps that cost thousands. Cost does not equal quality, or suitability to a particular application. Cost can often be an indicator of a piece of equipment that is robustly built or that it is made by a company that provides excellent customer service, but even this is not a rule. I know one manufacturer of very expensive studio monitor speakers—whose speakers I love to work on, and whose speakers are definitely robustly built—but whose customer service is absolutely abysmal. I will not toss them under the bus here by naming them, because I do love the sound of their equipment, but suffice it to say, cost is not a true predictor of anything but cost. Companies develop reputations in the industry, and if you want to know if a manufacturer has good customer service or builds robust equipment, ask professionals, check Internet forums (*take them with a grain of salt*) and trust your own experience.

CHAPTER 7

Microphones (Wireless)

DEFINITION

> **Wireless Microphone**
> A microphone system that connects a mic to a portable radio transmitter, which sends signal to a receiver, which then outputs wired signal to a sound system.

Generally, there are two reasons to want a mic to be wireless:

- Ease of movement (of the performer, the mic, or both)
- Invisibility (of the mic)

Two of the most common uses of wireless mics are as follows:

- Concert systems, where performers love to have either handheld or headset wireless mics, so they can move and dance around a stage without being tied up in cables
- Theater systems, where actors are frequently mic-ed with hidden sub-miniature mic elements connected to hidden transmitters so that their voices can be amplified without *appearing* to be amplified (in theater, the illusion of "natural" performance is quite often desired)

What is the basic shape of a wireless mic setup, and how it will interface with your overall system?

In Figure 7.1, we see a conceptual diagram of the signal flow of a wireless system.

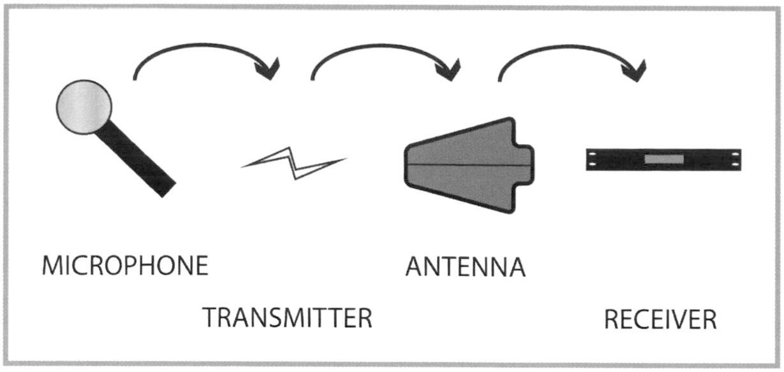

FIGURE 7.1
Signal Flow of a Typical Analog Wireless Mic System

Any wireless mic system consists at its base of the following four parts:

- Microphone (the transducer itself)
- Transmitter
- Antennae
- Receiver

We see a handheld mic (technically, its transmitter would typically be built onto the end of the mic, but it is shown here with a "transmitter" represented by the drafting symbol for wireless transmission—a lightning bolt). A wireless mic either has a built-in radio transmitter (as with handheld mics) or connects to one (as is the case with lavalier/sub-miniature mics that plug into beltpack format transmitters). That transmitter will send the microphone's signal as radio frequency over the air to antennae, which collect the radio signals, and connect via wires to the receiver unit. The receiver processes the radio transmission back into usable audio and sends the audio signal via cables into your mixing console or other signal processing device. Every wireless mic system will operate with some variation of this basic design. Components will vary (and in the case of digital mic systems, clocking will need to be considered), but all of our wireless mics will feature the aforementioned four components.

What do we need to know about wireless mics in order to make proper use of them in a system design? We need to understand their common form factors, some principles of radio transmission and antennae, and why we would choose one type of mic over another for specific common purposes.

CLASSIFICATION

Besides the usual microphone categories (explained in the previous chapter, all of which apply to wireless mics as well), wireless microphones are further classified by the following:

- Form factor (Figure 7.2): wireless mic systems come in one of three basic subcategories:
  - Handheld: these are standard handheld-style mics, with a transmitter built into the mic body. Aside from the radio transmission (which we will describe in detail later in this chapter), the general rules for any handheld mic apply: What is the mic engine? What is the polar pattern? How sensitive is it? Et cetera.
  - Lavalier/beltpack: these mic systems consist of tiny (usually sub-miniature) mic elements, typically hidden on a performer (we will discuss mic mounting a bit later in the chapter). These mic elements connect via very thin wire to a radio transmitter known as a "beltpack." These are small rectangular devices that can be worn on a belt, hidden in a pocket, or squired away in a costume in a custom hiding place. Special attention is usually paid to both the polar pattern and sensitivity of these mics. Most of these mics are condensers (though occasionally you will find a dynamic).
  - Plug-end: the least common form factor of wireless, plug-end mics are simply radio transmitters that plug directly into the XLR connector of regular (wired) mics. Inexpensive units of this type do not support

Tales From the Field

I made a discovery about double mic-ing it that I didn't succeed in fixing in the first go-round. When we did *Hadestown*—we now double mic basically everyone because we're so paranoid about losing the mic and it's gotten to the point where my RF order is twice the size of my cast, plus spares. So we were double

FIGURE 7.2
(Left) Handheld Wireless Mic (Shure SM58 Wireless); (Top Right) Lavalier Mic (DPA 4060); (Center) Sennheiser SK300 G3 Beltpack Transmitter; (Bottom Right) Plug-End Transmitter (Lectrosonics HMa Digital Hybrid Wireless Plug-On Transmitter)
Source: (Left) Courtesy Shure, Inc.; (Top Right) Courtesy DPA; (Center) Courtesy Sennheiser; (Bottom Right) Courtesy of Lectrosonics, Inc.

mic-ing everyone, and I had everyone sounding really good, except our tenor—who was our lead—and for some reason, with his voice, it sounded like there was this crazy DeEsser on him. And we tried everything, I mean I literally was like ReEssing him at the console, I basically made an expander I'm trying to solve it this way and that way, I'm dialing his high-end up . . . he was wearing, on a boom, two DPAs right next to each other—one above, one next to, depending on how it slightly rotated in reality. We tried everything to change that, we changed him from Point-Source to DPA, we moved the mic placement forward and backward, I did all this crazy stuff in the console. I made it tolerable by the end, but I was never happy with that. And I was like, "Well, maybe it was just this guy's voice. Maybe there's something about this particular singer's voice that we never notice until we close-mic him. Maybe all the recordings of him had been on a handheld up until that." So I was at my wit's end and just sort of gave up. And then, a few months later, I was doing another musical, two actors in this other musical, again double mic-ed: the woman, identical mics, identical mic rigs, literally physically identical to each other at the end. And the woman sounded fine and the *guy had the same problem*, and I was like, "oh my god, I'm going to kill myself." Literally, I was like, "I don't understand!" At first, I thought there was something wrong with my system design. I thought there was a phasing problem in the system in certain frequencies. Finally, I actually put it out to one of the listservs—I don't know if it was the theater sound list or the one of the Facebook groups—and I said here are the things that I've narrowed down: either there's something wrong with the way that mic is being rigged, or the surfaces of the mics are phasing off each other for this type of [voice] . . . like the vocal frequency is causing a problem . . . and a couple people wrote back. And someone sent a note about the fact that these particular DPA [lavalier elements] might have an actual left and right side to them that I had no idea about, so we checked that, my A2 checked that and she was like, "Oh my god, yes." These little tiny holes are oriented the wrong way! The other thing was that we moved the second mic back by, like, two fingers, by, like, an inch, and it was gone. The problem went away.—Robert Kaplowitz

phantom power (and thus do not support condenser mics), though high-end plug transmitters do provide phantom.

- Radio band: typically, professional wireless mics all operate on an FM (frequency modulation) basis of transmission. Almost all wireless mics transmit either in the VHF (Very High Frequency) or UHF (Ultra High Frequency) radio bands,[1] with most professional-use mics transmitting in the UHF range.
- Analog vs. digital: for decades, the vast majority of wireless mics have been analog. Now, however, there are some very strong offerings arriving in digital wireless mics (e.g., Shure's Axient Digital series), that convert the mic signal to digital information before transmission, and then (depending on your system's configuration) either send that digital audio directly into your system as data, or re-convert to analog in order to connect to your system.
- Antenna type: there are a wide variety of antenna types for your receiver system. Depending on the range/ distance the mic signals need to cover, the shape of your venue, and other factors, you will find one or another type more desirable for your task.

FORM FACTOR

Choosing the overall form factor of your wireless mic system should be the easiest part of your wireless planning. It comes down to simple questions of intended use. Whether you need a handheld, a lav/beltpack, or a plug-end should be relatively clear to you based on what the performer(s) need to do with the mic. That said, within each sub-category of form factor (particularly within the lav/beltpack category), there are a range of options to consider that will impact the usability of your mic for the given application.

HANDHELD MICS

Handheld wireless mics are most commonly used for stage productions or events where it is acceptable for the mic to be seen, and where the performer or speaker wants to move around stage unencumbered by wires, but also does not want to wear a headset-style mic (which we group in the lav/beltpack category). Conferences, seminars, and concerts are the events where handheld wireless mics are most commonly found. Within the category of handheld wireless, we really have only two other options from which to select:

- Dynamic vs. condenser: most handheld wireless mics are dynamic, but condenser models are available. Considerations for which to select are identical to considerations for any mic selection between these two engines. See the previous chapter for information on how to select between these two types.
- Analog vs. digital: this will be addressed later in this chapter, and the choice will apply to any wireless mic you select (not just handheld).

Besides the earlier options, the usual rules about polar pattern, frequency response, etc., apply as they would for any microphone.

If you are choosing a handheld model, the other choice is really whether to select a dedicated handheld wireless mic (with transmitter built into the body), or a plug-end transmitter that can connect to any wired handheld mic. If the application is a standard handheld mic use, typically designers tend to choose dedicated handheld mics, since they are less unwieldy, and don't have the danger of the transmitter being disconnected from the mic during an event. I will address the specific cases in which a plug-end transmitter might be advantageous next.

LAVALIER/BELTPACK MICS

Lavalier microphones (often referred to as "lavs," "body mics," or other shorthand terms based on their mounting form, e.g., "headset mics") are a very useful category of microphone and present the designer with the widest range of choices to make in selecting units to suit their application. From the mic element itself, to the mounting type, to the transmitter, there are a host of factors to consider.

MIC ELEMENTS[2]

When selecting a mic element for a lav/beltpack setup, there are a range of factors to consider. While you want to think about the range of options available for any microphone when selecting a lav element (several of which we will address next in the "Specification" section), there are specific considerations for choosing a lav that you must attend to:

- Size: lav elements range from as large as nearly ½" in diameter[3]—which won't be terribly easy to hide but will likely be inexpensive and can be found on television chat shows where hiding the mic isn't terribly important—to as small as 0.1" in diameter[4]—which may be more expensive but will be a world easier to conceal on a performer. This is one area where *size does matter*. Particularly when it comes to theatrical work where mics almost always need to be invisible to an audience, the smaller the element, the better.
- Brightness: many mic manufacturers sell lav elements in different tonal "brightnesses." For example, Countryman's B3 lav mic can be fitted with three different caps (that snap over the capsule), that change the mic from "flat" to "bright" to "very bright." In their specific case, the caps are giving the mic additional boost at 15 kHz, which might be desirable if the mic needed to be hidden in a hat or wig that might make the voice sound a bit muffled otherwise.
- Color: most wired mics come in one color. Wireless elements, however, are often sold in a range of colors, to match a variety of skin tones (thus making them easier to conceal), or in stark white or black (to conceal in wigs/costume pieces/hair). If you are designing for a single production, you want mics that suit the performers and costumes/hair/makeup of that production. If you are designing a system for a venue, you probably want to select a range of colors. However, if you need to change the color after the fact, this is possible. Mics are colored by spraying them with shoe paint (which is designed not to crack when a shoe flexes, making it ideal for coloring mic wires), or using certain types of paint pen (whose paint is also flexible after drying). Care must be taken to protect the capsule and particularly the diaphragm from getting painted directly, though, or the diaphragm may get stuck in place, rendering the element useless (typically the capsule is wrapped tightly in plastic while the cable is sprayed).
- Connector type: each manufacturer of wireless beltpack transmitters makes them with a different, proprietary, connector type. Each of these connectors requires different electrical impedance from the element, as well as different *pinouts* (wire terminations). Therefore, when selecting lav elements, you must ensure you select those that will work with your given or selected transmitters. Re-wiring these is no simple feat, so choose wisely the first time.
- Mounting: depending on the intended application of your wireless mics, the mounting may be something that you create in a custom configuration for the production (as is often the case with theatrical designs, where the mounting will depend on the costumes, hair, etc.), or it may be something you want to order pre-configured. If it is pre-configured, the most common options are either a regular clip (used to mount the mic element to a jacket lapel or necktie), or a headset (that places the mic capsule at the end of an armature near the performer's mouth). The full range of professional practices for custom lav mounting is beyond the scope of this book, but some very common tools (shown in Figure 7.3) used for custom mic mountings include the following:
  - Medical tape—used for affixing mic wires to skin to run down the back of a performer's neck and back to the transmitter. Note that the use of tape is often paired with the use of surgical skin-prep wipes, which are used to remove oils from the skin before affixing the tape, making for a stronger bond. Also note that many medical tapes are made of latex, and some people have latex allergies. Always check with your performers before using latex products on them (non-latex alternates are available that, while not usually as good at bonding to skin, do not irritate those with allergies).
  - Wig clips/barrettes—used to secure mic elements in hair or wigs. Often in combination with floral wire or elastic thread. Sold in a variety of sizes and colors.
  - Floral wire—used to shape custom earpieces, and to help secure elements and wires in hair and wigs. Used because it is stiff enough to hold a mic, but not *too* stiff, and because it is coated in plastic. However, though floral wire is safer to shape around an ear than, say, piano wire, it can still cut the skin. Floral wire is, therefore, often used in combination with rubber tubing called Hellerman sleeves. Hellerman sleeves are thin plastic tubes in short lengths. Using a Hellerman tool (a device that stretches the sleeves around the mic capsule, then allows them to shrink back around the cable and floral wire), any part of the floral wire that needs to rest directly on skin can be insulated by the sleeves, thereby preventing injury.
  - Elastic thread—used to make "halo" mounts that wrap around a performer's head (typically hidden in hair or wig), and to connect mic cables to very small wig clips.

71

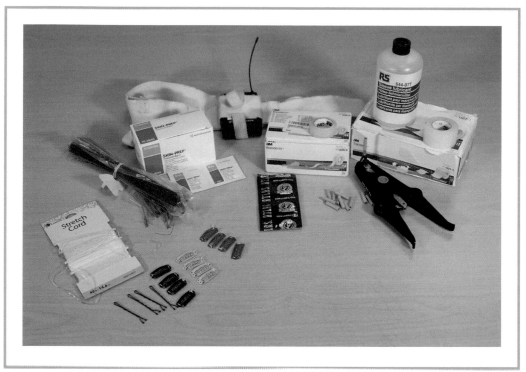

FIGURE 7.3
Mic Element Mounting Supplies: (Clockwise From Bottom Left) Elastic Cord; Floral Wire; Skin-Prep Wipes; Mic Belt (With Beltpack); Blenderm Medical Tape; Transpore Medical Tape; Hellerman Sleeve Lubricant, Tool, and Sleeves; Unlubricated Condoms; Wig Clips; Bobby Pins

BELTPACKS

When choosing beltpacks, there are a few things that you will need to consider, first among them being that beltpacks are designed as part of a paired system with receivers. Assuming the beltpack you select will work with your desired receiver, the form factor of the beltpack you select only presents a couple of options:

- Size: some beltpacks are large, some are very small. From almost 4" on either side of the pack, down to roughly 2" on either side, size is a major consideration. Generally speaking, the smaller the pack, the more expensive the purchase (assuming your pack still delivers professional performance). If you are placing a beltpack on a rock and roll musician, chances are it doesn't need to be particularly well hidden, so size isn't terribly important, but if you are placing a mic on a nearly naked theater performer (which happens more often than you might think), a tiny pack will of course be easier to conceal.
- Mounting: while beltpacks all come with essentially the same mounting option (a built-in metal belt clip), we don't always want to mount a pack visibly to a belt (or our performer may not be wearing a belt at all). Therefore, it is worth mentioning that there are a couple of alternate supplies you may need for mounting the beltpack on a performer:
  - Mic belts—these can be custom made or ordered pre-made, but essentially these consist of elastic fabric with Velcro fasteners at the ends (in lengths suitable for fastening around the waste/torso of a performer, or around the thigh, which is common for under-skirt mounting) attached to a fabric cradle for the pack itself.
  - Unlubricated condoms—used to isolate packs from the sweat of performers. This is a common tactic, but also a double-edged sword. While in some positions, a pack wrapped in an unlubricated condom can prevent moisture from gathering in the unit, in other positions (for example, with the condom's opening facing up, pack mounted in the small of a performer's back), this can be a recipe for *pooling* sweat directly around the pack. Caution must be exercised. Sometimes, a very small hole can be poked in the tip of the condom to allow pooling moisture to escape, but this also risks completely rupturing the condom and rendering it useless.

Tales From the Field

I used to work as the Sound Supervisor for a very well-known regional theater. When I started my position there, one of my very first purchases was a restocking of the wireless mic dressing kit, which included an order of 1,000 unlubricated condoms (enough for the whole season). I guessed that my predecessor in the position had not used condoms for mic mounting, as the purchase was immediately flagged by the accounting office. I received a very sternly worded email from the accountants and was forced in short order to bring mic packs and condoms into the office to give a live demonstration of the technique to defend my purchase. I found myself giving a 10-min condom tutorial for an audience of five giggling financial office employees. After having defended the need of these in our production stock, I realized that—my predecessor's practices aside—the folks in our office just thought it would be funny to make me come in and unroll condoms for them. Regardless, my purchase was authorized. I could have been defensive or combative when approaching the request to defend my purchase, but instead I approached it with good humor and left with allies (who wouldn't question whatever weird stuff I ordered for productions thereafter). Keeping good humor about you is always an asset!—J.L.

PLUG-END MICS

There is only one form factor option to consider: size. The same size considerations that apply anywhere else also apply here.[5]

RADIO BAND

When selecting a wireless microphone system, careful attention should be paid to the radio band of the equipment. Before getting into a discussion of transmission frequencies, we should take a quick detour so that we understand the radio principles at work in a wireless system.

Wireless microphones transmit by means of FM (Frequency Modulation). While a comprehensive description of how FM works is beyond the scope of this book, the short version is that the broadcast consists of the audio signal, which is modulated by (combined with) a *carrier wave*. The carrier wave is a signal at a frequency far higher than the audible frequency range (in the RF, or radio frequency, range). The combined signal is sent by the transmitter, and then at the receiver end, the RF portion of the signal is filtered back out, leaving the audio.

Early wireless mics were very susceptible to signal loss over the air (particularly of the quietest parts of a signal), so a technique was developed to prevent this, called *companding*. While we will get further into the mechanisms of dynamic effects later in the book, the essence of this technique is a combination of *compression* and *expansion*. The compression is on the transmitting end, and it reduces the dynamic range, making quiet parts of the signal louder, and thus making loss of those quiet parts during transmission less likely. The expansion widens the dynamic range on the receiving end, so that the sound isn't horribly over-compressed (and so that the program material won't lose all of its native dynamics). This technique isn't perfect, so every analog wireless mic system evinces *slightly* lower dynamic fidelity than comparable wired mics, but the trade-offs in mobility and lack of stage clutter are often worth the slight reduction in dynamic range.[6]

Wireless mic broadcast is susceptible to interference from surrounding radio devices (terrestrial TV and radio stations, wireless devices, and—if you are working somewhere with multiple productions in close proximity, like Broadway—other wireless mic systems), so planning of transmission frequencies (the carrier waves) is important.

Mic systems are sold with stated transmission frequency ranges (e.g., 554–590 MHz). These frequencies listed are the *carrier* frequencies. Care must be taken to select mics in ranges that will suit your location. Most manufacturers have tools on their websites that allow you to search the television and other broadcast stations in your ZIP code, to ensure that when you plan your production, your mics are in a range that will work. Each wireless mic you plan to use will take up no more than 0.25 MHz for its transmission, though professional practice dictates that you space your transmissions by at least 0.5 MHz to help avoid interference.[7] Each mic in your system needs to be assigned its own dedicated carrier frequency, with no overlaps—mics assigned to transmit on the same frequency will create what is known as *intermodulation distortion*. This sounds horrible and is to be avoided.

Finally, most professional wireless mic systems function with what is called *antenna diversity*. This means that each mic receiver has two different antenna connections (usually "Channel A" and "Channel B"). The antennae each receive the transmissions from the mic (or mics—some receivers are multi-channel and receive two or four mics in one unit). Whichever antenna has a stronger signal at any given moment is the one that is heard at the output. Transition between the antennae is seamless, and this structure ensures more robust signals.

ANALOG VS. DIGITAL

Until recently, this wasn't a choice we faced in wireless mic systems. All wireless mics were analog, and that was that. Now, however, we are starting to see some digital systems on the market.

Digital wireless mics have some advantages and some disadvantages.

ADVANTAGES OF DIGITAL MICS

- Digital mics convert the signal to data before transmission, and therefore there is no need of the companding that is used in analog systems. This means that digital systems can reproduce the full dynamic range of the program, as opposed to analog systems.
- Digital mics often transmit in a *frequency diversity* configuration. Much as antenna diversity means that there are multiple antennae for one mic, and the strongest signal is heard, frequency diversity mics transmit a single mic signal on multiple carrier frequencies simultaneously. This means that if there is momentary interference on one of the frequencies, but not on another of the frequencies, the clean frequency will be heard. This ensures the clearest possible transmission.[8]

74

DISADVANTAGES OF DIGITAL MICS

- As with any digital audio system, the conversion between analog and digital takes a bit of time. This time, called *latency*, means that there is a tiny lag in the mic-ed signal as compared to the live one. Some singers, especially those who use *in-ear monitors*,[9] find this latency to be distracting, and prefer the immediacy of an analog system. That said, the latency is extremely low, and most contemporary performers either do not notice it or can work around it.
- The frequency diversity system, which provides such clarity of signal, takes up enormous bandwidth. If you are in an area where there is a lot of competing signal (on Broadway, or in an area with a lot of digital television

Message in a Bottle

[On what skills young designers need] I would say, to be really honest, at this time it's *being an active learner*. I mean, there's so much happening and so much change. You can, and you certainly should, learn the fundamentals, and those are unchanging. You know, physics is absolutely still real. The very fundamental notions of why feedback happens, and how speakers work, and all that stuff. That's not changing. But what I find sort of challenging and delightful is that every aspect of the implementation is changing incredibly, incredibly rapidly. I literally just sat with a couple of guys—I just did this gig at the Constitution Center, and the house guy was complaining about the fact that they installed AVB rather than Dante. And to be able to have a conversation about that means that I have to have been thinking about—not just thinking about but *learning*—what is the difference between them and why. And I think, in 3 years, it's going to be another giant shift. I mean RF is becoming more and more digital, [but] how is that going to be a real thing once the spectrum is gone? At this point to be a system designer is to be able to learn and to constantly be chasing that knowledge down.—Robert Kaplowitz

FIGURE 7.4
Wireless Mic Antennae: (Left) Shure UA400B Omnidirectional; (Center) Shure UA874 Log Periodic; (Right) Shure HA-8089 Helical
Source: All Courtesy Shure, Inc.

stations, *or both*), this can be a problem. If you have a production that needs 30 mics, you might have enough space in the radio bands for analog mics, but not enough for digital mics working in frequency diversity mode.

ANTENNA TYPES

There are three basic types of wireless mic antenna (shown in Figure 7.4):

- Omnidirectional antenna—aka "whip antenna," this is the standard type included with wireless mic receivers, consisting of a wire encased in plastic for protection. They come in different lengths, either one-fourth wave or one-half wave antennae (which are one-fourth or one-half the wavelengths of the transmission frequencies). One-fourth wave antennae are only suitable for mounting to the body of the mic receiver. One-half wave antennae can be mounted remotely and connected to the receiver via cables.
- Log periodic antenna—aka "paddle antenna," after the shape. This is a unidirectional antenna. This type of antenna can provide an enhanced signal compared to a whip antenna, but only in the direction it faces. From the rear, it can provide rejection of signals, which is useful if you need to reduce ambient RF interference, or detrimental, if you have performers who might end up on the wrong side of the paddle.
- Helical antenna—aka "dish antenna." This is a unidirectional antenna, consisting of a wire ribbon wrapped around a plastic or fiberglass cylinder. This is the strongest collector of transmitted signal, but with the narrowest coverage. These are commonly used over long distances.

Antennae can be connected directly to receivers, or their signals can be strengthened by antenna distribution amplifiers. A distribution amp will take the signal from two antennae and distribute it to multiple receivers. While this can be useful—if you have two well-placed paddle antennae, those might work better than four pairs of whip antennae, depending on your venue and configuration—it can also cause problems. If you have a system that is receiving interference, a distro amp will amplify the interference along with the signal. See Figure 7.5 for an example of the typical configuration of an antenna distro system.

SPECIFICATION

When selecting wireless mics, all of the typical microphone specs apply, though when selecting lavalier elements, there are a couple of categories that need special attention.

- Sensitivity: mic sensitivity isn't inherently different in lavalier mics (and is still listed as "x mV/Pa"); it is worth noting that lav elements are often sold in "hi-sensitivity" or "lo-sensitivity" models. The mics will be identical in all other ways, generally, but the sensitivity is very different. If you are mic-ing actors who need to be able

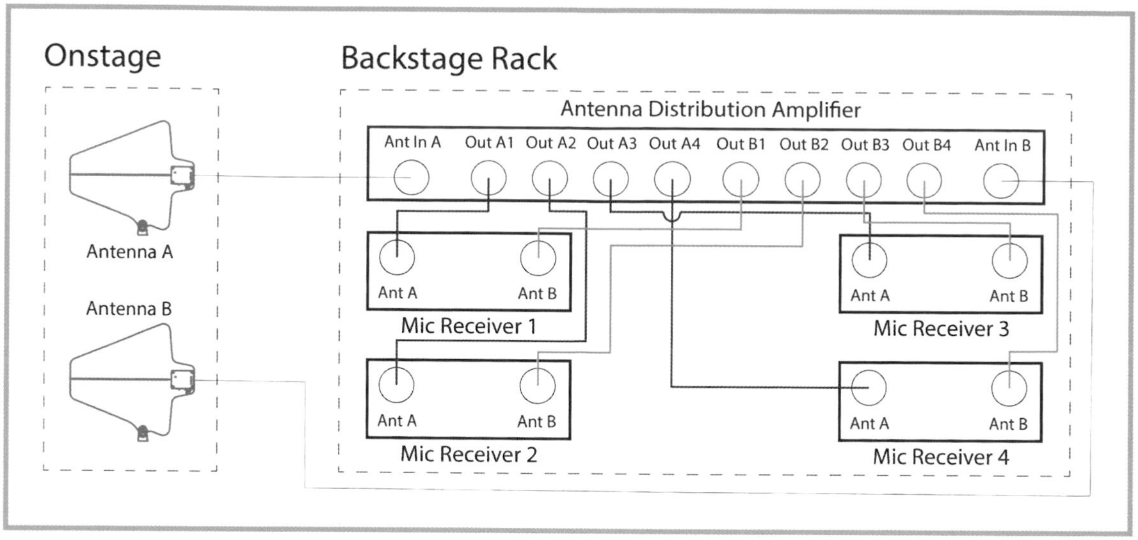

FIGURE 7.5
Configuration of an Antenna Distro System

to be heard while whispering, the hi-sensitivity model may be a good choice. If you are mic-ing a singer who screams through his or her whole show, lo-sensitivity is probably a better option.

- Polar pattern: again, the normal terms and rules apply, but when placing a mic on an actor's head—particularly for theater, but also for concerts—proximity effect can be a beast if you select unidirectional elements. Additionally, if the mic mounting falters, and the capsule faces away from the mouth, a unidirectional mic can result in lack of sound. Omnidirectional elements are almost always preferred for lav elements, with the exception of some headset mics, which are sometimes desirable in directional models. Generally, the only two choices for lav elements are unidirectional (typically a semi-cardioid, which is almost exactly like cardioid, but with a *touch* more coverage at the rear null point) and omnidirectional.

Beyond lav elements, the other important specifications are either identical to specs for any mic or are associated with the transmitter and receiver:

- RF carrier frequency range: what frequencies does the mic system transmit on?
- Effective range: how far can the mic transmit and expect to be received? Transmitters are sold at different power levels, make sure yours will cover the distance you expect by reading the details (in specs or in a user manual) before selecting.
- Receiver audio output level: does your mic system output mic-level signal (which would connect to a preamp) or line-level signal (which does not require a preamp), or both?
- Dynamic range: since analog wireless mics reduce program dynamic range, you need to understand this figure. If a mic system can produce dynamic range of 100 dB, that is roughly 30–40 dB less than human hearing. If that is acceptable, great! If not, look for a system with better dynamic range.

For advanced applications, there are other specifications that might be relevant to your work, but for basic system design, a solid understanding of the aforementioned will be enough to guide you.

CONSIDERATION

What questions must a designer ask (and answer) in order to select wireless mics as part of their system?

1. Do our mics need to be wireless? Much like designers who specify mics where no amplification is needed, some designers will specify wireless mics without thinking through whether or not they actually need to be wireless. Wired mics are more reliable, cheaper, and easier to configure, so if you don't genuinely need

wireless mics, don't specify them. In order to determine if our mics need to be wireless, use the questions from the start of this chapter: do we need the performers to be able to be mic-ed while moving through a wide range of space? Do we need the mic to be hidden? If the answer to one or both is yes, we might need wireless mics.

2. How many mics do we need, and do we have enough units that will work on discrete carrier frequencies? If you are mic-ing actors in a stage show, do all of them need mics to keep the tone consistent? Only that one actor who can never project their voice? Only the soloist for that one song that can never be heard over the chorus? If you are mic-ing concerts or other events, this becomes more straightforward. (Who is singing? Who is speaking?)

3. Do we have enough budget for backup elements? Elements are much more easily damaged than transmitters, and professional productions will budget for at least two elements for every single lav/beltpack system in their design.

4. Do we have enough budget for batteries? Professional productions use fresh batteries at the start of each rehearsal or performance. If your system is for a production, you need to account for this number. If your system is for a venue installation, does the venue have a good sense of what the operating cost will be of the units you specify (how many batteries you need in stock in a given week)?

5. How will we be mounting our mics? This needs to be planned well in advance, so that our system budget can accommodate all needed mounting supplies. Same notes for production vs. venue apply here (from number 4).

6. Where will we place our antennae? Do we need extra cable to get to far-flung antenna positions? What type of antennae do we need?

Wireless mics can be confusing for novices but are ultimately not terribly difficult to utilize. That said, in addition to the mics, transmitters, antennae, and receivers detailed earlier, many contemporary systems also feature computer network connections (at the receiver). By connecting these data ports (via a network switch) to a central control computer (or computers), crews can monitor the status of mics from any position in a venue (away from the receivers). These software programs show info including audio signal level, RF signal strength, battery level for transmitters, carrier frequencies, and more.[10] Networks aside, once you've mastered the finicky details, wireless mics are just more microphones for your system—at the end of the day, you put them in front of a sound source and manipulate and process the signal that comes out.

Well, now that we've got wired mics and wireless mics figured out, we have all our live sources for our system in place, right? Almost. We've got one more type of live input to our systems before we get to playback—DI boxes, which allow musical instruments to be connected directly to a mixing console without being mic-ed, and which are the subject of the next chapter.

Notes

1 There is much controversy over radio band usage in the pro audio world in the United States. The FCC, which allocates frequencies for device usage by regulation, has twice in the last decade decided that frequency ranges previously allocated for wireless microphone use will no longer be available for mics (they have been auctioned off at very high prices to companies like Google, AT&T, and the like for connected devices). Both of these occasions have triggered massive shifts in the industry, as mic manufacturers must discontinue mics tuned to those frequency bands, and users must discontinue use of mics in those ranges (under penalty of law) and replace their stock with mics in the new radio bands. I'd love to say that this time, the FCC has guaranteed a space for mics that won't change, but that's not true, and the industry continues to struggle every time they change these regulations.

2 In addition to mic elements, manufacturers also sell wireless instrument jacks, that are designed to plug into electric guitars, basses, and such, and connect them directly to wireless beltpacks. These are used frequently by musicians in concert, so they can move around stage easily without cables hanging them up (the receivers connect to their amps, or directly to the console via DIs, which are covered in the next chapter). There are no relevant specs or choices to be made with these units. For a given beltpack, there will typically be only one instrument cable available, and they are all essentially interchangeable.

3 For example, the Sennheiser ME 2-II, which has a capsule that is 0.4" in diameter.

4   For example, the Countryman B6, which bills itself as the "smallest lavalier in the world."

5   Plug-end transmitters are not always connected directly to microphones. I worked on a production in which a realistic lunchtime scene (people eating from brown bags), had to turn into a game show. Inside one of the lunch bags, we placed a long, thin, Bob Barker-style mic, connected to an XLR cable, which itself connected to the plug-end transmitter. Thus, a performer could walk in with their "lunch," and at a moment's notice, pull a functional "wired" mic out of the bag, without having to have a false bottom or a crew member connecting something during the scene.

6   Digital wireless mic systems do not have this problem.

7   Again, some digital systems are different. For example, Shure's Axient Digital system transmits on multiple frequencies at once (so the receiver can pick whichever is strongest and freest of interference), which makes the bandwidth necessary for each mic wider. See your specific system instructions for more information.

8   Some Broadway shows, instead of (or in combination with) using frequency diversity systems, will mount two wireless mics and transmitters on a performer, so if one goes out ("sweats out" where an element becomes overly moist and ceases working, or suffers another problem), the other is available as a backup.

9   Discussed in Chapter 17.

10 These setups will be discussed in Section 6.

CHAPTER 8

Direct Inputs (Instruments, Turntables, Etc.)

If our sound system requires live inputs (as opposed to just canned playback), there are only three component types that fall into this category. The first two (microphones, wired and wireless), were covered in the preceding chapters. The third is DI boxes, known as "direct inputs."

DEFINITION

> **DI Box**
>
> An electronic device used to convert high-impedance, usually unbalanced, instrument-level signals to low-impedance, balanced microphone signals. DI stands for "direct inject," as the device allows users to inject an instrument signal into the microphone signal path directly—though many people refer to these as "direct input" boxes.

Analog audio signals come in basically four different electrical levels:

- Mic level—the lowest level of the four. It is expected that any mic-level signal needs to be boosted by a preamplifier in order to make it strong enough for use in a mixing system.
- Instrument level—the second lowest level of the four. Instrument-level signals vary wildly in intensity since there are level controls on electric instruments (like the volume knob on an electric guitar). Instrument-level signals are too hot for mic inputs (without a DI, they will overload the preamp and distort), but too low for direct line input (with rare exception—some keyboard instruments can approach true line level with volume controls and effects, and some keyboard instruments feature both instrument and line-level outputs).

- Line level—the second highest level of the four. The standard nominal operating level of playback devices such as CD players, computer interface outputs, etc. There are technically two standard "line levels," expressed typically in dBu or dBV; dBu is a scale where 0 dB = 0.775 V, and dBV is a scale where 0 dB = 1 V. "Professional" line-level devices send signal at a strength of + 4 dBu, whereas "consumer" line-level devices send signal at a strength of −10 dBV.[1]
- Speaker level—the highest level of the four. The level of signal after it exits a power amplifier, hot enough to drive loudspeakers.

Mixing consoles and other processing devices have input connections at two standard levels: mic and line (usually "pro" line by default, but many have the ability to handle both pro and consumer). Thus when we want to connect an electric instrument (like a guitar or bass) to our mixing console, we must use a conversion device to send the signal. This is the DI Box. We typically want to step the signal down to mic level, because this allows us to connect the line directly to our mic inputs and make use of the preamp and other channel processors on the signal.

But, if we can mic up the amp onstage, why do we want to use a DI at all? What is the advantage?

Imagine a ten-piece band. Drums, three guitars, bass, keyboards, two trumpets, trombone, and tuba. They are crowded onto a tiny stage. When you try to mic up Guitar #1, the amp is so close to Guitar #2 that you can't isolate the signal for mixing; you get both guitars in the mic. The bass is so close to the drums that you can't get isolation there either. The horns are upstage (away from the audience) of the rest of the band, so their parts bleed into every other onstage mic. In order to get some cleaner, isolated signals, we might want to take DI inputs on the bass and guitars.

A DI allows us to get a clean signal directly from the instrument itself, while still sending the instrument signal out of the DI box to the amp onstage. If there are any effects pedals, we take our signal after the effects pedals but before the amp, so we get all the sonic texture the musician intends (see Figure 8.1 for typical DI setup). The musician gets to control his or her stage sound, we get a clean signal, and everyone is happy. Right?

Well, mostly. You might notice I said in the last paragraph that we get "all the sonic texture" of the instrument, but this isn't quite true. Amplifier choice is a very particular thing, and for many musicians, the amp itself (and the settings on the amp) becomes a crucial part of their sound. By taking a DI signal before the amp, we lose the tone of the amplifier itself, so it is very common to take both a DI signal and to mic up an amp on a single instrument. By doing this, we have good clean sound on the DI path that allows us to ensure that the instrument is heard clearly in the mix, but we also have the amp tone in a mic that we can use to ensure we don't alter the character of the musician's work too much.[2]

At any rate, DIs are an essential part of a system tool kit, particularly in any venue where music is played (from concert halls to studios). See Figure 8.2 for images of DIs.

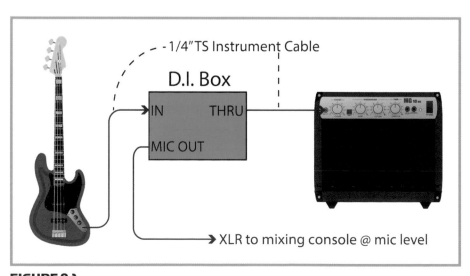

FIGURE 8.1
Connection Diagram for Bass DI

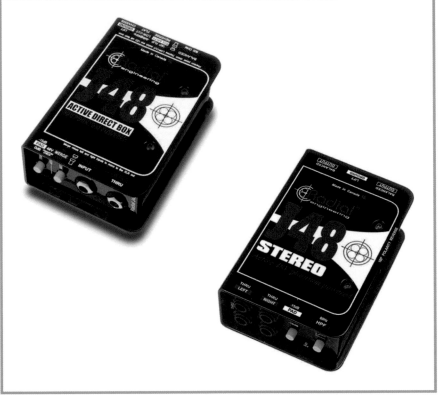

FIGURE 8.2
Radial J48 Mono DI (Left) and Stereo DI (Right)
Source: Courtesy Radial Engineering

CLASSIFICATION

There are only a couple of choices for a designer to make about DI box types:

- Active vs. passive DI
- Single-channel vs. multi-channel
- Pad vs. no pad

ACTIVE VS. PASSIVE

An active DI box requires a power source (either a battery or phantom power) in order to operate. A passive DI box does not require any external power. In terms of sonic quality, an active DI box can handle a wider range of electrical input signals, as the active nature means it usually has a pre-preamplifier built in that will ensure a particularly quiet signal at the input stage, and will still attain proper mic level at the output. A passive DI, on the other hand, has no pre-preamplifier, and is just a transformer with fixed value, which means that if the input signal to the passive box is particularly quiet (a guitar turned way down, for example), it can only boost according to the fixed electronics inside, and the output signal might still be a little quieter than desired. On the other hand, active DI boxes can sometimes add more circuit noise to the output signal from their power source (a good active DI that has been well maintained should not do this, but cheaper units, or units that have seen a lot of wear and tear, are more likely to add noise). Because passive DIs don't require external power, these can be good if you don't have a budget for batteries, or if you have a console that turns on phantom power in banks of channels instead of individually. If your console can only turn on phantom power for all channels or no channels (which is common for inexpensive consoles), and you need to use some ribbon mics in your system, then you might want passive DIs, so you don't have to engage the phantom at all. That said, nicer consoles have channel-by-channel phantom power switches, so this isn't an issue.

SINGLE CHANNEL VS. MULTI-CHANNEL

This distinction is as simple as it sounds. Are you using a mono (single-channel) DI, or one that has two or more channels in a single device? Stereo DIs are common for use with instruments like keyboards, which usually produce both a left and right signal. Additionally, since we often use DIs for electronic instruments that produce true line level rather than instrument level, stereo DIs are common for connecting to those as well (DJ mixers, laptops, etc.).

PAD VS. NO PAD

> **Pad**
>
> A switch that engages a fixed attenuation (reduction) of level on a signal device or path. Pad switches usually reduce input signal by 15 or 20 dB, and are commonly found on large-diaphragm condenser mics, DI boxes, input channels on mixing consoles, etc.

Why do we need a pad on a DI box? Instrument level varies wildly, especially once you account for all the effects pedals that might be in a guitarist's signal chain, for example. If the path is overloaded and distorting (not guitar pedal distortion, but too-hot-signal in the channel path distortion), we might want to pad the signal down so that we have some room to play with our preamp (which gives us more control over dynamics at the fader, and allows us to use a nice-sounding preamp to color the sound). In addition, when running line-level signals through a DI box (as from a computer or DJ mixer), sometimes a pad is needed in order to ensure that the line-level signals *act* more like instrument-level signals, and thus step down properly to mic level for sending to our console. Without a pad, some line-level signals can overload the DI box.

Note: there is one kind of specialty DI that we find a need for at times, which is the *turntable DI*. Turntables have specific impedance, grounding requirements, and frequency response that demands a dedicated input type distinct from any other device. Most turntables in production systems are found in DJ rigs and thus are usually run through DJ mixers, which take care of this and send line-level signals to the console. However, should you need to interface a turntable directly with a console (with no sub-mixer in between), you will need a *turntable DI*, such as the Radial J33, for that purpose.

SPECIFICATION

There are very few specs we need to pay attention to when choosing a DI box:

- Active vs. passive: covered earlier. If the DI is active, it should specify if it requires a battery, phantom power, or can be powered by both.
- Frequency response: should always be at least 20 Hz–20 kHz.
- Dynamic range: the wider the better. A good professional DI box has a dynamic range of at least 100 dB.
- Maximum input: this will state the electrical level at which the DI will overload. While standard line level is (at hottest) +4 dBu, a good figure for a professional unit is +10 dBu, which means it should be able to handle any instrument or line-level input, without necessarily needing a pad.
- Pad: if the device has a pad, this spec will tell you what the fixed attenuation amount is. Often this is either −15 or −20 dB.
- Channels: in a multi-channel device, this should be listed. If it is not listed, it is probably a single-channel device.

CONSIDERATION

What questions must a designer ask (and answer) in order to select D.I boxes as part of their system?

1. What signals do we want to DI? Or, if designing for a venue and not a specific production: how many DI signals do we want to be able to use simultaneously? In a single production, it is very common to DI bass,

guitars, keys, etc. If you are working on a Broadway-style musical, DI sources for these instruments are extremely common, as sometimes the pit area for the orchestra or band is so small that instrumentalists forego using onstage amps at all, and so rely on DI boxes for getting their signals into the system (a tactic which helps reduce in-pit volume as well, giving the FOH mix engineer more room to play with dynamics and achieve a "natural" sound).

2. Do we want both DI and mic for the sources in question or just DIs?
3. Do we want active or passive DIs? Or is it (as is true in many cases) not terribly important which kind we use?
4. Do we need any multi-channel DIs, or will single-channel units suffice?

DIs are a fairly simple but very useful part of our input tool kit, and any venue or facility that expects to engage with electric or electronic instruments should have some in their stock.

Now that we understand mics and DIs, we've covered the range of live input devices. The next major category of input devices is the playback system, and once we've covered those, we're on to the second stage of our system: processing!

Notes

1 When calculating actual voltage numbers, we use the same 20-log equation as we do for the inverse square law, except instead of inserting distances, we insert voltages. So in order to find the true voltage for the professional line-level signal,

$$+4 \; dBu = 20 \log_{10} \frac{x \, volts}{.775 \, volts}.$$

Since we know the value in dBu (+4), that takes the left side of our equation, and we are solving for x volts. The first step here is to divide out the 20 from each side:

$$+.2 \; dBu = \log_{10} \frac{x \, volts}{.775 \, volts}.$$

Now, in order to take this a step further, we must understand that the log equation is the inverse of a standard exponential equation. Thus in order to remove that pesky log, we need to convert it. The relationship between log and exponent equations is as follows:

$$\left[y = \log_b x \right] = \left[b^y = x \right].$$

If we plug in our values to the left equation above, we get

$$.2 \; dBu = \log_{10} \left(\frac{x}{.775} \right).$$

Then, in order to dispense with the log, we invert it to become an exponent:

$$10^2 = \frac{x}{.775}.$$

We then rationalize the left side to a decimal number:

$$1.58\ldots = \frac{x}{.775}$$

Then multiply both sides by 0.775:

$$1.228\ldots = x$$

Therefore, the actual voltage of a +4 dBu signal is 1.228 V. I have only once needed to do this math in a practical situation (and that was a very unique application), hence this equation being relegated to an endnote rather than in the body of the main text. For 95% of applications, knowing that +4 dBu is a hotter voltage than −10 dBV is enough info for the system designer.

2 Some guitar and bass amps have built-in DI outputs. While using this signal for the instrument does mean that you get the sonic character of the instrument *amplifier circuitry* in your DI signal, you still lose the character of the *speaker* in that stage amp, so it's not a perfect solution either. And, if the instrument amp is low-quality, sometimes taking a pre-stage-amp DI signal is preferable to the sound you will get from a built-in instrument amp direct output.

CHAPTER 9

Playback Devices and Systems

For the final chapter in the "Inputs" section, we leave live sources behind and turn to the *canned* (pre-recorded) sources, to examine playback devices and systems. The complexity of these systems ranges wildly. On the simple end, this might be just a CD player, or your phone plugged in via a headphone jack and streaming music from Spotify. On the complex end, this might be a dedicated rack-mounted server device, responding to a custom-programmed code sent from a show controller. In the middle we find regular consumer computers running software that again varies in complexity from something like iTunes to software designed specifically for running live shows, like Figure 53's QLab. We will examine all of these options.

DEFINITION

Playback Device

A device that stores pre-recorded audio files and is used to play them back on cue during a production.

Playback System

A subsystem consisting of the playback device and any controllers (hardware or software), interfaces (for D/A conversion, or straight digital transfer, of a signal), and associated devices.

Playback systems and devices store audio files. Audio files used in a production or installation can be stored in a wide variety of file formats, so we will detour here for a moment to examine some of the most common.

- Uncompressed (WAV or AIFF)—Uncompressed audio files are full-quality files that do not compromise signal quality. Whenever possible, it is generally desirable to use this type of file for two reasons: 1) the higher quality the pre-recorded file, the better it will sound; 2) unlike compressed formats, uncompressed formats can stream audio directly without the added step of "unwrapping" the format.[1] For most devices, there is no functional difference between WAV and AIFF (WAV was created by Microsoft, and AIFF by Apple, but they are nearly identical). Both file types can be burned to CD, both can be played by *almost* any device. The one exception is Android tablets and phones. The Android operating system cannot play back AIFF files, but it can handle WAV files. Both file types are forms of what is known as PCM audio (Pulse Code Modulation). It is beyond the scope of this book to explain how PCM works but suffice it to say these are high-quality files.
- Compressed—file formats that have sacrificed some sonic quality in order to shrink file sizes. The lower the file size (for an equivalent length piece of audio), the lower audio quality the file will be. MP3s (and some other compressed formats) are generally listed by *bitrate*, which is different from the *bit depth* we associate with uncompressed files. Bitrate tells us how much data is streaming across the processor per second of audio and is expressed in kbps, or kilobits per second. The highest-quality MP3s will be 320 kbps, but they can be encoded as low as 32 kbps.

  - MP3—the godfather of compressed audio files, these will play from most devices
  - M4A—Apple proprietary format for music downloaded from iTunes Store, also compatible with most devices
  - OGG—Ogg Vorbis files. These do not play on all devices and should generally be avoided (or converted) for production use
  - FLAC—the highest-quality widely available compression standard. However, like OGG, FLAC files do not play on all devices and should be avoided (or converted) for production use
  - MQA—a relatively new format that requires a paid license to use, this is truly the highest-quality compressed audio format, in some instances reproducing tracks from hi-fidelity studio files that are higher quality than CD files. Not in wide use because of the fee.[2]

While in the 21st century, most every playback device we will use in a system will be digital, the category of playback devices also includes older analog tech like turntables, tape machines, etc. These devices send analog audio signal through their outputs to the console. While turntables have specific requirements for their signal path,[3] other analog playback devices typically output line-level signal and can be interfaced with a mixing console directly. However, given that we live in the 21st century, our sound systems will generally be planned around digital playback devices, so that's what will be addressed for the remainder of this chapter.

CLASSIFICATION

Playback devices and systems come in three basic shapes:

- "Dumb" stand-alone playback devices—simple equipment whose system purpose is playing back sound files—e.g., a CD player, iPod, or the like. While there may be minimal programming options available in these devices, they generally need a human operator to control them (assuming they are not tasked simply with playing long playlists without interruption).
- Computer-based playback systems—setups using a computer as the central playback device, that may or may not require additional audio interface(s) to send a signal to a console or overall system.
- Server-based playback systems—setups centered around dedicated playback servers, which are computerized devices dedicated solely to playing back media files on cue, and are designed to be extremely robust and reliable hosts of content that may be able to run 24 hr/day, 7 days/week, with no interruptions. These devices can operate in a stand-alone mode with no operator or can be controlled by sophisticated programming and triggering.

"DUMB" STAND-ALONE PLAYBACK DEVICES

If your production or venue does not have very complex or elaborate playback needs, these simple stand-alone devices can be a great choice for your playback system. Many nightclubs, for example, only really need to be able

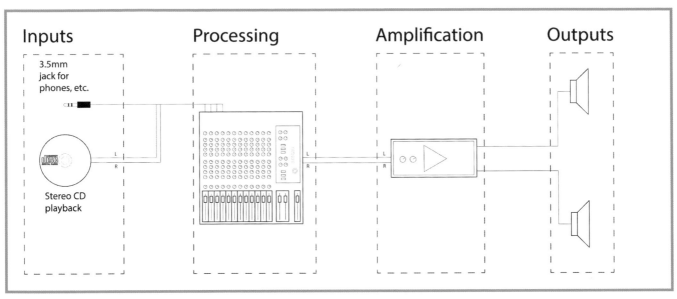

FIGURE 9.1
Basic Playback System Featuring CD Player and 3.5 mm Plug for External Devices

to play CDs (though the days of physical media are limited and decreasing by the moment) or to have a 3.5 mm cable available to plug in a laptop, iPod, iPhone, iPad, or another device of some kind. If all you require is stereo playback—usually of music files—these solutions may be good for you.

In systems of this nature, the playback device is usually connected directly to a mixing console, which then sends the signal out through the system and into the speakers. In Figure 9.1, we see an example of a very basic playback system.

COMPUTER-BASED PLAYBACK SYSTEMS

While the above simple system type can connect to a laptop, via the built-in 3.5 mm (headphone) jack, your playback needs may be more complex than a simple stereo system can handle. In theater design, for example, it is extremely common to create content and systems intended to feed over many discrete channels.[4] In order to send more than just simple two-channel audio (or six channel in the case of devices with HDMI outputs that support 5.1 surround sound), you will need an audio interface connected to your computer.

Audio Interface

A device—usually hardware-based, but sometimes software-based—that connects to a computer and allows a range of I/O (input/output) options. Depending on the interface, this can range from connecting a host of microphone-level inputs for recording, to connecting a range of line-level outputs for playback. Hardware interfaces perform the added function of A/D and D/A conversion, while software interfaces will usually only allow the input and output of digital signals (typically over a network connection).

Let's consider a simple example. We are designing a theatrical production, and we want to have speakers in the four corners of the theater, surrounding the audience. In order to send content from a computer to each of these discretely (which is to say, different signals to each speaker, individually addressed, not just sending one identical signal to all four), we need an interface.

In Figure 9.2, our computer (that hosts our audio files) connects to a hardware interface via USB[5] cable.

87

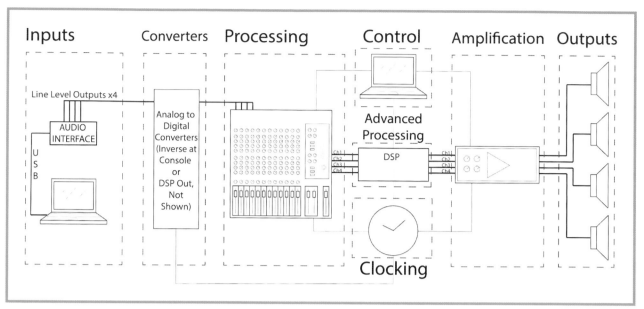

FIGURE 9.2
Computer With Hardware Interface Sending Signal to Mixing Console

This hardware interface, in turn, has at least four analog outputs, which we connect to our mixing console, and then route through the rest of the system to our speakers. Depending on the system configuration, and whether or not you need to change the audio with the processors in your console, you could also connect directly to the DSP (digital signal processor), skipping the console completely. Regardless of whether you connect to the console or the DSP, the principle remains the same, you connect discrete outputs from your interface to the discrete signal paths for your speakers, and thus can place content wherever you like among them.

If you are working with a computer connected to a hardware interface (whether external or installed in a PCI(e) slot on a desktop computer), your output channel count is only limited by two factors: the count of output jacks on the hardware interface itself, and the processing power of your computer. Working with a software interface can increase your available output channel count, but this only works if the console or DSP that your playback connects to speaks the same digital audio "language." A good example of this is Audinate's Dante Virtual Soundcard. Dante is a very popular digital audio transmission protocol. Dante Virtual Soundcard runs on your computer and acts like an audio interface. It sends digital audio in and out of your computer from any audio program, and if you connect the signal to a standard network switch, any other Dante-capable devices connected to that switch can send and receive audio to and from your computer.

Dante

A proprietary protocol for transporting audio digitally over network cables and infrastructure. Depending on the configuration of the device, Dante traffic can send/receive 64 channels of high-resolution digital audio, with very low-latency, over a single network cable (e.g., "Ethernet" cable, Cat5 or Cat6).[6]

In this type of system, if you have either a mixing console or a DSP device outfitted with Dante connectivity, you can send many channels of audio directly over a network cable from a computer to the rest of your system. Figure 9.3 shows an example of a system with playback via computer running Dante Virtual Soundcard (the software that runs as a virtual "interface" for the computer).

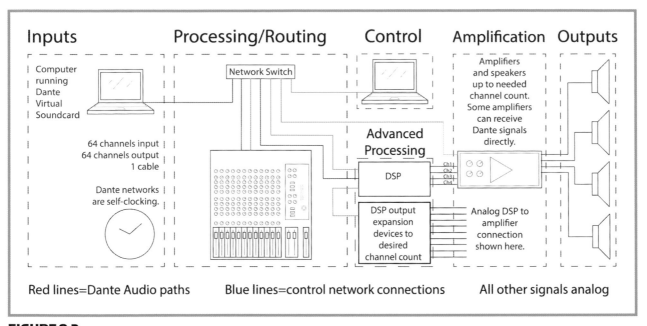

FIGURE 9.3
Computer With Dante Virtual Soundcard Connected to a Console With Network Cable

In one configuration or another, the computer-based playback system has become very common in systems of all types. For theater, concerts, and mid-level installation jobs, this is a suitable solution. However, at the top end of contemporary system complexity, a consumer-computer-based system is not robust enough. If you need to send sound to some 6,000 different speakers in one production, as was the case with Jonathan Deans' design for Cirque du Soleil's Beatles-based show *Love*, a laptop running QLab isn't going to cut it. If you need interactive sounds for a theme park environment that can receive cue information from 100 different contact points and send different content to those points depending on what message the system receives, likewise, even a powerful desktop computer is generally not the best solution. This is when we start to get into the next level of complexity.

SERVER-BASED PLAYBACK SYSTEMS

Now, it is a bit of semantic trickery that I have named the previous level of playback system "computer-based," when this level is "server-based." A server is definitely a computerized device; however, the most common usage of the word "server"[7] is in relation to Internet servers, which are computers that host websites and feed them over the Internet to our personal devices. This is not the kind of server we are talking about here—we are referring specifically to a class of devices called *media servers*.

> **Media Server**
>
> A computerized hardware appliance dedicated to hosting and delivering media content (audio and video files) across a system. Media server appliances are designed to be robust in performance, to be able to run nonstop for long stretches of time and to be able to respond in real time to customizable control messages (usually sent to the server from a *show controller*).

> **Message in a Bottle**
>
> On a playback show, the fact that even our standard playback system, QLab, has parametric plug-ins on the outputs now [means] I can tune any system. So I can walk into a system that literally is [just] a computer and speakers and no mixer, and still tune it. That's a godsend.—Robert Kaplowitz

> ### Show Controller
>
> A system (based on software that is run either on a conventional computer, or on dedicated show control hardware devices) that contains all the coded instructions for a given multimedia program and that feeds these instructions to various devices in the system (media servers, amplifiers, DSPs, etc.) according to either a programmed timeline, or according to interactive messages received by the controller from switches, buttons, sensors, and other such inputs situated throughout the system installation.

Media servers are used for sound systems that demand a great deal of complexity, reliability, and interactivity. Since they are designed solely for the task of hosting and delivering media on cue, no processing power is wasted on the kinds of background tasks that are frequently occurring in regular consumer computers (disk defragmentation, network status checks, etc.). There are many manufacturers of media servers, from Alcorn, whose Binloop devices are standards of the theme park design industry, to QSC, whose Q-SYS DSPs can be outfitted with a module for up to 128 channels of discrete audio playback, over 512 output paths,[8] to Meyer Sound's flexible D-Mitri system, with Wild Tracks (their playback server engine) and Space Map (a three-dimensional panning layer that allows for the most complex spatial programming of sound sources available).

A media server-based playback solution needs to contain at minimum the server appliance itself and a show controller (whether internal or external). Many media servers require that the media files be stored on high-speed flash memory cards, others feature all-internal memory, or swappable hard drive bays (akin to a desktop computer). Some systems (like D-Mitri) are fully integrated solutions covering the serving of tracks, processing, panning, etc. Figure 9.4 shows an example of a server-based playback system.

Servers can be more cumbersome to set up (and certainly more expensive) than a standard computer-based solution, but when your venue or production requires the added complexity and reliability, they are the best option.

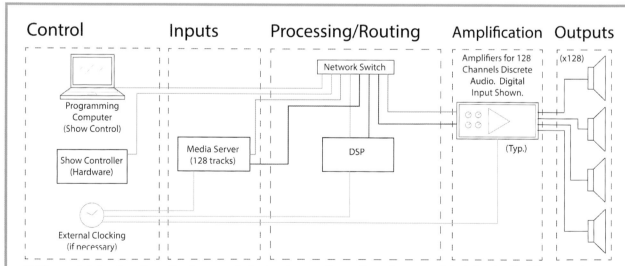

Red lines = digital audio paths   Blue lines = control network connections   Green lines = word clock signal   Black lines = analog
Show controller runs program for when media server plays tracks. Programming computer interfaces with show controller, server, DSP, and amplifiers. While only one signal connection is shown from server, more cables (at least two) may be necessary for channel count. External clocking is shown, though if audio is Dante, it may not be needed.   (Typ.)=Typical of other similar units in design.

FIGURE 9.4

Server-Based Playback System With Show Controller

A Note on Redundancy

In professional systems, wherever possible, we design playback to be performed by *redundant* devices. What does this mean? If, for example, we are using QLab on Mac computers, we will have two playback computers, running the exact same playback program ("workspace," in QLab terminology). We will normally hit "GO" on cues from an external controller (often a MIDI-based piece of hardware). That MIDI controller will be connected to a MIDI splitter, which takes our programmed GO command and sends it in real time to both playback computers. Only one of the playback computers is made audible at the start of the show, but should that computer encounter an error, crash, or other problem, we can switch over to the other computer. The signals running from QLab to our system might be, for example, running out as Dante digital audio signals routed through the Dante Controller software. In the event of Computer 1's failure, we can recall a new preset in Dante Controller, switching our inputs to the overall system from Computer 1 to Computer 2 (and unpatching/muting Computer 1). Similarly, if you are running QLab via hardware interfaces, you can either run them through a console, with associated mute groups for switching between computers, or through a DSP, with preset recall or live mute/unmute controls programmed into your ladder logic[9] system. Regardless of the topology of connection, running redundant playback is very desirable. It is the best insurance against playback system failure. Playback systems don't fail often (if they are suited to the scale of the project and well-programmed), but they do still fail, and having a backup that can operate in real time is essential to a professional system.

SPECIFICATION

When selecting a playback system, there are a range of specifications you can examine. It should be assumed that any professional playback system is capable of delivering full frequency range, ideally the dynamic range of at least 96 dB (CD quality) and at least stereo playback.

- Channel count: how many channels of audio is the device capable of reproducing simultaneously? Does your device need additional slot cards (PCI-style, if not actually PCI) or other hardware in order to reach the maximum channel count (as is the case with many media servers)?
- Sample rate/bit depth: what quality of audio file is your device capable of hosting. Often, this is interrelated to channel count, in that devices sometimes support very high sample rates and bit depths, but with a lower channel count than when they run at lower sample rate/bit depth.
- Output type(s): does your device or system feature analog outs, digital outs, or both? What form do those outputs take?[10] Does that format match the receiving devices in your system (whether your mixing console, DSP, or in some cases simply the amps themselves)?
- File types supported: playback software and media servers will have lists of what file types can be used in their programs. If you need to serve a ton of AIFF content, and your playback system only supports WAV, be prepared to spend a lot of time converting your media or find a playback system that will support your AIFFs.
- Digital audio protocol: if your device or system outputs digital audio, what protocol does it use to transmit this audio? We have mentioned the (increasingly common) Dante protocol already, but there are many others (CobraNet, AVB, etc.), several of which are proprietary and not inter-operable with other protocols. If you have a playback device sending CobraNet audio, you must have a receiving device that speaks the CobraNet language.[11]
- Storage format: do you need to buy external storage cards or drives in order to make your system work? If you have selected a computer-based playback system, keep in mind that it is *always* advisable to keep the media you are playing back on a separate drive from the system drive (the one that contains the software and operating system you are using). If you store the media on your system drive, it can slow performance down unacceptably. If you are using a desktop computer, or even some laptops, you can have more than one drive inside the chassis, so this doesn't necessarily mean an external drive for media, but if you are using a new, slim laptop, chances are you will need a high-speed external drive for the media.[12]

- Clocking: does your device/protocol require an external clock source? Can it clock internally (and/or serve as master clock for other digital devices in your system)?
- Timecode: many playback devices (especially servers) can be made to trigger sounds at a particular time measurement (either in real time, based on a so-called wall clock, or based on a fixed preset timeline similar to those found in film editing software). If you need your sounds to trigger on a fixed timeline relative to the start of the program, you need a timecode generator somewhere in your system (probably in your show controller). You also need your other devices to be able to read/interpret timecode values. Does your device support timecode? If so, what kind? SMPTE: standard time listed in Hrs: Mins:Secs:Frames—based on film timings (hence "frames," from frames of film per second), and named for the Society of Motion Picture and Television Engineers, who set this standard? LTC (Linear TimeCode, a version of SMPTE encoded for specific devices)? MTC (MIDI TimeCode, another version of the SMPTE standard that sends messages in the MIDI—Musical Instrument Digital Interface—language)? If you need timecode to run between show controllers and playback systems, make sure they can each handle the same format(s).
- Latency: if you are using a computer-based system with software interface, there will be some system latency. If you are using a hardware interface with D/A conversion for output, there will be some conversion latency. Find out what the latency is, and if it is short enough to suit your needs.
- Control language: playback systems have become increasingly flexible in terms of what control languages they speak. Software like QLab can receive (and send) commands in MIDI, OSC (Open Sound Control—a more flexible programming language than MIDI that is used for sound and other production devices), and Apple Scripting formats. Alcorn's Binloop line can receive UDP commands[13] sent by a show controller, or MIDI commands. And so on. There are many different languages and configurations for show control,[14] so just ensure that whatever your playback system can read can be delivered by your show controller. In a smaller system, where there is no show controller (and there is likely a human operator), this is less relevant.
- Electrical: any device that plugs into a wall needs to be examined to ensure that it works with your local power system. Most contemporary electronic devices are designed to work with either US or international power, but the United States has different power specs than the rest of the world (we run our power at 120 V, 60 Hz, where much of the rest of the world is at 220/240 V, 50 Hz). If your device is designed for only one locality's power specs, you need to know this. *This note remains relevant for every subsequent device and chapter covering any device that plugs into the wall, but it will not be repeated in subsequent chapters.*

CONSIDERATION

What questions must a designer ask (and answer) in order to select playback devices or systems as part of their system?

1. What is the nature of the content to be played back? Is it all stereo music files, being played back before a one-time concert? Is it complex theatrical sound cues with 24 layers each, sent to discrete speaker destinations, over a 6-week production run? Is it theme park content being sent to hundreds of discrete speakers, running all day, every day, for years? The nature of your content will be the first determinant of the complexity of your playback solution.
2. How many discrete channels do you need to be able to route content to? If all I am doing is playing stereo music, that would be two channels of output from my playback system (even if they get sent via a console matrix to many other destinations). If I am designing a complex themed installation with multiple rooms, action areas, and layers of sound, this could be dozens if not hundreds of channels. How many discrete channels you need will also determine how robust and complex your playback solution must be.
3. Will this be operated by a human, or programmed to run automatically with no human operator? Human-operated systems (e.g., concert interstitial music, most theatrical productions, etc.) are a different order of complexity from those with no human control. A theater production may run for 2 hrs per day, 6 days a week, with a human controlling the operation. A museum installation may run all day, every day, with no changes, so the device needs to be capable of handling this playback schedule with no interruptions (which indicates a server solution). A theme park installation may need to run longer hours than a museum and may have layers of interactivity (with the ride guests, the ride vehicles, or other inputs) that makes this a more complex system than either the theatrical or the museum.[15] Always ask yourself, does this system need to be

able to be controlled or changed by a human operator during operation, or does it need to be self-operating, or some combination of the two?

4. What format is my content? Does my playback device solution support my content?

5. What is the destination device of this playback content? If your playback system is sending to a console, directly to a DSP, or elsewhere, ensure that your playback device will interface properly with the destination device(s) (in terms of protocols, sample rate/bit depth, and cable/connector types).

6. How much content do I need to play back? Will it fit on a 512 Mb drive? Or is it 3 Tb worth? What is the storage capacity of my playback solution?

So! We've examined mics (wired and wireless), DIs, and playback systems. These four pillars encompass the entirety of the input stage of our system. Now that we have a basic understanding of inputs, we move onto Section 3: Processing. First up is the heart of most concert and theatrical production sound systems: the mixing console!

Notes

1. All digital audio formats play back via the mechanism of streaming the first part of the stored data into the device's RAM (Random Access Memory). The RAM is the active memory area of the digital device, and the audio we hear is coming from the RAM through the converters. Uncompressed file formats stream directly into RAM. Compressed formats (like MP3), have an additional "wrapper" around the file, which means that the machine needs to "unwrap" the audio and *then* stream it to the RAM. While this is usually fairly instant, if you have a very complex show programmed in software like QLab, and are trying to trigger 100 audio files at once, the delay of having to "unwrap" 100 files at once before streaming might mean (on a slower computer) that there is a slight delay from pressing "Go" on the cue to actually hearing it. Depending on the production, this might be an acceptable delay, or it might be a problem. Uncompressed audio files will fire on time, though they do take more storage space.

2. As of this writing, MQA files are only available for listening via Tidal streaming service (tracks marked with an M will be in this quality). They do sound fantastic.

3. See Chapter 8 for information on turntable DIs.

4. A small theater system for my own design work will usually have somewhere in the neighborhood of 30 discrete audio paths—speakers (or sets of speakers) to which I can send content individually—while a complex system might have *a lot more* than that.

5. Universal Serial Bus—a standard data connection type for computers.

6. These protocols will be covered in more detail in Section 6.

7. Besides the person who brings your appetizers to the table.

8. Q-Sys is also capable of acting as its own show controller, via the powerful Lua scripting language that is part of the software platform. These are powerful—if expensive—devices.

9. Ladder logic is a programming system based on conditional statements, laid out in a "ladder" structure to determine what happens in a given production (or industrial) control system when certain conditions (e.g., a trigger being activated, a sensor reporting data of a certain value) are met. Some DSPs feature ladder logic, where others use different control protocols, but the principles remain the same—conditional triggers for desired actions.

10. We discuss connector/cable types in detail in Chapter 24.

11. There is one exception, and that is the newly emerging AES67 protocol. AES67 is designed to be an inter-operable protocol that can accept Dante, Q-LAN (QSC's protocol), and others. It is fairly new as of this writing, and isn't available in many devices, but is quite promising. One of the advantages of the analog era that we have largely lost in the digital era is the ability to send a signal from one device to another easily without proprietary code and cable (a TRS ¼" cable is TRS no matter who makes it). AES67 is an attempt to return to this ease of configuration in the digital age. While there are some limitations (for example, Dante native can also send control, clocking, and program information along with audio, whereas Dante over AES67 only passes the audio itself), I hope that this, or another inter-operable standard like it, takes root in the industry to help simplify device connections and planning.

12. A note on drive speeds: for pro audio work, a mechanical drive (non-solid-state) must have an RPM of at least 7,200 (many standard low-end computer drives operate at 5,400 RPM, which is not fast enough for audio work). If you are working with solid-state drives, they should be so-called "high speed" solid-state drives. Do not count on serving media from a flash drive (USB stick), whether it is called "high speed" or not—USB stick flash drives are not designed to handle real-time media speeds.

13 UDP (User Datagram Protocol), is a version of internet communication language that, unlike TCP/IP (which is used for email and such), prioritizes the time domain in messaging over accuracy. Where a TCP message, like an email, sends packets of info and will not appear at the destination until and unless *all* of the packets arrive, a UDP message is received as soon as it is sent, even if some packets get lost along the way. Sound and other media devices tend to use UDP-based transmission protocols, because operating in real time (even with some small errors) is more important than delivering error-free content that is delayed.

14 Discussed in greater detail in Chapter 19.

15 This is an oversimplification. Some museums are now doing very exciting work with interactivity in system design, and the most complex theatrical systems (Cirque du Soleil, for example) may be run by a human operator but may *also* have a range of automated behaviors that take data from sensors and switches around the set and trigger actions based on those messages, independent of a human operator. As we get to the Chapters on networking, show control, and triggering, the choices you need to make when designing a system of this type will become clearer.

Preamplifiers and Mixing Consoles

Now that we've reviewed the fundamental components of the input stage, it's time to talk about what we do with them. In most systems, we will be wrangling a range of input signals, whether simply a short list of wired mics for a small live concert or a collection of wired and wireless mics, DIs, and analog and digital playback devices for a complex theatrical system, all of those input signals have to go somewhere so we can manipulate them. Whenever microphones are present in a system, preamplifiers are required, and wherever preamplifiers are needed, one of the most common places to find them is in a mixing console.

DEFINITION

We have already provided technical definitions of preamplifier,[1] and of mixing console,[2] so instead of repeating those, let's use this definition section to expand our understanding of the functions of each.

First, let's look at a fairly small mixing console and a basic concert-style sound system connected to it. By examining this setup, we can see how in many systems, the mixing console really acts as the nerve center for all operations.

In this setup, we have both wired and wireless mics, DI boxes, and a CD player as input sources. They connect to the mixing console's inputs. We have main left and right speakers playing sound to the audience—these are connected to power amplifiers, which are connected to a Digital Signal Processor (DSP), which is itself connected to the console's main outputs. We have stage monitor speakers (for the band to hear themselves, self-powered, so they don't need external amplifiers) connected to aux sends.[3] We have an outboard reverb unit, and one outboard compressor (effects). This system is drawn in Figure 10.1 as *plan view*, which shows a basic stage layout for the main items (though smaller items are not shown on such a drawing). This drawing, like any plan view drawing, is essentially a "bird's eye" view of the venue and major equipment. In Figure 10.2, we show this system as a technical drawing (a signal flow diagram, also known as a *block diagram*) to begin to introduce visual standards of showing system layouts that we will examine with more depth later in the book. Figure 10.2 looks a bit like the system flow

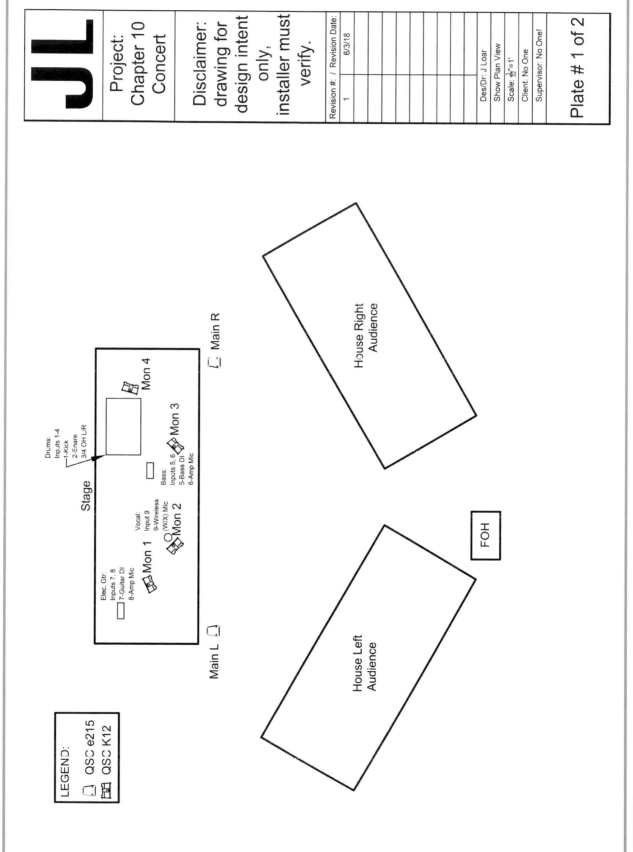

FIGURE 10.1
Small Concert Sound System—Plan View

diagrams we've been using so far, but instead of pictures representing items like the mixing console, we are now using drawing standards that would be found in professional system documentation.

This system is small but has a lot of features. These features are representative of the features of larger systems, and the functions of this small mixing console are good examples of functions (and the broader principles behind these functions) that you will encounter in any mixing console, anywhere.

Let's begin by examining a very common and simple analog mixing console—a console that could be configured just as the one drawn here—as all of the controls and features we find on this device will be replicated and expanded on more advanced consoles. Thus, if we can grasp these principles on the simple device first, we can see that advanced consoles (that sometimes look as intimidatingly complex as spaceship control panels) are really just variations on an essential set of concepts.

Figure 10.3 shows a very common 16-channel mixer, the venerable Mackie 1604.[4]

It is a workhorse small console, found in small venues around the nation, and it is a good model of the standard operating sections of a console. In Figure 10.3, we have drawn borders around parts of input 1's *channel strip* and numbered those parts to match the following descriptions.

Channel

A signal path in a processing device. A single microphone is plugged into a single channel on a mixing console, and that channel provides preamplification and other processing options for that microphone's audio. This is an input channel. A mono system takes all of the input signals and routes them in combination through a single output channel. A stereo system routes audio through two output channels (one for left, one for right). Effects processors are often classified by their channel count (e.g., "stereo reverb," "single-channel compressor," etc.).

97

Channel Strip

The set of controls in a signal processing device that applies to a given input, patched into a given channel. On analog consoles, the channel strip is usually a literal vertical strip of physical controls.

The numbered borders in Figure 10.3 correspond to the following list of sections, which we will find in one form or another on all mixing consoles. Before we get to the numbers, though, let's examine another set of numbers: 1604. Many mixing consoles are referred to by a combination number indicating the number of input channels and the number of output channels (e.g., "a 48 by 8 console"). The number 1604 hints at this. There are 16 input channels, and while there are more than 4 outputs, the number 4 is still relevant on the output side. There is the main stereo L/R output path, but additionally, there are four sub outs and four aux send controls (that actually send to six aux send outputs). We will identify and define those terms next.

1. Preamplifier control: aka "gain," "head amp," etc. (Figure 10.4)

   When routing a signal into a console, the very first thing we need to do (after we plug the cable in) is to set the gain of the preamplifier. The 1604, like many consoles, has discrete input connectors for mic signals and for line-level signals for each channel. Each of these hits the preamp at a different preset level of gain. Looking at the numbers on the gain knob, there are two ranges expressed, one for mic signals, one for line signals. At the bottom of the mic signal range is the number 0. This means the mic signal is turned all the way down, the channel is resisting it to the fullest. In the 1604's case, there is 60 dB of available gain above that point. Looking closely, we will find there is a "U" on this range for the line-level signal, which is 15 dB up the

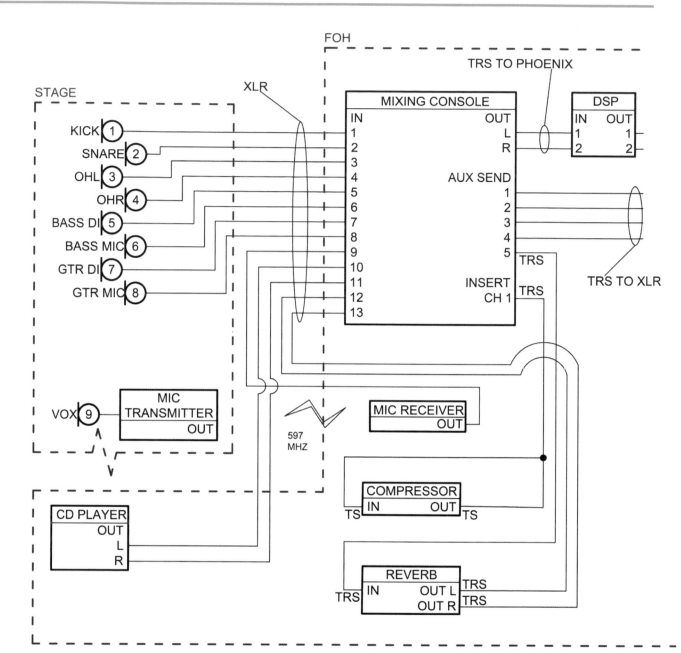

NOTE: MICROPHONE MODELS WOULD BE EITHER SPECIFIED HERE IN THIS DRAWING, OR IN AN I/O OR HOOKUP SHEET, WHICH WE COVER LATER IN THE BOOK. I HAVE SPECIFIED SPEAKER MODELS HERE IN SERVICE OF DRAFTING A PLAN VIEW (FIG. 10.1) WITH ACCURATE SCALE. A PLAN VIEW OF A SYSTEM LIKE THIS WOULD NOT USUALLY SHOW MICS. MICS WOULD BE SHOWN ON THEIR OWN PLOT, OR (MORE LIKELY IN A SYSTEM WITH THIS FEW INPUTS) ONLY IN A LIST--NOT DRAWN AT ALL. I HAVE ALSO SHOWN THE DIs AS MICS, WHERE I COULD HAVE DRAWN THEM AS DEVICES (RECTANGLES). I HAVE SHOWN THEM THIS WAY FOR SIMPLICITY'S SAKE, AS THEY ARE MIC LEVEL SIGNALS CONNECTING TO MIC CABLES, BUT SHOWING THEM AS DEVICE BLOCKS (THE FORMAL NAME FOR THESE RECTANGULAR DEVICE SYMBOLS) WOULD ALSO BE AN ACCEPTABLE CHOICE.

FIGURE 10.2

System Block Diagram of the System From Figure 10.1

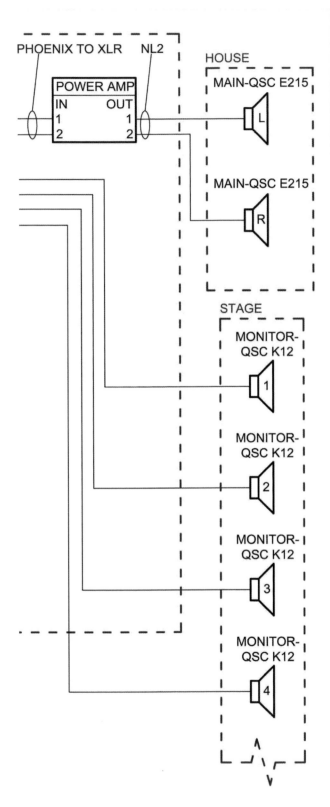

PHOENIX TO XLR   NL2

POWER AMP
IN   OUT
1   1
2   2

HOUSE
MAIN-QSC E215
L

MAIN-QSC E215
R

STAGE

MONITOR-QSC K12
1

MONITOR-QSC K12
2

MONITOR-QSC K12
3

MONITOR-QSC K12
4

JL

Project:
Chapter 10
Concert

Disclaimer:
drawing for
design
intent
only,
installer
must verify.

Revision #: /	Revision Date:
1	6/3/18

Des/Dr: J Loar
Block Diagram
Scale: N/A
Client: No One
Supervisor: No One!

Plate # 2 of 2

FIGURE 10.3
Mackie 1604 Pro Mixing Console
Source: Courtesy Mackie/Loud Audio

curve. U stands for unity gain, which means we are neither boosting nor cutting this signal. It is assumed that occasionally we will be sent a line-level signal that is too hot for use in our signal path, so there is 15 dB of available attenuation to the line input below unity. There is 45 dB of available gain above unity for line signals. While many consoles handle line and mic signals a bit differently (and the gain ranges will vary depending on make and model of the console), the principle will repeat itself. Gain is the first setting for an input signal.

Now, I want to take a moment to address a misconception about setting levels that I see in entry-level sound students over and over again. Many sound students come to me believing that the way to set mic levels on the preamp is to put the fader at the bottom of the channel up to its own unity gain position (the notched mark that is most of the way up the fader path), and then raise the gain knob until the signal is loud enough. This is incorrect![5]

It is here we need to introduce an important pair of concepts:
• Signal-to-noise ratio
• Gain staging

Signal-to-noise (S/N) ratio is a measurement that we will find over and over again in gear specs, and it is also a concept we need to understand when sending a signal through any piece of processing equipment. A S/N ratio measurement, usually expressed in dB, is a measure of the difference between the level of desired signal vs. unwanted noise in a given signal path. Any electronic device, no matter how nice or expensive, generates some small measure of "self-noise" just by virtue of operating. For example, the 1604 lists its "main mix noise," with faders and gains at unity, as −84 dBu. This is a vanishingly small amount of voltage and under normal circumstances should be totally inaudible. Now, when setting up a signal to run through a system, we want to ensure that we get the *most signal* and the *least noise* from our inputs through to our outputs. In order to do so, we want to maximize (without overloading) signal gain at the first part of the signal path, where the signal is most direct, and minimize signal gain at the last part of the path (toward output), where the signal has had the chance to build up noise from all of the processor circuits.

In a channel strip, the signal hits the gain control first and then runs (usually in visual order) down the channel strip processors out via the channel fader to the output paths.[6] This means that we want to maximize the signal at the gain control so we get the clearest mic (or line device) sound without overloading before we go on to process the signal in additional ways.

FIGURE 10.4
Preamplifier Gain Knob
Source: Courtesy Mackie/Loud Audio

This brings us to the concept of *gain staging*. The gain of a given input signal works in successive stages. The first is the preamp (with direct "gain" control). From there, it enters processors—for example, an equalizer (frequency control). If you turn up a certain frequency in an equalizer, you are adding more gain. From there, it goes to the fader (which sends to the main outputs). The fader can be pushed above unity gain (on the 1604, there is about 10 dB of gain available at that stage). All of these gains can be useful in certain circumstances, but the cardinal rule is always that the more gain you can safely add early in the signal path, the cleaner the sound. If I don't add enough gain at the preamp stage, the signal will be too quiet when it hits my fader. If I turn up my fader too far in order to compensate, instead of raising the preamp gain, the channel fader is not just boosting the mic signal, but all the circuit noise from all the processors between the preamp and the fader itself. This means a noisier, less clear signal. Gain staging is important, and as you build a system, you always want to be clear about where your stages of gain are coming from and to be sure they are set up in the proper order (which becomes more critical the more complex your system becomes).

2. Aux sends: "aux," short for auxiliary, which means "extra" (Figure 10.5)

Consoles are designed to send your signal from any given channel strip to the "main outs" (though as we will see, you still have to assign your channel to those outputs). However, we often want to send signals to more than one location, so console designs provide a range of options in this respect, mainly in the form of aux sends and in the form of so-called sub outs (also known as group outs or bus outs on different consoles). Let's imagine a small nightclub. A band is playing onstage, and in the house, we have main speakers and subwoofers. The main speakers are being run in mono (even though they are flanking the stage at the left and right edges). Onstage, the band can't hear each other naturally because it is so loud and clattery in the room, so each musician has at least one speaker (on the floor) aimed at them, and that speaker is providing a different mix of the band, to each musician's taste, that helps them hear each other and play in time and in tune. The drummer hears a lot of the bass player and some of the guitarist. The lead vocalist hears a lot of the guitar and even more of his/her lead vocal.[7] These speakers for the band are an essential part of any professional concert system, and they are known as *monitor speakers*.[8] Now, in large concert systems, mic signals are split before they reach the consoles, and there is a mixing console for the FOH (Front of House) mix (what the audience hears) and a separate mixing console for monitor mixes.[9] However, in small systems, it is very common for the monitors to be fed from the same console that is controlling FOH. Monitor signals are thus sent from the channel strip out the aux sends.

FIGURE 10.5
Aux Sends with Pre/Post Switch on Aux 1 and 2
Source: Courtesy Mackie/Loud Audio

Aux sends on any mixing console typically have one major setting to select, the "pre/post" switch. On the 1604, only aux sends 1 and 2 have this option, but in general, this is an important concept. The pre and post in question are pre-fader or post-fader. The fader in this title is the channel fader at the bottom of the channel strip, which (as we have mentioned) controls the overall level at which the channel is sent to our main mix. If an aux send is engaged in *pre-fader* mode, this means that the signal sent out over this path will be sent independently of the channel fader. Thus, if I have a kick drum in channel 1 (as is commonplace), and I feel like it is too loud in the house mix, I will turn the fader down. If my aux send is sending that kick drum onstage to the bass player's monitor, I want that monitor mix to be in pre-fader, because that way, my reduction of the signal in the house *does not* turn it down onstage for the monitor mix. Monitor mixes are ALWAYS run in pre-fader mode.

In *post-fader* mode, by contrast, if I turn the kick drum fader down in the house, that aux send is also turned down proportionally. This would be a disaster for stage monitors, but it is very suitable to the other main use of aux sends, which is connecting to *outboard effects*.

> **Outboard Effects**
>
> Sound effect units that are external to our mixing console. These are usually standard 19" wide rack-mounted devices situated in an equipment rack near the mixing console, so the engineer can make use of them in the mixing process.

Instead of a kick drum, let's now imagine a snare drum. It is very common when mixing to apply a bit of effect to a snare drum—often some reverb or delay.[10] If you are using any board without digital components, there won't be a reverb or delay available in the console itself, so you must route the signal out of the console, through an effect unit, and back into the console (the "return" signals from effects units of this type are often patched to their own channels, or to dedicated "aux returns," which can add the effect signals to the main mix but don't provide the full set of processing controls that a regular channel would). The send to these outboard effects is accomplished via an aux send in post-fader mode. Why post-fader? That way, if my snare is sent to my choice reverb and I get the balance between the clean "dry" snare sound and the reverbed "wet" signal (returning to other channels) just how I like it, I can then turn up or down the snare channel fader, and the send to the effect (and thus the effect level overall) remains proportional. Also, at the end of the set, I can fade the snare channel all the way out, and the reverb fades out too!

In addition, the 1604 features a toggle switch near aux sends 3 and 4, that allows the path to be switched to send to aux sends 5 and 6 instead. A single channel can send to *either* aux sends 3 and 4, *or* aux sends 5 and 6, but not both.

A *brief note on effects:* there are certain types of effects that we want to send to via an aux send, and those are defined as any effect that we want to use *a bit of* on an input signal, to blend with that input signal. Obvious choices here are reverbs and delays, but we might also send to a compressor via an aux send such that we don't compress the whole signal, we only create a compressed copy that we blend with the uncompressed signal in order to create a subtle dynamic effect.[11] However, for any effect where we want to *replace* the original signal with the wholly effected signal (instead of blending the clean signal with the effected), there is a better way to send the signal and return it than via aux sends. This is via a connection path called an *insert*.

> **Insert**
>
> A connection in the signal path of a channel that sends the entire signal to an outboard processor and returns the entirely effected signal at the same point in the signal path of the console.

So, say we want to compress our snare drum, and we don't want a subtle blend, we want to replace the clean uncompressed snare with an entirely compressed signal. We can patch the compressor to the insert point on our snare channel, and whatever we do to the snare in the compressor unit will be the whole of the signal from that point forward. In some consoles, like this one, the insert is a single-point send/return patch,[12] and in some consoles, this will physically consist of a connector for the send and another connector for the return. Because the insert point replaces the whole signal, there are no controls for it, just physical patch points.

3. EQ: equalizers are frequency controls (Figure 10.6)

We can turn up or down certain frequency ranges on our particular channel signal. There are a wide variety of equalizer types and controls, which we will examine in detail in Chapter 11, but for now, suffice it to say that we can adjust which frequencies we are focused on hearing for any given signal. If you have someone playing a flute onstage next to a bass amp, chances are the flute mic will pick up some of that bass. It would be a natural use of EQ to reduce low frequencies on the flute mic since that won't change the sound of the flute itself but will just remove some of the bass amp bleed, giving you better control of the flute as an individual instrument. EQ is the most commonly used effect, which is why just about any console will give you at least some equalizer control for each channel. EQ is referred to often by how many "bands" (or different distinct alterations) you have on each channel. The 1604 gives you a three-band EQ, one set at high frequencies, one set at low frequencies, and one with selectable frequency in the midrange. EQ is used to shade or color sound, to clean up signals, to emphasize tones, and more.[13]

FIGURE 10.6
EQ Controls
Source: Courtesy Mackie/Loud Audio

In the 1604, additionally, there is a fixed filter switch, which cuts out frequencies below 75 Hz, at a slope of 18 dB/octave. This is referred to as a "low-cut" or "hi-pass" (these are synonyms). Also note that the EQ, while physically below the aux sends on the console surface, is actually *before* the aux sends in the signal path of this particular console (which you can learn by reading the signal flow diagram provided in the instructions—any decent console worth its salt will give you documentation of the signal flow, so you can find out this kind of info). It is important to know this, because then you know that the EQ you are doing on the lead vocal that sounds so good in the house is also changing what the vocalist hears in the monitors onstage, and the vocalist may or may not like how it sounds up there. More advanced consoles will allow for separate EQing of these paths.

4. Pan pot: aka "assign pot," "balance knob" (Figure 10.7)

A pan pot (short for "panoramic potentiometer," a variable selector that sends electrical potential to one or another spot in the panorama—left/right field—depending on how you set the knob) is typically thought of as controlling one thing: how much signal is sent to the left or the right in a stereo system. While this is a function of the control, the knob also performs additional tasks. In a moment, we'll be addressing the assign buttons, which control where the signal gets sent (L/R mix, or subgroups 1–4). Signal is sent to pairs of output destinations, and while a traditional L/R system uses this knob as a straight pan control, the subgroups can perform differently. A pan knob set all the way left will also send a signal only to subgroup 1 and 3, and not at all to subgroup 2 and 4 (and vice versa with the knob set to the right). We'll address the uses of subgroups next.

FIGURE 10.7
Pan Pot
Source: Courtesy Mackie/Loud Audio

A note on pan vs. balance controls. On a mono channel, the pan is truly a pan control, meaning that it sends the mono input signal to the left or right sides proportionally based on the setting. However, some mixers feature stereo input channels (for connection to a stereo CD player, for example). These channels may also feature what appears to be a pan knob but is, in fact, a *balance knob*. Where a pan sends one signal in varying amounts to two destinations, the balance control is always on a stereo input source—which has already gotten a full signal sent to the left and right by default. Rather than sending both L/R input signals all the way left, for example, turning a balance knob all the way left just means that you have attenuated the right signal entirely, so all you are hearing is left. Balance sets how much of each of a pair of signals we hear in their default two destinations, where pan sets how much of one signal we send to two places.

5. Mute: a control that silences the signal running through the channel (Figure 10.8)

FIGURE 10.8
Mute Button
Source: Courtesy Mackie/Loud Audio

Typically, a mute will also mute any post-fader aux sends, but will NOT mute pre-fader aux sends (so that you can have instruments muted in the house prior to a show, but still be sending sound to the band behind the curtain onstage, to get their mixes set up before they start playing).

Advanced consoles with many input channels will often offer a feature called *mute groups* (this is most commonly found on digital consoles). If you have 96 inputs to your console, and you want 50 of them to mute at the end of a large musical number (since those 50 actors are going offstage), it will be pretty hard to individually press 50 buttons all at once, so with a mute group, you can assign all 50 of those channels to "Mute Group A," and when you press the control for Mute Group A, all 50 will mute simultaneously. A very handy feature for Broadway-style musicals, concert festivals, or any other situation where you need to be able to quickly mute or unmute a large number of channels all at once.

6. OL light/-20 light: over-limit light, aka "clip light"/signal present light (Figure 10.9)

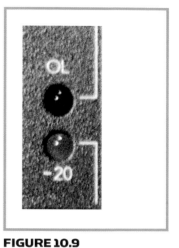

FIGURE 10.9
Over Limit and Signal Present Indicator Lights
Source: Courtesy Mackie/Loud Audio

The over limit or clip light is a small LED (usually red) that turns on when a signal in your channel is too hot and is either about to distort or is already distorting ("clipping"). This is a very valuable tool when working with lots of channels. That tuba player that you haven't been paying attention to might suddenly burst out with a loud BLAT! on one song, and you can see the light indicate the overload sometimes even faster than you can hear the problem. The OL light doubles (on the 1604) as the MUTE indicator, which will illuminate when a channel is muted, and otherwise be dark unless the channel is peaking.

The signal present light will flicker in time with an input signal to indicate that it sees signal passing through the channel. In this particular console, the signal present light doubles as the SOLO indicator (solo is addressed next). In more advanced consoles, these two lights have been replaced by full-range meters, so you can watch the signal levels in real time as they vary from weak to strong and everywhere in between.

FIGURE 10.10
Solo Switch
Source: Courtesy Mackie/Loud Audio

A note on analog signal and peak indicators. What frequency does the indicator register? All of them? In any digital console, the answer is yes—the meter is monitoring every frequency in the audible range. However, in many analog consoles (and pieces of outboard equipment) the indicators are actually only responding to a single frequency: 1 kHz. 1 kHz is in the center of a lot of sounds, and it is present in many others, so if you can only look at one frequency, this isn't a bad one to select, but this can cause problems particularly when working with low-frequency instruments. That tuba? It might not actually light up your peak indicator, even though it might be audibly overloaded and distorting, because a standard tuba tops out around 350 Hz. If your indicators are only seeing 1 kHz, they might never light up at all for your tuba mic! It is best to find out what your meters are showing before trusting your indicators (and as always, the ears are the real test).

7. Solo: solo switches are used to isolate a channel in order to hear it separate from the main mix (Figure 10.10) A concert mix engineer often uses headphones (or a dedicated speaker, aimed at the mix position) in order to hear the work she is doing. When setting mic levels, the proper procedure is to solo that mic channel and listen

(through phones or "cue wedge" which is what we call the speaker aimed at the mix position) to the signal as you boost it. When you solo a channel, three things happen:

a. The solo light turns on
b. On this type of console, which only has full-range meters on the main mix, the meters will switch over to be a full-range display of the soloed channel, so you have a good indication of level.
c. The soloed signal will be the only thing heard in the headphones or cue wedge (when nothing is soloed, the phones/cue wedge hear the main mix).

The solo control will not solo the signal in the main mix. We want to be able to solo a channel during a show so that we can identify, say, a mic cable that is having some static interference, without altering the main mix while doing so. There are two primary solo modes: PFL and AFL. AFL is further broken into regular AFL and SIP.

PFL stands for pre-fader listen. This is the mode you want to use when setting mic levels. Pre-fader listen is basically giving you the preamp gain sound and nothing else. AFL stands for after-fader listen. This is a mode you may use when checking problems during a show. If you soloed in PFL and didn't hear the issue, but then soloed in AFL and heard the problem, the problem is somewhere after your preamp—it could be your EQ, your insert effects, etc.

SIP stands for solo in place. Typical AFL solo is still mono for a mono channel, while SIP is a stereo solo mode that represents your pan setting as well as all the other channel strip processing.

8. Assigns: this is a series of switches that, when engaged, send a signal to a range of different output paths (Figure 10.11)

FIGURE 10.11
Assign Buttons
Source: Courtesy Mackie/ Loud Audio

On the 1604, there are three switches, sending to paths in pairs. One switch, LR, sends the channel to the main left-right mix output. This is known as the "main outputs." The other two switches are for sub 1–2 and sub 3–4. Subgroup outs (also known as bus outs or group outs on other consoles) are additional output paths. Why aren't they called aux sends if they are additional output paths? The short answer is that they are differently named to distinguish their use. Sub outs are typically alternate outs sending the main mix to speakers heard by the audience, often configured as different zones in a venue.

I have worked in many concert venues where the main performance hall was adjacent to a stylish bar in a separate room. That separate room still needed to get the mix of the concert, but at a different signal strength (at some venues, the two rooms only share signal during pre-show music playback, and the side room turns off once the concert starts). In order to send a copy of our main fader mix to another place, sending it via a subgroup out is a common tactic. These mixes take their balance from the fader settings, just like the L/R mix, so the balance of instruments or voices or whatever is identical to the main mix, but you have another master control with which to set the overall level.

In addition, subgroups can be sent *to* the main mix. Let's imagine a band onstage with guitar, bass, drums, and voice. Each of the instruments except drums might only have one mic. The drums, however, might have a mic on the snare, one on the kick, one on each tom-tom, one on the hi-hat, and stereo overheads to capture the rest of the cymbals. For a standard five-piece drum set, that's a minimum of eight mics. Now let's imagine that we have spent time at soundcheck getting the balance between those eight mics just right, and the set sounds amazing, but once the band starts playing together, the overall set is too loud. Rather than trying to turn each fader down by 3 dB, you can assign each of the drum mics to sub 1–2, but NOT to LR. Then assign sub 1–2 itself to LR. As long as only the drums are sent to sub 1–2, the master faders for sub 1–2 are now drum set master faders, and when you want to turn the set as a whole up or down, you can just raise or lower that pair of faders, instead of controlling 8 individual faders. This is very handy! In that it saves your hands![14]

9. Fader: finally (in terms of the channel strip, at least), we come to the fader itself (Figure 10.12).

The fader, as we have already intimated in previous descriptions, controls the level of the input signal after everything we've done to it in the channel strip. Once we've gained the preamp, set the EQs, sent the auxes, and whatever else (advanced consoles have some other options per channel that we will discuss next), the fader sets the level at which that processed signal is sent to the main mix(es). If all is set properly during soundcheck, the mixing of levels is "performed" primarily with the faders. Faders usually range from -∞ at the bottom (which is to say: off), through a range of negative numbers, to a U (unity gain, sometimes shown as a 0 instead of a U), and to a range of positive numbers at the top. At unity, the signal is neither being boosted nor cut from the processing applied in the channel strip and is being sent at full blast to the main mix. Below unity, we are attenuating the signal (we might only need a little of it in the mix). Above unity, we are boosting it further (usually up to 10 or at most 20 more dB—the 1604 gives 10 dB of boost at the top of the fader)—we might need just a little more level compared to other signals, and if the preamp is already gained fully up, the fader is the place to get it.[15]

FIGURE 10.12
Channel Fader
Source: Courtesy Mackie/
Loud Audio

This is our channel strip. Advanced console channel strips will also feature some expanded options: more EQ controls (or finer-grained EQ controls), the addition of dynamic effects (compressors, limiters, and gates, most commonly), additional aux sends, and more routing assigns, but the basic paradigm of available controls as listed above repeats. Large analog consoles appear to have tons of controls, but they are mostly just variations of the above themes, repeated many times. Digital consoles may not have a physical channel strip but may work on a "channel select" paradigm. In these systems, each channel has a "select" button. By pressing that button for a given channel, that channel's "channel strip" controls become visible on the main display and can be adjusted from there.[16] We have covered all the basic functions of a channel strip: gain, effects, routing, panning, muting, soloing, and leveling. These are present in any mixing console, anywhere in the world, from nearly any era.

From the channel strip, the signal is routed to our *master section* (Figure 10.13).

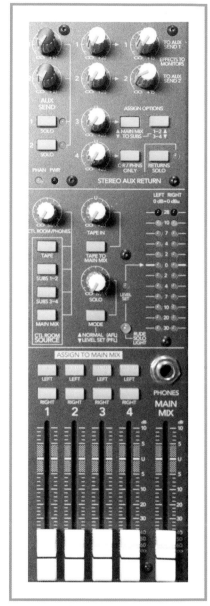

FIGURE 10.13
Master Section of Console
Source: Courtesy Mackie/Loud Audio

The master section of a console is so named because it is the home of all of our master controls: the master faders and level controls for main mixes, subgroups, and aux sends, and any other *global* controls and settings that have an impact over the entire console. The master section of the 1604 has only a few sub-sections, which—again—are exemplary of those you will find on just about any console. Note that one thing you find in the master section on this console is unique to this and other small-format boards. There is a phantom power ("PHAN") light. This light is on when phantom power is engaged. This is in the master section because on this console, there is one master phantom power switch for all of the inputs, and it is either on or off. On more advanced consoles, you will find a 48V indicator on each individual channel, as you can toggle phantom on and off on an individual channel basis.

1. Aux masters (Figure 10.14):

Aux masters are simply master faders (or in the case of the 1604, pots) for the aux send paths. If you turn up aux 1 on a range of channel strips, the extent to which each is turned up determines how much of that particular input signal will be sent to aux 1. Aux 1 master determines the overall level of that aux send. So, like our drum subgroup from earlier, imagine we have ten mics all sent to aux 1 in a perfect balance, but then we want to make aux 1 *overall* louder or softer, this is where the aux master control is useful.

2. Aux returns (Figure 10.15): (labeled "stereo aux returns")

   Aux returns are extra paths into the main mix for the return signals of outboard effects units. They are usually configured in pairs because the type of outboard effects that we send to via aux sends typically have stereo returns (reverbs, delays, etc.). All of the aux returns can be sent to the LR mix (by default use of their level controls), which is a typical routing for aux returns on any console. On this particular console, there are also options to route Returns 1 and 2 to aux sends 1 and 2, respectively (to add, say, reverb to a stage monitor mix), to send Return 3 to subgroups instead of the LR and to send Return 4 to either the main LR, or to the control room/headphone outs only (for preview).[17]

3. Control room/headphone section (Figure 10.16):

   This section assigns sources to the headphone/control room mix. If you are using the 1604 (or any console) as a studio device, you will have speakers connected to the control room outs in the patch panel. Those speakers will act just as headphones would for a live engineer (conversely, if you are using a cue wedge, you can connect it to the control room out, even in a live setting, and it too will act identically to the engineer's headphone send). In this section, you determine what the headphones/cue wedge/control room speakers will play when not in solo mode. On the 1604, your options are Main Mix (LR), Subs 1–2, Subs 3–4, or "tape." To the right, you will see a knob labeled "tape in," with a switch that, when engaged, sends the tape signal to the main mix. This is for a stereo input called a "tape return." Tape returns are artifacts of old studios, where it might be common to have a cassette or other two-track tape machine patched in for listening in the

FIGURE 10.14   **107**
Aux Masters
Source: Courtesy Mackie/Loud Audio

FIGURE 10.15
Aux Returns ("Stereo Returns")
Source: Courtesy Mackie/Loud Audio

FIGURE 10.16
Control Room/Headphone Section
Source: Courtesy Mackie/Loud Audio

control room, so a musician could (for example) bring in a demo and play it easily. Since you can patch this into the LR mix, this also can effectively act as another aux return.

Below the tape in, you find the solo level (which sets the overall level of the soloed signals in the headphones/cue wedge. You can also select the overall solo mode (AFL/PFL).

We also have our main meters here, which—when nothing is soloed—reflect the levels of the main mix. When something is soloed, the "level set" light turns on, and the meter acts as a meter for soloed channel.

Finally, in this section, there is a "rude solo" light. This is a feature that lights up whenever any channel(s) are soloed. On a 16 channel console, it is not that difficult to find a soloed channel without looking at the rude solo light, but on a 96 channel console, the rude solo light can be very helpful (additionally, on advanced digital consoles, you will often find a "clear solo" button, so you don't have to hunt for the channel you accidentally left in solo mode, you can just switch all channels out of solo mode with one button press).

4. Master faders (Figure 10.17):

Here we find the master faders for subgroups and the main output (additionally, because the 1604 allows the subs to be routed to the mains as described earlier, there are assign buttons above each subgroup fader). Many consoles (though not this one) also have a feature found in the master fader section called either VCAs or DCAs. In large-format analog consoles, VCA stands for Voltage Controlled Amplifier (on contemporary digital consoles, these are known as Virtual Control Amplifiers or Digital Control Amplifiers). These faders can be assigned a range of input channels for which they serve as a master fader. The difference between a VCA and a subgroup is that a VCA does not have its own signal path. A subgroup is a dedicated output path. It can be assigned to the LR, but it can also be sent out its own physical output connector at the patch bay, separate from the LR. A VCA is only a master fader; there is no output path. All the individual channel assigns still work as originally configured when they are part of a VCA; the VCA just gives you extra grouped level controls.

FIGURE 10.17
Master Faders
Source: Courtesy Mackie/Loud Audio

So! These are the basic functions of not only our small sample console but also any sound console you will encounter. Counts will expand, extra meters will be added, the "scribble strip" (the area where on an analog console you might lay a piece of white tape and label your channels with a pen) becomes a digital scribble strip where you can program the channel names, and so on.

If this is what all consoles do, what is the advantage of a digital console system? What are the disadvantages?

ADVANTAGES OF DIGITAL CONSOLES

- Effects: digital consoles usually provide not only full-featured EQs for every channel but also dynamic effects for every channel, and a range of built-in effects of other types (reverbs, delays, etc.), which means that you save space and labor because you no longer need racks of outboard effects, you just need your console. Does this mean we never want outboard gear when we have a digital console? No, some effects are very specific, and may only exist as outboard gear, and thus still need to be hardware-patched. Additionally, while not technically an "effect" some people will still want specific outboard preamps, boutique units that have a particular sound. This is more common in studios than live, but you will find it in some live systems as well.

- Programming: one of the signature advantages of digital consoles is programmability. Most digital consoles feature some version of savable "scenes," which allow the user to save whatever parameters they desire across the entire platform. This can be enormously useful. In concerts, it is common to save a scene for each opening act and for the headliner, so that when the stage changes over between bands, the mix engineer doesn't have to look at detailed notes from soundcheck (and the engineer doesn't need to have taken those very notes!). The engineer just hits "scene recall" and is ready for the next band. In theater, it is very common to save "scenes" according to the actual scenes in the show so that when new groups of actors come on or offstage, you don't have to remember which mics to turn on or off, the engineer just hits "scene recall" and the appropriate program is recalled. Additionally, digital consoles can send and receive messages (MIDI, OSC, or other types), which can allow a console to, say, trigger a playback system, or vice versa.

- Digital audio: digital audio can be easily copied, pasted, and routed to multiple locations in a way that in an analog system would require miles of extra copper wire. For example, a console working on the Dante audio protocol can connect to a network switch and via that path connect to any other Dante-enabled device. So, instead of patching, say, 30 copper-wired cables to the outputs of an analog console and then to the inputs of a DSP, you can send all 30 of those channels over one network cable. The same 30 signals can be sent simultaneously to a Pro Tools system to record the signals in an HD multitrack format, with the addition of simply one more network cable. Digital audio is thus flexible in a way that analog audio never can be, and in today's complex systems, this is a huge advantage.

- Stage Boxes: while small digital consoles still feature I/O connections on the console frame itself, medium and large-sized consoles host most of their physical I/O via external *stage boxes*. These are just rack-mounted external hardware pieces with a bunch of signal connection ports on them. Some come pre-configured (like Yamaha's RIO series), and some are configurable via card slots (like Avid's live console series). The advantage here is primarily one of cable runs—if you've got a band onstage with 64 mics, we don't need 64 balanced lines (192 copper wires) running all the way from stage to FOH, we just need to run to the side of stage where the stage box is situated, and then the stage box can feed the console usually with just one or two cables (often network cables). We can also connect more than one stage box, so we can mic most of the orchestra in the Broadway pit and then the extra horn section playing backstage[18] with a second box.

DISADVANTAGES OF DIGITAL CONSOLES

- Latency: as we mentioned with digital wireless mic systems, digital consoles have a measure of latency for the A/D and D/A conversion. This latency is vanishingly small, as processors get faster and faster inside our consoles, but it is still larger than the *zero* latency of analog consoles. For this reason, there are some musical performers specifically who still demand analog consoles for their monitor mixes (particularly when they use in-ear monitors), even though they allow digital consoles at FOH.[19]

- Cost: while costs of digital consoles have certainly come down in recent years, digital consoles are still typically more expensive than analog consoles with equivalent channel counts.[20]

Tales From the Field

The worst time I've ever had was—I was doing something at the Kennedy Center. We had an A-list line up, and for whatever reason, the [then-new] Avid console kept dumping the patches. Couldn't figure out why. We had to hold the whole show and that was pretty bad news. It was just that one night which is part of the reason that you have to be careful with the new-fangled stuff or a software update. I think we came down to [the fact that] it was a software update, and that goes back to the point: try to do as much in the shop as possible, to avoid those types of issues in the field.—David Kotch

In the professional systems world, analog consoles are only used for three reasons: 1) the venue is small, the operators are non-professionals, and the budget is tight; 2) the latency issue mentioned earlier; and 3) it is a studio production system, and the high-end (often vintage) sound of a particular analog console is desired.[21] Otherwise, digital consoles are now the industry standard for all professional sound systems that need mixing consoles.

CLASSIFICATION

When we are planning to select a console, there are infinite minute varieties in software platform, included effects, and other such details, and to examine these in-depth is beyond the scope of this book. However, in broad strokes, when selecting a console, there are a few basic categories that form the anchor of our classification system.

- Channel count (inputs): this is the first consideration that must be settled when selecting a console for a system. Does the console have enough input channels for the number of input sources that need to run through it?
- Channel count (outputs): does our console have enough main outs (if it is an FOH console) or aux sends with pre-fader mode (if it is a monitor console), or both (if it is a dual-duty console) to serve our needs?
- Effects: what effects are included? Any console provides effects, and digital consoles usually provide a robust complement. Does this console include the effects you need for your system?
- Analog vs. digital: as discussed earlier.
  - If digital:
    - Protocol: Dante? AVB? CobraNet? Something else?
    - Sample rates/bit depth
    - Stage box vs. onboard connections: most analog consoles have all their cable connections built into the back of the console itself. Most digital consoles have only a few connections on the back of the console, included among which are connections to their *remote stage boxes*. As mentioned earlier, remote stage boxes are rack-mounted sets of physical inputs and outputs, that connect to the console system via trunk lines (often network cables, or coaxial cables, or occasionally proprietary connector types). What kind of stage boxes are available for your console (and what connections are available on those boxes)? It will vary by model, so make sure you know what options exist as you plan your console system.
    - Additionally, there are a host of potential variations in the software and program features of a digital console that may influence your decision-making. Some consoles are designed around speedy use in concert with easy interfaces; some are designed to be live consoles also feeding a broadcast truck, which may offer additional features but more complexity in setup and operation; others are designed around theatrical performance, with "understudy" features to easily swap one actor's mic settings for another's throughout an entire play's worth of cues. Additionally, some console software will just feel "more natural" to some people than to others, and this is really a matter of taste.[22] What features you need depends on your specific production or venue. These options can be overwhelming, but many brochures, manuals, and spec sheets are available, so do your research, and you will end up with the console that suits your particular needs.

That's basically it. We typically classify consoles by channel counts, effects available, and whether they are analog or digital. The devil is in the details, though, so we'll examine specifications a bit more extensively.

SPECIFICATION

While many of the following specifications will apply to either an analog or a digital console, we will examine a list appropriate for a digital console. Assume anything specifically related to computers (like Ethernet control) or digital audio (like sample rates) will not be applicable if you are selecting an analog console.

- Input channels: number of inputs available in this console. This does not always correspond to either the number of input connectors on the console (you may need a stage box or multiple stage boxes to make use of the full number of inputs the console is capable of processing), or the number of faders on the console (in fact, most digital consoles have fewer faders than input channels; they make use of a system called "banking" channels, whereby, for example, 24 faders selected on bank A will be controlling inputs 1–24, selected on bank B will be controlling inputs 25–48, bank C will be controlling inputs 49–72, D for channels 73–96, etc.).

- Output channels: main outs, sub outs (or bus or group outs), and aux sends. Like inputs, be careful to understand the distinction between output paths and physical output connectors.
- VCAs: does the console feature them? How many?
- Mute groups: does the console feature them? How many?
- Internal effects engines: this is a count of how many non-channel-strip effects can be used simultaneously. For example, a console featuring 16 effects engines can have 16 different reverbs, or a reverb, a delay, 2 pitch shifters, a ring modulator, etc., etc., up until you have installed 16 effects.
- Internal presets: aka show automation, scenes, etc. How many "scenes" can be stored in the console? A good limit is somewhere in the 500–1,000 range, which makes a console very useful for the most complex productions.
- Sample rate/bit depth: some consoles only operate at one sample rate and bit depth. Some operate at a range of rates and depths.
- Dynamic range: expressed at different points of the signal chain. Often identified at the A/D conversion, or at the D/A conversion. Measured in dB, this should be a *minimum* of 100 dB, and the higher the better.
- Latency: measured in milliseconds. This should identify what exact signal path the latency refers to. For example, a console that has both inputs on the console frame and also an external stage box will likely have shorter latency for the path from internal connection through output than from the inputs at the stage box through to outputs. As with any digital audio device, the smaller this number is, the better.
- Connector types (this spec may be listed for the stage boxes as separate hardware items, but still needs to be considered when planning your console system): while mic ins are always XLR (as are many line inputs, though TRS is also very common), outputs almost always come in a range of configurations. TRS, XLR, AES3 (digital over XLR), etc.
- Phantom power: per channel, in banks, or console-wide.
- Total Harmonic Distortion + Noise (THD+N): This is usually listed for a specific part or parts of the signal path. For example, you may have a THD+N spec for the mic input channels. This number should be tiny, ideally less than a tenth of 1%.
- Frequency response: should be at least 20 Hz to 20 kHz.
- Input impedance: while for the most part, we can take impedance for granted (as the input impedance will be higher than the mic output impedance, always), this is still listed. When using vintage mics with unusual output impedances, this may become relevant. If the signal from your vintage ribbon mic is too low, sometimes you may want an external preamp with variable impedance control, so you can drive the mic with lower impedance resulting in a stronger signal. Most of the time, even in the case of vintage mics, this is an unnecessary consideration.
- Output impedance: even less relevant, typically, than input impedance, as professional equipment will be designed to work together.
- Power: it is important to know if your device is set up to work in your country. While US power is 120 V, 60 Hz, European power is 240 V, 50 Hz. Most professional consoles have switching power supplies that can work with either source, but especially when working with older consoles, you may need to add a voltage transformer to your system if you plan to use US gear in the United Kingdom (for example).
- Maximum levels: what is the maximum voltage of an input signal before the input channel will distort? Usually expressed in dBu. Often in the +20–30 dBu range. Since line-level signals are technically +4 dBu (nominally, they actually vary by the intensity of the content), this should accommodate any line-level signal.
- Computer control: is this via Ethernet? USB? Both? Can you attach external keyboards or mice to the USB ports? Can you connect the Ethernet port to a wireless access point, and control the mixer via iPad (additionally, does the mixer feature an iPad control app?).[23]
- Additional effect specs: what type of EQs are available in the console, and what are the frequency ranges for each band? What kind of dynamic effects are available? Et cetera.

There are other specs you will see listed on consoles, but for 99% of your applications, they can be functionally ignored.[24]

CONSIDERATION

What questions must a designer ask (and answer) in order to select a mixing console as part of their system?

1. Should the console be analog or digital? Assume digital unless compelling reasons to the contrary are found.
2. How many channels do I need to run through the console (in and out)? All mics, DIs, and, depending on the situation, playback channels must go through your console (note that it is increasingly common for playback systems to use digital audio protocols like Dante, and to send the signal directly to the DSP, skipping the console altogether). Do you need to run mains? Delay speakers? Monitor mixes (or do you need a second console for monitors)?[25]
3. What kind of programming do I need, or would be advantageous? It's hard to specify this until you get a sense of what is available—read manuals and brochures, and you can learn a lot about what options are available to a contemporary system designer.
4. What kind of effects do I need? If there is no compelling reason to use outboard effects, find a console with appropriate built-in effects.
5. Does my console need stage boxes? How many does it need in order to achieve my desired input/output count? Some console systems have fixed stage boxes that have a set number of ins and outs and can't be customized. Some systems, however, allow customization of the stage boxes. It can be helpful, for example, to create one box of all mic ins to go in the orchestra pit and one of all line outs to feed the amp stacks.
6. How large is the console vs. the space in which it will be operated? This seems obvious, but sometimes a beginning designer will specify a console that is simply too large for the FOH or monitor mix position available in their venue. This definitely needs to be considered.

OK! Mixing consoles are complex. This was a long chapter, and we still only scratched the surface of the kind of advanced options available in contemporary digital mixing consoles. An entire book could be written just on the variations in this one component category. That said, if you grasp the information noted earlier, you will have enough ammunition to approach the selection of a console intelligently and to end up with a model that will suit your needs.

As we have mentioned, many consoles have a range of built-in effects. What kinds of effects are available in our industry, and what do we use them for? This is the subject of our next chapter.

Notes

1. In Chapter 6.
2. In Chapter 1.
3. For simplicity's sake, we've skipped the DSP on the monitors, but in reality, we would still want a DSP between the console and the monitor system.
4. Product photos by author.
5. With one exception, which is if your system is so overpowered for the space it is covering that by following the proper gain setting procedure you can't help but feed back, in which case, some variety of this "wrong" setting process might be useful. However, in a system properly scaled to the room, use the process as identified in the next section of the main text.
6. Again, an oversimplification, we will address other paths as we elaborate our description of the console and channel strip's functions.
7. At the risk of being controversial, lead vocalists always want to hear *a lot* of themselves in the monitors. This is partially because it makes it much easier to stay in tune and hear how their voice is interacting with their mic, but also because lead vocalists just *like* to hear themselves a lot.
8. Except in the United Kingdom, where they are typically known as *foldback speakers*, as in "the sound of the musicians is folded back to them via these speakers."
9. For the splitting of analog mic signals, we run the mic cables into a dedicated splitter, which is a device that not only splits the signal from one wire down two wires to run to two consoles but runs those splits through transformers that are set up such that they isolate each copy from the other. This ensures that the FOH engineer and monitor engineer can each set their own preamp gain levels on their copy of the signal. This is preferable to sharing one preamp gain, as each engineer can set strengths as appropriate to the output level of their particular system. It is fairly common for monitor engineers to run with slightly lower preamp gain compared to the FOH

engineer, because signal to noise is not as important in monitor engineering as is the prevention of feedback. A monitor engineer may need to have more range of control of the signal at their channel fader than would be possible with full preamp gain, and thus they might need to reduce the preamp. The FOH engineer will still usually prioritize the clarity of the signal. Some digital console systems skip the transformer-split, taking one copy of the mic signals and sending a virtual digital copy to the other console, but that second console then has no control of the gain. They do usually, in that case, have a supplementary "trim" control which can produce the same effect of lower signal at the input stage for more range at the fader, but the risk is that if the engineer managing the preamps sets a level either too low or too high, the secondary console engineer can't fix this on their own—and thus has to live with either circuit noise or overload distortion in the signal that they cannot change.

10 Discussed in Chapter 11.

11 Again, effects will be discussed in detail in Chapter 11.

12 This is explained in Chapter 24.

13 Detailed examination of EQ in Chapter 11.

14 I'm starting to groan at *myself* for some of these jokes. . .

15 It is worth noting that many professional mix engineers strive to avoid using the + side of the fader whenever possible. If your system has enough power for your venue, you should be able to get all the level you need at unity or below, and if you need a bit more of one channel, you should first turn down the others. That said, I have worked as a mix engineer on systems I did not design, and when the system is underpowered for the room, I was pushing faders into the + side on the regular.

16 Note: some companies, in an effort to cater to old-school engineers who prefer to have channel strip controls laid out visibly for each channel, and to have controls for multiple channels accessible simultaneously (as opposed to the "channel select" model, where only one channel can be modified at a time), have designed digital consoles with vertical strips giving at least basic controls for each channel all simultaneously. Soundcraft consoles are notable in this respect. However, these consoles still do not allow full feature control in the physical channel strip and some only provide readouts of settings on each strip and still require the settings to be controlled via channel select.

17 There are also solo buttons for both Aux Send Masters 1 and 2, and for the aux returns as a group.

18 This is a thing. A friend of mine played in the orchestra as a sub on trombone for one of the very popular Disney Broadway shows (that will remain nameless). He played in a room backstage, with a video monitor of the conductor and a speaker playing the rest of the orchestra. He was in the room with only other horn players. He never saw or met the conductor or the rest of the orchestra for that gig.

19 Notable artists in this category include Fleetwood Mac and Sting. Artists who make this request are typically baby-boom or older generationally and are typically very high-priced artists who can demand such things. Digital consoles are the standard for FOH and monitors, around the world, for the vast majority of professional systems for concerts and theater and any other live events.

20 Though, when it comes to boutique studio consoles, some analog consoles are more expensive than a large house.

21 We will cover some studio-specific console considerations in Chapter 45.

22 For example, I find Avid consoles' software very easy and intuitive, but there is one manufacturer of digital consoles (who shall remain nameless), whose consoles are really wonderful sounding and filled with interesting features, but which I avoid specifying whenever possible, because I find their software to be overly complex and difficult to use for no apparent benefit in operation.

23 While there are a wide range of available tablet computers, the industry has (for now) standardized around the iPad as the preferred tablet for external mixer control. I have yet to see a console offering Windows or Android tablet control, but many models offer iPad control, which is an enormously useful feature. A mix engineer can move around an audience space, tweaking the mix to ensure it is consistent and good for all of the guests, which is especially helpful when the mix position is horrible and unrepresentative of what the audience is hearing (which is often the case in small venues, theaters, etc.).

24 I know a couple of console designers who are probably mad at me for saying this, but I have been designing sound systems around the world and have never needed to look in great detail at any specs other than the ones listed in this chapter.

25 A note on monitors: some digital console systems feature "personal mixing systems," which are often used on Broadway. These are systems whereby instead of sending aux sends to monitor speakers, you configure them as stems of the musical program. So, Aux 1, for example, instead of being the "drummer's monitor mix," might instead be a mix of the drums themselves. Aux 2, instead of being the "bassist's monitor mix," might just be "bass." These mixes all run to small personal mix stations (panels set next to each instrumentalist that are usually no larger than an iPad). The instrumentalists connect headphones to the personal mix stations, and on the stations are usually 16 knobs, each controlling one of the auxes so that each instrumentalist can create her own blend in her headphones. We will address these systems more in Chapter 17, but when specifying a console, it is important to know that this is an option—if your production or venue would benefit from these systems, choose a console with native integration of such a system as it is much easier than configuring one externally.

CHAPTER 11

Effects—Part 1

Creative and Mixing Effects

In mixing practice, sound engineers and designers use a wide variety of effects to adjust and alter the content of their sound programs. While this is not a mixing textbook, we will examine the standard categories of effects (and their most common representatives) in just enough depth that you should be able to understand when one or more of these effects needs to be specified as part of your system. This chapter focuses on effects as used for creative sound modification, and thus positioned in the system in or around the mixing console.[1] Effects used in system calibration will be examined in Chapter 13.

DEFINITION

Mixing effects can be broken down into seven essential categories:

1. Spectral: effects that alter the frequency spectrum (e.g., EQs, filters)
2. Dynamic: effects that alter dynamic range (e.g., compressors, expanders, limiters, gates)
3. Drive: effects that create distortion via the overloading of a circuit (or the digital approximation thereof—e.g., overdrive, distortion)
4. Time-based: effects that alter sound by virtue of changing the timing of the signal (e.g., delay, algorithmic reverb, chorus, and other modulation effects)
5. Convolution/modeling: effects that use a digital technique called "convolving" to combine an input signal with a reference signal to create a new tone (e.g., amp-modeling effects, any plug-in that is modeled on old outboard hardware)
6. Analytical: effects that examine sound and represent or interpret it graphically and mathematically (e.g., Waves PAZ analyzer)
7. Hybrid: effects that combine two or more of the aforementioned types to create something new (e.g., Logic Pro's spectral gate, dynamic equalizers, Auto-Tune)

Let's examine each category.

SPECTRAL EFFECTS

Spectral effects are listed first for two reasons.
1) They are the most commonly used effects.
2) Historically, they are the first effects, and our understanding of them goes back to early analog technologies.

Spectral effects are any effects whose chief action is on the frequency spectrum, altering what frequencies we hear in a given piece of audio. Let's imagine we want to change the amount of 1 kHz in a signal. A basic analog equalizer[2] operates by splitting an input signal along two paths.

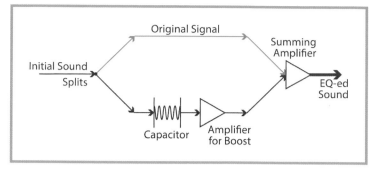

FIGURE 11.1
Basic Analog Equalizer Signal Path

The first copy remains unaltered until it is recombined with the second. The second is routed into a tuned filter (essentially a capacitor—or series of capacitors—with a width between plates that will resonate at the analogous *electrical* frequency of our desired *acoustic* frequency). This acts as a rudimentary *bandpass filter*, only allowing the desired frequency (and those immediately around it) through. The second copy can then be delayed a tiny bit so that when it recombines with the first copy it will cancel out the frequency in question.[3] On the flip side, the second copy can be boosted with an amplifier in circuit, which means that when the two copies are rejoined, there is more of the given frequency (in this case, 1 kHz) in the resulting signal. We see a simplified version of an analog EQ circuit in Figure 11.1.

Digital EQs are essentially performing the same task, except via mathematical calculations that simulate the physical activity of electronic circuits.

The basic component of an equalizer is a filter. In today's mixing systems, you will find bandpass filters in such forms as parametric EQs, graphic EQs, etc. (more on these types next). There is a simple filter type that is commonly used that is sometimes included in EQ components, but is also sometimes found separately, and this is what we know as a *high-* or *low-pass filter (aka low or high cut)*.

High-Pass/Low-Pass Filter

A filter with a fixed slope that is either situated at the bottom or the top of the frequency range, and that allows all frequencies either above or below its own cutoff frequency to pass through. High-pass filters (Figure 11.2) are often used in live systems to remove unwanted stage rumble from sources that don't have much low-frequency content—e.g., most female vocals. Slope is defined as a value of dB/octave, normally expressed in 6 dB increments. Cutoff frequency is not the frequency at which the attenuation begins but is typically the frequency by which attenuation has reached 3 dB (at which point it begins to be noticeable).[4]

Equalizers (aside from high- and low-pass filters) come in four basic shapes:

- Shelving: these EQs, like so-called hard filters noted above, have a cutoff frequency beyond which all sound is altered. However, where a high-pass filter rolls off everything below the cutoff frequency at a steady rate marching into infinity, a shelving EQ instead attenuates or boosts everything below (in the case of a *low shelf EQ*) or above (in the case of a *high-shelf EQ*) the cutoff frequency by an equal amount. So, we could, for example, set a low shelf at 250 Hz, and reduce everything below that point by 10 dB. In a digital model, we can typically select the frequency and gain (up or down) of this EQ, and sometimes we can even adjust its (sometimes listed as "Q"), but in analog models, the frequency and slope are normally preset, and we can simply adjust gain. Figure 11.3 shows a low-shelf EQ.
- Parametric (Figure 11.4): a bandpass EQ (operating on a certain band of frequencies) that allows for the user to select three specific parameters of operation: center frequency of the band to be altered, gain (up or down),

and Q—also known as bandwidth[5] of the filter. A wide parametric setting is sometimes known as a bell EQ, for its shape, and, likewise, a very narrow parametric setting is sometimes known as a notch filter. Parametric EQs are widely used in mixing (and in system tuning, which we will address in Chapter 13).

- Semi-parametric (Figure 11.5): similar to a parametric, except without the ability to alter bandwidth (Q). Often found in analog equipment.
- Graphic (Figure 11.6): a bandpass EQ with a range of filters at preset frequency centers, with preset Q, where generally the only adjustable parameter is gain (though in software, sometimes we get "paragraphic" EQs, which offer a graphic-style interface with more parameter control). Often, the frequency centers are spaced by one-third octave each.

These are our basic EQ types.[6] In the digital age, there are some hybrid EQs like the aforementioned paragraphic EQs and dynamic EQs (a hybrid effect that combines the action of a compressor with the action of an equalizer). However, most EQs for mixing use fall into one of the categories noted earlier.[7]

While some people might consider pitch/tuning effects (like Auto-Tune) to be spectral effects (since their impact is chiefly on the frequency spectrum), I have instead classed these as hybrid effects, since the action is a spectral alteration based on digital analysis of audio input and complex mathematical adjustment to the signal, as opposed to EQs and filters, which are much simpler. Auto-Tune (and Melodyne and others like them) involves spectral/analytical effects and thus falls in the hybrid category.[8]

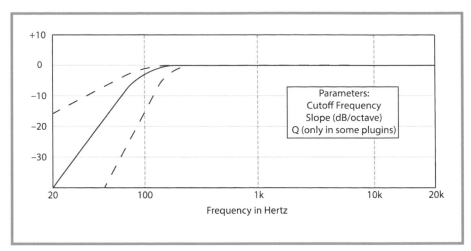

FIGURE 11.2
Graphic Representation of a High-Pass Filter

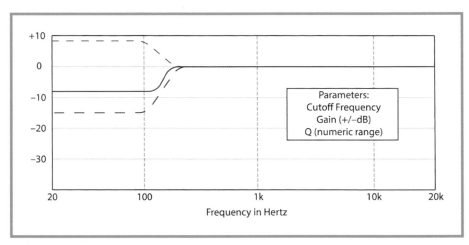

FIGURE 11.3
Graphic Representation of a Low-Shelf EQ

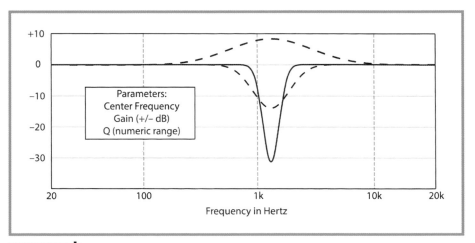

FIGURE 11.4
Graphic Representation of a Parametric EQ

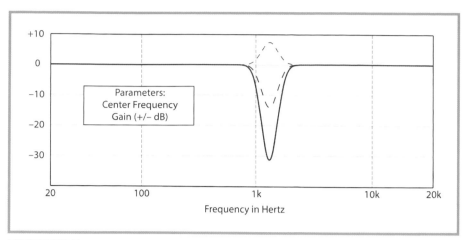

FIGURE 11.5
Graphic Representation of a Semi-Parametric EQ

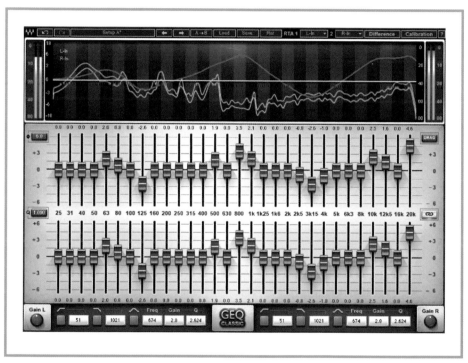

FIGURE 11.6
Waves GEQ Graphic Equalizer
Source: Courtesy Waves Audio Ltd.

DYNAMIC EFFECTS

Sound consists of vibrations at a given frequency and loudness. While spectral effects handle adjustments to the frequencies, dynamic effects exist to help us manipulate the loudness. Between the two categories, we have the most commonly used mix effects, bar none.

Dynamic effects seek to do one of three things to an audio signal:

1. Reduce dynamic range
2. Increase dynamic range
3. Make use of dynamic differences to "clean up" the signal

FIGURE 11.7
Logic Pro X Compressor Display
Source: Logic Screen Shots Reprinted Courtesy Apple, Inc.

The first class of dynamic effects, those that reduce the dynamic range, include *compressors* and *limiters*. Figure 11.7 shows a compressor with built-in limiter.

Compressor

An audio effect that both reduces the loudest parts of a signal and boosts the quietest parts of a signal, to narrow the incoming dynamic range.

Limiter

An audio effect that reduces the loudest parts of a signal.

Imagine a band playing, with a lead singer in front. The lead singer has a problem. During the quiet parts of her vocal performance, the lyrics get lost in the music; her voice is too quiet to be heard. When the band kicks into high gear, and she starts screaming the lyrics; she is so loud that we start to distort the channel (assuming we set our input gain based on the quiet parts of the song). A mix engineer could spend the whole show chasing her level moment by moment, or we could apply a compressor. When her vocal signal enters the compressor, it is sensed by the circuit. We set a *threshold* control. This is a fixed number, whereby any sound louder than the threshold will be reduced in level, and any sound below the threshold will not be reduced. The *ratio* control has to be set to a number greater than 1:1. The ratio determines how much the signal will be reduced once it exceeds the threshold. At 1:1, no reduction is taking place. If we set the threshold so that the singer's voice only goes over the threshold when she is singing her loudest parts, and then (for example) set the ratio to 2:1, we will see reduction on our gain reduction meter. At a ratio of 2:1, a signal 1 dB over the threshold will be reduced to 0.5 dB over the threshold at the output. A signal 10 dB over the threshold will be reduced to 5 dB over, and so on. If we increase the ratio, we increase the reduction factor. If our ratio is 5:1, a signal 1 dB over will result in a signal 0.2B over at output, if input is 10 dB over the threshold, we get 2 dB over at output, and so on.

Now, how fast does our compressor need to act? How quickly should it stop compressing? These are our attack and release settings. Attack determines how long the compressor waits to reduce the level after sound has begun to be reduced in level (usually expressed in milliseconds). If we have long, slow, legato notes (say, slowly bowed cello) we can usually use a fast attack, since we just want to suppress level whenever it happens to tip over the limit. If we have quick staccato notes (like a snare drum), using a fast attack will often kill the first impulse of the transient, thereby dulling the drum sound. If we want a punchier drum sound, we want an attack just slow enough that we compress the ring/sustain out of the drum hit, but not the initial strike—this kind of setting cleans up the sound but leaves some crunch in the track. Release time, on the flip side, determines how long the compressor will wait before stopping the reduction (this is not keyed to the threshold, just to sound being gain-reduced). A fast release means that the sound will only be attenuated briefly, which can be good for things like drums, where you want the compressor to have released before the next transient, so that you preserve some *sense* of dynamics, even as you tame them. For our long legato cello, we may want a longer release, so that the suppression of level gently fades back into the distance along with the note itself, rather than shutting off quickly, which would be very noticeable in a continuous bowed note.

Finally, we have the output gain . . . this allows us to raise the overall level of the track, and since we're holding down the highest levels, this effectively boosts the lowest levels in particular, and thus we have narrowed the dynamic range. Some compressors have further controls (knee—which determines how rigid the threshold is; punch—a tone control; etc.) as well.

Compressors are also sometimes used as *parallel* effects. This means that compressors would be applied, not to a main signal, but to a copy of the signal in an additional channel. This technique allows a mix engineer to easily turn up the compressed copy of the signal when it needs to be more consistent (say, during the fist-pumping chorus of a good bop) and turn it down during less intense sections (say, the moody middle eight). While this technique is most common in studio mixing, it has become quite common in concert engineering as well.

In digital mixing consoles, it is assumed that each channel has a compressor built-in and available.

Limiters (Figure 11.8), on the other hand, are simply concerned with reducing the peak levels and ensuring that they do not go into overload. Typically, digital limiters have fast attack, and a gradual knee (set by default and often unchangeable), such that some reduction starts as a signal approaches the threshold, and the reduction increases the closer to the threshold it creeps. This helps a "hard limiter" sound less choppy and keep smooth sound while

FIGURE 11.8
Waves L1 Limiter Display
Source: Courtesy Waves Audio Ltd.

reducing exaggerated peaks. While limiters do get used in creative mixing, their main use is in system calibration, and we will address these in more detail in Chapter 13.

The primary effect designed to increase dynamic range is called an *expander*.

Expander

An audio effect designed to widen the dynamic range of a given input signal by reducing the level of already quiet material (shown in Figure 11.9).

Where a compressor's threshold sets a limit above which sound will be attenuated, an expander's threshold does the opposite. When sound is *quieter* than the threshold, an expander reduces that sound in level even further, widening the difference between loud and soft for a given input signal. Expanders are used sometimes in mixing, but not as frequently as compressors. While a compressor is a very easy tool for narrowing dynamic range, in practice, an expander is harder to use successfully—the fact is that when we have signal with too little dynamic range, hitting the exact spot for the threshold where the sound we want to make quieter gets quieter and the rest stays loud is very challenging. Expanders are, as identified in the wireless mic chapter, essential to the operation of wireless mic transmission.

Our third category is of effects that clean up a signal by making use of dynamic range—this is where we find our friend the *noise gate*.

Noise Gate

An audio effect with a threshold control, wherein signal hotter than the threshold remains audible, but the signal that falls below the threshold is silenced completely, often used on drum mics to gate out bleed from other instruments (shown in Figure 11.10).

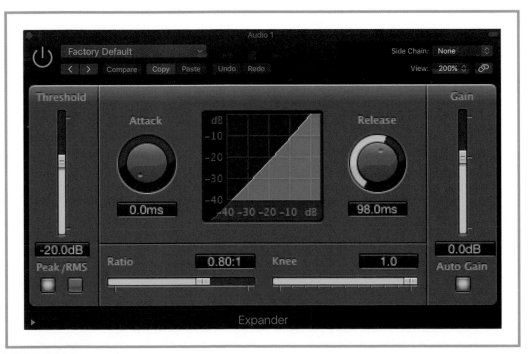

FIGURE 11.9
Logic Pro X Expander Display
Source: Logic Screen Shots Reprinted Courtesy Apple, Inc.

FIGURE 11.10
Logic Pro X Noise Gate Display
Source: Logic Screen Shots Reprinted Courtesy Apple, Inc.

122 Imagine a farmyard. In this farmyard are two kinds of animals: cows (loud sounds) and chickens (quiet sounds). There is a fence around the farmyard, with a big heavy gate. When the cows want to leave the farmyard, they are heavy (loud) enough to push the gate open and walk on through. The chickens, on the other hand, are light (quiet) enough that on their own, they can't push the gate open, and they are trapped inside the farmyard. If, however, a chicken decides to sneak out under a cow's legs, it can get through the gate that way. If the farmyard inside the fence is total silence, and the animals outside the gate are what we hear when a noise gate is triggered open, then you have a basic idea of how this effect works. A threshold is set, such that the sounds we desire are louder than the threshold and thus open the gate, allowing the sound to be heard. Sounds below the threshold do not open the gate and are unheard, unless they are playing at the same time as other loud sounds.

A snare drum is a good example. Our snare top mic is aimed to minimize bleed from other instruments, but we might still hear a little bit of the kick, a little distant electric guitar, and other such. So long as those other sounds are significantly quieter than the snare, a gate can help us clean up our signal. We set our threshold so that the snare hits trigger the gate to open, and all the other sounds don't, and we've mostly isolated the snare in our channel. I say mostly because we still have the chickens hiding under the cows' legs to deal with. Once the snare triggers the gate to open, any bleed during that opening will still be audible, so we can't avoid it entirely, just *mostly*. Given that the instruments causing bleed are typically audible via their own mics, we don't tend to notice this little bit of bleed, and we have a lot more control of EQ, dynamics, and other aspects of the snare sound now that we've gated it.

Gates are very common. Most digital consoles will also provide the option to gate each channel, alongside the EQs and compressors. Some gates have the option to change the amount of gain reduction performed when the gate "closes." When you set reduction to anything less drastic than "completely," a gate ends up acting sort of like an expander. Gates have attack and release settings, and often have a third control for *hold*, which determines how long the gate will stay open after it is triggered (regardless of whether sounds in this stretch of time are keeping it open). Manipulating attack, hold, and sustain times allows for the smoothest possible gate action.

Many effects (gates included) can also be *side-chained*.

Side-Chain
The principle whereby an effect placed on one channel (e.g., bass) is triggered by the signal from a different channel (e.g., kick drum) such that the action of the effect processes the first instrument based on the playing of the second.

Side-chaining is very common in dynamic effects. Imagine a compressor on the aforementioned bass. If it is side-chained to a kick drum, that means that each time the kick hits, the bass compresses (and ideally reduces level) at that moment. This is a common trick for helping bass and kick sit together in a mix. Side-chaining can be used in a wide variety of situations—a *ducker*, for example, is a gate where the "reduction" is actually set to a boost. Imagine a TV commercial. There is background music playing. When the announcer speaks, the background music gets quiet, and when the announcer stops speaking, the music swells back up. Instead of having to chase a fader every time the announcer speaks, you set a ducker on the music track, side-chained to the announcer's voice. When the announcer speaks, the threshold is triggered on the music, which reduces the signal, and when the announcer stops speaking, the music swells back up.

DRIVE EFFECTS

Drive effects are a very simple category. In an analog drive effect, an input signal is turned up until the circuit overloads. The overloaded circuit distorts the sound (we often picture this as a waveform reaching the limits of a circuit path and having the tops and bottoms of the wave sheared off—or *clipped*). In Figure 11.11, we see a clean waveform next to the same signal that has been clipped via overdrive.

This kind of distortion is usually found in guitar setups, either as an overdrive setting in a guitar amp or as outboard guitar pedals. We also use these effects in creative mixing, but when they are used in live systems, they are typically controlled by the performer rather than the mix engineer. As such, while we do find these effects as plug-ins in digital mixing consoles, they are not in use very often (far more common in the studio).

TIME-BASED EFFECTS

These effects typically operate on a copy of a main input signal, alter the timing of it in some way and then recombine it with the original.[9] Whether this is a delay effect with a wet/dry control (the dry being the original signal and the wet being the copy that has been delayed for effect) or a chorus effect that by definition takes the copied signal, varies it in time compared to the original very slightly (sometimes with a shifting delay offset), and recombines them to create a shimmery summed signal, all of these effects do essentially the same thing.

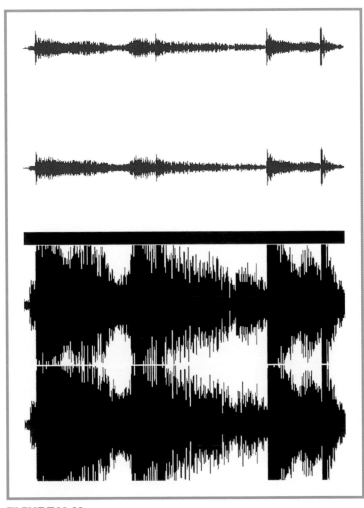

FIGURE 11.11
A Clean Waveform; the Same Waveform Clipped With Overdrive

There are three main categories of time-based effects:

1. Delays
2. Reverbs
3. Modulation effects (chorus, flanger, phaser, etc.)

Delays: delays are pretty straightforward. We use them frequently on musical content. It is very common to hear a lead singer with just a little bit of short "slap delay" on the voice to give it a sense of size (or to fit a certain vintage recording style, or the like). Psychedelic acts (e.g., The Flaming Lips) make use of long and unconventional delays to draw sounds out and create a swampy, druggy sound world. A delay consists of *taps*, which are the individual delay points in the program. A delay can be single-tap, meaning that the copied signal is delayed by one fixed amount (e.g., 150 ms), and played at the output. Or a delay can be multi-tap, which means that a signal is delayed at multiple different lengths and those different delayed signals are then all combined at the output—for a bouncing, sequenced delay trail. These multi-taps can be spaced evenly, or unevenly, depending on the desired effect. Figure 11.12 shows a delay with a wide variety of controls for customizing the sound.

> **Reverb**
>
> An audio effect meant to mimic the reverberation of an acoustic space (or of a physical "reverb chamber").

There are two kinds of digital reverb, the first of which is time-based and the second of which (and the generally nicer sounding, with notable exceptions)[10] is convolution based. The time-based reverbs are known as *algorithmic reverb* (shown in Figure 11.13). Natural reverberance has been broken down into component parts and those

124

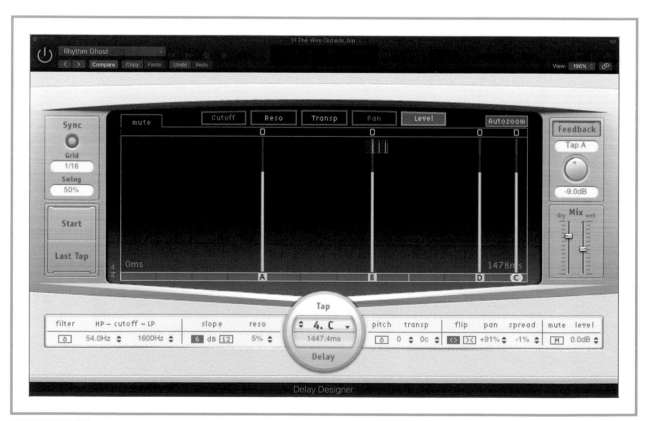

FIGURE 11.12
Logic Pro X Delay Designer
Source: Logic Screen Shots Reprinted Courtesy Apple, Inc.

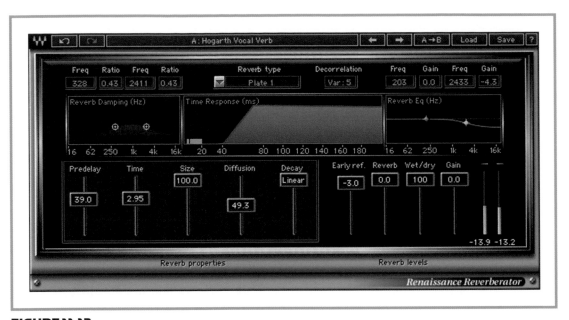

FIGURE 11.13
Waves Renaissance Reverberator Plug-In
Source: Courtesy Waves Audio Ltd.

component parts consist of a series of time-delayed copies of a signal summed together to create a simulation of a reverberant space or piece of equipment. You will often find controls like "predelay" (time between input signal and start of reverberance), early reflections (close-by small sound reflections that are the early wave of a reverberant signal), density (how many reflections the software calculates for a given stage of the reverberance—the higher the number, the greater the smoothness of the sound), and overall size (the size of the space being mimicked, calculated as a distance in time between input and reverbed signal).

MODULATION EFFECTS

Modulation

The technique of altering a signal by combining it with another signal.

Modulation effects is the category under which phasers, flangers, and chorus effects are found. All of these effects are variations on a single theme. An input signal is split, and the copy is varied in time compared to the original. A phaser typically copies the signal at a very short delay (5–20 ms) and sums the copies together. The just-out-of-time copies sum together creating peaks of summation and valleys of cancellation, resulting in a swirly sound that is sometimes desirable. A flanger does the same thing, but sometimes with a slightly longer time window and varying the time of the offset copy such that the swirling is enhanced. A chorus effect is typically just like a phaser, but with a longer time window. These are very frequently used on guitars (Andy Summers of The Police almost always has a chorus effect on his guitar), and sometimes on voices and other instruments. Figure 11.14 shows a common guitar chorus pedal.

FIGURE 11.14
Boss CE-2W Chorus Pedal
Source: Courtesy Roland

125

CONVOLUTION/MODELING EFFECTS

Convolution

The mathematical process of combining an incoming set of information with a preexisting set of information and creating by combination a third set of information. In audio, this means combining an input signal with a preexisting impulse response (taken either in a venue or through a piece of equipment) to create a combination signal that makes the input signal sound like it is in the venue or equipment in question.

Impulse Response

1). The process of capturing the acoustic signature of a venue or piece of equipment by exciting it with a calibrated test signal, recording the results, and running them through software that removes the initial calibrated signal, leaving only the signature of how the space or gear reverberates. 2) A calibrated test signal used to excite an acoustic space or the signal path of a piece of equipment or system, and used for system balance, calibration, and optimization (aka "tuning").[11]

Many of the most sophisticated effects we use are convolution effects, otherwise known as *modeling* effects. These effects capture the signature sonic performance of a place or a piece of equipment and allow us to run our signals through them. There are two main categories of modeling effects:

1. Spaces
2. Gear

Space modeling effects take samples from real venues and allow us to use them as reverb effects in our mixes. This can be very useful. Figure 11.15 shows a convolution reverb.

In a convolution reverb, you select an impulse response, and then adjust parameters from there. In some convolution reverbs (e.g., Logic Pro's Space Designer, Figure 11.15), you can also record and upload your own impulse recordings (so, if you love the way your garage sounds, for example, you can capture it to use in your mixes!).

Gear modeling effects take the same principle and apply it to gear.[12] When you see plug-ins mimicking a guitar amp or a piece of outboard gear, they are typically created by recording impulse responses for each conceivable combination of settings in the equipment and using those as the basis for the effect. Broadway orchestras use amp-modeling effects rather often because they often want an amplified sound without the noise in the room from an actual amp (and because space in Broadway pits is at a premium, and they may not have room for an amp). Figure 11.16 shows a very popular hardware amp modeler.

ANALYTICAL EFFECTS

These are not *effects* per se in that they are not typically altering the sound, but they are important processors in our system and are nearest to effects in functioning, so they are covered here. Analyzers typically take input signals and represent them for us graphically. The most common version of this is a simple meter (Figure 11.17), representing relative loudness by intensity and color in a linear path.

A note on metering scales: digital audio workstations often use a decibel range known as dBfs, or decibels full scale. On a dBfs meter, the very top of the meter will be the value of 0, and all values below will be negative numbers. Digital mixing consoles, however, more commonly display a 0 value as unity gain (as this corresponds with analog mixers). Make sure you know what

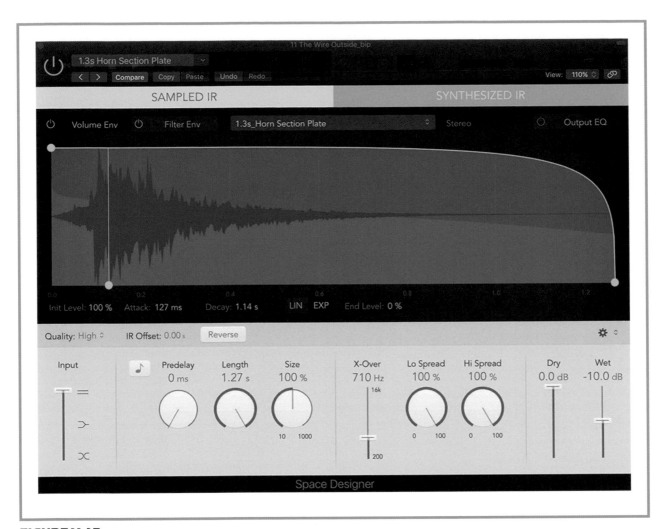

FIGURE 11.15
Logic Pro X Space Designer Reverb Display
Source: Logic Screen Shots Reprinted Courtesy Apple, Inc.

FIGURE 11.16
Line 6 POD Pro X HD
Source: Courtesy Line 6

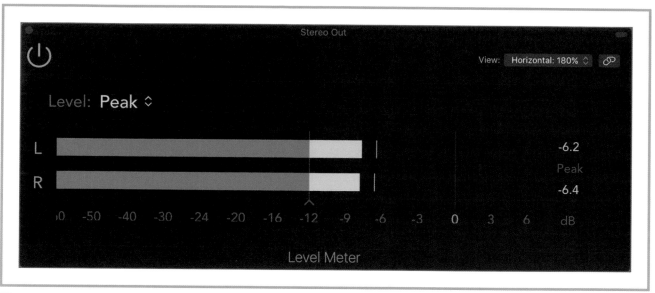

FIGURE 11.17

Logic Pro X Level Meter

Source: Logic Screen Shots Reprinted Courtesy Apple, Inc.

kind of meter display you are working with, as a 0 value in dBfs is too loud and about to distort, where a 0 value in a fader system with +12 dB of headroom above it is totally acceptable. Logic Pro X's meters are dBfs meters on the channel strips, but the Level Meter shown in Figure 11.17 has 6 dB of headroom above 0 to show you how far above your maximum the signal is, so reduction can be performed appropriately. Whatever equipment you work with or specify, it is important to understand what kind of meters your gear will possess.

As our meters get more complex, we have analyzers that will show us relative intensity over the frequency spectrum. These are called *real-time analyzers* (RTAs—see Figure 11.18).

These can be very useful installed on channels in a mixing system, so you can watch the signal as it changes both with input variation and based on what you are doing as a mix engineer.

Beyond RTAs, we get into advanced analyzers that make use of the *Fast Fourier Transform* function (FFT) to give us phase data, information about coherence between two copies of a signal and more. We will examine these in-depth in Chapters 12 and 13.

HYBRID EFFECTS

This is a catch-all category that includes any effect that makes use of two or more of the aforementioned categories in a signal effect. For example, Auto-Tune—which is becoming increasingly common in live systems in its rack-mounted version—is a hybrid effect combining analysis (of the incoming signal) with a map of ideal spectrum values to alter the spectral response of the signal for output. Subtle Auto-Tune is used all the time to help singers keep on pitch. Drastic Auto-Tune . . . well, have you heard of a singer named T-Pain? 'Nuff said.

CLASSIFICATION

When selecting effects for a system, there are basically two categories:

- Onboard/plug-in effects
- Outboard effects

Many effects will be available either built-in to the digital console you select or (as is the case with studio effects) as a plug-in.

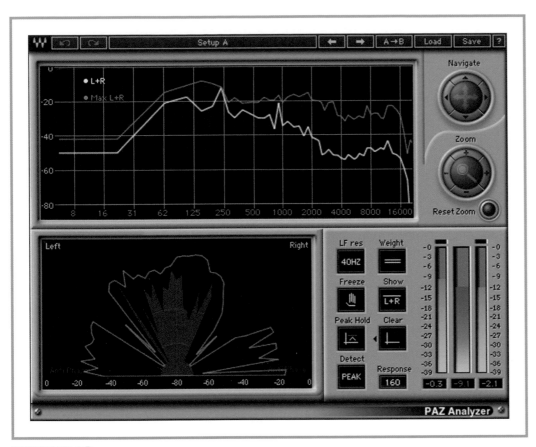

FIGURE 11.18
An RTA Display (Waves PAZ Analyzer)
Source: Courtesy Waves Audio Ltd.

Plug-In

A software effect made to work inside another software's architecture as a sub-routine. Plug-in audio effects can be purchased from vendors other than the console's (or digital audio workstation's) manufacturer, and so long as the software format is compatible, they can be installed for use and can thus extend the capabilities of a console beyond its original stock effects.

Some effects are only available as outboard gear, and, of course, if you are using an analog console, all of your effects (beyond basic EQ and possibly comps/gates) will of necessity be outboard.

Otherwise, the classification of effects really is by type and use, as described in the "Definition" section of this chapter.

SPECIFICATION

It is generally assumed that effects work as designed. Beyond a few options, specs are not as relevant for effects as for most components. Here are the few specs you will want to consider when choosing effects.

EQs:
EQ types—parametric? Semi-parametric? Graphic? Shelving? Low/high pass? Fixed frequencies and slopes or adjustable? How many bands available per instantiation?

Compressors:

    Gain reduction amount—at least 30 or 40 dB desirable

Limiters:

    Adjustable attack or no?

Gates:

    Hysteresis or no? Hysteresis is a secondary threshold control. If the main threshold sets the level at which the gate opens, the hysteresis sets the level at which the gate closes so that a sound at −15 dB might open the gate, but it will stay open till the sound drops below −18 dB (if the hysteresis is set at −3 dB).

Drives:

    What type of distortion sound is it producing? Analog overdrive? Something else?

Delays:

    What delay controls are available? What is the minimum and maximum delay time?

    Single-tap or multi-tap or both?

Reverbs:

    Convolution or algorithmic? If convolution, can you use your own impulses?

    Min/max reverb time?

    What controls are available?

Amp models/gear models:

    What gear is being simulated, and what controls are available on that simulation?

Analyzers:

    What types of graphic readout are available? RTA? FFT? Goniometer (angular measurements that give physical representation of sounds in a stereo field)? Et cetera.

Hybrid:

    Too many to count. There are so many hybrid effects that the specs are well beyond the scope of this book.

CONSIDERATION

What questions must a designer ask (and answer) in order to select creative/mixing effects as part of their system?

1. What type of program will be presented? Concerts and other musical events need more creative and mix effects available than event halls hosting corporate board meetings and nothing else.
2. What effects do we need? We always need EQ, we always *want* compressors and gates, but beyond that, this is really a choice that needs to be based on the desired action in the system. Do you need Auto-Tune? If you are designing a subtle reinforcement system for a jazz club, probably not. If you are designing a touring system for Mariah Carey, probably so.[13]
3. Can we get the effects in our console, or do we need outboard?
4. How many channels of the given effects do we need? Consoles have limits on how many plug-ins can be instantiated at once. Racks have limits to how many pieces of outboard gear they can hold (and budgets provide the hardest limits). If I need amp models, do I need them for all three guitarists, or just for one? Do I need Auto-Tune for a lead vocalist and two guests, or just the lead? Et cetera.

While effects are ever present, always expanding and becoming more complex, choosing effects is actually pretty simple. If I want an outboard compressor, I know that just about any professional compressor will do the job, but I also know by virtue of experience that a Universal LA2A is going to sound different and have a different operating character than, say, a DBX166. Either will compress your signal, but like mic preamps, what *flavor* of compression you want is up to you—I can't teach you how to navigate these differences in a book. As always—read reviews, talk to professionals, and listen to every piece of gear you get your hands on until you have a mental catalogue of what different pieces sound like and why you might (or might not) want them for your system.

As system designers, a list of needed creative and mixing effects is often given to us by the content designer or mix engineer (if we are not also performing those jobs). We will be more concerned with the effects built into our DSP unit, that we use for tuning the system overall. Tuning a system is the core subject of the following two chapters.

So how do we tune a system? What does that even mean?

All of our outputs run through a DSP, and then to the amplification stage, and it is inside the DSP that we work to turn a lumpy system with horrible delay blur into a coherent, crystalline listening experience.

Notes

1 Systems with creative mixing effects usually feature a mixing console or some controller/digital facsimile thereof. Systems that only feature a central DSP unit and no console typically are not systems designed to allow creative mixing choices to be made in the system itself—in these systems, any creative use of effects is typically done to the pre-recorded content being played through the system, in studio, before the content is assigned for playback.

2 Fun historical fact: Equalizers are so-called because they were originally invented to help avert signal loss in high frequencies over old-school telephone lines. They were called equalizers because they boosted HF content in order to equalize the signal on the receiving end of the phone line with the signal on the transmitting end. A lot of sound system technologies that we use every day were originally designed for telecommunications purposes and only later adapted to entertainment.

3 Phase cancellation is complete when a copy of a signal is $180°$ out of polarity with the original, but partial cancellation happens at different phase values between original and copy.

4 High- and low-pass filters come in different types: Butterworth, Linkwitz-Riley, Chebyshev, and others. It is beyond the scope of this work to examine these in detail, but good information on them is available in other texts—for example, Davis/Patronis' *Sound System Engineering*.

5 Q stands for quality factor, which relates to the physics concept of the spring constant. It is beyond the scope of this work to explain the workings of the spring constant and how it applies to audio filters, so just know that Q is the bandwidth setting—how wide a range of frequencies do you want to alter with your parametric EQ?

6 In studio work, there is one more option for EQs, which is to select a conventional EQ versus a *Linear Phase EQ*. Since conventional EQs work by making use of time offset between the copied signal and the original, there is some phase distortion necessarily implied in their action. A linear phase EQ holds back the original signal as well as the copied signal and re-aligns them at output (as near as possible, aside from the actual altered frequencies). This can result in a slightly cleaner/clearer output signal, but due to the mechanism needing to delay the original signal (only a tiny bit, but still), linear phase EQs haven't really caught on in live work. In studio, linear phase EQs have their devotees. If you are doing creative work, though, there are times when the phase distortion of a conventional EQ may be a desirable sound.

7 Graphic EQs were for many years the standard tool for system tuning/speaker leveling, but in the digital age, they have fallen out of favor, and in their place parametrics have risen, often used with wide curves and subtle gain reduction, which can level a speaker system's frequency delivery more smoothly and with fewer audible artifacts and feedback nodes than a traditional "surgical" graphic EQ.

8 As with any "category," the purpose of so classing an effect is just to better be able to understand and discuss it. Don't get too hung up on which bin I place something in, just make sure you understand what it does and how it works.

9 This is in the mixing/creative domain. Things work a bit differently in the system tuning domain . . . see Chapters 12 and 13 for more.

10 Lexicon, in particular, makes wonderful-sounding algorithmic reverbs.

11 See Section 10 for more info on system tuning.

12 Though some amp modelers also rely on algorithmic imitation rather than true convolution.

13 With apologies to Mariah Carey. I don't know if she uses live Auto-tune, but I simply chose a pop singer for this joke. For all I know, she sings every note perfectly! Please don't sue!

131

DSP

Digital Signal Processing

Digital signal processing devices are critical to creating a sound system that is professional quality. DSP devices are often the most difficult standard component for new sound students to understand—most small sound systems (and particularly semi- or non-professional systems) skip this step and connect amplifiers directly to the mixing console. Since many students come to this topic with no experience of what a DSP is or does, it can be challenging to grasp why they are necessary—yet they are absolutely necessary.[1]

So, what is a DSP, and what do we do with it?

DEFINITION

A DSP can take several forms. Most commonly, DSPs are stand-alone, rack-mounted devices that take signals from a mixing console's outputs (and/or directly from playback systems), and route signals to all applicable system outputs. Stand-alone DSPs vary considerably in complexity, from simple devices with only a few inputs and outputs, to sophisticated frames that can handle hundreds of channels. Some of these devices have very elaborate controls built in that allow the user to recall different system presets at the touch of a button (changing as much or as little of the system configuration as the user determines).[2]

DSP

A device that is placed before the amplification stage in a sound system, that calibrates the signal paths for both routing and system tuning/optimization purposes. Common uses include equalizing output paths to ensure that a set of loudspeakers is delivering the flattest frequency response possible, providing limiters on the output paths to prevent overload damage to speaker drivers, and time-aligning speaker outputs such that main and delay speakers are properly summed in acoustic space (rather than canceling one another out due to phase irregularities in the delivery).

DSP, as a function, can also be found in some power amplifiers, and at the output stage of the rare mixing console. Such DSP programs are typically less full-featured than dedicated outboard DSPs.[3]

In this chapter, we will examine the overall range of features available in typical stand-alone DSP units. Be aware as we continue that every DSP system will have unique features that make it more or less suitable to your particular system task—it is beyond the scope of this book to examine the variations in the whole existing range of DSP devices.

The simplest DSPs feature a fixed number of input and output channels, with a standard internal configuration for routing inputs to outputs. For example, let's consider a very simple concert system. In this system, we have L/R main speakers (full range) and L/R subwoofers (for low frequencies). Typically, when we pair full-range speakers with subwoofers, we don't want them to overlap too much in frequency coverage—if they are well-matched for SPL delivery, then in the overlap range we won't have flat delivery; we'll have a boost since all speakers are delivering the same content. How do we take the L/R output of a mixing console and split it to send the highs and mids to the full-range speakers (known as *tops*, when paired with subs), and the lows to the subs? In analog days, we used an outboard hardware device known as a crossover. The full L/R signal from the console would be sent into the crossover, which would then split the signal by frequency and send high/mid through one physical output and low out another. In today's digital world, we still use a crossover, but it is the software crossover built into a DSP frame.

In Figure 12.1, we see the analog crossover in action.

The outputs on the back of the crossover unit are typically labeled "high" and "low" (or "high," "mid," and "low"). Now, in a DSP unit, the crossover is a digital processing object in a custom signal flow that we have designed. Most professional DSPs come as neutral devices. In order to route signals through them, we have to create a custom program that identifies the input signals, routes them through processing objects that we select, and then out via output paths that we assign to physical outs on the device. Figure 12.2 shows a sample signal flow from inputs through crossovers to outputs for a system with both top and subspeakers.

While the degree of programmability varies considerably from model to model, most professional DSPs demand at very least that you establish which of the device's inputs are assigned to which of the device's outputs. In advanced DSPs, not only will you often select what kind of inputs and outputs to use (digital, analog, via network?) by virtue of selecting expansion slot cards that feature different types of physical connections, but you will also select every detail about the signal paths between input and output. While our simple concert system from Chapter 10 might

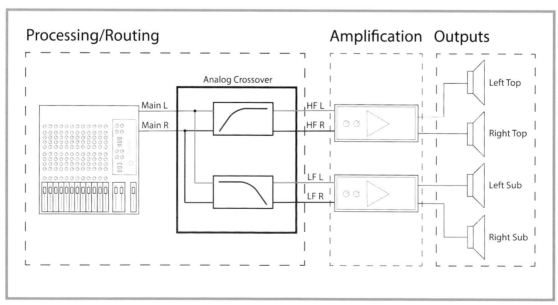

FIGURE 12.1
Conceptual Diagram of Analog Crossover Splitting L/R Mains Into Top/Sub Signals

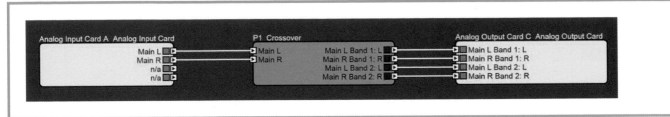

FIGURE 12.2
A Simple DSP Signal Flow From Input Through Crossovers to Outputs for Tops/Subs (Shown in Harman's Hi-Q Net Audio Architect)
Source: Courtesy Harman Professional Solutions

not need a great deal of flexibility in its design (though as we shall see, when we add delay speakers, more stage monitors, a feed to the venue's lobby, and so on, complexity definitely increases), there are other systems that use a DSP less like a linear effects path before speakers, and more like the central brain of a large distributed network. Such systems are very common in places like theme parks and airports, where the number of live input signals is usually small and/or doesn't require mixing with other sources, and the number of outputs (and the need to flexibly route the signal to different targets at a moment's notice) is vast. An airport, for example, has no need of a mixing console—there is probably not a live band playing in the food court—however, each gate has a paging microphone, so staff can make announcements and contact passengers and so on. A staff member needs to be able to select the zone for their announcement: does it go out only to the gate at which the staff member is standing? To all gates for a certain airline? Through the whole terminal? The whole airport? This kind of flexibility is one of the chief advantages of professional DSP systems. By virtue of their customizable programming, which can allow for a range of devices to send and receive control messages to and from the DSP, we can create custom interfaces (touch panels, numbered keypads, etc.) that act as our control platform for our sound system. At the touch of an onscreen button, we can send the signal to a variety of destinations, mute signals, trigger playback cues, and a wide range of other actions.

Let's briefly examine a system of middling complexity: a theater system with playback, live mics, and a wide range of speakers (plan view shown in Figure 12.3, block diagram shown in Figure 12.4). How does a DSP interact with such a system? Let's assume our playback system is software-based and is sending the signal via Dante over a network cable (ten channels of playback, in this case). As such, it doesn't need to go through a mixing console, but can connect directly to a DSP (well, "directly" via a network switch so that designers can also access the Dante setup via their own computers). Let's also assume our digital mixing console (necessary for balancing the live mic levels) is *not* sending Dante out but is sending analog audio. Mics will be routed to three main speakers (left, center, and right mains), and playback audio needs to be able to reach all of the speakers in the system. Each speaker has to connect to an amplifier channel (as none, in this example, are self-powered), and each amplifier in turn connects via analog signal wire to the DSP. First, let's look at the physical layout of the imaginary venue. This drawing shows each speaker in place, the FOH mix position, and the location of the equipment racks housing the DSP, amps, and playback computer.

Now, let's take a look at the signal flow of such a system. Remember in examining this that Dante carries multiple channels over a single network cable.

We see where the DSP rests physically in the signal path. Now—what is happening inside the DSP? Each input is fed along a custom-created path, through processing objects/effects that we select for our particular use in this system and then out the various output paths as we have configured them.

In Figure 12.5, we see the signal flow within the DSP software that would be typical of this system.

The matrix router allows both the vocal mics and the QLab playback to send to the central vocal speakers, and otherwise, QLab is channel by channel into the speakers. Each channel features standard effects related to system tuning—as will be discussed later in this book in more detail. This only scratches the surface of what this particular DSP is capable of (the BSS BLU-806DA is a very powerful device—its software Hi-Q Net Architect is shown), and this is only one DSP among many available on the current market.

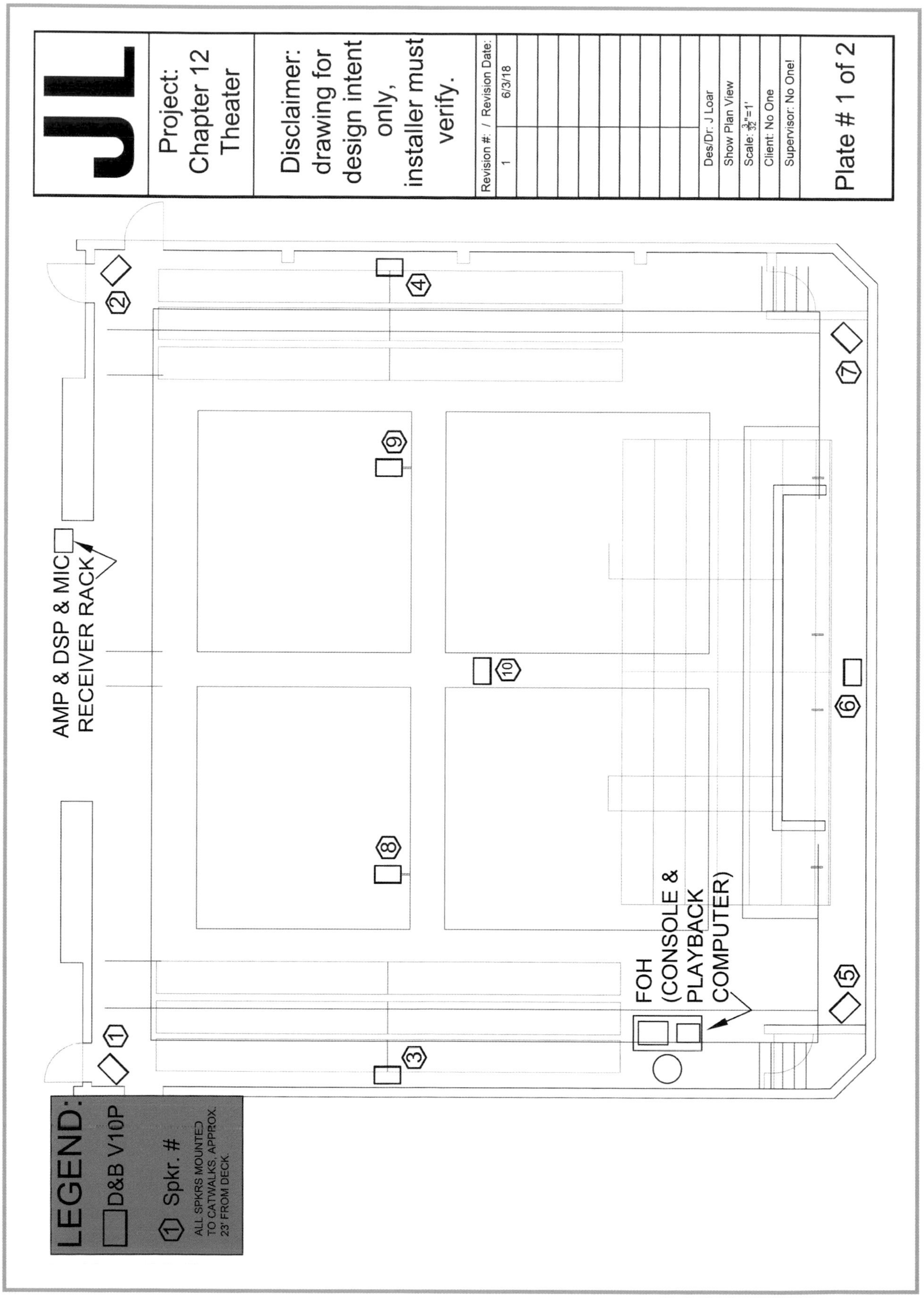

136

LEGEND:

☐ D&B V10P

⬡① Spkr. #

ALL SPKRS MOUNTED
TO CATWALKS, APPROX.
23' FROM DECK

AMP & DSP & MIC RECEIVER RACK

FOH (CONSOLE & PLAYBACK COMPUTER)

Project:
Chapter 12
Theater

Disclaimer: drawing for design intent only, installer must verify.

Revision # /	Revision Date:
1	6/3/18

Des/Dr: J Loar
Show Plan View
Scale: $\frac{3}{32}$"=1'
Client: No One
Supervisor: No One!

Plate # 1 of 2

FIGURE 12.3
Plan View of Simple Theater System (in a Black-Box Theater)

With such a wide range of features and configurations available, we need an understanding of the range of features typically or frequently on offer, as well as the form factors associated with such features, before we can hope to select the appropriate DSP for our task.

CLASSIFICATION

There are both physical and digital features that we must consider before knowing whether a given DSP unit will suit our system. We begin with the physical—while there are still a range of options available in the physical configuration of DSPs, there are fewer physical options to consider than digital ones.

- Form factor: the first consideration is whether your unit(s) is stand-alone or built-in. As mentioned earlier in the chapter, for some small systems (or even large systems with very little complexity), using only built-in DSP may be appropriate. However, for most applications with any degree of complexity, a stand-alone DSP will be preferable. Built-in DSPs will have fixed input and output types (and usually channel counts as well). If you are using DSP built into a two-channel power amp, you can be certain that your physical input connectors are limited to the connectors on the amp and that you will have at most two channels of processing available in that unit.
  - Stand-alone DSPs: these units are typically rack-mounted devices. They come in sizes ranging from 1 RU to 4 RU.[4] With rare exception, physical inputs and outputs on stand-alone DSPs do not come pre-configured (they do often come pre-configured with network ins and outs, both for control and for digital audio over Ethernet cables, but we'll get to those in a moment). The physical inputs and outputs are typically configured by the user, by virtue of *slot cards*.

> ### Slot Cards
>
> In computers, a PCI slot is an internal slot in which a processing card can be installed (e.g., a graphics card, a soundcard). In DSPs and mixing consoles (and their stage boxes), slot cards function similarly to PCI cards, but the cards themselves typically contain either a range of physical connections for expansion of I/O, or dedicated processors for third-party effects (e.g., a Waves plug-in card for a mixing console).

- The user can, therefore, select what combination of analog and digital inputs and outputs they desire. For example, in the BSS BLU-806DA processor (Figure 12.6), Dante digital audio connections are built-in, but there are four card slots available for further expansion. These could be configured as half inputs, half outputs, all inputs, or all outputs—all analog, all digital, or a combination of the two. The choice is up to the user, which allows a DSP to interact with a wide range of equipment in a wide variety of configurations.
- The larger your DSP frame is, in RU, typically the more card slots are available. This isn't a hard and fast rule, but a general guideline—always examine the specs of the unit in question for info about the available connections and slots on your selected device. When selecting a stand-alone DSP, knowing how many physical inputs and outputs it can feature is important. Keep in mind that some external hardware units are also available that can be associated with devices in a given product line that can expand the I/O count without needing more card slots. For example, the BLU-806DA shown in Figure 12.6 can also be configured in a system with what BSS calls "break-in" or "breakout" boxes, which are external pieces of hardware with more physical connections that can route signals in and out of the main DSP over their proprietary BLU-link network digital audio protocol.
- It is a rare circumstance (though certainly not unheard of) that the saved space between a 4 RU DSP and a 1 RU DSP would be the deciding factor for selecting a unit, but it is something to consider. More important are the connection types discussed earlier.[5] In addition, many DSPs feature network-based digital audio connections, but these vary in type. Just because there is an Ethernet jack on the back of your DSP for audio connection does not mean that it can handle your particular type of audio signals. A DSP with AVB ins/outs over Ethernet will not handle Dante ins/outs, and vice versa—so while the physical connector here looks the

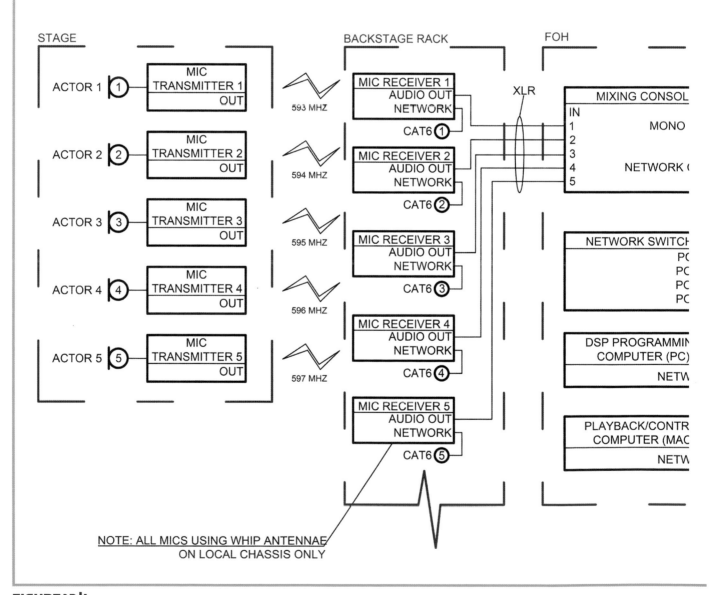

FIGURE 12.4
System Block Diagram of Simple Theater System From Figure 12.3

NOTE: IN A PROPER DRAWING SET, THE TITLE BLOCK WOULD NOT MOVE FROM ITS ORIGINAL POSITION. I HAVE TURNED IT ON ITS SIDE IN SERVICE OF SHOWING THE BLOCK DIAGRAM AT A LARGER SIZE IN THE SCALE OF A TEXTBOOK. THE SMALLEST THIS DRAWING WOULD BE PRINTED ON A REAL PROJECT WOULD BE 11"x17".

Des/Dr: J Loar
Block Diagram
Scale: N/A
Client: No One
Supervisor: No One!

Plate # 2 of 2

139

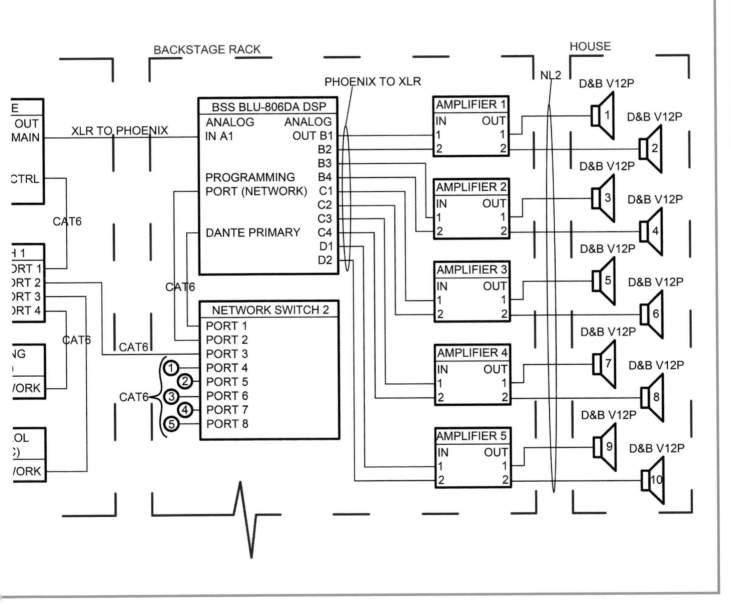

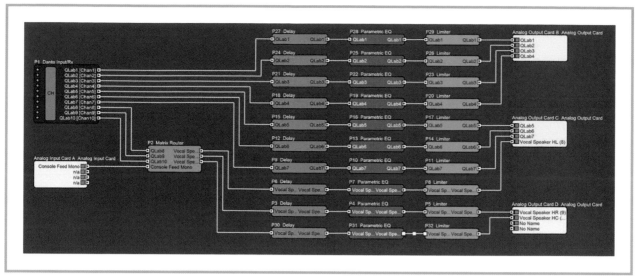

FIGURE 12.5
DSP Internal Signal Flow of Simple Theater System (Shown in Harman Hi-Q Net Architect)
Source: Courtesy Harman Professional Solutions

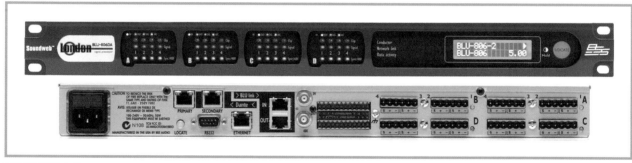

FIGURE 12.6
BSS BLU-806 Front, BLU-806 Back with Card Slots
Source: Courtesy Harman Professional Solutions

same, you need to make sure the internal software engine can process the particular types of signals you will be sending it.

- Peripherals: another physical form issue to consider is what peripheral devices are available to use with the DSP.
- As in the airport example earlier, many systems require peripherals in order to properly function. If I am designing for an airport system, I want the paging stations to be able to send a signal to the destinations I desire at the touch of a button. Whether that is a physical button or a virtual one on a custom-designed touch-panel interface, those features need to be available. Do you want physical buttons on the station? Or would you prefer a touch-panel? Each has its advantages and disadvantages.[6] If your paging mic will only ever broadcast to one destination, perhaps you don't need a paging *station* itself, but just a dedicated paging mic with a switch that opens the channel and triggers the override to make sure the page signal goes out over any content currently playing in the appropriate channels (this would be a common setup for, say, the operator of

Peripherals

Devices external to a central processor or computer that expand the functionality of that central device. A printer is a peripheral to a computer, where DSPs may have peripherals that include paging stations, touch-panel interfaces, extra I/O boxes, etc.

a roller coaster, who typically only needs to speak into the paging system in order to make an emergency announcement, and doesn't often need to flexibly talk to specific areas on command).[7]

- Additionally, depending on what signal types you are using, you may need network peripherals (switches, for example) in your system, but those will be addressed in Section 6 of this book.

▪ Digital features: once you have a sense of the desirable form factor for your DSP, you will turn to the critical subject of what internal digital processing features your device offers (or doesn't). Again, the range of available options on the market is virtually limitless and expanding every day, so this writing will be far from a comprehensive examination of the variations possible. Instead, we will here endeavor to simply identify the major categories of consideration for DSP selection.

  - Processing engine (often referred to as *processing power*). One unit may offer 128 x 128 channels of processing (meaning it can simultaneously handle 128 inputs and 128 outputs, across a range of form factors both physical and network-based), where another unit that is the same RU might offer 512 x 512 channels. Some speaker manufacturers offer units that I refer to as "semi-DSPs"—these units are often marketed as "loudspeaker management systems" (LMSs) instead of as DSPs.[8] These often offer some of the features of a standard DSP (EQ, limiting, sometimes delays), without the flexibility in configuration and channel count of true DSP units. LMS units often have vastly reduced channel counts compared to true DSPs (e.g., two inputs, six outputs). If we are merely setting up our simple concert system from earlier, with two tops and two subs, an LMS unit may provide all the processing we need (at typically a much lower price than full DSPs). However, if we are instead designing a theater system with some 75 individual speakers (all of which need to be able to receive unique playback signals), and mic signals for singers and orchestra mixed through a console, we will need a real DSP with high channel count. This is not even to mention airports and theme parks and casinos and the like—systems that are inherently very complex and always require sophisticated DSP in order to properly distribute signals.

  - Interoperability with other system gear: most professional DSPs feature not only physical wired connections and network-based audio connections (e.g., Dante, though some smaller DSPs do not have this functionality); they also feature proprietary signal transport via network cables. What this means is that, for example, QSC's Q-SYS DSPs can send a signal to devices via analog or digital wired connections, via network protocols like Dante (depending on which Q-SYS device you choose), but also via a proprietary signal type called Q-LAN (short for Q-SYS Local Area Network).[9] If you are selecting a DSP that has to interact with other existing components, you will want to ensure that the signals you pass in and out of DSP can make use of any preexisting proprietary communication types if available. For example, if you are designing a theme park ride system, but that system has to interact with a preexisting parkwide sound system, and that parkwide system is all Q-SYS, you will probably want your new DSP to also be a Q-SYS device, so you can make use of Q-LAN signal connections and other such proprietary features (custom trigger messages, for example).

  - Sample rates/bit depths: some DSPs can handle a wide variety of sample rates, some are limited. If you are selecting a DSP for a live concert system, but that concert system also has to send the digital audio to a Pro Tools recording rig, you need to make sure the DSP and the recording rig can operate at the same sample rates. If your DSP is limited to 48 kHz, but your recording rig is default configured at 96 kHz, you might want to consider a different DSP

  - Programming: different DSPs will offer different levels of programmability. From the LMS end of the spectrum, where you might be able to send a control message that mutes or unmutes channels or performs some other basic actions, to the most complex systems (e.g., Q-SYS) that can not only act as DSP but also as playback server and as show controller for the AV system and for a wide variety of objects (even controlling animatronics, pyro effects, and other such).[10] It is, once again, well beyond the scope of this book to examine every possible programmable option available in the DSP market, but here are a few examples of items to consider:

    - Processing object types: it is assumed that every DSP should be able to provide EQs for frequency calibration, dynamic effects (particularly limiters) to help prevent overload and speaker damage, crossovers (for sending full-range signal to tops and subs, or splitting further to HF drivers, MF drivers, LF drivers, and subs), and delays, for use in time-aligning your system. That said, there are some features that are available in certain DSPs but not in others that might be critical to your work. For example, Peavey's DSP systems include a processing object called a *ramp*, which is a timed boost or attenuation to a signal path, with customizable time and level settings. This allows a system to trigger channel mutes that take, say, 0.25 sec instead of 0 sec. A 0-sec mute often makes a clicking or popping sound that is undesirable, where a 0.25-sec ramp down avoids this pop but still effectively mutes the channel instantly. In BSS DSPs, the shortest ramp time (in the *gain-timed* object) is 1 sec. For most applications, the difference between 1 sec and 0.25 sec isn't a deal breaker, but for some calibrated theme park installations, that difference might be critical.

> **Scripting**
>
> A software platform that supports scripts, which are programs and commands designed to operate in a real-time environment (known as a *runtime* configuration). Scripting allows a complex series of actions to be automated.

- Logic: logic programming allows the user to create customized action paths that are triggerable via single points of control. If we are designing a system for a multimedia classroom, for example, we might want one button on a touch-panel to switch a stereo system to a 7.1 surround system, enable input from the Blu-Ray player, and disable all other signal paths. With logic programming, we can do this, by assigning a trigger message to the touch-panel button and assigning that same message value to a logic source in the programming that then feeds through other logic objects to trigger the various changes we desire. Different DSPs have different amounts and styles of logic programming available. In addition, DSPs have different logic connections in the physical domain. If all your logic will be triggered via network cables, that is different than logic that will be triggered via two-wire electrical relay switches (generally into GPIO ports), and you need to make sure your device can send and receive control messages as desired.
- Scripting: some DSPs feature built-in scripting programs that allow the user to create highly elaborate software programs that allow their systems to perform complex actions easily.
  - Users of Mac computers may be familiar with the fairly simple scripting platform Apple Script that is built into those machines. This is a simple example of scripting language (and even Apple Script can be used to control QLab playback software). Q-SYS DSPs, for example, support the Lua scripting language, which can enable the user to run very complex instructions via the DSP itself, without the need of an external show controller.

SPECIFICATION

The specs for a DSP are in one sense innumerable—if you examine the types and classes of all the internal processing objects, the list could be very long. However, in another, more hardware-oriented sense, there are relatively few specs we need to really track when choosing a DSP

Chassis: is the unit a standard 19" rack-mount unit (most, but not all, will be)? How many RU high is the device?

Frequency response: it should be assumed that any DSP provides a minimum of 20 Hz to 20 kHz frequency response, with no significant peaks or valleys (we do not want a DSP to color our signals if set in default mode—i.e., if we are not EQing signals on their way through the device). Frequency response is often expressed as a range with a given tolerance (e.g., 20 Hz–20 kHz +/− 3 dB). In a DSP, +/− 3 dB would be *way too much* variation in frequency. A good spec would be 20 Hz–20 kHz +/− 0.5 dB. Some DSPs won't even list frequency response, and while normally we would be suspicious of a device omitting a crucial spec like this, in DSPs, it is just generally assumed that the frequency response will be flat and full range, so the omission is acceptable so long as the unit is from a reputable manufacturer.

Input dynamic range: should be at least 100 dB. Some DSPs will not list this spec either, and again, we typically assume that dynamic range should be at least 100 dB. That said, this can get tricky. If we are creating a system that is supposed to gently reinforce classical music, we need the widest dynamic range possible, whereas if we are designing a system for rock music, we can survive with much less dynamic range.

Phantom power: does the DSP provide phantom power, and if so, is it switchable by channel or by bank/card slot? There are situations where running a mic directly to the DSP without going through a console is highly desirable—for example, a theater system where the stage manager's GOD mic (a mic used to make announcements to backstage for calling actors to places, and/or to the house in case of emergency announcements) is connected directly to the DSP and not the console, a fairly standard configuration. In these systems, the mic used might be dynamic, or it might be a condenser, and in this case, it is good to know what the phantom power options are on your given unit.

Sample rate(s), bit depth(s): again, this can be very important. Some DSPs will only operate at one sample rate, some will have selectable rates. You need to ensure all your digital audio devices are operating at the same rates and depths.

Clocking: if you are using digital audio signals, where is the clock source? Is there a physical word clock connection on the DSP, or does clocking travel over the digital audio signal cable (some digital audio signal types can also carry clocking info, some cannot)?

Physical input maximum counts: how many channels can physically connect (beyond any networked signal paths)? Do you need extra cards in order to make these connections, or are they built-in?

Custom or fixed I/O: is this a unit that can take card inserts to change the physical connections, or is it a unit with fixed I/O ports that you cannot change?

Analog inputs: does the unit feature analog inputs? How many, and in what form factor?

Digital inputs: does the unit feature digital inputs? How many, and in what form factor?

Analog outputs: does the unit feature analog outputs? How many, and in what form factor?

Digital outputs: does the unit feature digital outputs? How many and in what form factor?

Network I/O: does the unit feature the ability to handle network-audio ins and outs? What protocol (Dante, CobraNet, proprietary, other)? How many channels? Bidirectional over single cable? Can that protocol travel over conventional network switches/architecture, or does it demand direct connection to other signal devices?

Maximum channel count: both in terms of how many channels the unit can process and how many it can physically connect to for I/O. It is very common for units to process more channels than they can physically connect to, especially with units that take networked audio.

Configuration: via software, via buttons on the chassis, or both? PC, Mac . . . Linux?

Display: does the unit have any kind of LCD status display or other lights?

GPIO: general purpose input-output connections. Does the unit allow these? These are connections to relays and other electrical devices that allow you to wire custom triggers for actions or presets within the software. If GPIO is present, how many connections are available?

RS-232: a serial control language used by many PCs (especially older computers). Does the unit have a connection for this and understand the language?

Processor speed: like any computer, processor speed is listed in MHz or GHz. The faster the better.

Alternate connections: some DSPs offer telephone line connections (commonly used in television studios for dialing through a system to control rooms or other parts of a facility); is the unit part of a product line including other devices that make it more useful (custom touchscreens, paging mics with on/off switches, etc.)?

Playback: some DSPs feature the option of including an audio player/server inside the DSP to simplify your system. Does your unit feature this?

CONSIDERATION

What questions must a designer ask (and answer) in order to select a DSP as part of their system?

1. How many signal paths do you need to process? A system that is merely sending L/R concert signals out of a console needs much less processing power than a theater system with 50 discrete channels of playback sending to 50 individual speakers, in addition to a microphone mix sending in stereo to L/R mains and monitor

speakers onstage for the performers. Think through all the inputs in your system. Do they get summed before they go to the outputs (as in the case of the stereo concert system), or are they discrete one-to-one connections (as in the theater system)? Make spreadsheets identifying these counts, and this will help you plan your needs.[11]

2. What types of signals will you be connecting to the DSP? Analog? Digital? Networked? A combination?

3. What kinds of actions do you need the DSP to perform? Is it just acting as a simple system calibration device—providing EQ, delay, and limiting for a small set of speakers? Or is it managing cross-flowing communications among a wide range of inputs and outputs (as in an airport sound system)?

   a. What kind of peripherals do you need in order for my human operators to properly interact with the system and to perform the actions required?

4. What equipment will you be interacting with? Are there other proprietary connection/signal types you need to consider?

5. Do you need the system to respond automatically to sensor or relay input, or can any changes to the system be triggered by a human operator? If a human is operating all changes, you may not need robust GPIO, for example, because you could operate the system with online software monitoring and change setups via the software itself. If you need everything automated, you need to make sure whatever trigger/input/control data feeding into the DSP will be recognized so the DSP can respond as you desire—whether it's coming from a show controller or directly from triggers into the DSP.

6. How much time is available for programming and testing the system? This is always a consideration for any system component, but it is more important for DSPs—with all their custom programming options—than for, say, a power amp. You may have the desire for highly elaborate programming in your DSP system, but if your programming team (which might be *you* on smaller productions or projects) doesn't have time or budget to allow for programming and testing of the custom program, then you may need to simplify your plans. If you simplify the planned needs of the system, you may then be able to select a less-expensive and less-complex DSP, so this should definitely enter into consideration when selecting this component.

We will be discussing DSPs and their functions in more detail in the next chapter (examining the use of effects within a DSP for system processing), in Part II of this book (examining the setup, testing, and calibration of sound systems), and in Part III of this book, as DSPs play a critical role in the structure of different types of systems. All of which is to say that if you are still a little unclear on some of what a DSP does, we'll be coming back to this topic again and again, and by the time you reach the end of this book, the purpose and function of DSPs should be clear.

Most DSP manufacturers make their configuration software available free via the Internet. If you are curious about how to configure signal paths and what kinds of processing objects, logic, and scripting are available in a given platform, download their configuration software, and you can explore this before settling on which DSP to select.[12]

Our DSP is the heart of the system. All the arteries and veins run to and fro in this central spot, and without its work, the system does not function as desired. DSPs aren't the *sexiest* part of a system (mics, consoles, and speakers all have them beat on that front), but they are essential to creating a system that works as intended, sounds professional, and is easily controlled.

It is helpful, when planning a system, to have a good idea of the effects we tend to use within DSPs. In the next chapter, we will discuss these effects—many of which are also used in different configurations for creative mixing, and some of which are decidedly NOT used for creative mixing (such as all-pass filters).

Notes

1 Note that in computing, DSP can refer to the actual algorithms inside digital audio devices that calculate the A/D and D/A conversions, and actually process the digital audio data. While that is a vitally important field (if no one was working in that realm, we would have no digital audio), that is not how we will use the acronym DSP in this book. Whenever you see DSP in this text, we will be referring to digital signal processing devices, placed before the amplification stage, unless otherwise identified.

2 The custom programming of these controls is often known as the *logic* programming. Whereas the actual signal paths are the *audio* programming. More on this later in this chapter.

3 There are exceptions—for example, the speaker manufacturer d&b audiotechnik makes speakers that are designed to be powered only by their dedicated proprietary power amplifiers (in fact, if you power them with

other amps, you violate their warranty), and those amps have very sophisticated DSP built into them with presets designed to work with their specific speaker models. That said, there will still be times when you might need to feed these amps with an outboard stand-alone DSP if you need the flexibility in system programming and control that the stand-alone unit would allow. If you only need the DSP to calibrate timing, EQ, and limiting, the built-in amp DSP on the d&b unit is more than sufficient.

4 RU = rack unit. This is a standard measurement of the height of rack-mounted devices. 1 RU = 1.75", and equipment racks are sold by height as measured in RU (e.g., a 44 RU rack, a 25 RU rack).

5 In most DSPs, there are no XLR connectors, but a series of custom-wired connections known by the generic terms *Phoenix block* or *terminal block* connectors. These will be addressed in detail in Chapter 24. Just be aware that, for example, to wire XLR cables into a Phoenix block, you will need to make custom connectors. This is not difficult.

6 For example, with a touch-panel, there is a slight added risk that the software will crash, rendering the system inoperable (though with professional control equipment, this risk is very, very slim). With a button-station, there is potentially a need to label each button with tape or a label-maker . . . labels that can be removed or changed by accident rendering the system less clear to an operator. Additionally, a physical button that is in use all day every day will eventually wear the mechanical switch out, where a touch-panel does not have that risk. If your system needs to operate in the dark, a button station might be preferable to a touch-panel, as it would give off less light. These are just a couple of the many options to consider when selecting these kinds of peripherals.

7 Though again, this is a generalization, and on a large ride with different track sections, there might be an advantage to being able to talk only to passengers in one car at a specific part of the ride, rather than to everyone all at once.

8 For example, dbx's DriveRack series of processors.

9 A LAN is a standard show network configuration, that will be detailed in Section 6.

10 For more on show controllers, see Chapter 19.

11 Spreadsheets and other production documentation will be discussed in detail in Part IV of this book.

12 Much professional DSP configuration software is PC-only. I am primarily a Mac user, but I still own a PC in order to run two types of software: DSP configuration software that isn't Mac-compatible and AFMG's EASE® (speaker prediction software, discussed in Chapter 34). If you only have a Mac, you can run Parallels or other virtual machine software that allows you to install Windows on a part of your drive, and thus can still create DSP configurations in these programs—a little annoying, for sure, but since the high-end corporate AV market is still heavily PC-oriented, not terribly uncommon (since the high-end sound content creator market is very Mac-oriented).

Effects—Part 2

System Calibration Effects

Digital signal processors are only as useful as the signal paths we establish within them. While some simple DSPs (often known as LMSs—loudspeaker management systems) have pre-configured signal paths with stock processors in each, true DSPs are fully customizable to the needs of the individual system application. In this chapter, we will focus on the effects (and some other processing objects that are not truly "effects," but can be lumped into this general category for the sake of understanding) used in DSP configurations to calibrate and otherwise manipulate the signals traveling through the DSP on their way to amplification. Like mixing effects, this is a category of processing components that is virtually endless in specific detail but can be generalized by use of several basic categories. This chapter does not pretend to be a comprehensive illumination of every available effect, but instead a general presentation of effect types as commonly found in most sound systems.

DEFINITION

Most commonly, sound systems find themselves in need of some of the following 4 basic categories of processing object within DSP channel paths:

1. Spectral effects: EQs, used for "flattening" speaker frequency response—which is to say, removing anomalous peaks in certain frequency ranges, which might be the result of the specific speaker's design, or (more commonly) the result of the combination of a given speaker with a given position in a given venue
2. Dynamic effects: limiters are the most common dynamic effects used in system processing, simply as a preventive measure to avoid overloading either amps or speakers later in the signal path. Gates and duckers are occasionally used (for example, ducking all other sound sources when the stage manager activates the god mic), and even more rarely, compressors might be used (typically when feeding an auxiliary feed out to a recording or broadcast setup).
3. Time-based effects: delays are the most commonly used time-based effects in DSP-land. In a sound system where more than one speaker is carrying identical copies of a signal, and those speakers' areas of coverage

overlap *at all* (or are contiguous), the signals from those speakers need to be time-aligned for signal coherence (a detailed description of this principle is given following this list). Also in this category is a type of effect common only to system calibration via DSPs, called an *all-pass filter*. An all-pass filter is one that does not truly change the gain of any frequency range but tackles a selected frequency range and delays it by a fixed amount, usually measured in degrees of phase. This is a very refined time-alignment tool that can take a signal that is routed to two otherwise time-aligned speakers and correct remaining blur of signal that may occur in one or more frequency areas but not across the averaged signal as a whole.

4. Routing/leveling objects: mixers, matrices, and other such devices that allow a signal to be sent from one input path to more than one output path—sometimes flexibly, allowing that input signal to change its destination path upon recall of a new setting—and allow said signals to be boosted or attenuated (or muted altogether) according to the programming logic of the DSP.

Of the earlier four categories, number 4 (routing/leveling objects) contains the most variety, as different DSPs will offer different variations on the same theme and each has slightly distinct modes of operation. We will remain generic in our examination of these processor types.

Before continuing to the "Classification" section, there are two fundamental concepts about DSP and system configuration that we must grasp. The first is the generally preferred signal order of processing objects in a DSP channel path. While there are variations to the "rules" I will lay out here, there are some broad-stroke ideas that it is important to understand. The second is the principle of time-aligning speakers within a sound system. Why we do this—and how it is done—is an essential process to understand before setting up a DSP system.

SIGNAL ORDER OF PROCESSING OBJECTS

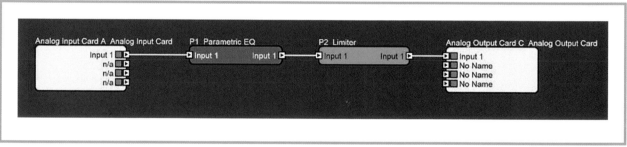

FIGURE 13.1
A Lone Signal Path From Input to Output in a DSP (Shown in Harman Hi-Q Net Architect)
Source: Courtesy Harman Professional Solutions

As always, we begin at the beginning, with inputs. Some DSP configuration programs open as a blank slate, which means you do not even have input objects in place on your screen. If this is the case, the first step to configuring a signal path is to create the input object within the software. Programs vary, but you will need to follow whatever software protocol your system provides for creating objects and create an input object that matches the signal type you are planning to use. In Figure 13.1, our input object is an analog audio input, for simplicity's sake—however, this could just as easily be a digital signal input, a Dante networked audio input, or the like. Once we have an input object that corresponds to a literal signal input to the DSP, we need to route the signal somewhere. Typically, DSP software allows us to route signal simply by creating the next object in the signal path and drawing a line between the output of the "input object" and the input of the next processor. In a very simple signal path (shown in Figure 13.1), where an input is routing directly to an output, without splitting and being sent to multiple destinations, we will generally want at minimum an equalizer and a limiter in the signal chain.

An equalizer, by its very nature, changes signal levels. Whenever we use equalizers (whether for mixing or for system calibration), it is always desirable to attenuate unwanted frequencies first, rather than boosting wanted frequencies. The reason for this is simple—reducing unwanted frequencies creates fewer *artifacts* than boosting desired frequencies.

When we reduce frequencies in a mixing effect, we are deemphasizing parts of a sound we don't like, or don't need in our mix. When we reduce frequencies in system calibration, we are reducing peaks in a speaker's performance as measured at a given position (or ideally, an average of given positions) in the audience area.[1]

Typically, the first processing we want to engage with in a signal path as simple as the one in Figure 13.1 is equalization. We want to achieve flattened frequency delivery to our audience before we do anything else to the signal. *Why* do we want flat frequency response in a sound system? Don't we sometimes want speakers to add "flavor" to our presentation? To answer the second question first, NO, we do not want "flavor" from our speaker system. A good sound system, whether for mixing in a production studio, delivering a live concert, or exciting theme park riders (or any other use) delivers a flat frequency response (or as close to flat frequency response as possible) because this allows the creators of the sonic content to be assured that what they heard in their own production facility when creating that content is what will be delivered to the intended audience via the sound system. If I have spent 20 hrs mixing instruments in a single music cue for a stage show, and I believe I have achieved a perfect mix, I want the sound system in the theater to deliver exactly what I heard in my (flat frequency response) studio when I was creating it. Any production sound system should strive for flat frequency response, because this allows the "flavor" to be entirely baked into the content being delivered, without the content creators having to account for unpredictable "flavor" from the system of delivery.

Additionally, when we are routing a signal through processing objects or components, whether in a DSP or via hardware, it is vital that any effect or process that can alter the gain of the signal in question be placed *before* any limiters in our signal path. Even if we only reduce frequencies with our EQ, the truth is that a reduction in one range can cause boosts in adjacent ranges.[2] Since a limiter's job in a DSP is to provide a hard barrier against overloading amps and speakers down the signal path, it needs to be the very last processing object in the signal chain that has the possibility of altering gain. Thus, even a reductive EQ needs to be placed before the limiter in the processing order.[3]

Let's now make our system just the slightest bit more complex, and we can see how this impacts our signal order. In Figure 13.2, we have taken a single signal input (such as the mix of actor mics in a stage show) and split it to two sources that need time-alignment (e.g., main vocal speakers and under-balcony fills placed farther into the house).

While we will discuss the principles of time-aligning speakers later in this chapter (and in Section 10), for now, assume that we need to delay the under-balcony fill speakers so they are aligned with the mains. In Figure 13.2, we can see the input signal split.

In Figure 13.3, we see the basic layout of the speakers we are discussing in this scenario.

If we think about this closely, we will quickly realize that the main speaker and the underbalc will need to be EQed differently in order to achieve a flat response. This is so for two reasons: 1) they are different models of speaker, so they will have different natural frequency response, and 2) they are positioned in vastly different locations, and their signals will interact with those locations creating different response. Even if they were identical speaker models

> ### Artifacts
> New and typically undesirable sonic material introduced to a signal by virtue of a processing object or device."

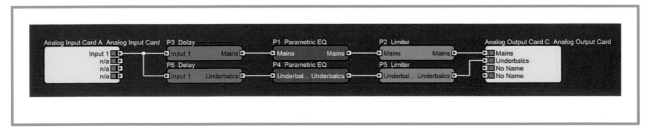

FIGURE 13.2
One Input Split to Two Time-Aligned Destinations in a DSP (Shown in Harman Hi-Q Net Architect)
Source: Courtesy Harman Professional Solutions

149

150

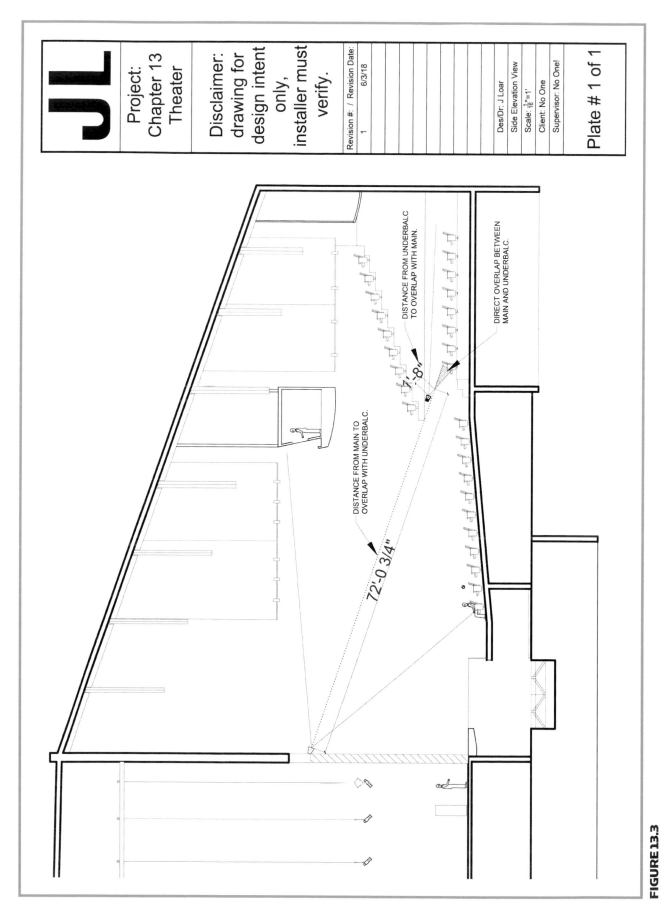

DISTANCE FROM UNDERBALC
TO OVERLAP WITH MAIN.

7'-8"

DIRECT OVERLAP BETWEEN
MAIN AND UNDERBALC.

DISTANCE FROM MAIN TO
OVERLAP WITH UNDERBALC.

72'-0 3/4"

FIGURE 13.3

Proscenium Theater With Ma n Vocal Speaker and Underbalc Fill (Side Elevation View) (which would be very rare in the case of mains vs. underbalcs), the acoustics of the front of stage are different from the acoustics of the underbalc area. By virtue of being placed under a balcony, the underbalc speaker will create reflections/diffusions of a signal off of that balcony that sum with the original direct sound in a way that will be different from the main.

> **Proscenium**
>
> An arch framing the opening between the performance area and the audience area in some end-stage theaters. End-stage refers to any theater or venue where the performance area is facing one direction toward the whole audience (with or without a proscenium frame), as opposed to two-sided audience in a "runway" configuration with performance area between the audience sections, three-sided audience with a "thrust" performance area nestled in the horseshoe of audience areas, or four-sided "round" performance areas with audience surrounding the performance area.

At any rate, because the two speakers that are both receiving the same input signal will perform so differently in the acoustic realm, we want to have the equalizers AFTER the signal split. However, as we see in Figure 13.2, we have not placed the EQs first in line after the split, we have placed the delays first. Why?

We tend to process our systems in signal flow order. The human brain likes order, and it is easiest to think about the work that needs to be done if it is all laid in a line. Now, in our system, as shown in Figure 13.3, will the main speaker's signal be completely silent for the audience that is directly in line with our underbalc? In other words, if we turned the underbalc off, would the audience under that balcony hear anything from the main speaker? Of course, they will hear the main, just not as directly or as loudly as might be desirable (which is why we added the underbalc in the first place). Thus, our two speakers playing at the same time will combine acoustic signals in any area where their coverage overlaps. When those signals combine, if they are far out of time-alignment, those time effects will cause frequency artifacts (remember comb filtering from a few chapters back? This is one possible artifact of this combination). Those time artifacts will turn up in our frequency measurements (time problems often show as frequency problems when we are only looking at frequency). Thus, if we don't time-align our speakers *before* we EQ them, we will see artifacts that we think we should be fixing with EQ that, in fact, we need to fix with delay. Additionally, because our underbalcs will be combining with some part of the main speaker signal, we need to not just EQ the underbalc so that it performs flat on its own, but so that it delivers a flat response *in combination with* the main speaker.

Since the zone of overlap between the two speakers is going to happen only where the underbalc covers along with the main, and not for the area that the main covers without the underbalc, we can EQ the main first, and then use the underbalc's EQ to properly blend the two while still delivering good response to any areas that the underbalc covers where the main doesn't reach. In practice, *there may not be* any area where the main doesn't reach at all, so we EQ the underbalc after we've got the main working as desired.[4]

At any rate, this is why we want the delays to be the first objects after the split—so that we can avoid mistaking time artifacts created by poor alignment for frequency artifacts that we need to EQ away.

THEN, we place the EQs and then our limiters.[5]

TIME-ALIGNMENT OF SPEAKERS

What does time-aligning speakers mean? In a given sound system, whenever we have more than one speaker that is delivering identical content (a good example is the dialog mix for a mic-ed theater show or a concert mix in mono), we have to consider how long it takes for the sound to travel from the speakers to the audience. If we've got a concert stage, and our only speakers are the L/R mains (playing in mono), chances are we don't need to do any time-alignment: while an audience member on the left of the audience section will get the signal from the

151

left speaker earlier than from the right, it is assumed that the left speaker's copy of the mono signal will be loud enough for that audience member (because the right signal will have decreased while traveling through air) that the two won't fundamentally compete in any meaningful way. In addition, in a concert system (especially a system for music that is typically heavily amplified—rock, pop, rap, electronic . . . anything other than acoustic jazz or classical, mostly), we are usually OK with hearing the sound from the speakers as if this is the original source, rather than trying to hear it "from the stage."

This brings us to an important concept: *localization*.

Localization

In sound system design, localization is the principle of determining what sound source an audience member hears first and thus where in space the audience member(s) perceive that sound to be traveling from.

If we have our rock concert system, and the main L/R speakers have no delay on them, then because they put sound out instantly and they are loud, the audience will naturally *localize* sound to those speakers. Even if the stage is loud, so long as the speakers are louder, the speakers are closer to the audience than the stage itself, so the signal from those speakers will arrive at the audience before any natural sound from the stage itself.

Now, if that concert system is for an audience area that is very long/deep (audience stretches back from the stage a great distance), there is a good chance that in order to have our main L/R speakers quiet enough to not totally deafen the front row of the audience, there will come a place as we stretch away from stage that the SPL of the mains will be too low (because of the inverse square law) for our coverage goals. One of the principal tasks of the sound system designer is to determine how loud coverage needs to be (at minimum) throughout the audience. In the United States, for example, large concert systems often specify a minimum of 120 dB SPL available through all audience areas.[6] If by the time the audience is 100 m away from the stage the SPL being delivered is down to 100, this is not acceptable. Rather than turning the mains up, we place what we call *delay speakers* in the crowd, covering the audience so that as soon as the sound from the mains begins to dip below 120, the supplemental delay speakers pick up and boost the signal back up to 120.

Now, why do we call them *delay speakers*? Because we always need to delay the supplemental speakers to time-align the signals with the mains, both to avoid nasty comb filtering, and to preserve localization back to the front of stage.

In Figure 13.4, we show a side elevation view of just such a system.

For simplicity's sake, I have drawn *coverage cones* that loosely represent the basic coverage shape of our two speakers (though, as we will discuss later, coverage cones don't hold their shape across the frequency spectrum, and actually change quite a bit from low to high). Sound does not move very fast. Unlike light (which moves at 186,000 miles/second, and therefore feels "instant" to us), sound moves at about 1,130 ft/second. Now, onstage we have a drummer. If all we have in our system is the main speakers, when the drummer hits the kick drum, the speakers immediately produce the sound. Let's imagine that as an audience member, we are 280' from the stage (which, at a big concert festival, is not unexpected). That means it will take nearly a quarter of a second for the sound from the mains to arrive at our ears. If we also then turn on our supplemental speaker (because the main speaker is too quiet by the time it reaches us 280' away), but we don't delay the signal, the supplemental speaker will *also* fire instantly. Because the supplemental speaker will then be both louder, and the signal from it will arrive at our ears sooner, we won't localize to the mains/stage, we will localize to the supplemental speaker. Not only that, but a quarter of a second is easily enough time for us to hear the delay of the main sound, so we will hear *two* kick hits, where one was played—the first from our supplemental speaker and the second from our mains as the signal travels acoustically to us. This is unacceptable. We want to localize to the stage, and we want to only hear one kick when one was played. Therefore, we need to delay the signal to the supplemental speaker so that it lines up with the signal from the mains.

Now, if you're thinking ahead, you will already realize that the time difference between the two speakers isn't going to be consistent over the whole audience since audience members are at different distances from that speaker. So how do we pick a point to align the speakers? Typically (though there are exceptions we will cover later in the book), we align speaker signals at the point where their signals overlap so that the transition between them is smooth.

If, as we see in our drawing, the distance between the main speaker and where the supplemental speaker's coverage overlap is 233', then we need to calculate the delay time for the supplemental speaker like so:

$$\frac{233\,ft}{1130\,ft\,/\,sec} = .206\,sec.$$

Since most digital delay units set their times in milliseconds, rather than seconds, we then convert our seconds to milliseconds (1,000 ms/sec) and get 206 ms. We then set the signal sent to the supplemental speaker to a 206 ms delay, and then when the drummer hits the kick, it starts in the mains, arrives acoustically (and electronically) at the supplemental speaker 206 ms later, and then to the audience 280' away directly thereafter. This prevents the double-kick hit and causes the audience to still perceive the sound as traveling from the stage, even though level has been boosted at the supplemental speaker.

While we will be discussing the principle of system delay in great detail later in this book, there are two other points to clarify right now:

1. Sometimes we do need delays on main speakers. In theater, for example, it is very common for the designer and director to not want to localize actor voices to the main speakers, but to the actors onstage. Therefore, sometimes we put small (and sometimes changing) delays on the main voice reinforcement speakers, to help the audience localize back past the main speakers to the performers themselves. We'll cover this more in the theater chapters later in the book.
2. You only need to delay supplemental speakers if they are carrying the same signal as main speakers. Again, in theater, we often design systems with a ton of speakers surrounding an audience (for immersion in effects, among other uses). There is no need to delay a speaker that is behind the audience for delivering ambient cricket sounds to the main speaker delivering actor mics. They are not delivering the same content, so time differences between them are not relevant. Delaying speakers is *only* relevant when they are delivering identical (or at least, largely the same) content.

CLASSIFICATION

There are a wide range of processing objects/effects available in DSP units, more than we can describe in detail here. Instead, we will touch on the major categories of effects, their main subcategories, and basic reasons why you might select one versus another.

- EQs: for system tuning and calibration, the standard for many years was graphic equalizers, because with one device, a designer could modify frequencies across the spectrum. However, as system design has become a more refined discipline, increasingly advanced measurement tools have allowed us to see that the kind of notch-filtering that is a common feature of graphic-tuned systems creates too much phase incoherence in the signal. These days, the most commonly used EQs for system calibration are parametric EQs. Parametrics allow us to adjust frequencies as subtly as possible, and designers often try to make wideband (wide-Q) adjustments of as few dB as possible, so as to keep the system in the best possible time-alignment while still creating a relatively flat frequency response. Any decent DSP should provide the option for parametric EQs in your output paths, but a salient question is "how many bands are available?" If your signal path only allows one band of parametric EQ (which is an exaggeration, most would allow more than this), it won't provide enough flexibility to really calibrate your signal paths. However, if you can instantiate endless eight-band parametrics, you have plenty of processing power in hand (and in fact the risk becomes that you will over-EQ the system and thus introduce unwanted anomalies in performance).
- Filters: there are a wide variety of filter types available in DSPs. They are often referred to by technical names (Butterworth, Linkwitz-Riley, Bessel, Chebyshev, etc.). In addition, filters are classified by their "order." Filters

154

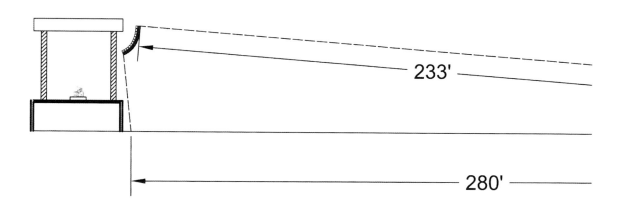

FIGURE 13.4
Concert System With Mains and Delays (and Distances Listed)

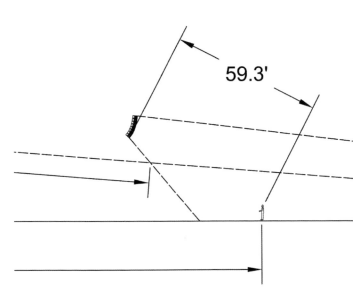

59.3'

Project:
Chapter 13
Concert

Disclaimer:
drawing for
design
intent
only,
installer
must verify.

Revision #:	/	Revision Date:
1		6/3/18

Des/Dr: J Loar

Side Elevation View

Scale: $\frac{1}{32}$"=1'

Client: No One

Supervisor: No One!

Plate # 1 of 1

are typically used as crossovers in a system (some DSPs have both filter and crossover objects as distinct instantiable devices, the difference being that a "crossover" device assumes more than one output path, where a "filter" device just operates on the single signal passing through it, with no split). Filter order refers to the amount of attenuation per octave from the "cutoff" frequency (normally 3 or 6 dB down from flat response). A first-order filter has a slope of 6 dB of reduction per octave, a second-order has 12 dB/oct, third is 18 dB/oct, fourth is 24 dB/oct, and so on. In terms of the different filter types, they exhibit different performance with regard to amplitude flatness and phase distortion (group delay). A brief summary of the major types is here:

- Butterworth: very flat amplitude response, minimal and consistent group delay
- Linkwitz-Riley: Butterworth filter(s) cascaded into other Butterworth filter(s) to create second-, fourth-, or eighth-order filters. Where a first-order Butterworth crossed over to top and sub will sum at +3 dB at the center frequency of the crossover, a second-order L-R will create a flat sum across the center frequency, which is often desirable in system performance. L-R filters are some of the most commonly used crossover filters in system calibration. Same group delay response as Butterworth (just increased by a factor of the order).
- Bessel: maximally flat group delay/phase response, but with more amplitude ripple. Not included in all DSPs.
- Chebyshev: specialty filters, that sometimes evince flatter response in the *passband* (range of audio let through the crossover) than other filters but have more *stopband* (range of audio being suppressed) ripple, or vice versa depending on their particular topology (*inverse* Chebyshev is the former, regular Chebyshev the latter). Not very common in systems, not included in all DSPs.

- Additionally, there is a type of filter called an *all-pass* filter. As mentioned earlier in this chapter, all-pass filters do not directly alter frequency (in the usual gain/reduction sense), they select frequency ranges and alter their phase response directly (forward and backward in time). These are very frequently used in refined installation system calibration, but less frequently for one-off system installs like touring shows or concerts. We will talk more about all-pass filters in Section 10.

- Dynamic effects: the most commonly used dynamic effect for system calibration is the limiter. Sometimes, we will actually use two limiters in a signal chain, one that is a *peak* limiter for fast transient materials (to prevent over-excursion of a speaker driver and thus rupture of the cone), and one that is an *RMS*[7] limiter to prevent overheating from consistent levels that are too hot. The process of setting up these limiters can be complex, especially in very fine-tuned systems.[8] However, a basic process can be followed that will get you to workable settings (that, while not as precise as possible, will protect equipment in 99/100 situations).
  - Assume that we have a sound system with speakers and amps that are well-coupled for use together. What does this mean? The amplifier channels can supply at least the amount of maximum power that the speakers can handle and ideally can supply *more* power than the speaker can handle, so we never worry about distorting the amp. Set up your system, put in place any EQs or crossovers you need, get time-alignment working. Put the limiter in the signal chain but bypass it. Now, in order to set your limiters, follow this procedure:
  - Gather your supplies. You will need a voltage meter that measures both peak and RMS values (and that has a high maximum voltage), a pink noise[9] source, and the specs for your loudspeaker(s) in question.
  - Disconnect the loudspeaker from the amplifier.[10]
  - Connect the voltmeter to the output terminals for the amp channel you are setting.
  - Using the speaker specs for the unit you have just disconnected, calculate the maximum voltage it can handle.[11]
  - Connect pink noise source to the channel in question. Run it through to the amp. While watching the voltmeter, turn up the level of the pink noise source until you are measuring a voltage 2 V above the max speaker spec. When doing this, all signal channels should be started at a nominal line-level gain, so the level is coming from the pink noise source itself, not from a preamp on your console or something. Most pink noise sources are computerized and contain their own level controls.
  - Now, engage the limiter, and reduce the threshold control to the point where your measured voltage is the stated maximum of the speaker (or ideally, just a hair below the stated max, for safety).
  - Now you have set the RMS limiter!
  - You can repeat these steps using program material (e.g., recorded music if this is a concert system, recorded speaking if this is a lecture system) to set a peak meter, watching for the highest levels hit in the program. This can be tricky if you don't have a meter that records peaks, but many meters will register high peaks and store

them so that you can examine them after they have passed. By doing so, you can set your limiter numerically without the transient test source playing and then play the program content again to check your work.

- Generally, you won't need to use many other dynamic effects in system calibration. The exception here is a ducker (or inverse gate). Duckers allow a signal set to be attenuated by a preset amount when another signal above the threshold is present. These are very commonly used for systems feeding backstage performer reference monitors, where in general the program of the show is being fed backstage so that the performers can keep track of the show for entrances and exits, but the signal may be fed through a ducker with the trigger signal being a stage manager's mic. Therefore, when the SM talks and gives commands to the backstage team, the show program attenuates (ducks) under the SM's voice, and it is easy to make out her instructions. Once she stops speaking, the program material swells back up to its normal reference level.

- Routers/matrices: there are a wide variety of routing objects and styles in DSPs. Typically, they come in a couple of basic shapes:
  - Matrix: a simple patch system, where for example signal input 1 can be routed to outputs 1–48, as desired, with on/off patch switches. Each output in a device like this can only have one input source.
  - Matrix routers (Figure 13.5): a crosspoint setup whereby several signals can be routed to a single output (at full signal level).
  - Matrix mixers: like matrix routers, except with variable level controls for each crosspoint send.

- Fade/leveling/gain objects: these can take a wide range of shapes. From simple gain objects to raise an anemic signal in the system (or attenuate a too-hot signal), to objects that have customizable triggers and time parameters (e.g., to mute a whole system with a 0.4 sec fade at the input of a touch-panel button). These are more common in distributed systems, where signals need to be turned on or off in different locations throughout the operation day, as opposed to concert/simple event systems where the DSP flow stays more or less in a steady state through the event.

- Delays: usually listed in ms of time, though sometimes also available in feet or meters (for system time-alignment). Most delay objects simply delay the whole signal passing through them, though there are some available that can act as crossover and delay, and set different frequency ranges to different times from within one delay processing object.

- Scripting/logic/control programming: different systems have wildly different amounts of control options. For example, Meyer's Galileo/Compass systems are more LMS than DSP (even though they are definitely performing digital signal processing), whereas their D-Mitri system features some of the most advanced programming options on the market. Compare with QSC's Q-SYS, which incorporates a powerful Lua scripting engine that allows it to act as a full-scale show controller of other devices as well, where BSS's BLU processor series falls in between, featuring ladder logic programming (in a visual tree style), but not the powerful show control scripting. Figure 13.6 shows BSS ladder logic. Your needs in this realm will be based on the complexity of your system, and it is well beyond the scope of this book to go into great detail on control programming for one or more DSPs.

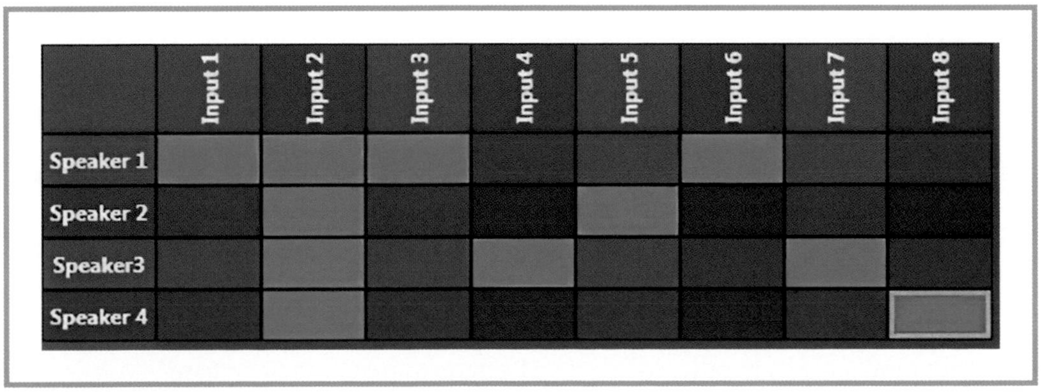

FIGURE 13.5
Inside a Matrix Router (Shown in Harman Hi-Q Net Architect)
Source: Courtesy Harman Professional Solutions

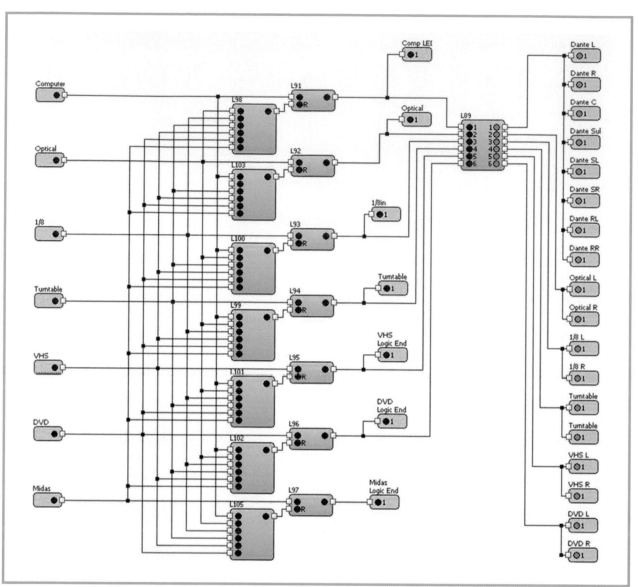

FIGURE 13.6
Ladder Logic (Shown in Harman Hi-Q Net Architect)
Source: Courtesy Harman Professional Solutions

SPECIFICATION

There really aren't specs for the DSP effects independent of the categories listed earlier in the "Classification" section of this chapter. Yes, you need to know what kinds of filters, EQs, etc., are available in your unit, but these aren't published as "specs." The best way to find out what is offered in a unit is to download the free control software and explore it, looking at the settings within each device you will need.

CONSIDERATION

What questions must a designer ask (and answer) in order to select DSP effects as part of their system?

1. What processing is necessary on my signal path? Are you just tuning frequencies and protecting from over-loads? Do you need time-alignment? Does routing need to change at the touch of a button?

2. What is the best processing object for the job within the system I am using? Does each EQ have the same parameters, or is one better suited to my particular task than another?

3. As you build your system in the DSP software: is this the simplest way to accomplish my task? It is easy to over-complicate things by adding too many processing objects when a simpler path or an object with multiple functions will handle the job and reduce the load on the DSP's processors. The lower the processor load, the smoother everything will function (DSP software always has a DSP load meter to show you how hard you are hitting the physical processors).

That's it for the processing stage of a system! Now we need to make things loud. Chapter 14 will take a quick break from the "Definition," "Classification," "Specification," and "Consideration" system of sections to analyze power and electrical math important to our work. Then in Chapter 15, we'll be back to this structure to look at power amplifiers themselves.

Notes

1 Two notes: 1) mixing EQ is much more likely to feature boosts than system calibration EQ. While a professional mix engineer still usually begins by removing the unwanted, sometimes the engineer will *then* boost other ranges, in part because some EQs have sonic signatures that introduce desirable artifacts to the signal. 2) System measurement and calibration practices are discussed in detail in Part II of this book.

2 It is beyond the scope of this book to get into the detailed physics explanation of why this happens. However, for example, if I reduce (in a very tight notch filter) 200 Hz, there is a likelihood that this might cause a boost at harmonic intervals of 200 Hz (400 Hz, 600 Hz, etc.). The narrower the Q of your filter, the more likely you are to create this kind of artifact (which is a signature reason that in system tuning we seek to use EQs with the broadest parametric Qs possible).

3 In mixing systems, this is slightly less straightforward, though the same general rules apply. For example, if I have an input channel on which I want to place both an EQ and a compressor, I want my EQ generally to come first. I want to reduce/remove unwanted frequencies before the compressor boosts quiet parts of the signal. If I compress first and then try to remove unwanted frequencies, it will be less successful.

4 Note that this may still involve some alterations to the EQ of the main—but we want to try to get the main working for the majority of its coverage area first, and then alter it as little as possible in combining it with the underbalc.

5 In some complex configurations, we will place two stages of EQ. These could be before and after the delay—which I know contradicts what I've just been saying about time-aligning first. However, if we know there are huge frequency problems in a given speaker's performance in a given position, we may address those, then work on timing, and then do fine EQ. There are innumerable complex approaches to difficult system problems, but for the most part, we are sticking to general rules in this chapter (and, unless absolutely necessary, in practice as well).

6 Europeans, by the way, think this is insane; 120 SPL is way louder than most people actually want a concert to be, and in Europe, national noise laws (while varying by country) typically would *prohibit* a system from getting this loud. Of course, just because a system is designed to deliver 120 doesn't mean it will be run at 120—there is a principle called *headroom*, which will we get into in more detail in Part III of this book, that states that we typically want a system to be able to deliver at least 12 dB more than we plan to use, so that at no time are we at risk of peaking and distorting our signals. At any rate, for this example, we'll use 120.

7 Root mean square, an averaging of signal strength that corresponds both more closely to how the human ear hears but also to the long-term energy flow in a system.

8 See www.prosoundtraining.com/2010/03/21/using-limiters-to-help-protect-loudspeakers/ for a good description of advanced frequency-based limiting of output signals.

9 Pink noise is a standard test signal for audio systems. It is a signal with randomized frequency content delivering equal energy per octave.

10 This process assumes passive loudspeakers with external amps. In the case of powered loudspeakers, where the amp is built into the cabinet, limiters are often built into the speaker itself already—though certainly check manufacturer's data on this front. If you find yourself using powered loudspeakers with no limiters built in (a true rarity at this point), you will need to alter the limiter setting process as explained earlier to catch the signal before it enters the amp. The essential procedure remains the same, but you will be working with a lower-voltage range as you will be examining line-level signals rather than speaker-level signals.

11 This requires a little Ohm math, which we'll get into in detail in Chapter 14. In short, speaker manufacturers don't list maximum voltage handling but do list peak wattage handling. Using Ohm's wheel, we will find that if we know the wattage and the impedance, we can calculate voltage. So, for example, the JBL CBT-50 column loudspeaker states a maximum wattage of 150, with 8 Ω impedance. The equation for deriving the voltage of this signal is

$$V = \sqrt{W\,I}.$$

Thus, we multiply 150 by 8, getting 1,200. The square root of this is approximately 34.6 V maximum this speaker can handle. I have used 150 W in this example because the specs for the CBT-50 list this as the "2-hr" power rating, which is long enough for me to want to set my RMS levels there (even though this is the "peak handling" spec for the speaker). If you set two limiters, RMS and peak, you would use 150 W as the peak reference for calculations and 100 W (the stated long-term power rating of the speaker) for the RMS calculation. I am indebted to SLS Audio for the succinct explanation of limiter setup.

Power Math

Electricity is very important to our work: without electricity, our jobs as sound system designers would disappear (or we'd just be amphitheater designers, using the acoustics of the stage shell to our advantage). We examined the fundamentals of electricity in Chapter 4. If you need a quick refresher, jump back there before reading this chapter.

When planning a sound system, we need to understand how speakers and amplifiers fit together, and why to choose one amp for a given speaker over another. Generally, we will choose the speakers first, and then figure out how to power them (unless they are self-powered, in which case the amp is built into the speaker enclosure). This chapter examines how loudspeaker and amplifier impedance works, how to match wattage values between the two categories of device, and the different types of speaker/amp circuit layouts we find in professional audio systems, from series and parallel to 70V.

SPEAKER SENSITIVITY VS. LOUDNESS

You will often hear people describe speakers by their wattage. They will say something like, "That's a 300-W speaker," but in a vacuum, that really doesn't mean very much unless we also understand the speaker's *electrical sensitivity*. A speaker that lists a sensitivity spec of "93 dB SPL @ 1 W, 1 m" is going to be substantially louder when driven by a 300-W amp channel then a speaker that lists "82 dB SPL @ 1 W, 1 m." This spec, sensitivity, is usually listed in the form just shown: "x dB SPL @ 1 W (1 watt), 1 m (measured at a distance of 1 meter from the enclosure's grill or face)."

Sensitivity is a measure of how much sound pressure a speaker will create when powered by a standard value, which is always 1 W. How often do we power speakers at 1 W? Very seldom, but because this is a simple value to plug into calculations, we use this as the baseline so that we can easily determine how loud our speaker will actually get.

The equation for figuring out max SPL at a given wattage is simple, and it is a variation on the same equation we used for level falloff over distance. Where that one (the inverse square law) is referred to colloquially as the "20-log" equation, this one is the "10-log" equation. It is identical to the 20-log equation, except that we use 10 in the place of 20 in the operations.

$$dB\Delta = 10\log_{10}\frac{P_1}{P_0}.$$

In English, this reads, "Decibels of change for a new power value are equal to 10 times the log (base 10) of power 1 (the given power of the amp to the speaker) over power 0 (our reference power from the sensitivity spec, aka 1 W)."

So, let's take our speaker that listed 82 dB SPL @ 1 W, 1 m. If the speaker is rated to handle 300 W of power (it will say so in its specs), how loud will the speaker be when powered by 300 W?

$$dB\ \Delta = 10\log_{10}\frac{P_1}{P_0}$$
$$dB\ \Delta = 10\log_{10}\frac{300}{1}$$
$$dB\ \Delta = 10\log_{10} 300$$
$$dB\ \Delta = 10(2.477...)$$
$$dB\ \Delta = 24.77$$

Add that to our original 82, and we'll find that the speaker will be producing 106.77 dB SPL when powered by 300 W, measured at 1 m. If we run the same equation for the 93 dB SPL speaker, of course the answer in *dB* Δ is identical. The starting sensitivity is only relevant at the end. The change between powering *any* speaker at 1 W versus powering the same speaker at 300 W (again, assuming the speaker is rated to handle 300 W) will always be 24.77... dB. So, the speaker with 93 dB SPL sensitivity would be producing 117.77 dB SPL at 300 W, 1 m, because we just add the dB of change to the original sensitivity figure.

Before we move on from this equation, it is worth our time to do one more example. If we have a speaker being powered at 150 W, that is producing 100 dB SPL at 1 m, and we double the power, what is the SPL change?

$$dB\Delta = 10\log_{10}\frac{P_1}{P_0}$$
$$dB\Delta = 10\log_{10}\frac{300}{150}$$
$$dB\Delta = 10\log_{10} 2$$
$$dB\Delta = 10(.301)$$
$$dB\Delta = 3.01$$

A doubling of power *always* results in approximately 3 dB of SPL gain. This is true from 1 W to 2 W of power, and it's true from 10,000 W to 20,000 W of power. When we double the power, we gain 3 dB. The human ear doesn't tend to *hear* this as double the loudness, though. If we go by the inverse square law (which, again, tells us that for every doubling of distance from a sound source we lose roughly 6 dB of intensity), we might think that 6 dB *sounds* like double the loudness to us. It can, but usually the human ear responds to an increase in sound level as having doubled when it is louder by between 6 dB and 10 dB. Thus, when the promoter of your concert comes to you and says, "I need it twice as loud!", you can't just double the power; you may need to double it at least twice (to get to 6 dB louder) or go even farther.[1]

SPEAKERS AND AMPS IN COMBINATION

So, we know how loud our speaker will get with a given wattage of power from an amp. Now we need to select an amp that is appropriate for delivering this power. Before we even decide how much power our amp needs to provide versus how much our loudspeaker can handle, we need to ensure that the impedance of our selected amp works with the impedance of our selected speaker.

Let's imagine that we are working with a single speaker, whose specs tell us it is rated to handle 300-W continuous power, at an impedance of (the very common) 8 Ω. What kind of amplifier do we want to power it with?

Most professional audio power amplifiers will feature a range of operating impedances. A very common spec for an amp is an impedance rating of 4–16 Ω. This means that for any *electrical load*,[2] so long as the impedance remains between 4 Ω and 16 Ω, the amp will deliver power to the speakers as desired (assuming a good wattage ratio, which we will address in a moment). Amplifiers are generally specified with a range of impedances with which they are compatible.

Most professional loudspeakers will feature unit impedance of somewhere between 4 Ω and 16 Ω as well, with 8 being the most commonly found figure (it is listed as one number in speaker specs). However, sometimes we want to power more than one speaker from a single-amp channel—this will change the impedance of the load, and *how* that impedance changes will vary based on how you wire the speakers: in series, in parallel, or in series-parallel.[3]

Before we leap into the situation of multiple speakers on a single channel, let's address the question of wattage. If we are powering a single 300 W speaker, how powerful should our amp channel be? We have three options:

1. More powerful: for a 300 W speaker, an amp channel capable of delivering over 300 W of continuous power
2. Matching: for a 300 W speaker, an amp channel capable of delivering maximum 300 W continuous
3. Less powerful: for a 300 W speaker, an amp channel capable of delivering under 300 W continuous

It should be self-evident that in an ideal system, option #3 should be ruled out. In large part, we select speakers based on how loud they will be when covering a given audience section. This loudness is predicated on wattage vs. sensitivity, as we've seen earlier. If we expect the speaker we have selected to deliver its full range of power/SPL, we can't power it with an amp that is weaker than what the speaker can handle. If we do, we risk pushing the amp into distortion and heat overload when we try to push the speaker to the max SPL we calculated it would deliver for us. If the amp distorts the signal, that distorted signal can also damage speaker drivers because of its jagged shape— even though we won't be overloading the speaker, the difficulty of producing the sawed-off distorted waveform may still cause problems with a speaker cone.

That said, there are circumstances (particularly when working with a venue's preexisting stock of gear to create, for example, a theatrical sound effects system), where you will not ever need the max SPL of a speaker. If I have a 300-W speaker, capable of producing 110 dB SPL at a distance of 3 m, but all I am using it for is subtle crickets and wind sounds during a very quiet play, I might never need to run that speaker hotter than 85 dB SPL (if that). In that case, use the sensitivity 10-log equation noted earlier to calculate how much power you will need in order to achieve the desired level, and you may be able to settle for an underpowered amp. If, however, you have the chance to specify the gear you desire, under-powering your speakers is a poor choice and will leave you without *headroom* with which to reach full SPL free of distortion.

Headroom

In audio systems, the concept of providing a system capable of at least 12 (and ideally 20) dB of level to the venue *above* the desired operating SPL. Thus, if you estimate your loud rock concert will operate at roughly 105 dB SPL, you want a system capable of delivering at least 117, if not 125, dB SPL. This method of design ensures that you can run the system as loud as you desire without ever risking overload distortion in the power chain, which not only makes for an easier mixing experience and more pleasant listening environment but prevents heat damage to amps and speakers.

Option 2 (matching power) can certainly work; however, there is mild risk involved in that you are trusting the specs of both amp and speaker manufacturer to be consistent with one another. Additionally, if your amp and speaker have matching continuous power ratings, this still doesn't necessarily mean they have matching *peak* power ratings. Peak power is defined differently per manufacturer (some will say continuous power is for 10 hr of operation, where peak is for 2 hr; others will say continuous power is for 6–8 hr of operation, but peak is only for *momentary transient peaks* like a loud band crescendo that only lasts 30 sec or so—check the specs to clarify), but if your amp has lower peak rating than your speaker, you can still overload the amp.

The ideal amp to speaker power ratio is one in which the amp is overpowered compared to what the speaker can handle. Thus, you can set gain limits in your DSP to prevent speaker overload but will never be at risk of overheating or damaging the amp itself. This will prevent signal distortion, allow maximum use of the speaker's SPL range, and generally work out better for everyone involved. Whenever you have the chance, using an amp with more power than a speaker can handle is ideal—however, you must be diligent in setting limiters to prevent the destruction of speakers.[4] You also don't want an amp that is *drastically* overpowered for the speaker, as it becomes too challenging to set appropriate power levels going to the speaker. An amp with moderately more power capability than the speaker connected to it is the best scenario.

Let's return to the idea of multiple speakers on one amp channel. Why might we want this, anyway? Imagine a mid-sized outdoor concert, with only the main left and right speakers. We're not using line arrays, but large, wide-coverage, *point-source* speakers.

> ### Point-Source Speaker
> A single speaker covering a wide range of audience area, as opposed to a line array, which divides a coverage area among several coincident speakers working together.

If we want to mix the show in mono (which is very common in concert, so that the whole audience hears all of the same sounds), we might want to power the two speakers from one amp channel to simplify our setup.

Or, imagine a grocery store with speakers in the ceiling. They are all playing the same canned music and ads, so why not power them from the same amp channel?[5]

Whatever the reason, if we want to send the same signal to multiple speakers and we don't need to time-align them, we may want to connect more than one speaker to a single amp channel. In this case, there are three ways (70V aside) to connect them:

1. Series circuit
2. Parallel circuit
3. Series-parallel circuit

In Figure 14.1, we have shown an amp channel connecting to four speakers in series. A series circuit is one in which the wiring is as drawn here, with the positive terminal of the amp channel connected to the positive terminal of the first speaker, the negative terminal of the first speaker then connected to the positive terminal of the next speaker, and so on, until the negative terminal of the final speaker connects to the negative terminal of the amp.

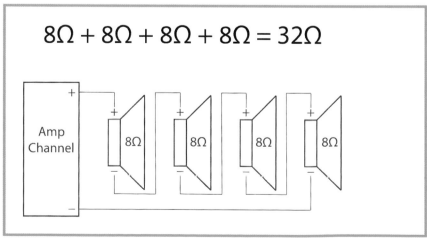

$$8\Omega + 8\Omega + 8\Omega + 8\Omega = 32\Omega$$

FIGURE 14.1
Speakers in Series

In a series circuit, we calculate impedance by summing the impedances of the speakers connected.

$X_1\Omega + X_2\Omega + X_3\Omega\ldots$. = Series impedance

Therefore, in our earlier circuit with (4) 8 Ω speakers, we have an amp channel load impedance of 32 Ω. This is too high for our amp's operating range of 4–16 Ω. What happens when the speaker impedance is too high for the amp channel? The amp cannot drive enough power into the speakers, which has two negative effects:

1. The speakers cannot reach full power/SPL.
2. The amp will face strong backflow current from the added electrical resistance and overheat, possibly causing damage.

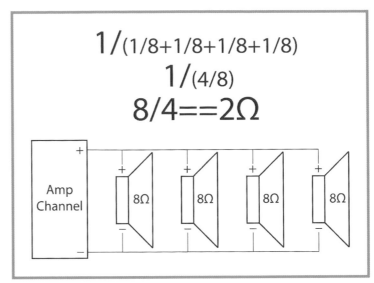

FIGURE 14.2
Speakers in Parallel

There is another notable downside to series connection of speakers. Imagine our four speakers each have individual impedance of 4 Ω instead of 8 Ω. In that case, the impedance would be 16 Ω and would work with our sample amplifier. However, if the first speaker in the chain is a bit older and more worn than the others, it will be more susceptible to overheating damage. If we run the system at its hottest edge, it is possible that this first speaker might burn out and overheat. As we'll discuss in Chapter 16, overheating a speaker for too long can cause the internal mechanism to melt and thus stop the cone from moving entirely. If this occurs, that speaker provides a new impedance to the amp, which is effectively an *infinite* impedance, because it no longer passes electricity at all. If the first speaker in the chain dies in this way, the signal will fail to reach *all other speakers in the chain*. It's like old-fashioned Christmas lights, where one bulb going out kills the whole string. For this reason, series speaker wiring has fallen out of fashion in systems use.

The next method of connecting the speakers is in a parallel circuit.

In a parallel circuit, the wiring works differently from series. As shown in Figure 14.2, in a parallel circuit, the positive terminal of the amp is connected to the positive terminal of every speaker in the chain, and the negative terminal of the amp connects to the negative terminal of every speaker in the chain. This has the advantage that if the first (or any) speaker in the chain dies, the others are unaffected, as the power travels right past the dead speaker to the others in the parallel chain. Professional passive loudspeakers (using standard Speakon NL2 or NL4 connectors) often have two signal ports in their enclosures—one is for input from the amp and the other is for "daisy-chaining" to another speaker on the same amp channel. In 99% of these speakers, the "daisy-chain" wiring defaults to parallel connection.

To calculate impedance of a parallel circuit, we need a slightly more complex equation:

$$\frac{1}{\dfrac{1}{x_1} + \dfrac{1}{x_2} + \dfrac{1}{x_3}\ldots} = Parallel\,impedance.$$

Thus, in our example of the four 8 Ω speakers, we would calculate impedance like so:

$$\frac{1}{\dfrac{1}{8} + \dfrac{1}{8} + \dfrac{1}{8} + \dfrac{1}{8}} = Parallel\,impedance,$$

which becomes

$$\frac{1}{\left(\dfrac{4}{8}\right)} = Parallel\,impedance,$$

which becomes

$$\frac{8}{4} = 2\Omega\,.$$

Thus, we have a 2Ω impedance for this circuit. This is also a problematic impedance. Not only is it outside of our amp's range of 4–16 Ω, but it is an unsafe load that we try to avoid in professional systems in general. This can be confusing because there are some professional power amps that state a safe operating range of between 2 Ω and 16 Ω. If our amplifier states that it can operate down to 2 Ω, why do we still want to avoid setting up a channel at that value? Remember that impedance, unlike resistance, is frequency dependent and fluctuates based on the input signal. While this can mean that the impedance temporarily rises higher than 2 Ω, it can also mean that impedance temporarily drops to below 2 Ω. As the impedance approaches a zero value, we will essentially be shooting lightning bolts through our equipment. This can damage the amplifier but is very likely to damage the speakers. If they are no longer resisting the electrical current much (or at all), they are much more likely to suffer heat death and become totally nonfunctional. Thus in professional systems, we try to limit our amplifier loads to a minimum of 4 Ω whenever possible.

So, with an amp rated for 4–16 Ω, and four 8 Ω speakers, is there any way to use all the speakers on a one-amp channel? There is, and it's called series-parallel.

In a series-parallel circuit (Figure 14.3), we are combining the best of both worlds. We configure speakers as shown below, in so-called series legs. Those series legs are internally totaled with the series equation (just adding unit impedance). Then, the total impedance of each leg is calculated in parallel. This means that we can increase values in the series domain and decrease them in the parallel domain, allowing us to use greater numbers of speakers combined on one amp channel. The wiring, as shown below, is a combination of the two methods. Each series leg is internally wired just as a series circuit would be (with the negative terminal of the first speaker feeding the positive terminal of the next speaker), and each series leg is then combined as a parallel system (with positive terminal of amp hitting positive terminal at the top of each series leg).

In the example we've drawn, we have two series legs, each with two 8 Ω speakers.

First, we calculate the series total of each leg—16 Ω.

Then we run the parallel equation with the series legs as each x value:

$$\frac{1}{\dfrac{1}{16}+\dfrac{1}{16}} = Parallel\,impedance,$$

which becomes

$$\frac{1}{\left(\dfrac{2}{16}\right)} = 8\Omega.$$

Look at that! We've got four 8 Ω speakers running at a combined 8 Ω! This means our circuit will work! Hooray!

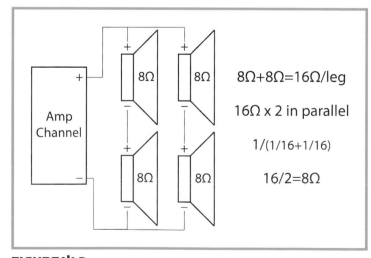

8Ω+8Ω=16Ω/leg

16Ω x 2 in parallel

1/(1/16+1/16)

16/2=8Ω

FIGURE 14.3
Speakers in Series-Parallel

The series-parallel circuit has also fallen somewhat out of fashion these days. It used to be quite common to power, say, the ten boxes on the left side of a concert stage from one massive amp channel, in series-parallel. However, contemporary advances (from self-powered speakers, line arrays that need individual DSP settings for each box, etc.) in technology have made this setup less common (though not unheard of).

When using any number of speakers greater than one on a single-amp channel, you also need to pay attention to wattage ratings, but this is a much simpler endeavor. Add the wattage values for each speaker, and that combined total is what your amp should be able to provide on the single channel.

70 VOLT SYSTEMS

Finally, we should address 70V speaker/amp systems.[6] When designing large distributed systems where high-fidelity sound is not imperative (e.g., the PA system of a grocery store), it is very common to use 70V systems, because they remove the impedance calculation from the equation of amp to speaker matching altogether.

In a 70V system, the speakers all have *transformer taps* in them, which (without getting too deep in the weeds with electrical engineering) allow for an endless count of 70V speakers to be connected to an amp (up to the wattage rating of the amp) with no impedance calculations necessary at all. The amp must support 70V mode (not all amplifiers are designed to handle this configuration), but this can allow large, mono systems to power many speakers with a single unit, very little rack space consumed, and less math than usual.

For example, the JBL Control 24 CT Micro is a ceiling-mounted speaker made for installation in shops and other such environments. It can operate in a normal impedance-matched mode, or it can be wired with either a 70V or 100V tap. It offers 70V taps rated at 9 W, 4 W, 2 W, 1 W, or 0.5 W. This means that, using the sensitivity spec, the speaker would be fed at these stock wattages (only one of them, depending on how you wired up the speaker—each of these wattages are specific wiring points, aka *taps*). If you have a 70V-capable amp with a 1,000 W channel, you could safely power 111 Control 24 CTs in 9 W mode, or *2,000* of these speakers in 0.5-W mode, from that single-amp channel.

If we can dispose of the impedance equations and power so many speakers, why don't we do this for every system in the world? Because sound quality is reduced in this kind of setup. The presence of the transformer in each speaker typically reduces sound fidelity by virtue of its own self-noise being added to the signal. Theoretically, if you designed the highest-quality neutral sounding transformer, you could avoid this problem, but the cost would be so high per unit that at that point we generally switch back to regular amp/speaker combinations.

So! Now that we have a grasp on power concepts, let's turn to the next chapter and examine power amplifiers themselves!

Notes

1 It is worth noting that promoters *always* want it twice as loud, except when they think it is *way too loud* (which generally only happens on jazz or classical events).
2 Defined as a component in an electrical circuit that consumes power. In this case, the load is the speaker or combination of speakers being powered by the given amp channel.
3 This does not address 70V transformer-tap systems, which we will discuss later in the chapter.
4 We'll talk more about "blown" speakers in Chapter 16.
5 In practice, this example is more likely to be handled by a 70V system, which we will discuss a bit later.
6 And 100V systems, which work identically to 70V speakers, but at a different voltage.

167

CHAPTER 15

Power Amplifiers

Plus a Note on Power Conditioners, Uninterruptible Power Supplies, and Distros

Power amplifiers are an utterly essential, and (apologies to amplifier designers) often boring, piece of equipment. That is, *good* power amplifiers are boring, because they do their jobs, alter the sound traveling through them very little except for level, and function endlessly without much attention. A *badly designed*, or poorly chosen, power amplifier can make your life a living hell, though, so it is definitely important to understand these components fully in order to design a robust sound system. Figure 15.1 shows the standard block diagram symbol for power amplifier.

DEFINITION

There are only two basic types of power amplifier that we will encounter (though, as with any component, there are seemingly endless variations on the types).

1. External—for use with passive speakers, these are typically rack-mounted units designed to send an unbalanced speaker-level signal to speakers.
2. Internal—these are amps built into speaker enclosures, and when paired with the speaker, they make up a so-called active or self-powered speaker system.

Additionally, there are two important variations to consider (each of which can be found in either external or internal amp systems):

1. Analog—these amps take in an analog line-level signal and output analog speaker-level signal.
2. Digital—these amps take in a digital audio signal and convert it to send an analog signal to speakers; often, these amps have some form of internal computer control, whether for amp monitoring via software, or for minimal internal DSP processing, or some combination of the two.

Power Amplifier
A device that takes input signals at line level and boosts them to speaker level.

CLASSIFICATION

Power amps actually are sorted by literal amplifier classes, A-H, which we will address here.

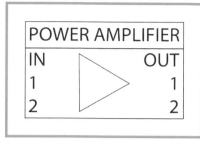

FIGURE 15.1
Block Diagram Symbol for Amplifier[1]

- Class A: these devices are also known as "always on" amplifiers. This means that all parts of the signal chain are producing 100% of the signal at all times, all 360° of phase. The circuits and components internally never switch off. What this means is that the audio quality is the highest possible, because the device is always ready for changes, new transients, ramp-ups in level, etc. It also means that the device uses more power than any of the other amplifier classes. As such, it is not terribly common to find Class A power amplifiers, particularly in a live sound setting where the power consumed can be enormous. It is VERY common (and desirable) to find Class A electronics in boutique external preamps and the like.

- Class B: these devices cut power use in half by having the signal components inside produce only 180° of the given signal at any given time, trading off which 180° is being handled as the signal passes through the unit. While this does reduce power use quite a bit, the crossover between one component handling the first 180° of a signal's rotational phase and the next component handling the opposing 180° is rough. The resultant *crossover distortion* means that this is not a terribly good-sounding amp topology. As such, pure Class B amplifiers are also fairly rare in production systems. These devices are still "always on," but because of the push-pull action between internal components, the device isn't 100% *"always on"*, as Class A would be.

- Class A/B: these devices make use of the concept of Class B amplification but extend each component's coverage to between 181° and 200° of the signal's phase. Thus, the crossover region is being handled by each component as it hands signal off, which makes for far fewer problems in the crossover region. A great many professional power amplifiers are Class A/B, as they produce cleaner, clearer sound than Class B, but still offer substantial power use savings over pure Class A. These devices are still "always on," but because of the push-pull action between internal components, the device isn't 100% *"always on"*, as Class A would be.

- Class C: *these are not used in audio applications.*

- Class D: these devices are not "always on." In Class D amps, control voltages (or digital control algorithms) are utilized to switch off powered components when they are not needed. As such, these are by far the most efficient amps in terms of electricity usage, but they often risk weak transient response (if the device has switched off internally, and a sudden sharp transient signal comes through, the rate at which the amp can respond—known as *rise time*—may not be fast enough to represent the transient accurately). In addition, because they use far less power, there is less need for *heat sinks* (physical escape paths for heat built into the amplifier casing), which means that Class D amps are often substantially lighter weight than their higher-quality compatriots in Class A/B. There are a large number of professional power amps available in Class D, but they are to be used where weight and power are the primary concerns, not where sound quality is of paramount importance.

- Classes G/H: these devices are modified versions of the Class A/B paradigm. In Class G, the addition of voltage rail switching, and in Class H, of rail modulation, allows them to operate at still lower power than standard Class A/B devices, while providing higher audio quality than Class D amps. Some amps will actually list their class as "A/B—H," to explain that these are sub-classes. G and H amps usually utilize more signal components than straight A/B amps. Where an A/B amp can get away with one pair of components along the path to enact its switching, G and H amps often have four or more multiple pairs of components, and the complex switching patterns among them allow for the lower overall power use. Because of the addition of extra components, these devices are often more expensive than straight A/B amps, however, because of the complex combination of signal paths, oddly, these can be designed to be smaller than standard A/Bs and as such are found in a fair number of small active speakers.

SPECIFICATION

A power amplifier features a host of relevant specs to understand. Please note that the following are written in reference to external amps (for passive speakers) and the options available therein, but internal amps feature all of these specs, it's just that if you want a given active speaker, you have no choice about the amp. Amp specs (especially things like rise time, which relates to transient response, particularly at high frequencies; see the following) should still be on your radar when you choose an active speaker system.

- Channels: amps come in a variety of channel counts. Most common is the stereo (or two-channel, if running two mono signals) amplifier. That said, four-channel and eight-channel amps are widely available as well. Rarely, you will find six-channel amps.
- Power per channel: expressed in watts. A good amplifier manufacturer will provide a range of specs in this category. Most amplifiers can be run at more than one impedance. An amp that can power an 8Ω speaker as well as a 4Ω speaker will generally deliver higher power to the 4Ω speaker than the 8Ω, simply because lower impedance means less push-back on what the amp is trying to deliver. Additionally, some amp specs will show power with a broadband 20 Hz–20 kHz signal as well as with a 1 kHz signal. Why is the latter relevant? If you are designing a system meant primarily for human speech (e.g., a lecture system), 1 kHz power will be closer to what you can actually produce in your system. However, for any full-range use (music, theater, film, etc.), the 20–20 measurement is more useful. Wattage is generally listed in continuous/long-term values for power amps but check what your specific manufacturer is listing.
- Bridge mode: bridging is a mode that can be enabled on some amps that allows the user to combine two channels as one for greater power. In theory, this mode gives the user four times the wattage, but in practice, it is usually around three times the available power that is now usable for the bridged channel pair. This is typically used for subwoofers that need a lot of power. Impedance changes in bridged mode, and, in general, it is unwise to bridge a pair of channels on an amp that doesn't offer a specific "bridge mode" switch.
- Frequency response: should be at least 20 Hz to 20 kHz, with minimal variation or distortion. And +/− 1 dB is a decent spec, and less than 0.1% THD (Total Harmonic Distortion) is ideal.
- S/N ratio (signal to noise): expressed as a dB figure (e.g., 106 dB), this is the difference between signal strength and device self-noise in a channel powered by broadband signal at full line level. This can loosely be equated to the dynamic range of the device. If you are counting on full 24-bit signal recreation (which would have a dynamic range of up to 144 dB), an amp with 106 dB S/N ratio won't cut it (in practice, even a lot of "24-bit" gear doesn't actually perform at 144 dB of dynamic range, but that's a story for a different section).
- Input sensitivity: the voltage at the input that will cause the amp to reach full rated power. Usually listed in reference to one of the stock impedances the amp will handle (e.g., "1.2 V @ 8 Ohms"). Professional line-level signals are nominally 1.228 V, so ideally, your amp will reach full power at or near that value. If the input sensitivity is lower, and you plan to drive your system fairly loud, you risk overloading the inputs with a well-balanced console/DSP send. If the sensitivity is higher, you need to ensure your delivery device can send a signal that is properly rated to this input. A weak DSP or console signal to a high sensitivity amp means that you won't get as much effective gain from the amp as is theoretically possible, because you're not driving the input at full.
- Maximum voltage gain: listed in volts, this is how much a given input voltage can be increased by an amplifier. If your input sensitivity is well-matched to the sending device, the voltage gain is simply an expression of how much louder the signal will get. Voltage gain will scale with the input signal, which means that if you are not driving inputs at full, you won't get the full rated wattage, and the voltage gain figure is provided so you can plug these numbers into Ohm wheel math and figure what your wattage will be with a given input signal. This becomes more relevant in large installed systems for permanent use, but in production systems, we (honestly) usually gloss over this spec in favor of simpler wattage and impedance measurements.
- Impedance: if not already stated as part of the wattage spec (it usually is), the impedances that the amp can handle should be listed. Note that while some amps rate their channels down to 2 Ohms, we never want to actually configure a system that way; 2 Ohms is too close to 0 Ohms, and given the frequency-dependent nature of sound equipment impedance, we put our gear in peril if we configure channels to run at 2 Ohms.
- Class: this is discussed earlier in the "Classification" section of this chapter.
- Damping factor: a passive loudspeaker has a mechanical resonance peak at the frequency whose wavelength equals the diameter of the cone. So, for an 8" driver, that resonant peak is at about Hz, the frequency with an 8" wavelength. If powered with *no* damping, when the speaker hits its mechanical resonance peak by being excited by signal at that wavelength/frequency, it will overproduce that frequency. That over-resonance thus produces a higher electrical impedance at that frequency, because the driver is operating on its own, and is

thus generating magnetic/electrical variations of its own back down the speaker line. Damping factor is the measure of an amp's ability to sense that backflow current and keep it from interfering with the speaker's performance. By controlling that backflow, we can ensure that a speaker driver delivers a flat response, even at its resonant peak. This number is a ratio between load impedance and output impedance, but we don't usually need to perform this calculation ourselves. Instead, a guide: a damping factor of 200 is considered the minimum value for professional amplifier performance, with 500 or greater being desirable. The higher this number is, the more controlled the coupling of amp and speaker (particularly at low frequencies), and thus, the better for us.

- Rise time: The time it takes for an amp to rise from 10% power to 90% power when driven by changing input. This is another figure that has to do with transient response. This figure is measured in microseconds, typically. The lower the number, the better.
- Slew rate: measured in volts per microsecond, this tells us how much power the amp can rise to in one microsecond. This is particularly relevant to high-frequency performance. In short, slew is the ability of an amp to power a given signal quickly enough. Generally, a slew rate of 40 V/μsec is a minimum standard.[2]
- Input impedance: much like a console's inputs, amp input impedance is very high, so that it can be shut off by turning down amp level controls. This is not a spec that generally informs amp selection.
- Seventy-volt/100V taps: as described in the previous chapter, 70V and 100V systems can save a lot of hassle for large, monophonic, distributed systems. If your amp offers 70V or 100V taps, they will be listed in the specs, with the associated wattages available. A typical 70V spec might read: "70V taps: 10 W, 5 W, 2 W, 1 W." This means that you can wire 70V speakers (with their built-in transformers) to this amp channel, set to any of these values, and using the 10-log equation you can figure out how much SPL you will get at that setting.
- Other options:
  - Cooling: what kind(s) of fan are built in? What triggers them (are they heat-sensitive or on all the time)?
  - Class D level threshold: some amps that shut off power when the signal is absent also shut it off when the signal is very low. While this might make sense in a corporate boardroom system where a tiny trickle of a signal is not likely an intentional signal, this is murder on, say, theater sound design, which is known for playing really quiet subtle sounds.
  - Control: does this amp feature computer control? Via what kind of connection?
  - DSP: is this a digital amp with built in DSP?
  - Connector types: this is self-explanatory.
  - RU: almost all professional amps are 19" wide for rack mounting, but how tall in RU is the device?
  - Depth: How deep is the amp physically? Where are the heat sinks? Racks come in different depths, and some amps are quite deep—make sure yours will fit into your racks.

CONSIDERATION

What questions must a designer ask (and answer) in order to select power amplifiers as part of their system?

1. How many channels of amplification are needed? If this is a complex theatrical system, with 60 speakers, each of which needs to be individually addressed from the playback system, you will need 60 amp channels in order to keep those signals discrete. If this is a supermarket, which might also need 60 speakers, but which is sending the same mono signal to all of those speakers, with no concern for time-alignment, you might get away with a single amp in 70V mode. This must be assessed based on the needs of the system itself.
2. How much power can my speakers handle? What is their impedance range? Does it feature bridging? 70V/100V taps?
3. What class of amplifier performance is suitable for this task? Does it need to be high-fidelity Class A electronics (as we might find in a high-end post-production house), or can it be power-saving Class D bashers (as we might find in a punk bar)? Or somewhere in between?
4. Do we need any extra features? Computer control/status monitoring? Built-in DSP?
5. How much heat does the device give off? This is a tricky one. Heat is not always listed in the specs, but it is measured in BTU (British Thermal Units) per hour, and for permanent installations (or any large installation indoors, even temporary), this is something you must consider. Amps give off a ton of waste heat. We can find out how much by taking into account how much wattage we will use. 1 Watt of power

is equal to 3.412142 BTU/hr. Any additional watts are a straight multiplication of that figure. So, for example, an amp run consistently at 350 W is giving off approximately 1,194 BTU/hr. This calculation will be used to figure out how much HVAC (particularly the AC part) is needed in your amplifier room (known through many parts of the industry as an EER, or *electronic equipment room*, which houses amps, DSPs, control computers, etc.). If you don't properly cool your gear, it will overheat and die. No one wants this.

6. How many electrical circuits are needed for your amps? A single amplifier might draw 7.1 amperes at full power, but for most calculations, we assume one-eighth power measured with pink noise, as this is roughly equivalent to music usage. However, that doesn't mean we just divide 7.1 by 8. It's best to seek actual measured current draw ratings provided by the manufacturers (which many release). As we know, US circuits provide 15 Amps of current. If you have an amp that draws 7.1 amps at max continuous power, you don't want to plug more than one of them into a single power circuit because peaks are just around the corner. While running at one-eighth power (2.8 amps draw),[3] we would assume we can stack five amplifiers safely on this circuit, we don't know when we'll drive into overload, so best practices demand that you provide power that would allow all amplifiers to hit peak draw all at once without tripping the circuit breakers. This becomes especially relevant in new installation projects, where you are often responsible for telling the electrical engineers how much power you need in your equipment room, but is also very important in live entertainment with outdoor/temporary setups, where you need to ensure enough power circuits can be fed from the generator or venue to deliver your desired amplifier performance.

7. Do you need more RU than the space literally taken up by amps? Because of varying heat sink design, some amplifiers state in their operating manuals that they run best with an empty RU (or two) above and below them in the rack. You want to observe this suggestion; otherwise, you can easily overheat and damage your amps. This may mean you need larger equipment racks than you were initially planning, but pay close attention, as this is important.

Power amplifiers will never be the most *scintillating* part of a system—for my money, the most exciting part is our next chapter, on speakers—but understanding how amps work and how to choose them can mean the difference between an anemic, distorted system, and a crisp, clear, powerful one.

Before moving on. . .

A NOTE ON POWER CONDITIONERS, UNINTERRUPTIBLE POWER SUPPLIES, AND DISTROS

Power conditioners are devices that are generally rack-mounted and feature—at the most basic—a row of electrical sockets and a power cord. As such, to many first-time designers, they seem like glorified power strips, and young designers often leave them out of their system plans. A decent AC power conditioner[4] connects to the wall socket and uses its own active electronics to smooth the current coming from the wall, stabilize it (such that any surges will be prevented from reaching gear, and current served to attached gear will be within safe operating range), and help to reduce electrical noise within the audio system. Whenever possible, always include power conditioners in your systems—every device in a system should ideally be plugged into a power conditioner, not directly into the wall (or, worse, into a dime store power strip, which probably won't even trip as a surge protector). There are several reputable manufacturers, Furman and Tripp-Lite being two of the most common.

Tales From the Field

We did a tuning for a system outside of Pittsburgh. and solved an amazing amount of problems and were at the point of signing off on the system. And all of a sudden, this huge hum comes on through the system. The [general] contractors tied their refrigeration system ground with the audio ground—it ends up being a huge problem. Value engineering that goes on, you see when people ignore improving the room or giving you the right positions, and they maybe value engineer out that the treatment of the room to make it a place to listen to stuff. I've never seen that *not* come back to haunt you.—Jamie Anderson

Uninterruptible power supplies are basically large-scale battery backups, designed to prevent critical systems from shutting off immediately in the middle of operations if the wall power goes out. Depending on the size, scale, and cost of your UPSes, these may provide 5 min of backup, all the way up to several hours of backup. Whenever possible, we also want our systems powered via UPSes, such that if the power goes out, we have *at least* protected mission-critical gear. We would plug a device like a console into the UPS, which would then plug into the conditioner, and then to the wall (though some UPSes also act as power conditioners). At very least, we would want any life-safety or emergency announcement systems in our system connected to a UPS, recording equipment in studios, and any other particularly important or fragile gear (desktop computers without internal batteries, for example). Ideally, your whole system is run through UPSes, so we can just safely shut off the equipment after directing audiences to do whatever is necessary in an emergency, but when budgets are tight, prioritize the most important gear. In the event of power loss, some UPSes automatically sound an annoying loud klaxon, and some flash lights in addition to or in place of the noise. Make sure, if you are using UPS devices for a sound environment, that you can either disable the klaxon or that there is no klaxon and only warning lights. There is nothing slick about being able to save the take your band was playing if the take is interrupted by a loud blaring alarm.

It is beyond the scope of this book to go too far into further power system design details, but there is one other common power consideration worth mentioning: *power distro*. Frequently, performance venues, or outdoor generators for concert events, will not deliver power directly in usable 120-V/15-A form. Often, power will enter a building or come from a generator in what is known as *three-phase*. Three-phase power is high voltage and sends current over three "hot legs," all referencing one common neutral and one common ground. Typically, each leg is 120° out of phase with the others. Such power needs to be distributed, so we use devices and systems simply called *distro*. Such devices take in the high-voltage electricity (generally via twist-locking connectors called *camlocks*) and split it into signals we can use—120 V/15 A, or 30 A, or 40 A signals (in the United States, of course overseas these numbers are a bit different). What kind of distro you need depends on how many amplifiers you are powering, what their current draw will be, how hard you will be driving them, and so on. It is worth being familiar with the concept of these devices, as they will recur throughout our work.

Now, onto arguably the *most* important part of any sound system—the speakers!

Notes

1 Note that this very symbol is the cause of controversy. Many designers have abandoned the triangle all together in block diagram drawing, finding it to be useless. However, a small group of designers are *very* committed to keeping it as an identifier of amplifier power. I have no horse in this race and have added the triangle, or left it out, on various projects as the project demanded.

2 It is beyond the scope of this text to get into detail about slew rate and how it works. In practical reality, you just want the highest voltage per microsecond possible.

3 This value (7.1 vs. 2.8) based on specs from a QSC amplifier.

4 "Power" here is a misnomer, as we're really dealing with electric current, rather than "wattage."

Speakers

How important are speakers to the success of a sound system?

Really, think of your answer. I'll wait.

OK—if you answered anything other than THE MOST IMPORTANT, you are off-base. As this book is hopefully detailing, a sound system is made of many component parts, all of which are important to consider, and all of which perform their roles. However, if you have not properly considered your speaker coverage needs, and selected speakers that will meet those needs and deliver the appropriate content at the appropriate level to the appropriate section of audience—in other words, if you chose the wrong speakers for the job—your system will not succeed. You can have the nicest mixing console, the most advanced DSP, and thousands upon thousands of dollars' worth of boutique mics and preamps, but if they are fed through inadequate or inappropriate speakers, all that money and effort will be at best frustrated and at worst totally obscured—worthless.[1]

It has become a truism that when selecting loudspeakers, unlike the process of selecting mics, we generally want to select speakers that will give us the flattest, most "reference-quality" response possible. This is true a vast majority of the time (and is *certainly* true any time we're designing a speaker system that live mics will be fed through in real

Message in a Bottle

Speaker positions mean as much as speaker choices. It's one of the things where your ability to get a job done comes down to getting speakers in the right position, and sometimes you're just completely hampered by [the fact that] there isn't a speaker that's actually gonna get the job done for you since you weren't given a place to *put* those speakers to get the job done. Far too often, people spend their time working out the drive of the system, all of that stuff, and don't necessarily find the place to put stuff to sound good. That's an insidious thing.—Jamie Anderson

time), but there are exceptions, particularly when selecting *specials* (individual speakers for coverage of particular areas, or for reproduction of limited and particular effects—like a speaker built into a refrigerator prop, meant to reproduce the sound of the compressor and chiller) that are associated with particular effects. In this chapter, we'll operate with the standing assumption that we are attempting to create a speaker coverage system that is capable of flat frequency delivery and even coverage wherever possible and will note any exceptions to that rule as we go.

DEFINITION

Loudspeakers (or "speakers" for those not tied to the pedantry of the industry) are generally complex devices that are made up of several components. While occasionally (particularly in theater) we will use so-called *raw drivers* as the speakers for a given application, generally, speakers are a bit more complex than that. Speakers are most typically made of the following:

> **Loudspeaker**
> A transducer that converts electronic signal to acoustic sound.

- Drivers (Figure 16.1): speaker drivers are the cone, coil, and magnet assembly that vibrates and produces sound.[2] Many speakers contain more than one driver, often of different sizes, intended to produce different frequency ranges from one another. High frequency (HF) drivers are colloquially referred to as *tweeters*. Low-frequency (LF) drivers are often referred to as *woofers*. Ultra-low drivers (meant to only reproduce very low bass frequencies) are referred to as *subwoofers*. Midrange drivers don't have a cute nickname and are usually called *mids* or *midrange drivers*.
- Drivers consist of the following:
  - Cone: the vibrating membrane that actually generates the acoustic pressure waves we hear emanating from the speakers.
  - Voice coil: a wound metallic coil, coupled to the inside diameter of the speaker (via a shaping piece, called the former, which prevents the coil from warping). This coil, when energized with audio signal, vibrates in and out of its magnetized home position, and thus moves the cone, which generates sound.
  - Spider: the accordion-folded material that helps hold the cone and coil in lateral place, and also allows them to flex and move as needed to generate sound.
  - Magnet: the permanently charged magnet between whose poles the voice coil floats. Like in a dynamic mic, there is a pole piece run through the center of the coil (polarized one direction), and a surrounding ring magnet (the opposite pole).
  - Basket: the metal frame that holds the whole driver assembly in place and allows it to attach to an enclosure.
  - Surround: the ring of connective material (usually a form of plastic) that connects the wide outer diameter of the cone to the basket, holding it in place, also flexible (like the spider), but not accordion-folded.
  - Dust cap/phase plug: the central cover, whose primary role is to prevent dust and other particulate material from entering the driver assembly. In some speaker drivers, the central dust cap may serve a dual role (and may be shaped specifically for the secondary role), as both preventer of particle entry and as a phase plug, which is a piece that helps guide waves (particularly HF waves) out toward the listener rather than allowing them to cause destructive interference close to the driver, before sound has reached the listener.
  - Terminals: these are posts for wire connection.

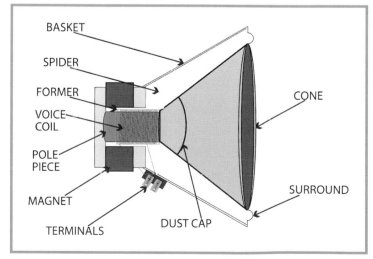

FIGURE 16.1
Speaker Driver Diagram (With Parts Labeled)

■ Enclosure: The "box" in which the drivers are situated. Speaker enclosures are often referred to as *baffles*. These come in a wide variety of shapes, sizes, and materials (including absorption inside the cabinet). Some enclosures are designed to deaden the backflow of sound from the rearward motion of the driver, some to tune it to enhance bass response, and some to fall somewhere in between. The range of enclosure varieties is beyond the scope of this book to cover in full, but we will here outline the most common types.

> **Baffle**
>
> An object designed to reduce airborne sound. In the context of speakers, a baffle refers to the enclosure in which drivers are situated, which is intended to help prevent out-of-phase signals generated by the open rear of drivers from combining with the on-axis front sound and causing destructive interference.

● Sealed/infinite baffle: a sealed box, also sometimes called an infinite baffle,[3] is an enclosure with no holes, ports, or vents to the outside air. This is the best design for time coherence in a signal, as the sound given off by the rearward motion of the driver is typically absorbed before we (as listeners) can hear it outside the box.

In order to properly understand the benefit of a sealed box, let's imagine a raw driver hanging in open air (Figure 16.2).

In this example, our in-phase sound is generated from the front of the driver, but as soon as the first cycle rarefacts, the cone pushes the opposite direction and generates a half-cycle, out-of-phase signal off its backside. In this scenario, it is very easy for these two signals to combine in acoustic space, causing destructive interference and phase blurring. For high-performance audio, this is far from desirable, so we must place the driver in a baffle—an obstacle which will prevent that sound coming off the back of the driver from interfering with the front.

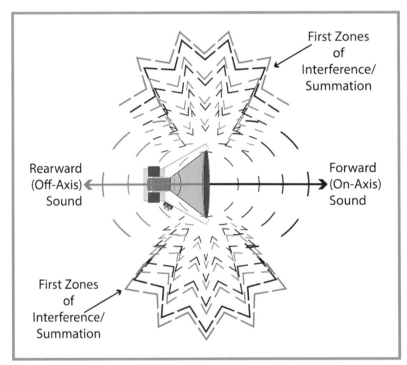

FIGURE 16.2
Raw Driver With Sound Waves Generated by Both Directions of Motion

177

Now, in a theoretical world, the perfect baffle would be an infinite wall (Figure 16.3).

In Figure 16.3, the wall in which that driver is set goes on in all directions for infinity. As such, the waves generated by the out-of-phase rarefaction motions can't ever recombine with the front waves, and the driver's signal is clean and clear. Of course, this assumes that our infinite wall is acoustically dead and doesn't itself become a sympathetic vibrating membrane—which is a big assumption, as the choice of baffle material will greatly impact the acoustics of that speaker's delivery.

At any rate, we want to prevent rear-facing interference, but the practicality of setting a driver in an infinite wall is, obviously, nil. So how do we get around this? We fold our infinite baffle in on itself until we've got a (usually) cuboid box with no openings, with a driver (or multiple drivers) set into the face.

As such, this is what we refer to either as a sealed box or an infinite baffle enclosure (Figure 16.4).

These speakers often have very tight and controlled phase response, particularly if the interior absorption (used to prevent or reduce resonance and standing waves in the box itself) is well-planned.

- Ported/vented box (also known as "bass reflex" enclosures, see Figure 16.5): these enclosures have holes in them. Simple enough, right? But, if we have established that one of our primary goals in designing an enclosure is to prevent interference from the rear-facing side of our drivers, why would we then open our boxes back up, theoretically allowing such interference? Well . . . the answer generally is MORE BASS.[4] If you hadn't guessed from the fact that the other name for ported or vented boxes is "bass reflex" boxes, it should be evident now—the reason for adding a port or a vent to a speaker enclosure is to accentuate bass response by providing a tuned resonant opening that adds to the sound. This is most commonly found in one of two speaker types—subwoofers—and in small enclosures that otherwise would have inadequate bass response (such as very small nearfield mixing monitors).

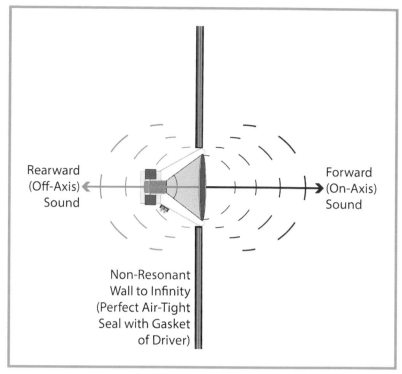

FIGURE 16.3
Driver Set Into Infinite Wall

While the terms are sometimes used interchangeably, technically, a "ported" speaker is one with a round opening, while a "vented" box is one with an opening shaped like a slot. These openings are tuned via the principles of Helmholtz resonance (the same principle that allows you to tune the note you hear when blowing across the open mouth of a Coke bottle by changing the amount of liquid inside the volume) to ensure that they are resonant at frequencies that add to the speaker's performance in a controlled and desired way. Open enclosures of this type do evince poorer phase performance, particularly in the

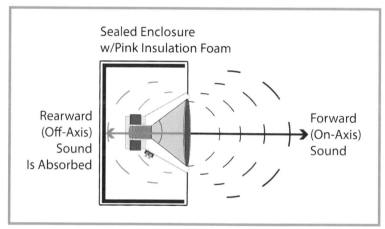

FIGURE 16.4
An "Infinite Baffle," or Sealed Enclosure

low frequencies the ports/vents are designed to accentuate. However, the human hearing system has a much harder time identifying phase issues in low frequencies than it does in high frequencies, so the trade-off is frequently deemed worth making in exchange for the additional bass response.

- Passive radiator: an enclosure type in which there is a primary active driver and a secondary passive driver that is not wired to amplifier power, but merely exists to be vibrated by the pressure waves generated by the primary driver. These enclosures can increase bass response by virtue of the array effect,[5] with the benefit of not requiring a hole in the box. If properly designed, this type of enclosure system can both improve upon (though not eliminate) the phase issues inherent in ported/vented designs, and also can avoid another problem some poorly designed ported boxes face, which is a kind of "whistling" of air through the opening at high SPL (also referred to as "port chuffing").[6] However, there is much calibration needed in order to ensure that the passive radiator resonates as desired, and the added cost of the extra driver is sometimes prohibitive.

This is not a terribly common design type in live production speakers but will still be found on occasion.

- Dipole: these enclosures don't really close. The essential idea is to mount a driver into a fascia of limited dimension, thereby reducing (though not eliminating) front-back interference and radiating sound both forward and backward. True dipoles are sometimes used in theatrical sound effects systems, especially in smaller venues where space is limited and splashing sound in two directions may be desirable. Most ribbon-driver speakers (like most ribbon mics) are naturally dipoles and as such are used for locations where a speaker needs to be concealed in a thin space.[7] The most common speakers sold as "dipoles" are not true dipoles but are dual-sided speakers made to be used as extremely wide-coverage cinema surrounds. In Figure 16.6, we see a version of this kind of "dipole," which does not throw sound literally in two 180° opposed directions but does feature two main axes of coverage.

- Horn-loaded (Figure 16.7): speaker horns are among the most iconic images associated with sound production. Most often used with HF drivers (though also found in specific shapes with LF drivers), a horn is, very simply, a *waveguide*. This is a physical object that induces acoustic sound to travel along a specific path. In the case of the horn, the mechanism is quite easily understood.

The driver, often a dome-shaped tweeter, is situated in a frame called a *compression chamber*. This can generally be equated to a sort of sealed basket, but there are many designs. At any rate, the chamber helps focus the sound forward. The driver is coupled to the narrow end of the horn, which is referred to as the *throat* of the horn. The throat provides an *acoustic impedance build-up*. What this means is that due to the narrow opening, there is a higher air pressure standing in the throat than on either side (the throat itself being inflexible). As the driver pushes sound into the throat, that impedance creates an obstacle requiring added

FIGURE 16.5
(Left) A Ported Enclosure—Focal Trio 6 BE; (Right) A Vented Enclosure—Focal Solo 6 BE
Source: All Courtesy Focal/Audio Plus Services

FIGURE 16.6
Klipsch RP-250-S
Source: Courtesy Klipsch Group, Inc.

force to overcome. That added force makes the sound that travels through to the other side more likely to flare out along the waveguide's sides, spreading evenly before exiting the horn at its wide *mouth*.

If you have a hard time understanding why the throat helps the horn action, think of trying to open a car door that is frozen stuck in the winter, as opposed to opening it in warm weather.[8] In warm weather, if you open the car door, it usually will swing easily open. If you are carrying groceries, you might open it and let it swing free while your hand travels away from the door handle, so you don't hit your bag of groceries. However, in the winter, if your door freezes closed, it's a different story. If you pull and pull, but fail to open it, you may start pulling

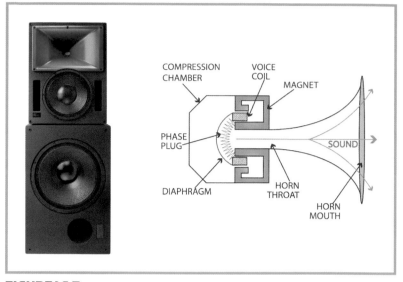

FIGURE 16.7
(Left) Meyer Bluehorn System (Courtesy Meyer Sound); (Right) Side Section Diagram of Horn-Loaded Compression Driver (With Parts Labeled)

harder each time. When the door finally pops open, the chance that your arm will swing out with the handle, tracing the arc of the door's swivel, is great—the extra force you needed to overcome the obstacle of the frozen doorway means that when it gives, your arm (the sound) travels quickly and without resistance along the given guided path.

This is the basic principle of the waveguide. By pushing sound through the obstacle, the lower pressure past the obstacle is effectively a void for the sound to fill, and because pressures want to equalize themselves (remember the balloon from the start of the book?), sound follows the shape of the horn quite naturally. It is very common for speakers to have one driver with a horn (usually an HF driver) and one or more without.

- Transmission line: a transmission line enclosure is one that incorporates elaborate maze-like chambers and a great deal of absorptive materials, with the ultimate goal of absorbing *all* rear-driver radiation of sound. In a so-called perfect transmission line, the mazing behind the driver is infinite and thus can fully attenuate the rear-facing sound (when sound energy is absorbed, it is mostly transformed to heat/friction, which then dissipates in air inaudibly). Like the infinite baffle, the infinite transmission line is not a practical reality, so in reality, the enclosures are designed with enough of a line to reduce the sound considerably. Care must be taken in these enclosures to not overfill the transmission line with absorptive material, as this can create a higher pressure than is generated by the driver and actually induce acoustic backflow out across the driver, which also results in phase/timing distortion.
- Crossover: in any speaker that features more than one driver, especially where those drivers are designed to handle different frequency ranges, it is common to find a crossover.

Crossover

An electronic device that takes a single audio signal and divides it into two or more output signals, split by frequency range, which are then sent to speakers or drivers designed to handle those specific frequency ranges. Crossovers are found in speakers, where they are used to send one signal split into high and low-frequency bands to tweeters and woofers (for example) and in DSP programming, where they are used to send one signal split into high/mid frequencies and low frequencies to top speakers and subwoofers (for example).

Speaker crossovers in passive speakers are designed to handle speaker-level amplified signal and split the signal into two or more frequency bands so that no driver is sent frequencies it is not designed to handle. Speakers very commonly are designed with two-way crossovers (split into highs and everything else) but can be three-way or higher (some mastering studios, for example, use five-way speakers). The electronics design of a crossover can have a huge impact on the sound of a speaker. It is beyond the scope of this book to go into crossover design, but there are some key points to understand.

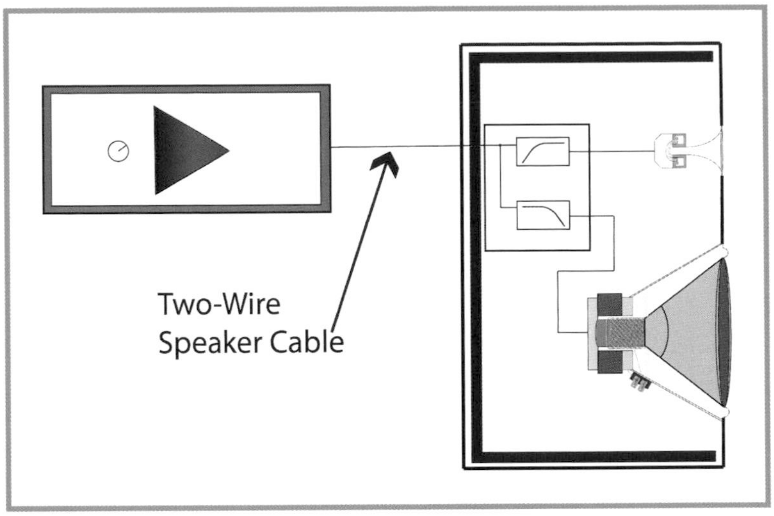

Let's imagine a two-way speaker (Figure 16.8).

FIGURE 16.8
Audio Paths in a Two-Way Passive Speaker

In Figure 16.8, we can see that our full-range audio is sent from amplifier to our speaker, and the crossover in the back of the enclosure splits the audio to our tweeter and our woofer. Now, if we are powering this speaker with a single amp channel, the limit to how much power we can send (and thus how loud the speaker will get) is set by the tweeter. The tweeter, being smaller and more mechanically fragile, will naturally distort at a lower wattage input signal than will the woofer. This means that even if your woofer is able to handle 375 W of continuous power, if your tweeter distorts after 100 W of input, you are effectively limited to 100 W of power on the whole speaker. This is, perhaps obviously, somewhat less than desirable. Even with sophisticated crossover design that can pad down some of the power level sent to the tweeter, this imbalance often remains.

So how do we fix this problem? We introduce a technique called *bi-amping* the speaker.

Bi-Amped Speaker

A two-way speaker which can be powered by two amp channels: one for the HF driver(s) and one for the mid/low driver(s).

Some speakers are designed to be bi-amped (Figure 16.9), which means that (usually via the flip of a switch on the enclosure's back panel) we can switch from powering the speaker via one amp channel for all drivers, to powering each crossed-over zone with its own amplifier channel (in the case of bi-amping, this is two zones, in a tri-amped speaker it would be three zones and amp channels, etc.).

The advantage of this technique is that by sending a dedicated amp channel to the high drivers and another to the other frequency bands, each can be powered to their full potential. This means we can now get a full 375 W out of that woofer and 100 W out of the tweeter without overload distortion.

If this is such a useful technique, why don't we do it all the time? Because it costs more. Both in the speaker design and manufacture, and also because it effectively doubles your amplifier and cable count, which can really add up. That said, it is a very common technique in systems that demand high-fidelity audio, as driving each set of frequency bands independently is more electrically efficient and cleaner overall.

Many self-powered—aka active—speakers (which have the amplifiers built in) are bi- or multi-amped speakers, but we don't think about them that way since we're now sending line-level signal to them and all the amplification/internal crossing over is performed out of our sight.

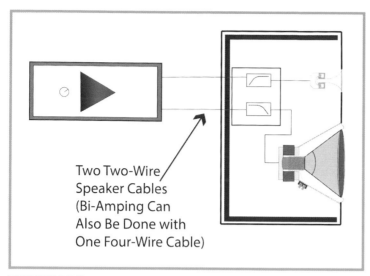

Two Two-Wire Speaker Cables (Bi-Amping Can Also Be Done with One Four-Wire Cable)

FIGURE 16.9
Audio Paths in a Bi-Amped Two-Way Passive Speaker

Before we move into "Classification," I want to offer a brief note on the destruction of speakers through "regular use." Any speaker that is run loud night after night, year after year, will become what we refer to in the industry as "tired." A tired speaker does not evince the same crisp transient response it once did, and as a result, the phase response (particularly in the HF range) suffers. This is because the surrounds, spider, and even the cone itself are just physically worn out. A good re-coning can bring such a speaker back to life. However, sometimes, in using speakers, we destroy speakers. Our goal, of course, is to never do this (see Chapters 12 and 13 on limiters in DSPs for more info on how to prevent it). That said, there will come a time in every sound person's life when they either cause or come across a speaker that has "blown."

A speaker can be "blown" in two distinct fashions. The first is the traditional tearing of the cone. Usually, this happens if a speaker has been temporarily fed peak levels far above what it can handle, causing the speaker to travel farther than it was designed to in its linear compression-rarefaction path (which is called *excursion*). If we exceed the designed excursion limits of the driver, the cone will tear, often directly at the surrounds, and come unmoored from its linear travel path. If the cone still remains connected to the voice coil, it will still produce sound, but the sound will be jagged, harsh, distorted, and overall quite unusable in any professional setting. This speaker can be re-coned and brought back to life. Sometimes, the voice coil will push through the speaker cone in the center, leaving the cone attached to the surround but not to the drive engine. The speaker will then evince a kind of low-level buzz or hum as the voice coil keeps moving but the cone produces nothing. Again, a good re-coning will fix this.

The second way we destroy speakers is through heat death. If we push the speaker, not against its peak limit, but against its continuous power handling limit—push it too hot, for too many hours—we can overheat the driver, which literally will melt the voice coil. The coil will fuse in place in the magnetic gap and stop moving altogether. This produces no sound, provides infinite resistance (which is pushed back through the circuit to the amplifier which also then starts to overheat), and means that the driver is toast. It's no longer useful. A driver whose coil has fused can't be saved, it must be replaced.

All of this is to say—set your limiters properly! Don't blow your speakers! They cost too much money!

CLASSIFICATION

There are many ways we classify loudspeakers. Because sound people have such intimate relationships with speakers, they often describe speakers in terms that are meaningful for their particular applications but may not apply to all design disciplines.[9] In this section, we attempt to address the major classification types that are applicable to the majority of professional loudspeakers.

Speakers are classified in three primary fashions:

1. Power type: active or passive
2. Form factor: shape, size, mounting type
3. Intended use: point-source, line-source, column-source, subwoofer, stage monitor, practical, etc.

Power type: this is a fairly straightforward determination—does your speaker come with a built-in amplifier? Or does it require an external amplifier? An active speaker comes with built-in amp, a passive speaker does not. Why would we choose one over another?

Let's consider an EAW JF60z speaker (Figure 16.10), a fairly common unit.

This is a passive speaker. I've chosen this as our example because, while EAW makes a range of speakers, the only amps it makes are for its own range of active speakers. Thus, if you want to power a passive EAW speaker, you will be selecting an amp by another manufacturer. On the one hand, this gives us enormous flexibility in design planning, as we can choose any amp we like, we can match impedance and power handling from amp to speaker in any ratio we choose, and we can (if necessary) use already in-stock amps with this speaker. However, there are some drawbacks. Speakers are notorious as the least power-efficient units in any given entertainment system, and particularly within a sound system. A *very efficient* speaker might use as much as 25% of the power sent to it to actually produce sound, the remaining being lost as waste heat (primarily). A speaker manufacturer recently introduced "the most efficient concert loudspeakers ever made," and they boasted that they made use of a whopping 45% of the power sent to them. Contrast this with LED lights (increasingly the choice for much stage lighting in concert design), which boasts upwards of 90% efficiency and approaches 100% in many units. Whenever we combine a passive speaker with a random amplifier, we are creating a relationship that will be inefficient. In addition, since power amplifiers have different damping factors, crest factors, and the like, we may not be getting the full performance from our JF60z with a given amp—the speaker might be capable of better performance than the amp we've paired it with is delivering.

FIGURE 16.10
EAW JF60z
Source: Courtesy EAW/Loud Audio

183

Now, an active speaker, such as the Meyer M1D (Figure 16.11).

The M1D will not only be more electrically efficient but also has been designed specifically to pair the damping factor, crest factor, and other amp specs precisely with the specs of the drivers and enclosure. This means that the active speaker will more likely sound crisper, cleaner, clearer, and more powerful (even at the same wattage used), because there is less muddying due to mismatched electronics.

If this is the issue, why don't we just match the external amp to the passive speaker based on ideal damping and crest factors, and all that other jazz? Primarily because it is nearly impossible to find information on passive speakers that would allow us to do so. Even if manufacturers gave us that info (e.g., "This speaker works best with a damping factor of 500–1,000."), there is no guarantee we could source an amp that met all the desired specs. Thus, it is easier to just use active speakers.

In addition, active speakers take line-level (instead of speaker-level) inputs. While we'll address balanced and unbalanced cables in Section 8 in detail, suffice it for now to say that line-level signals to active speakers are typically balanced, which are more noise resistant, and can run over very long distances, where unbalanced signals are more susceptible to outside noise.

If active speakers often sound better, why would we ever choose passive speakers if we have the

FIGURE 16.11
Meyer M1D
Source: Courtesy Meyer Sound

option to choose active? Active speakers are typically more expensive than passive. For the price, the EAW JF60z is a wonderful speaker, that I've specified many times, the M1D, (another of my favorites) is a good deal more expensive. Active speakers can't be housed in locations without ventilation, since that risks overheating the built-in amps (passive systems often house the amp racks in a dedicated equipment room with powerful HVAC systems). And active speakers are larger and heavier, since the amp takes up space. Sometimes we need to hide speakers in tight quarters, and in this case, passive speakers are almost always advantageous.

Form factor: the term *form factor* is often used in production systems of all kinds to refer generally to the shape and size of the items in use. This is a particularly important consideration for speakers. While there are many, many different shapes and sizes of speaker available, I will attempt here to break down *not* the different shapes and sizes, but the elements of shape and size that we frequently must consider when selecting a speaker. I am doing this for two reasons: 1) there are too many shapes and sizes to properly list, and 2) many of the classifications in our third item (intended use) tie in so closely with form factor that we will be addressing much about shape and size in that part of the chapter (which directly follows this one).

When planning loudspeaker form factor, here are some key elements to consider:

1. Size: how large is the enclosure you are thinking of using? Speakers take up space, and you need to know how much space you can allocate to speakers before selecting a unit. A given speaker might meet all the coverage angle, SPL delivery, and power requirements of your design, but if it is too large for the spot in which it must be mounted, you can't select it.

2. Mounting type: speakers mount in a wide variety of fashions. So-called line-array speakers[10] usually hang from a top bracket that itself can be suspended from 1, 2, 3, or more points, depending on the design of the particular bracket (aka "fly frame") and the available rigging points in the venue. Point-source speakers (single boxes meant to cover a broader area than the typical line-array box by itself) are often rigged with u-shaped brackets, attached to points on the sides of the enclosure, which then may be attached to a rigging pipe via a c-clamp, or bolted to a scenic structure, or mounted in other creative ways (Figure 16.12 shows fly- and u-bracket systems). Ceiling speakers typically mount to custom rings or plates set into the drywall or drop ceiling itself. Pendant speakers hang from either their speaker cables or long aircraft cables. These are only a few of the options available (in addition to which, for unique installations, designers often find themselves in need of custom mounting solutions, which they or the structural/scenic teams on the project may design from scratch). In some circumstances, mounting type can be flexible, but in many circumstances, it is very important to understand mounting type, as the speaker you want may not be able to be rigged in the position desired.

3. Weight: how heavy is your speaker? Will the mounting point at which you plan to install the speaker support that weight? This is a seemingly mundane, but terribly important, question.

4. Shape: this is both a practical and aesthetic question. Sure,

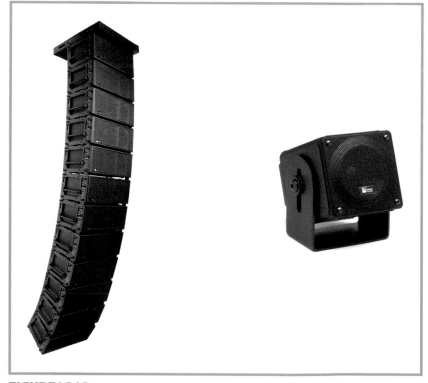

FIGURE 16.12
(Left) Meyer Lyon Array With Fly Bracket; (Right) Meyer MM4XP With U-Bracket
Source: Both Courtesy Meyer Sound

the eight-box line array might provide perfect coverage of your venue when hung as a center cluster, but if it blocks sightlines of the top of the designer's set, chances are the director is cutting your speakers or requiring them to move. This is related to size but is distinct in that two speakers may have similar dimensions (say, the enclosure height) but vary so much in depth/width as to make one a good choice and the other a poor choice for your application. A speaker with a convex face may not be desirable in a ceiling mount where the architects want the speaker to "disappear" visually, and a flat grilled speaker will do better in that case.

   Intended use: this may be the most important classification category. Different speakers are designed for (or are useful for, regardless of design intent) different applications. It is crucial that we know what task we need our speaker to accomplish before we select the unit that will serve that task. While, again, intended use-categories of speakers are much more numerous than we can go into here, the following is a list of the major use-categories.

- Point-source: typically, this is a single speaker, placed by itself (or at least spaced far apart from other units), intended to produce full-range (or close to full-range) sound over a broad area. Point-source speakers are often used as "main" speakers (the primary speakers in a given system) for small nightclubs where line arrays would be overkill, theaters where a single box is more easily concealed by scenery (or by shadow)

Tales From the Field

Love is the largest system I've done, because it has 16-line arrays, plus left, center, right for each seat. It's funny, I had no clue as to how many speakers there were. It's not like I sat down and said, "Look at all these speakers." I knew there were going to be three speakers on each seat and something like 2,000 seats, so that's a lot of speakers, but I'm only hearing three, so what does that mean? It doesn't mean anything. How many am I gonna hear? Why are they there? When it comes down to it, and I'm doing those kinds of designs, I'm not counting equipment. When I'm doing a production like that, it's obviously going to have a budget, so I'm working on what's needed for the production as opposed to trying to "one-up" myself, because that would be pretty sad. *Love* itself is such a unique show, because of the Beatles music and what it means to me—growing up in the same era as the Beatles and how important it was to me. They developed as I was developing as a person, so it became part of the soundtrack of my life, of my story. As soon as they broke up, it wasn't the same story, my story had to change. *Love* is a special thing, but no I don't think there's anything else with quite as many speakers, because it's in the round. *Michael Jackson One* and also *Ka* have quite a lot of speakers. *Ka* had only two speakers per seat, but a lot more seats. The thing is, students are freaking out about how many speakers there are, but it's funny. We did a press opening for *Love*, and someone came up to me, and there's a Cirque person there saying, "We've got a million miles of cable and the amp power is one trillion billion whatever." And I'm looking at the person who's saying all this thinking, really, is that so? We don't just turn that all on at once because we'd kill everyone. Why is this of interest? So this person is reeling out all this information that's totally useless. If I had known that at day one, [as a student], I would have shriveled up and died because that just seems ridiculous. I would have thought, why not just go back to two big bean cans and a piece of string because it'll be just as interesting. Different but just as interesting. A system is always a means to an end—you have to think about the beginning of why you're doing it and let it grow so it becomes something that is manageable, like your body, your wardrobe, anything else. These are things that grow and develop. Yes, it can go wrong, but it grows and develops. One of the reporters came up and said—what kind of format is this? Is this 5.1 or something like that? I went, "No. Do you know what 5.1 is?" He said no. I said, "Then why are you asking me that? Let me tell you what 5.1 is. It's 5 speakers that you're hearing around, left, right, rears, sub, and so on. So after this, I'm going to think about what I can hear sitting in the average seat—how many speakers I can hear." I worked it out, and *Love* is 26.4—that's the "format." That's totally irrelevant, and with reflections, it's probably more than that anyway. There are a couple of seats where you hear less and some where you hear more, but I had to think and work out what an average seat actually hears. It was just an idea that developed and grew, there wasn't any idea about it in advance.—Jonathan Deans

than line arrays, or venues where the coverage pattern of the particular point-source speaker matches more closely the shape of audience based on the available speaker mounting positions.

- Advantages: smaller than line arrays, simpler to tune/calibrate, fewer amp channels (or circuits, in the case of active speakers) than an array in the same position, easier to hide.
- Disadvantages: no availability of targeted gain shading (a technique used with some array designs), which means that instead of delivering uniform SPL to the whole audience, you will invariably deliver louder signal to the audience closer to the speaker and quieter signal to those at the back of the house. Often fewer drivers per enclosure than equivalent array boxes, which can result in less spectral control and thus (depending on design of the speaker and on system tuning) greater phase distortion.
- Line-source (aka "line-array"): these are speakers that are typically wide and not very tall, designed to be combined with other units of the same speaker type[11] in vertical arrays. A true "line-array" sees all of these boxes simply configured in a vertical column with no splay between boxes. Most common "line-arrays" seen in the wild have some form of curvature, to match the HF coverage of the boxes with the slope and spread of the audience being covered. A very common shape is the "J-array" which is mostly straight for the top few boxes, then curves into a J-shape at the bottom, in order to better cover audience closest to the stage. There are also "constant curvature" arrays, which have the same splay between each box to form a consistent arc in the air.[12] Figure 16.13 shows several array configurations.

It is important to understand that the "array effect" in any physically proximate collection of speakers serves to join those speakers together to effectively act as one larger speaker. If we imagine two point-source speakers of different sizes (Figure 16.14), this will help us imagine how and why an array joins together in this way.

A very small driver (like, say, the speaker in your cell phone), will generally be omnidirectional at all frequencies. This is because, barring intense horn/waveguide modification, speaker drivers become omnidirectional starting at the first frequency whose wavelength is larger than the diameter of the driver in question, and stay omni for every frequency below that. Imagine that the speaker on the left in Figure 16.14

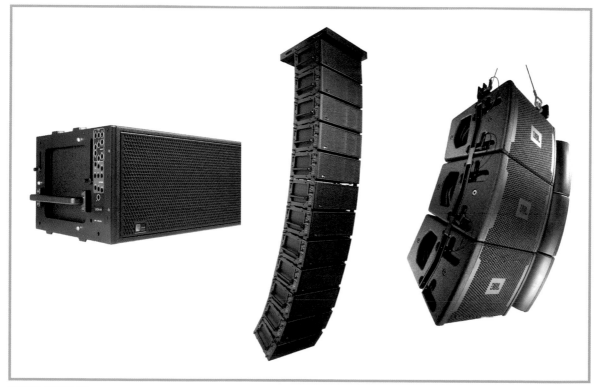

FIGURE 16.13
(Left) Meyer Lyon-M (a Single Array Box); (Center) Meyer Lyon in a J-Array; (Right) Constant Curvature Array (JBL VRX928)
Source: (Left and Center) Courtesy Meyer Sound; (Right) Courtesy Harman Professional Solutions

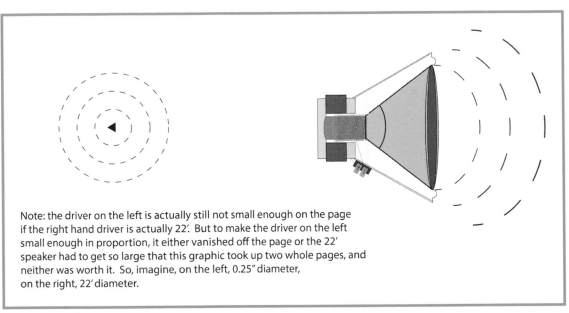

Note: the driver on the left is actually still not small enough on the page if the right hand driver is actually 22'. But to make the driver on the left small enough in proportion, it either vanished off the page or the 22' speaker had to get so large that this graphic took up two whole pages, and neither was worth it. So, imagine, on the left, 0.25" diameter, on the right, 22' diameter.

FIGURE 16.14
A Small Driver With Omni Coverage (L); A Large Driver With Directional Coverage (R)

is a 0.25" driver; this means that every frequency below roughly 4.5 kHz[13] will be omnidirectional. Why does this happen? Well, logically, if the wavelength of the sound is larger than the driver, it can only generate that tone by virtue of the sum of its vibrations in both directions along its axis, whereas a sound whose wavelength is smaller than the driver diameter will be driven forward, on axis to the in-polarity vibration of the driver. If the speaker on the right in Figure 16.14 is a 22" driver (yes, *feet*), this driver will exhibit directional behavior down to about 51 Hz and then turn omnidirectional.

How does this apply to arrays? When speakers are arrayed, which is to say placed next to each other at a distance no greater than one-half (and ideally within one-third) of the lowest wavelength the speakers are intended to produce in combination (and are driven by the same signal), the summation zone,[14] which is pronounced in the low-mid and LF ranges, effectively creates an extended diaphragm. Instead of, say, one 12" woofer, if we stack 4 of them in close proximity, we have effectively created a 48" woofer. This can steer low frequencies much more effectively than a single 12"! Note that this only applies to low-mid and lows. In the high-mid and HF ranges, tweeters are shaped with horns, and the intent of a good array design is that each box's HF driver(s) aims at its own section of the audience, with little or ideally no overlap with the other boxes in the array. Thus, line arrays combine the crisp individual targeting of good point-source systems in the HF range with the combined LF directivity/power only possible from a stack of speakers.

In so-called pure line-array theory (which has never been demonstrated in reality, but which drove original development of these technologies), the idea of vertically arraying these boxes was that we could array enough boxes together to create a true cylindrical coverage shape forward of the array—rather than the spherical coverage of a point-source with no enclosure or waveguide. This, again in theory, would have changed the 6 dB of loss per doubling of distance that is typical of other sound sources to a robust 3 dB of loss per doubling of distance, thus saving power. While well-designed arrays do somewhat improve the signal loss over distance from a true spherical source (and in any speaker, the 6 dB of loss is an average that varies with frequency), the literal 3 dB of loss and perfect cylinder have yet to be achieved. Just because real-world arrays don't perform true to "perfect" theory doesn't mean they aren't enormously useful for our work.

Line arrays have become the most commonly used FOH system speakers for large concerts, they are found in many indoor venues and theaters, and are used in a range of other contexts.

• Advantages: line-array boxes deliver very even and articulate coverage in the HF and hi-mid ranges (assuming you have splayed the boxes properly for your audience shape and size). For long arrays covering a deep audience section, boxes covering audience close to stage can be "gain-shaded" (level reduced) progressively, getting louder as the target audience section is more distant, so that as an audience member moves through the venue, the level experienced at any given position is the same.[15] Line-array boxes increase directional control compared to many point-source boxes.

- Disadvantages: long line arrays are large, and to many (especially in live theater) they are unsightly. They are difficult to conceal. If you are to have full control over the array, you may need individual channels of DSP and/or amps for each speaker in an array, making the processing and amplification end of array design very costly. They take some finesse (and often, good use of speaker prediction software, covered later in this chapter) to configure properly, and if configured improperly for the application and/or venue, can drastically increase comb filtering and other undesirable sonic artifacts.

- Column-source (Figure 16.15): these are essentially vertical line arrays contained in a single enclosure.

In the most basic of these speakers, we simply have a row of drivers, set vertically, in order to control coverage angles. Many columnar loudspeakers (like the Meyer CAL, Figure 16.15) can be *digitally steered*. Without getting too intense about the technical details, this means that by use of software, you can customize the aim of the speaker and sometimes the shape of the coverage. This technology is very useful in tight spots (the Meyer CAL is a favorite around the outsides of sports arena audience areas, because it can produce very even coverage at good SPL, without necessitating huge unsightly arrays). This type of speaker is most often used where a traditional line array's performance might be desirable, but there isn't space for a traditional array to be mounted.

- Advantages: thin physical profile, often useful coverage angles that aren't commonly found in other speaker types, digital steering. Excellent for public-address systems. Lighter weight than most traditional arrays.

- Disadvantages: unless the speaker is extremely tall with a ton of drivers, bass response is unlikely to be robust (these are very common for public PA voice reinforcement, rather than music production). Digitally steered units are costly and can be time-consuming to configure.

- Subwoofer (Figure 16.16): bass! A subwoofer is a speaker designed to handle frequencies at the low end of the spectrum, typically below 150 or 200 Hz. As such, the drivers and enclosures are quite large. Subs are a very useful extension of a speaker system any time full-range music is being amplified at a high SPL, and they are a requirement these days for cinema and television work (and for many video games). A classical concert hall that only does very light amplification may not need subwoofers, but a Jay-Z concert definitely does. Like line-array boxes, subs are often arrayed, sometimes vertically, often horizontally across the

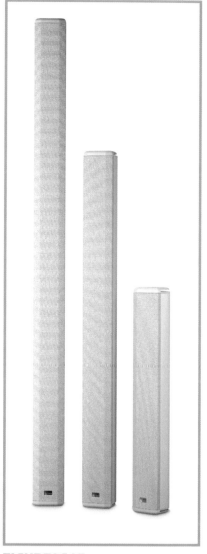

FIGURE 16.15
Meyer CAL (Column Array Loudspeaker)
Source: Courtesy Meyer Sound

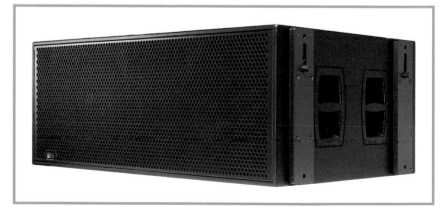

FIGURE 16.16
Meyer 1100-LFC (Low-Frequency Control)
Source: Courtesy Meyer Sound

front of the stage, sometimes in "ground stacks" (piles of subs at each downstage corner of the stage, under the L/R mains), and sometimes in interesting configurations like "end-fire" and other cardioid arrays.[16] In recent years, innovative engineers have created speakers like Meyer's 1100-LFC, which may be the most directional single subsystem I have ever heard.

- Advantages: in any system needing LF power, subs are generally a requirement, whether paired with point-source or array tops.
- Disadvantages: they are large, heavy, and unnecessary for most speech-only, classical music, or public-address systems.

- Stage monitors (Figure 16.17): speakers onstage aimed at performers (especially in amplified concerts), so they can receive custom mixes of the music that best help them hear their place in the performance. These will be

FIGURE 16.17
L-Acoustics 115XT-HQ Wedge
Source: Courtesy L-Acoustics

covered in detail in Chapter 17. Often found as so-called wedge floor speakers, but also in the form of side fills (often point-source) aimed from wings onto the stage, drum subs (large monitors just for the drummer to provide extra tactile thump during performance), etc.

- Advantages: some performance types will require stage monitors (any rock, pop, rap, electronic, or any other loud concert type other than symphonic or marching band). Allows artists to hear themselves.
- Disadvantages: unnecessary for some performance types. Can raise stage volume unnecessarily, leading to more feedback problems and a muddier, washed out mix in the house due to combination with overloud stage sources (in-ear monitors, also covered in the next chapter, are a common solution to this—Broadway shows rely increasingly on in-ears for the monitoring needs of their instrumentalists).

- Practical: a "practical" (a term of art deriving from theater) is a speaker installed on a set or in a location where it is to explicitly function as a visible speaker. In theater, a practical often carries the implication that an actor may interact with it (e.g., a radio onstage that an actor turns on and listens to). As such, a practical, more than almost any other intended use of speaker, will define itself first by the form factor. How large is the object or space in which this speaker has to be situated? A handheld radio from the 1980s will be smaller than a console-mounted radio from the 1940s, and so on. Once the form factor has been decided, consideration is paid to SPL and frequency response. In some cases (onstage radios are a good example), we may connect an existing period speaker to contemporary electronics in order to drive signal from our playback system into a vintage unit, with all of its associated vintage characteristics. Sometimes, this will be loud enough (and will possess, naturally, the desired frequency response and tonal characteristics). Other times, we need to fake it, by installing a higher-resolution, higher-powered unit inside a vintage-looking prop. Still other times, we need to do both: make use of vintage parts and supplement their sound with contemporary speakers hidden nearby. Practicals are almost always a custom installation, calling for detailed consideration case by case. Many times, we find ourselves working with raw speaker drivers (without crossover/enclosure) and building those into props such that sound can emanate from the objects, but the speaker remains hidden. In these cases, we crossover the frequency response in our DSP and consider the resonance of the prop object in which the driver will be situated in order to best plan for coverage. Prop materials can be challenging to model in speaker prediction software, so the reality is that as a designer becomes familiar with different drivers, acoustic responses of materials, and the like, their work in specifying drivers will become more successful.

- Advantages: when you need a real speaker onstage, especially one that actors or performers can convincingly interact with, practicals are a necessity.
  - Disadvantages: can be difficult to model in advance, limitations of period equipment may demand supplementation beyond initial plan.
- Specials: like practicals, except hidden in scenery. Speakers are often hidden in theater sets and props, in theme parks, and elsewhere. When we need sound to come from somewhere, but we *really* don't want to see a speaker, we need to hide a special. Form factor can vary wildly depending on what the speaker is trying to accomplish and in what context.
  - Advantages: fool the audience! Create an atmosphere of magic!
  - Disadvantages: hiding speakers in props is difficult, and also has a way of absorbing HF, which makes for a lot of customization in the content to offset this effect.

SPECIFICATION

A solid understanding of loudspeaker specifications will go a long way to helping your system designs achieve their goals. While there is no substitute for the gained experience of hearing lots of different speakers (as some live up to their specs and some don't), specs are still essential. Most designers will, at some point, find themselves forced to design a system using speakers they have never heard, and at that moment, specs are your best friend.

- Components: this specification will sometimes be listed under different terms ("drivers," "physical," etc.), but is a listing of the drivers involved. Unless specified, you can assume the drivers are dynamic (magnet-driven). Drivers are usually listed by size: *8" midrange driver with 1.25" compression tweeter*, for example. They are often referred to by frequency range in generic terms, as noted earlier.
- Frequency range (−10 dB): this is a general listing of the range produced by the speaker. Note that −10 dB is a drastic change in coverage, so by listing a spec this way, the manufacturer is able to state a wider range of coverage than will be useful in an ideal design (since we want the frequency response of our system to be basically flat). Also, some manufacturers are sneaky, and will include this spec with a "+/− 10 dB" qualifier. Without seeing the frequency response chart, this is a scary number, because it means you can have as much as 20 dB of variance in your speaker. That's a ton!
- Frequency response (−3 dB) (Figure 16.18): some manufacturers will list the frequency response as a separate spec than frequency range. Frequency response is typically listed with a maximum diversion from flat of −3 dB (though again, keep an eye out for those that list "+/− 3 dB"). This is a more accurate range of numbers for the useful frequency coverage of a given speaker. If a speaker lists frequency range (−10 dB) of 45 Hz–18 kHz, you might see a frequency response (−3 dB) of 95 Hz–15 kHz. When you have both numbers, always trust the −3 dB figure more than the wider range. If you only have the wider range, know that the speaker's response is probably not terribly flat in frequency.
- Coverage pattern: stated as a matter of so many degrees of horizontal and vertical coverage, or as one figure "conical" (both horizontal and vertical are identical). Coverage pattern is very important when selecting a speaker. A box that covers 10° vertical by 110° wide will perform very differently than one that is 60° conical. With some exceptions (practicals, speakers aimed at walls of a small theater for diffuse reverberant sound, etc.), you typically want to match the coverage angles of your speakers to the size of your audience such that everyone in the audience gets good direct sound, and there is minimal (ideally no) splash on side or back walls or other barriers. Of course, in reality we must always be aware that the speaker's directional coverage will only be applicable in the mids/highs, and that the lower the frequency goes the more omnidirectional coverage will be (with rare exception, such as the Meyer Sound 1100-LFC subs, which, again, have *astoundingly* good pattern control in low-frequency ranges). For most speakers, as we recall, the smaller the driver, the higher the frequency that will be omnidirectional.
- Sensitivity: for any passive speaker, this spec is very important (inactive/self-powered speakers, this spec is usually not listed—as with impedance, its relevance comes in matching the speaker to an amp channel, and in a self-powered device this work is already done for you). It is typically listed as "X dB SPL, 1 W, 1 m," which means that the speaker produces X decibels of sound pressure level when powered by 1 W, measured at 1 m. This is our standard sensitivity spec, so that however we power the device, we can calculate what pressure we will be able to generate.

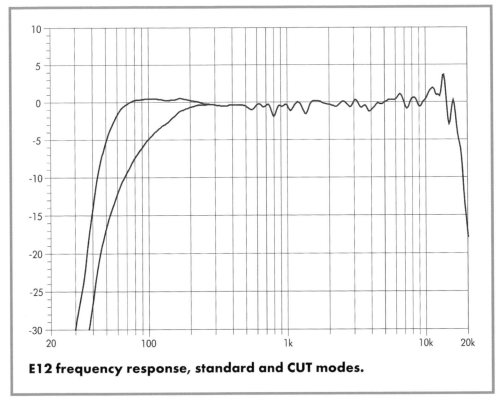

E12 frequency response, standard and CUT modes.

FIGURE 16.18
d&b audiotechnik E12 Frequency Response Plot
Source: Courtesy d&b audiotechnik

- Nominal impedance: impedance, as we know, varies with the frequency of input, so the impedance listed for a given speaker will seldom measure at that actual value under test. This is a standard figure that allows proper calculations to be done for powering the speaker via various amps. Most professional speakers have nominal impedance of between 4 Ω and 16 Ω.
- Power capacity: again, more relevant for a passive speaker than an active one—this spec should tell you how much power the speaker will use in two modes: peak and continuous. Note that manufacturers vary in how they define these particular terms. For some manufacturers, a peak value will be high-intensity sound over the course of not more than 2 hrs, where the continuous will be for 8–10 hrs of sustained use. For some manufacturers, peak might be true instantaneous peak value associated with loud transients (momentary only), and continuous might be 100 hrs. Regardless, the intent of these figures is both to suggest max SPL (via the calculations covered earlier in this book) for both types of content, and also to help calculate what amp best suits driving a given speaker.
- Crossover: in any speaker with a crossover (passive or active speakers), the specs should tell you what frequency is the crossover point between the two (or more) frequency bands. A speaker with a tweeter and woofer that is crossed over at 1.9 kHz will sound different from one crossed over at 3 kHz, as the former tasks the tweeter with more work, and the woofer with less. Depending on the driver components, either might be a better idea than the other, but it's important to know this: when selecting speakers, if you know that you will be using them for vocal clarity, having a speaker that crosses over right at 3 kHz might be more awkward than 1.9 kHz, since 3 kHz is right in the vocal clarity range, and there is often some "crossover distortion" or phasing around the crossover point. That said, speakers that are very precisely engineered have taken this into account and will usually have corrected for phase distortion around the crossover. The cheaper the speaker, the more you are likely to encounter crossover distortion. In addition, when tuning systems, it is helpful to know where your crossovers are set, since anomalies at those frequency points might be hard to fix with external DSP.

- Max SPL: many speakers will tell you their max SPL. If you have a sensitivity spec, you can calculate your own estimated value. A little variation from calculated to measured max SPL is common, as the theoretical limits based on the log equations might break down a bit at the edge of driver performance. However, a speaker whose calculated max SPL based on sensitivity and measured max SPL as listed in a separate spec widely diverge from one another should be approached with caution. Any time specs seem to contradict themselves, it is time to consider switching your selection for gear made by a more reliable manufacturer.
- 70V transformer taps: if your speaker is capable of being used in a 70V system, the transformer taps (different wattages at which you can power the speaker in 70V mode) will be listed here. Max SPL at different tap values can be calculated using the 10log equation. You may also find 100V taps listed as an option.
- Amps: if your speaker is an active speaker, it should list a decent range of amplifier specs (power, impedance, electronics class).
- Enclosure: what is the enclosure made of? A wooden enclosure sounds different from a plastic enclosure and knowing what your units are made of in advance is critical to understanding how your system will sound.
- Environmental/outdoor: any speakers designed for outdoor use will typically be rated on the IP scale. IP technically stands for "international protection," but is sometimes listed as "ingress protection," as it essentially measures how easily particles and moisture can get inside the unit. Typically, this code is listed as IP followed by two digits (IP-57, for example). The first digit on a scale of 0 (lowest) to 6 (highest) measures solid particle protection capability. The second digit on a scale of 0 (lowest) to 9k (highest) measures liquid ingress protection. Thus a speaker rated IP-22 is resistant to fingers or similar sized objects being inserted into it, but features no dust protection, and is resistant to dripping water at 15°, but not to full multidirectional rain. An IP-69k is protected from all dust and particle ingress and from high-pressure directional water jets, for up to 8 hrs continuously.[17] In truth, if a speaker is only rated IP-22, you probably won't find an environmental spec listed at all. You tend to only see this spec when the numbers get high enough for the unit to be considered for an outdoor installation.
- Mounting: the specs should describe the mounting options for the speaker.
- Power: if you are dealing with an active speaker, this spec should tell you the form factor of the power connection (PowerCon? Something else?) and should also tell you what power systems the device can handle (e.g., is it designed only for US systems at 110 V/60 Hz, or can it also handle European 220 V/50 Hz?).
- Dimensions: if you purchase a speaker that doesn't give you its dimensions, you deserve whatever is coming to you. Sorry, not sorry.
- Weight: often overlooked by beginning designers, but critical when determining how/where you will mount that speaker.
- Included accessories: does it come with a mounting bracket? Or do you have to buy that separately?
- Phase response: very nice speakers will sometimes tell you the phase response—how linear is its timing when we compare acoustic output to electrical input? A spec might read: "800 Hz–12 kHz +/−45°." And 45° isn't much phase variance (though note that "+/−" again, which actually means as much as 90° of variance possible across a single point). This spec is from a very nice speaker,[18] which has a frequency response of 75 Hz to 15 kHz. This tells us that in the 75–800 Hz range, we are more than +/−45° out of phase, and the same is true in the 12–15 kHz range. In the LF range, it is more difficult to hear this phase variance than it would be in the hi-mid speech range (where we hear most accurately at the lowest level), so this still sounds like a clean, clear speaker. However, if you consider up to 90° of phase variance from a very well-engineered speaker, consider how much phase blur you are likely getting from the vast majority of speakers that do not list their phase response at all.

All told, the most important specs relate again to loudness, frequency response/linearity, coverage angles, and mounting/size/shape/weight. If your physical location can hold the speaker you want to place, the speaker covers the desired area, with sound both loud and flat enough to suit professional delivery standards, you've got a winner!

That said, there are times when the specs will seem to lie. Some of this comes down to the old adage "you get what you pay for." If you find a speaker that seems to do impossible things based on its specs, and it is very inexpensive, there is sadly good reason to mistrust the specs as stated. You can always try to find out from the manufacturer under what conditions the specs were measured (which can be difficult when they are trying to hide something by writing the specs *around* their performance issues). The best way to avoid this problem is to listen to as many speakers as you can, wherever you go, and to learn about which manufacturers have a good reputation and which do not. It is not my place to spread rumors about poor manufacturing quality control, about misleading specs, or the like, in this book, but I assure you that in talking with other professionals in the business you can quickly get a sense of who makes quality equipment, who is middling, and who is not to be trusted.

CONSIDERATION

What questions must a designer ask (and answer) in order to select speakers as part of their system?

1. What area do I need to cover with sound? How far away are my speakers from that position, and will their angular coverage match this audience area, or is it too narrow/wide?
2. What content am I delivering, and is it full frequency range? If so, can my speakers deliver that range? If not, do my speakers perform well in the needed range?
3. How loud must coverage be? Do my speakers reach that level easily from the distance at which they are mounted?
4. What space do I have in which to mount speakers? How much weight will those points support?
5. Do I need my speakers to be hidden or can they be visible?
6. Are they indoors or do they need to be rated for outdoor use? If outdoor, what kind of elements are we facing? High and low temps? Dust storms? Rain? Snow?

Most of these questions have already been explored in the chapter leading up to this list. However, before we leave:

A QUICK NOTE ON SPEAKER PREDICTION SOFTWARE

In today's sound industries, speaker prediction software has become an indispensable design tool. Many companies issue proprietary tools (many free for download) for use in planning speaker systems within their brand ecosystems. From Meyer's MAPP to d&b's ArrayCalc, there are many varieties. However, one of the industry standards is AFMG's EASE® (Enhanced Acoustic Simulation for Engineers). Figure 16.19 shows an EASE plot of a single speaker in a relatively basic venue.

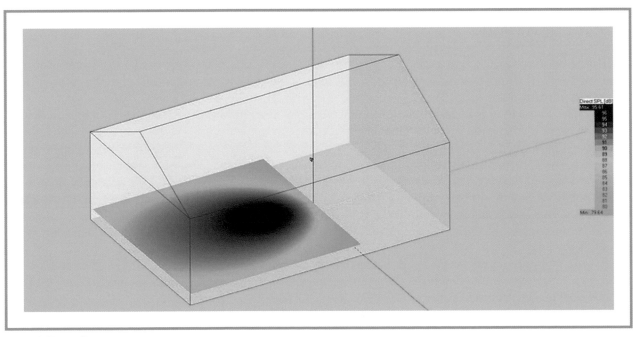

FIGURE 16.19
d&b audiotechnik E12 Direct Sound Plot via AFMG EASE® (EASE® is a registered trademark of AFMG Technologies GmbH—this applies to all references to EASE® throughout this book, of which there are several)

While it is beyond the scope of this book to go into EASE® in detail,[19] this tool is powerful. In it, you can build a venue in 3D; assign appropriate acoustic textures to every surface; insert loudspeakers that are pre-prepared with professionally measured performance data, supplied by all of the world's major loudspeaker manufacturers; and test positions, angles, timings, and models in your sound system. At the top end, most designers are using EASE® (or some variety of speaker prediction software) to estimate how their chosen system components will perform before ordering a rental/purchase, hiring in crew, or any of that. Modeling your system performance in 3D is now an essential skill for sound system designers to possess, both because it saves you time and money in the venue, and because you can generate colorful and attractive coverage plots for your clients to help them understand why you *really need* that extra array of 15 boxes, or whatever. In addition, in EASE®, you can model a wide variety of acoustic measurements, from Direct and Total SPL to standard acoustic values like STI/P, D/I ratio, Alcons, C7, C50, etc. It is again beyond the scope of this book to go into all these measurements in detail, but if you are designing systems for permanent installation and working with architects on a new building, you may be asked not only to understand, but to model, many of these values to ensure your works meets a variety of performance metrics.

SO!

That's enough about speakers. How about some more speakers? Ha!

In the next chapter, we examine monitoring systems, which do, yes, include stage monitor speakers, but also include in-ears, program feeds for backstage, etc.

Notes

1   Of course, this statement applies to those systems that are the central focus of this book: live entertainment, themed, and installation systems. If we are speaking of studios, it's a slightly different equation. Yes, good studio speakers are essential for doing accurate high-level work, but in that milieu, it can be argued that the mics and pres are as, or more, important than the speakers. If *I* am designing a studio, though, you'd better believe it's going to feature excellent reference monitor speakers.

2   In most drivers, anyway. There are ribbon drivers and piezoelectric drivers that work a bit differently, but the vast majority of loudspeakers are "dynamic" devices, much like dynamic mics.

3   Not entirely correctly. An infinite baffle, as we discuss further in the chapter, is truly infinite, where a sealed box in some ways mimics an infinite baffle, but in other ways cannot.

4   With due respect to my former student A. J. Diehl, whose famous rallying cry for years, no matter what the content, was always "MORE BASS!"

5   Discussed later in this chapter.

6   This sounds like the name of a British children's television show to me, but that is well beside the point.

7   I have used ribbon speakers to great effect strapped to the underside of a piano keyboard, to generate a very natural "performed" sound that resonated the soundboard of the instrument and localized to the "player," who in fact could not play at all.

8   With apologies to my friends in Southern California, who may never have experienced the freezing-shut of a car door. Go with me here, the example still works.

9   For example, I have often heard theme park designers speak of "backcan units." These are in-ceiling speakers, so-called because they are usually sold with the option of a metal or plastic "backcan" that covers the electronics and terminal posts that are situated inside a dropped ceiling. I've never heard designers, even in other architectural fields, call these "backcan units" (they are usually just referred to as "ceiling speakers" or the like), but if you hear this "backcan" term in the wild, you've been warned!

10 Which are often shaped into J-shaped or uniformly curved arrays, rather than just straight lines, hence my quotation marks.

11 And/or similar types (common varieties are boxes that have the same basic shape and configuration but a wider vertical coverage, say, for the down-fill portion of a curved J-array) . . . also with subwoofers from the same product line, typically mounted at the top of an array.

12 Note: for the remainder of this text, when I use the term "line-array," I may be referring to any of these configurations and will clarify which I intend if it is relevant to the issue being discussed. "Line-array boxes/speakers" will be used to refer to any speaker intended for use in a vertical array of similar or identical speakers.

13 Give or take a few Hz, depending on the temperature and how that's changing the speed of sound.

14 "Summation" here means combining of signals. It can be additive or subtractive, which is to say it can make signal louder or can cancel signal out, but either way, these phenomena are due to the summation of the acoustic signals.

15 Some designers think gain shading is unacceptable heresy, some think it's an essential tool. I am agnostic on the subject, I've worked on productions where it helped, and some where it was not appropriate, in my own estimation. One physical trait must be observed, which is that we don't just lose level at close proximity when we reduce the gain of boxes at the bottom of an array, we actually effectively steer the overall array's coverage higher. These are arguably two sides of the same coin, but it is important to remember that our "hotspot" for the combined frequencies (those experiencing the array effect) rises physically toward the back of the audience as we lower this gain level.

16 There are many phenomenal papers on different cardioid sub array techniques, rather than try to replicate them here, I will simply link them. If you need to make subs fire in cardioid fashion, these are very useful guides:

www.prosoundweb.com/channels/live-sound/the_end_fire_cardioid_subwoofer_array_made_visible/

http://eaw.com/docs/6_Technical_Information/White_Papers/Directional%20Subwoofer%20Arrays.pdf

www.dbaudio.com/en/support/downloads/category/download/7925.html

17 For more info on the IP-code scale, see https://en.wikipedia.org/wiki/IP_Code.

18 Meyer Sound M1D.

19 Though we examine speaker prediction in more detail in Chapter 34.

CHAPTER 17

Monitoring Systems

When most people think of monitors in a sound system, they think of wedge-shaped floor speakers, aimed at musicians in a band. While this is a common monitoring paradigm, it is far from the only way that performers, speakers, crew members, and others involved in production hear the event on which they are working. In this chapter, we'll examine the most common monitoring system types, and the basic considerations surrounding the use of each.

DEFINITION

There are two major categories of monitoring systems:

1. Live event monitoring
2. Playback monitoring

In a live event monitoring system, performers and crew members are being sent some version of the event's sound as it happens, such that they can keep track of what is going on. In a theatrical production, this often involves following the script by ear so that cues can be performed on time (entrances, exits, scene changes, etc.). In a concert setting, this often involves sending dedicated mixes to different performers (or groups of performers) onstage so that they may more clearly hear one another to stay in tune and in time with each other—a four-piece band playing on a festival stage 100' wide may not be able to easily hear each other's onstage sound because they

Foldback

A mix of a performance sent to a set of speakers or headphones, intended to guide performers and crew members by allowing them to hear themselves, other performers onstage, or other important information. Also known simply as "monitoring" systems.

are so far apart from one another (especially compared to the 20' x 20' rehearsal space they may be used to), but by mic-ing their instruments and sending a balance to each performer (balanced according to what they want to hear), we are able to help them stay calibrated over large distances. A varietal of live event systems is also found in broadcast studios and remote facilities, monitoring both the live feed of what is on camera, and often also the feed of another camera/mic that may be about to go live, such that the crew can ensure the next feed is ready before switching to it.

Playback monitoring is used in all the same kinds of productions listed earlier but is designed to feed pre-recorded or "canned" material to the performers or crew. In a Broadway musical, for example, it has become quite common to feed the conductor, and sometimes the orchestra itself, a click track for some songs that need very rigid timing. This playback signal will never be heard by the audience but needs to be clearly heard by the performers to help them synchronize their performances in time. A feed of playback sound effects cues might be sent backstage in a theater, so crew and cast can perform actions that are supposed to land at specific moments within the pre-recorded sound. In broadcast, of course, there is frequently a need to intercut between live field pieces and pre-taped pieces, and all of these must be monitored by the technical crew live-assembling the work in the control room.

Both live event and playback monitoring systems often use many of the same equipment types and frequently interact such that many systems offer both live and playback monitoring as needed.

CLASSIFICATION

There are essentially two different mechanisms we use for transmitting monitoring signals, with multiple variations in each main category. We will address the two categories, and primary subcategories, here. The two main mechanisms, perhaps obviously, are

- Speakers
- Earphones

Speakers:

- Stage floor monitors ("wedges") (Figure 17.1): these are large, wedge-shaped floor speakers, designed to aim up at performers' ears. These are found primarily in electrified concert settings, very seldom in musical theater orchestras, and almost never in classical music, spoken/seminar engagements, etc.
  - Advantages: they can get quite loud, they typically cover a wide area, so a performer can move a fair amount and still hear, they have some bass power that in-ears will never possess.
  - Disadvantages: they are large, and as such, are quite visible which means they are unsuitable for most theatrical applications, and undesirable for most speech/seminar type events (with notable exceptions).
- Drum sub: this is typically a 15" or larger single-driver speaker, often configured in tandem with a wedge, aimed at the drummer of an electrified band (usually rock/pop/hip-hop/reggae/country, not so often in jazz), to extend the bass response of the monitoring

FIGURE 17.1
L-Acoustics 115XT-HQ Wedge
Source: Courtesy L-Acoustics

system. Drummers often like to feel the bass—not just the bass guitar or bass synth, but even their own kick drum—as if it is a thunderstorm crashing on top of them.[1] A simple stage wedge might produce full-range audio, but still usually won't hit the performer in the chest like a cannonball. Enter the drum sub. So-called because 99% of the time, it is only the drummer who wants this speaker. A suitable drum sub is any subwoofer of great power and good low bass extension, with a footprint that will fit next to the drum set. Often, a standard point-source top might be set on top of the drum sub, aimed sideways at the drummer as their "wedge" since the elevation of the drum sub would mean a typical wedge placed on top of it might end up aiming above the drummer's head. Of course, if the drummer is on a riser, and the sub/wedge are on the stage deck, this can work out quite well for the angularity of coverage.

- Advantages: loud, bassy, powerful
- Disadvantages: large, too loud sometimes

▪ Side fills: typically point-source speakers, mounted at the side of the stage (in the wings, if there are any). These are sometimes used in electrified concerts as a means of giving the band an overall balanced mix of their work (particularly in combination with dedicated in-ear monitors giving more specific and detailed reference information). Care must be taken with side fills, as it is very easy to create feedback with point-source speakers aimed at all the mics onstage. These are more common in dance concerts, where there are no live mics onstage, and in musical theater, where the orchestra might be sent to side fills, but the live mics on the actors might not be, helping to keep the actors grounded with the score but avoiding immediate feedback issues.

- Advantages: broad coverage, blended field of sound
- Disadvantages: high feedback risk, not a custom mix for each performer

▪ Down-fills: monitors mounted above stage aiming at the deck, mostly used in musical theater, as mentioned in the side-fill example earlier.

- Advantages: see side fills
- Disadvantages: see side fills

▪ Backstage speakers: these might take a variety of forms and shapes. From production speakers mounted in a green room (cast/crew holding area and break room) to dedicated in-ceiling or in-wall speakers connected to an intercom system, these are often referred to as "program speakers" or "program monitors." The intent of these speakers and feeds is typically to keep offstage personnel grounded in the timing of the production enough that they can perform duties as needed. A cast member who only has one scene in a 2-hr play does not stand in the wings for 2 hrs waiting to enter, they sit backstage, listening to the program feed, and only move to the wings when it is approaching the time of their scene. Where onstage monitors each receive destination-specific mixes, this is not often the case for program feeds. Usually, a program feed consists of one or both of the following: a) a single mic, placed in the house of the venue (either hung or mounted near the FOH booth), transmitting the entire production as heard in the house and/or b) a general mix of all signals running through the console or DSP. Many theaters have intercom wall panels (Figure 17.2) in all dressing rooms and green rooms/crew areas. In that case, it is very common to connect a single program mic to the intercom system, and feed that to the speakers in all of these panels, providing a not particularly pleasant-sounding but serviceable reference for the timing of events onstage. Whether in the control room or backstage in the chorus holding area, these speakers must be loud and clear enough to provide a true reference for the work being produced, but also must be quiet/directional enough to not spill onstage or into the house. What units accomplish

Message in a Bottle

[On stage musicals] I'll never forget, it was the mid-1990s when we started saying, "I can put vocals in the foldback." Right? We still try to avoid it—I do it as minimally as possible. Everyone just said, "You can't put vocals in the foldback." So I had never put vocals in the foldback—that was the rule. *You can't put a wireless microphone . . .* Well, actually, as we discovered, you can. How productive is it? *Somewhat.* Sometimes more, sometimes less. But the reality is that even if it's just a low level, it can give actors what they need sometimes. And that's what it's there for.—Robert Kaplowitz

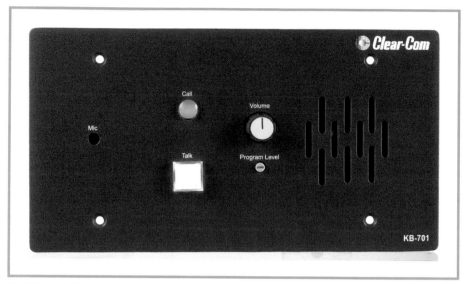

FIGURE 17.2
Clear-Com KB-701 Intercom Wall Panel With Speaker
Source: Courtesy Clear-Com

this will vary considerably based on the locations, materials, acoustic properties, and configurations of your backstage production areas.
- Advantages: facilitates necessary referencing backstage
- Disadvantages: sometimes low fidelity, possible bleed onto the stage or into the house

▪ Earphones—this class encompasses everything from custom-molded nearly invisible in-ear monitors to giant foam-padded over-ear headphones
  • In-Ear Monitors (IEMs, see Figure 17.3): IEMs are a very popular monitoring mechanism for concerts and theater alike. The best IEMs are fitted to the performer by virtue of a custom mold, suited to shape their exact physical form, which helps prevent the earpieces from falling out, and enhances comfort. IEMs are in-demand for a number of reasons: vocalists love them because it is much easier to stay in pitch when your reference is directly in your ear than when you have to depend on a stage wedge coupling acoustically to air and the venue stage; use of in-ears can help reduce overall stage volume (fewer monitor speakers=quieter stage), which in turn can make the house mix easier to make clear and powerful without being overloud; in-ears are often fed via wireless transmitter and beltpack receiver, such that performers are freed up to move around stage quite a bit more than if tied to a wedge formation.
    • Advantages: quiet, focused monitoring; wireless freedom of movement; well-molded units are comfortable and easy to use
    • Disadvantages: will never have the physical power of a speaker, if a performer needs the sensation of being "hit" with sound (again: drummers); poorly fitted IEMs can slip out, or be very painful to wear; poorly mixed IEM feeds risk very intense hearing damage (limiters are critical to IEM feeds); wireless signals are always susceptible to interference that a wired speaker may not be.

In-ear monitors come in a variety of system shapes. The simplest, as shown in Figure 17.3, is a transmitter mounted in a rack, a beltpack receiver, and a set of in-ear phones. Broadway and other orchestras that need in-ear monitoring often add a wrinkle to this equation, in the form of personal monitoring systems.

Personal Monitoring System

A personal mixer panel that is fed a number of mix stems from the main system (typically 8 or 16 channels), among which each performer can dial in their preferred level balance.

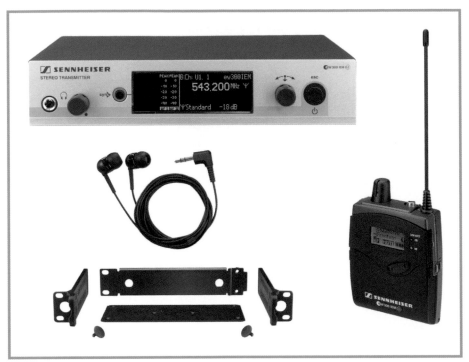

FIGURE 17.3
Sennheiser EW300 IEM G3
Source: Courtesy Sennheiser

In a Broadway orchestra, the stems might be something like 1) drums, 2) percussion, 3) bass, 4) guitars, 5) brass, 6) winds, 7) piano, 8) synthesizers, 9) reeds, 10) click track, and 11) conductor mic (for direct verbal cues to musicians). The musicians each can dial any of these sources louder or quieter in their personal earphones, which are connected to the personal mixer (often hard-wired). Earphones might be in-ears (often), or sometimes might be over-ear (depending on how much isolation from the acoustic room sound is desired, and the individual performer's preferences). Aviom (Figure 17.4) was one of the first companies to perfect the personal mixer system, but today there are many manufacturers who offer these, some of which are directly integrated into digital mixing console systems.

Then, there is the Assisted Listening System.

Assisted Listening System (ALS)

A system dedicated to making performance content more audible for those who are hard of hearing. Some systems work on radio transmission, much like standard in-ear monitors, and some work via induction loop, which involves a network of wires (lining the interior space of a performance venue) that generate a signal that can be received directly by many hearing aid systems.

In the United States, the Americans with Disabilities Act provides regulations on which venues must have ALS in place, and these regulations get updated from time to time. Courtrooms, for example, are required to have ALS, whether there is amplified sound or not in the courtroom. Venues with amplified sound are required to have a certain percentage of attendees able to use ALS devices. The exact regulations change, and around the world, other nations set their own standards. Check with the appropriate regulatory body to ensure your venue or production is providing ALS as required.

FIGURE 17.4

Aviom A360 Personal Mixer
Source: Courtesy Aviom, Inc.

Of course, there are other uses of earphones. Electronic News Gathering (ENG) teams typically use earphones in the field to hear the production audio being recorded, as do on-set sound recordists for film/TV productions. A musician in a studio session often needs earphones to hear what has already been recorded so they may layer their performance in time with the song as assembled. These are all monitoring systems of one kind or another.

When feeding monitor mixes, whether to musicians in a studio or to performers on a stage, we typically feed via the aux send of a mixing console. This might be a simple feed from the FOH console, where we send a generic blend back to side fills or backstage speakers via a single aux. Each channel of input will have aux sends, and to assemble the mix, each will be dialed into the desired amount.

In large concert systems in particular, it is most common to have dedicated FOH and monitor consoles. In this case, the preferred configuration is that all stage sources (mics, DIs, playback) are fed into a signal splitter. This splitter is *transformer isolated*, which means that the signals can be split to each console, and each console can set preamp gain as desired for their purposes, without any backflow (resisted current) traveling into the other console. A simple Y-splitter cable will not do this, and if you try to set the preamp gain with both an FOH and monitor desk for a mic split with a simple Y, you will run into interference and level issues. In situations where a transformer-split is not available, or where the FOH and monitor engineer can agree to share one gain setting, sometimes the preamp gain is set by one of the consoles, and the other console receives a (digital) copy, but has "digital trim" controls that allow the engineer to reduce or boost the signal *after* the preamp, but before it is sent into the channel strip.

SPECIFICATION

SPEAKERS

Specifications for speakers can be drawn from the same list in Chapter 16. The only added considerations for monitoring system speakers have to do with their interconnectability to your particular system (e.g., if you are using wall-panel speakers connected to an intercom system, are they compatible with your intercom system?). Special attention must be paid to coverage patterns, as monitoring systems often risk feedback.

EARPHONES

If you are specifying wired headphones of one type or another, the specs you need to pay close attention to are few: frequency response is the chief unit of concern. Beyond frequency, the main considerations are in form factor: are you selecting in-ear or over-ear phones? Do you need them to terminate in 3.5 mm connectors or in ¼" TRS? Do you need closed-back over-ear phones (which block more ambient sound but are slightly lower fidelity due to reflections within the cans themselves), or open-back over-ear phones (which are higher fidelity but admit more ambient sound)?

If you are working with wireless in-ear monitors, however, there are a few more items to consider, and they mainly relate to the transmitter/receiver pairs.

Carrier frequency: what frequency band does this particular unit operate on? Units will typically be sold with operating ranges of between 30 MHz and 60 MHz, but care must be taken to select units that are both legal to operate in your region[2] and that are generally as free of interference as possible. Check manufacturer's websites for search tools that can tell you what MHz bands are in use in your area by digital television and radio stations. Purchase a unit that is in the freest band possible, therefore reducing the chances of interference in your transmissions.

Presets/channels: a count of how many pre-configured channels are available (or in some cases, user-determined presets) for easy recall of carrier frequencies.

Frequency response: good IEMs should be nearly, if not completely, full range, though inexpensive models sometimes sacrifice very low and very high-frequency content, which will not suit high-resolution musical needs, but might suit the needs of an actor receiving prompts from offstage.[3]

S/N ratio: the larger the number, the better; 90 dB or greater is desirable.

RF output power: devices of this type are limited in their output power, so as to eliminate interference with critical safety radio transmissions in adjacent bands (police, fire, etc.). However, even within the limited power available, not all units will be equal, and some have multiple power modes. For example, a unit may offer 10/30 mW transmission, selectable. In the case of short distance communication (inside a small black-box theater, perhaps), a user might opt for the 10-mW setting, as it is less likely to generate intermodulation distortion (distortion caused by multiple units operating near one another). Higher power operating modes increase the risk of intermod distortion and often require users to space their transmission channels farther apart from one another. Over a longer distance (say, a football stadium), the higher power setting may be necessary, however, in order for a signal to reach the receivers.[4]

Operating time: if the device comes with a dedicated rechargeable battery, this is a very important figure to note. If the device uses third-party batteries (such as AA or AAA size), this number should be regarded with suspicion, as the actual performance will vary wildly with different battery models (and their attendant differing milliamp-hour ratings. While a higher milliamp-hour rating tends to add up to longer battery life, even this figure varies based on the actual chemical compounds used inside the particular battery). A good IEM system will last at least 4–6 hr on its batteries. A good rule of thumb for use of units with replaceable batteries is to use fresh batteries for every rehearsal or performance.

Dimensions: as with any wireless device, dimensions of the belt pack are important. Check to ensure the device you select is sized appropriately to your needs.

Water-resistant or waterproofing: how well will the unit perform in the rain? Water resistance is usually acceptable in light rain for an outdoor concert, while true waterproofing is only in specialty units used for events like Sea World shows where a trainer must dive underwater and then come out with electronics working.

Other specs, such as Input voltage range (how much voltage can be sent to the system before overload) may be relevant in specific cases, but on the whole can be assumed to be functional if one is selecting a professional unit for use.

PERSONAL MIXING SYSTEMS

The basic questions here are broad. What platform does signal travel under—is it Aviom, which uses the proprietary A-Net signal type? Dante, which is very common? Something else? How many channels are available? What kind of output jacks does it feature (for earphones)? What are the unit dimensions?

ASSISTED LISTENING SYSTEMS

As mentioned previously, ALS come in generally two main types: loop or RF. For RF systems, most of the relevant specs are the same as any RF system (mics, IEMs), with one general exception, which is compression ratio. One of the main mechanisms by which ALS make live content more easily perceivable by audience members who are hearing impaired is by compressing the dynamic range of the signal, ensuring that the quietest parts of the program are brought up to sufficient level to be easily understood. Different RF systems will offer different compression ratios, and in selecting a unit, you need to consider the available range. Units with selectable ratios are generally nicer than those with a fixed ratio, and the wider the available range of ratios, the better for properly suiting your needs.

Both loop and RF systems may offer multiple channels, such that Channel A could be transmitting the live program, while Channel B could (for example) be transmitting an additional channel of information, such as a narration channel for audience members who are blind, or the like.

Loop systems themselves will generally be complex custom installations, and it is usually wise to have a site survey conducted by a representative from a loop vendor or manufacturer who can assess the facility and its needs. Some buildings provide obstacles to good loop coverage that are insurmountable without a great deal of retrofitting, while others make loop systems easy to install. Rather than dive deeply into loop system parameters here, which is beyond the scope of this book, I will suggest that you see the bibliography for further reading and that you consult professionals for advice on systems of this type.[5]

CONSIDERATION

What questions must a designer ask (and answer) in order to select monitoring equipment as part of their system?

1. What are my monitoring needs? Who needs to hear performance audio, and does it need to be custom mixed, or can it be a general show feed (either from a mic in the house or from the FOH console)? Is the performance audio limited to playback, in which case feedback is no concern, or is it a live feed, in which case position and gain levels become much more important?
2. Do the listeners need control over general signal level? Over individual instrument/stream levels? Over more parameters?
3. Is it OK if we see monitor speakers onstage, or do they need to be hidden?
4. Am I worried about stage volume from monitor speakers? Will it overwhelm the house system? Does it risk being too loud for the genre of production I am supporting?
5. Does my venue require an ALS? If so, is the architecture compatible with a loop system?
6. If stage monitor mixes will be sent, do I need a dedicated monitor console (and thus a transformer-isolated mic split), or can we mix monitors from the FOH console?
7. Do our monitor feeds need to turn on and off in different preset modes for venue operation (as might be the case in a venue that produces several different types of stock events in rotation)? Are they show-dependent, and thus changing completely each time? Or are they just always on (as in a working production facility where a feed from a soundstage might always be on to the control booth)?

Monitor systems are a complex world, not least of which because so many different types of gear and needs fit under that blanket term of "monitoring." While this chapter is far from comprehensive, hopefully, you now have at least a sense of what general needs exist in typical production facilities and venues.

Now, we turn our attention to that mystical world of 1s and 0s. The computers!

Notes

1 I have played drums since I was 8 years old, I speak from personal experience! No insult intended to my fellow adrenaline-junkie percussionists!

2 In the United States, the Federal Communications Commission (FCC) regulates all wireless bandwidth, and regulations frequently change, so a unit that was perfectly legal last month might be totally illegal to operate this month (when these changes happen, there are typically buy-back deals from manufacturers to help users offset the cost of changing over to new gear in the new carrier ranges).

3 A less than ideal situation, but one we face once in a while, particularly with older actors.

4 At the time of this writing, power for wireless mics and IEMs in the United States is limited to 50 mW, but this could easily change. As little as 5 years earlier, devices could operate at 250 mW, so check with current regulations before selecting your units.

5 I assure you that this is not a cop-out. I have designed many facilities with ALS systems, and never once have I done so without consulting with the makers of the given ALS systems I am considering. It is a very particular field of discipline, and the manufacturers and vendors are generally up on the most current ADA requirements, technologies, and techniques in a way that I—as a sound, lighting, video, and control systems designer—am not. There is no shame in consulting with the pros before selecting expensive gear. In fact, I do this frequently— before using a new playback server I will pepper the manufacturer with a host of questions about response time from trigger cues and other such questions that aren't typically listed in specs or manuals, but that may be relevant to my system and its needs.

Networks and Connectivity

It is the 21st century. In your pocket, you probably have a computer (smartphone) with more computing power than the space shuttle that first brought humankind to the moon. This is a stunning development, and even more stunning (in a way) is that the extremely complex processes taking place within that computer are given a shiny, easy-to-understand user interface (GUI—graphical user interface), that allows the typical smartphone owner to click to their heart's content without ever having to manually configure a network. At most, you might have needed to set up your home network, but plugging in your wireless router to your modem, choosing a network name and password, and then selecting that name from your phone's WI-FI menu is a pretty simple operation.

However, in the world of event systems and production, we need to have a deeper grasp of computers, especially when it comes to networks and their functions. In contemporary sound systems, not only is much of our digital audio transported over computer networks from device to device, but we also design robust control systems wherein we can use computer networks to configure devices, monitor signals, and trigger actions.

The world of computer networking—at the highest level—is extremely complex and dense. By necessity, this section will be a glance at the most important terms and concepts, and not a deep dive. Every attempt will be made to address all the major points of consideration for a system designer, but inevitably I will leave something out that someone considers critical.[1]

In addition, while I am a professional system designer (a job that increasingly requires IT skills), I have never had any formal IT training. I have learned by doing, by reading articles and books, and by experimenting. When I began as a system designer, the idea of sending sound over a network cable was not yet a practical reality (at least not for any system other than the most expensive and cutting edge). Now, almost every contemporary system sends sound over networks and uses network control. As such, while I have the IT skills I need in order to perform my work, I didn't get them by taking college classes in IT. It is my goal that after reading this section, you will come away with at least enough knowledge to specify and assemble show systems (if not to do the most advanced control work). This is, after all, a *primer*, and not an upper division lesson.

Note: this chapter does not follow the "Definition," "Classification," "Specification," and "Consideration" pattern of the surrounding chapters, as we need to spend time on some elemental concepts before returning to the practical selection of gear.

WHAT IS A NETWORK?

In simplest terms, a network is a group of computers[2] connected to one another for the purpose of sharing information. These computers may be connected via a series of wires, or over a wireless access point, but they are connected—and are set up so that each can see the others, as necessary for the desired functionality.

Computers send information to one another in a basic format called "packets." Let's imagine that I am sending you an email, labeled "Big Jim's Surprise Party!" In that email, there is some information about Big Jim's party, the time, the date, the location, a picture of a cat in a party hat, and some terrible, terrible jokes.[3] Instead of sending all of this as one large bundle of info (that's one high-resolution cat photo!), the computer that is transmitting the message breaks this down into packets, which are reasonably sized chunks of data. At the other end, the receiving computer will collect all the packets and assemble them, and once they are assembled, Big Jim's party is yours for the attending!

This is a basic way of functioning that has been around for a very long time, and it is called the TCP/IP protocol (or Transmission Control Protocol/Internet Protocol). When sending a message via TCP/IP, it is generally assumed that the receiver wants the complete message more than just random pieces, so the transmitter sends the packets, and a list of what packets go in what order so that the receiver can tell for sure if the whole message has arrived before displaying it. Once in a while (much less often nowadays than in the early Internet era), a packet or three will get lost or corrupted along the way. If the network is doing its job, it re-requests the missing or damaged packet, which is cached (stored temporarily) in the machines through which the message has been routed (in the case of the open Internet, these are the service provider's computers). Therefore, the sender can turn the computer off, but the message can still get completed on the other end.

This is a plan that makes a lot of sense for any data that isn't time-sensitive. If I'm sending a draft of this book to my publisher,[4] yes it needs to arrive by the deadline, but whether it arrives at 9:00 am or at 9:15 am is not (usually) a big deal, so long as when it arrives, it arrives in its complete form.

However, when we are sending audio data over a computer network, time is of the essence. If a sound is delayed, even by a few milliseconds (not to mention a few minutes) in its transmission, it's really no good to us. We need sound to be transmitted in as close to real time as possible. Therefore, professional media transmission over computer networks typically uses an alternate protocol called UDP (User Datagram Protocol). In a UDP messaging system, the priority for delivery is given to the time domain. Certainly, we still want all the packets to arrive, but we assume that if something has to give, it is better to suffer a little digital distortion or noise from a missing data packet than it is to suffer an out-of-time signal. UDP is a platform with many advantages for media work, including the flexibility of its code base that allows us to send custom UDP messages from show controllers to media server devices.[5]

But how do computers know how to talk to one another? How does the message physically get from one machine to another, and how does the sending machine know which destination machine the message is intended for?

MAC ADDRESSES, IP ADDRESSES, AND NETWORK CONFIGURATION

To start with, each computer device with at least one network connection has a unique identifier, called a MAC address. No, this isn't because Apple got there first—it has nothing to do with their namesake line of popular computers. In this case, MAC stands for Media Access Control. This is a unique hardware identifier that only one particular network device in the world possesses.[6] Within a computer itself, there may be different ways of connecting to a network—for example, a desktop computer may feature two Ethernet ports and a WI-FI card. In this case, each of these three Network Interface Controllers (NICs, for short) has its own MAC address.

These addresses are used by the computer and the network to coordinate assigned communication paths, but as many of you are no doubt aware, when we talk about tracking a computer's actions on a network, we don't talk about tracking a MAC address, but about tracking an IP address.[7] IP (Internet Protocol) addresses are flexibly assigned addresses (one per NIC), that exist so that computers can join different networks at different times, and have access to only those parts of the network that they have permission to engage with. By creating IP address schemes (and later, subnets), we can combine a group of computers in such a way that Computer A might have access to all other computers in the group, where Computers B–F only have access to Computer A, but not to each other (or some such setup).

A typical IP address[8] consists of 4 "octets" (8-bit numbers, separated by dots, rendered in decimal—rather than binary or hexadecimal—form). A typical show system IP address might read: 192.168.0.10—though technically each octet can range in value from 0 to 255 (for addresses ranging from 0.0.0.0 to 255.255.255.255). The first address in the scheme (0.0.0.0) is the "default network," and so remains unused, and the last (255.255.255.255) is the "broadcast address" (which we will briefly address a bit later).

This leaves a pretty wide range of numeric values for network-enabled devices to use, but how do we assign these numbers, and more importantly, what do they mean?

Before we can understand how to assign addresses intelligently, we need to talk about the basic types of networks we will encounter in our work.

- WAN: Wide Area Network—this is the Internet and other such broad area networks. A WAN connects computers nowhere near one another over network devices and cables that form a telecommunications grid. In your home, your modem connects to the WAN via cable (from the cable Internet company, or from your satellite dish if you live way out in the boonies). The modem then connects to your home router, which distributes the WAN to your local devices. Local show systems do not use WANs in their work, however, in advanced themed installations a company may configure a production system so that it can be monitored from very far away (I've dialed into production systems across state and sometimes national lines via the Internet), and in the case of broadcast uplinks, connecting to some form of WAN is often requisite. Typically, this is done by connecting a node (individual computer device) in the local network to the Internet via a separate NIC, such that the node acts as a bridge from the outside world into the show system.
- LAN: Local Area Network—the vast majority of show systems are LANs. These are groups of computers in some semblance of physical proximity to one another, configured so that they can talk to and control one another as the production team desires. In a LAN, it may be the case that all computers can speak to each other, or the LAN may be further subdivided (subnetted) into sections so that, for example, lighting systems and sound systems aren't able to interact (though these days, it is becoming more common to want all of the systems to interconnect, such that a sound playback system can send a cue or trigger to the lighting system, or vice versa, in order to synchronize show events).
- WLAN: Wireless Local Area Network—same as a LAN, but wireless (or featuring wireless components). Home wireless networks are WLANs.

There are many different network configurations and shapes ("topologies") possible, but these are beyond the scope of this book.

Let's imagine a basic digital show system. We have a concert hall, and in that hall is a digital mixing console. That digital mixing console offers, as a companion to its internal software, an iPad app that allows the user to control certain parameters on the console in real time. In this case, the console has an Ethernet control port (using a standard RJ45 connector, usually wired in either Cat 5 or Cat 6 configuration).[9] That port will either connect directly to a wireless router, or to a router which itself is connected to a Wireless Access Point (WAP). The router will generate a LAN using a range of "private IP addresses" (about which we will discuss more later in this chapter). The iPad will connect to that wireless network, launch the app, and thus be able to control the mixing console from anywhere the wireless signal reaches. This is enormously helpful, especially in venues where the mixing console is in a less-than-ideal acoustic position, because the mixing engineer can take the iPad around the room and adjust sound from the audience's perspective, rather than from the less-than-ideal, under-balcony, away-in-a-corner spot where the console is located.[10]

In this example, there are three or at most four network devices: the console, the router (and/or router + WAP), and the iPad. That's it. It's a very simple network, and it is easy to configure, but already we've begun introducing terms like *router*, and we're getting ahead of ourselves.

To return to the question of IP addresses, in the above example, we must set up the devices such that they are all "on the same network," and to do so, we must assign them IP addresses. First, there are a range of so-called private IP addresses, which are reserved for use in LANs, and which we use extensively in show systems. Outside of this range, all addresses are used for WAN communication.

Private IP addresses:

- 10.0.0.0–10.255.255.255
- 172.16.0.0–172.31.255.255
- 192.168.0.0–192.168.255.255

Within show systems, all of our devices that need to speak to one another must be assigned IP addresses in a similar range. Different address ranges fall into different "classes" of address, and thus have slightly different rules for how they communicate.

In general terms, the first number(s) of the IP address is the network identifier, and the subsequent number(s) is the device identifier. In the 10.x.x.x address range, the first octet only (the "10") is the network identifier, meaning that this is a Class A network address, which leaves the subsequent three octets for individual device addresses. Therefore, a 10.x.x.x LAN can have over 16 *million* unique addresses in that range. This means that any devices connected to the same network infrastructure and sharing 10.x.x.x as the prefix (assuming no subnet divisions, which we'll address later) can all see and talk to one another over the LAN (assuming their internal settings allow sharing).

The 172.16.x.x—172.31.x.x IP addresses are Class B network addresses. Class B network addresses use the first two octets as the network identifier, therefore a 172.16.x.x address can talk to any other 172.16.x.x device, but won't talk to 172.30.x.x devices, as these are on different NetIDs, and thus distinct from one another.

The 192.168.0.x—192.168.255.x range is Class C, which means that the first three octets are the network ID. Thus a machine with address 192.168.0.10 can talk to 192.168.0.15, but not to 192.168.1.10, because this is on a different NetID. (Subnets introduce "classless" addressing, and again, we'll deal with that a bit later).

Each network device (NIC) must have a unique IP address in order to join a network. If your computer's wired Ethernet port has a manually assigned address of 10.0.0.2, and it tries to join a network in which your smartphone has already claimed that same IP address, your computer will not be able to join, and will be instructed to select a new IP address (or, depending on the computer's operating system, will just fail to connect).

So! How do we assign IPs? There are two primary methods: manually or automatically (via DHCP).

In a manually assigned system or device, the user enters into the network settings menu of that device and manually types in an IP address. In a lot of professional show networks, this was the preferred method for many years (and still is for some devices and purposes). By manually assigning addresses, you can ensure that no two devices attempt to claim the same address if/when they are temporarily knocked offline, and that addresses follow a scheme that you can pre-plan to make installation smooth and simple. However, manual setting takes a bit of time (or a lot, in the case of very complex systems with dozens if not hundreds of network-enabled devices), and some devices (an increasing number these days) don't really *allow* you to manually set their addresses. This is where DHCP comes in.

DHCP stands for Dynamic Host Control Protocol. Remember how the first octet (or two, or three) of an IP address are the NetID? The remaining octet(s) are the "HostID," and each NIC connected to the network is a single *host*. DHCP is a protocol by which devices are assigned addresses when joining a network. In a DHCP system, there must be a server (which provides each connected device with the addresses needed) and clients (the devices connecting to seek addresses). In a home wireless network, the DHCP server function is performed by your router. In a show system, if any devices need DHCP-assigned addresses, we need a server to be running, either as software on a show system computer, or as part of a hardware device (some advanced switches—so-called *managed* switches—can perform this task, as can any router—we'll examine the difference between routers and switches in a moment). The advantage of DHCP-addressed devices is that there is no risk that devices all being assigned by the same server from the same range will have trouble talking to one another (it eliminates the human error of mistyping a single digit in the address). That said, DHCP servers provide devices with a "lease time," which is the stock amount of time that particular address is usable. If you connect to your favorite café's WI-FI with your phone, each time you go into the café your device queries the DHCP server for that LAN and seeks an address. If the lease is set to 30 min, you find yourself kicked off the network after 30 min and have to reconnect. A DHCP server for a show system typically will set lease times to indefinite, or at least to lengths far longer than needed (a touring system that only stops at each location for a single day might set the lease time to two or three days, just to be safe).

So, we set up IP addresses, and connect devices. How do we actually, literally, connect them?

NETWORK HARDWARE

Q: *Why did the data enter the router?*

A: *To get to the other network.*[11]

What is the difference between a router, a switch, a hub, and a wireless access point? Who cares, anyway? Well, I know we (as sound people) are all about making things sound good, and this is veering dangerously into a whole different category of nerdery, but bear with me—the answers are actually not all that complicated, and they are important to grasp in order to plan (and configure) contemporary sound systems.

ROUTERS

A router is a network device that acts as a bridge between two networks. For this reason, it is often called a "gateway." The routers that most of us have experience with are our home network routers. As mentioned earlier, our modem connects to the WAN (Internet), and our router connects to the modem. It is the router's job to take the wide world of the Internet and translate it into data that we can access via our home LAN. The router sends packets of data (which are encoded not just with the data itself, but also with destination information, which is why you get your own email and not my email—unless you are a hacker and a creep, in which case, *cut it out!*) from one network to another, and ensures that the data is delivered. Most routers these days also contain features typical of switches, and often such other features as a DHCP server, DNS servers, firewalls, etc. Routers are used in show systems where some part of the system needs to communicate with the greater web (remember the example of dialing in across state lines to monitor show progress), or where two or more show LANs may need to interact. In the latter case, this might be a situation where sound systems are one LAN and video systems are another, but occasionally very specific types of data must cross between those otherwise isolated networks, so a router might be used for this purpose. A router allows devices with different NetIDs to communicate.

SWITCHES

A switch is a multi-port device (ports in this instance are the physical network cable connections) that allows multiple devices on the same network to talk to one another. In a switch, a computer connected to Port 1 can send and receive data to and from any other computer connected to the other ports, whether there are 8 or 24 or 48 of them. A switch keeps track of the MAC addresses of each device connected to it, such that Computer A connected to Port 1 can send data to Computer G on Port 7 without the switch having to copy the data to each individual port. Because the data itself contains the destination, the switch merely finds that destination and sends the data on its merry way. A switch will be listed by its communication speed ("gigabit switch" for example), and it must be able to serve that speed to all devices connected simultaneously with no loss in speed to any others.

Switches come in two basic types: managed or unmanaged. A managed switch allows the user to set priority for certain signals, such that in times of heavy traffic, the important data gets through in full quality at top speed. This QoS (Quality of Service) shaping can be critical in themed installations where some piece of the network is life-safety oriented, and others are less important. In this case, you might give priority to the emergency communication devices, and give less priority to the server playing the background music in the queue. Managed switches can also offer a great deal of detailed monitoring of network traffic and quality to and from all connected devices. Unmanaged switches offer none of these options, but just send a signal from place to place. Unmanaged switches have the advantage of being worlds easier to set up and use, so systems that don't need QoS or other such data tools usually use unmanaged switches.

HUB

A hub is a dumb version of a switch. In a hub, there is no tracking of MAC addresses, so the data is literally just copied to all ports. A hub also lists its performance by data rate, but that data rate is shared among the ports. So, if Port 1 is demanding a lot of traffic, Port 2 might slow down as a result, which does not happen in a proper switch. It's the 21st century, no one really uses network hubs in show systems anymore, but you might encounter one in a venue whose network systems are older.

WIRELESS ACCESS POINT (WAP)

A WAP is a device that gives a network wireless connectivity. In a home wireless network, typically the WAP is built into the router, which serves both the function of translating the Internet to your home network and of sending it over WI-FI around your house. In a show system LAN (without Internet access) the WAP is used to allow access to show devices from wireless devices (such as the iPad controlling the console in the concert example earlier in the chapter). In this case, the WAP is just creating a wireless connection to a wired router, and thus to any wired devices connected to it. The router connected to your WAP may need to have some form of DHCP server operating inside it (unless there is another already in your network) so that wireless devices can join easily. WAPs are extremely common in production sound systems.

These are the basic hardware components of a network. There are variations on all of them that we haven't covered here ("metro edge router" anyone?), but if you grasp the earlier concepts, you are ready to start most show system work.

Figure 18.1 is an example of a basic show sound system, featuring a digital console, DSP, playback computer, iPad controller, PC programming computer, and network-enabled wireless mics. The map in Figure 18.1 shows how these devices interconnect and talk to one another.

It is important to note that unless necessary, it is generally desirable to keep show systems disconnected from external (Internet) networks. This process is often referred to by the colloquial term "air gapping" (which pre-dates the rise of wireless networks). This means that you would literally unplug the Internet-connected cable from your show machines. The reasoning for this is that once a computer connects to the Internet, there are a host of background processes it runs to maintain and monitor that connection. These can bog down processors at inopportune moments, leading to errors in show software, which is undesirable.

SUBNETTING

Subnetting is an IP process whereby a secondary IP address can be used to subdivide the first. Without getting too far into the weeds, let's imagine a computer with IP address 192.168.0.15. If you have set this manually, below the IP address field will typically be a field for Subnet Mask. By default, a 192.168.0.x address uses the first three octets as the NetID, so the generic subnet mask for that address will be 255.255.255.0. This is because any octet with 255 listed (when matched with the actual IP) will be the NetID. However, if we wanted to use a 192 address, but only use the first octet as the NetID (in order to expand the available device count on our network, for example) we could use the same 192.168.0.15 IP address, with a Subnet Mask of 255.0.0.0, and suddenly only the first octet is the NetID and the other three octets are the HostID. Anywhere the 0 appears instead of 255 in the subnet mask, that octet is now part of the HostID. This only works in sequence—you won't see a Subnet Mask of 255.0.255.0— that would make no sense. Once you have subnetted the devices on the network, other devices must be in the same network (NetID) *and* mask in order to communicate without a router. Subnetting can be very powerful for very large systems, when you want devices on the same numbering scheme but subdivided in how they can access one another. For most small or medium (and even many large) show systems, subnetting is not necessary.[12]

NOTE 1: NETWORK CONNECTIONS SHOWN AS GREEN LINES.

NOTE 2: DANTE PRIMARY CONNECTION IS SHOWN, BUT DANTE SECONDARY IS NOT. DANTE SECONDARY CONNECTION FROM CONSOLE TO DSP WOULD REQUIRE A SECOND (DISCRETE) SWITCH PATH THAT NEVER TOUCHES THE PRIMARY PATH. IT HAS BEEN EXCLUDED HERE FOR THE SAKE OF KEEPING THE DRAWING SIMPLE, BUT WOULD BE IN USE IN A REAL SHOW SYSTEM.

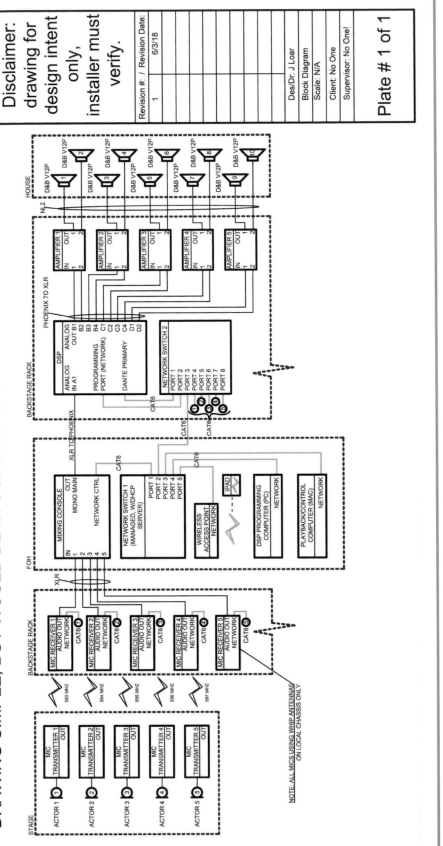

213

FIGURE 18.1
Sound System Network Topology Example

AUDIO OVER IP

In contemporary digital audio systems, many devices send audio over network connections. This is a very useful technique, as it means we can send large channel counts over single network cables, at high speeds, and with little or no error (and no worry about analog dropout or noise). There are some common protocols used by various systems for this kind of transmission. Before we can get into the differences between one protocol and another, we need a quick detour into what is called the OSI model of network communication (Figure 18.2).

OSI (Open Systems Interconnection) model is a structure of information transfer that divides different types of communication into layers.

Layer 1 is the physical layer. This defines actual physical characteristics of network connections (like, what is a Cat 6 cable, how many wires does it have, and how are they connected?).

Layer 2 is the data link layer. This is the link between two nodes, that defines how data passes from one to another at the elemental level. Ethernet and WI-FI are Layer 2 protocols.

Layer 3 is the network layer. This is the routing, addressing, and managing layer.

Layer 4 is the transport layer. This is a layer that performs data quality control, among other functions.

Layer 5 is the session layer. It establishes and maintains connections between devices ("sessions").

Layer 6 is the presentation layer, which translates between network format data and specific application formats needed for given software.

Layer 7 is the application layer, in which network data is finally translated and represented in a user-oriented format.

The aforementioned is a VAST oversimplification of each layer's functions, but in talking about audio over IP it is important to have some grasp of the model and its concepts. Some data protocols will make use of more than one layer—for example, many audio

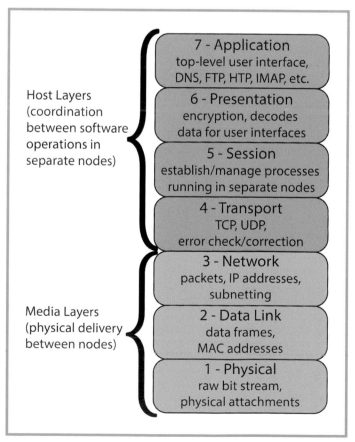

FIGURE 18.2
The OSI Network Model

protocols operate essentially on a layer 3 level but wrap the data in layer 4 packets. Thus layer 3 data can be sent via layer 4 forms, such as UDP.

COMMON AUDIO-OVER-IP PROTOCOLS

The following is far from a comprehensive list but addresses some of the most common network-audio types.

- DANTE (Digital Audio Network Through Ethernet)—a proprietary encoding type, created by Audinate. Dante is a layer-three protocol, which means it can send audio over standard wired LAN architecture (routers, switches, hubs, though not over the Internet or over WLAN). Dante devices vary in their configurations, with most able to transmit and receive 64 channels simultaneously at sample rates up to 48 kHz and bit depth of 24-bit. The Dante protocol is actually capable of sending and receiving a total of 1024 signals (512 each direction), at up to 192 kHz sample rate and 32-bit depth. Most hardware will not support this high channel count, though, and some devices feature a significantly lower count, even down to 16 channels at 48 kHz. Dante is very easy to configure. Most Dante devices are equipped with a Primary and Secondary port (for redundancy in case of failure of one of them). It is important to note that Dante Primary and Dante Secondary connections must be made over unique switches (and never interconnected). If Primary and Secondary are connected to the same switch (or to the same network path, even by interconnecting two different switches), the receiving device(s) will get confused, and many signal errors will occur. In many systems, Dante routing is configured via a computer running the Dante Controller software, which offers a simple crosspatch matrix for signal assignment. One Dante signal can be simultaneously sent to many receivers, though each receiving channel can only accept one transmitted channel. Dante also is used via the Dante Virtual Soundcard application, a software that allows a computer's wired NIC to become a Dante port, sending and receiving audio data over the network cable. It is very common these days to see QLab playback systems running via Dante Virtual Soundcard, as the transmission is high quality and low latency, and requires no hardware besides the computer itself (and maybe a network jack adapter if your computer does not have a physical network port). Dante is one of, if not the, most common audio-over-IP protocols as of this writing.
- AVB (Audio Video Bridging)—an open standard (non-proprietary), that in theory allows endless channels of 192-kHz/32-bit audio, scaled only to the speed and bandwidth of your network architecture. While the published standard does not have a channel count, in the field, practitioners have found that there are some practical limits (albeit very high limits). When AVB was introduced, it looked like it would take over the audio-over-IP world both because it is open (meaning there are no license fees to use it), and because it is theoretically limitless. However, between the announcement of the protocol and the actual publication of a usable standard, a large time gap elapsed, and Dante slipped in to take AVB's place. Some pro audio companies (notably Meyer Sound) are still working with AVB and hoping it will be the next big thing, but AVB devices are as of this writing relatively rare in the field. AVB is also a layer 3/4 protocol, meaning it can transmit over standard wired networks. AVB can also be used as a layer 2 protocol to transmit over the Internet. It has begun to be widely implemented in manufacturing plants, for quality-check cameras on assembly lines, and the like.
- CobraNet—this was the first successful audio over Ethernet format. It is a layer 2 protocol, but unlike AVB does not make use of the IP protocol, and as such cannot traverse routers (and thus cannot send over the Internet). It requires a complex addressing scheme, involving the selection of a transmission as either unicast or multicast (whereas a Dante signal can be flexibly assigned in that fashion at any time via the controller). CobraNet is not as reliable as newer standards and is generally being phased out of most professional system design.
- AES67—this is a promising new open protocol that is designed to make all other audio-over-IP protocols inter-operable. So, if you have a Dante signal, but the device that needs to receive it doesn't have a Dante port but does have an AES67 port, it can still receive that Dante audio. It provides translation for Dante, AVB, Q-LAN (QSC's proprietary Layer 3 protocol for use in its Q-SYS processors and its network-enabled power amps and other devices), RAVENNA (a high-resolution, high-bandwidth signal type designed for use in broadcast transport), and others. AES67 promises to return audio to the adaptability of the analog/copper wire age, where any signal could essentially be transported to any device. It sacrifices some of the proprietary data along the way (for example, Dante is capable of carrying its own clock signal and some control data as well as audio, and when it is sent via AES67, only the audio transmits), but it is an exciting recent development that is being found in more and more new equipment.

When configuring show systems that feature both audio-over-IP and control networks (as most do these days), it is often desirable (if not required depending on the protocols in use) to keep the control network and the audio network as discrete physical networks (or at least discrete subnets). This prevents confusion in setup and operation.

Computers are everywhere in production systems today. They are not going away, and, in fact, everything is getting more computerized by the moment. A good grasp on computers and networks is essential for the system designer. While I haven't (and won't) go into it here, it is assumed that as a system designer, you will have proficiency with the essential functions of computers themselves. Many designers like to stick either to Mac or Windows computers, but in truth, it will serve you best to understand both. QLab, the popular show sequencing and playback software, is Mac-only. The configuration software for all of BSS's DSPs is Windows-only. Logic Pro, the popular DAW, is Mac-only. AFMG's EASE® is Windows-only. If you don't know both platforms quite well, you will be hamstrung in choosing (and especially in configuring) the best gear for your system.

Now, returning to equipment, let's imagine a theme park ride. In such a large, complex system, is there a stage manager sitting in a booth, calling cues for the audio system (like there would be for a theater production)? No. We need to automate events in such systems (and even large theatrical systems, such as many Cirque du Soleil shows), so we turn our attention in the next chapter to *show controllers*.

Notes

1 For example, I do not give any depth of attention to QoS (Quality of Service) shaping, or network security, both of which are very relevant to large themed systems, but as an introductory guide, this book does not have space for these topics. See the Bibliography/Further Reading list for more info on these topics.

2 By "computer," I mean any device with a computational engine. This includes PCs, but it also includes digital audio consoles with Ethernet control, DSPs, and so on.

3 Just wait till I actually invite you to a party. The puns never stop!

4 Big ups, Focal Press!

5 For further discussion of triggering and show control messages, see Chapters 19 and 20.

6 At least, for the "universal," or burned-in, MAC addresses. There are ways to assign a "local" MAC address in certain network configurations, but this isn't relevant to our discussion here.

7 In non-network applications, MAC addresses are used to identify computers and peripherals so that they may operate together for certain tasks. Bluetooth communications, for example, use MAC addresses to identify senders and receivers. This is a form of communication that does not necessitate IP addresses.

8 All addresses used in this chapter will be IPv4 addresses, as IPv6 has not yet (as of this writing) made its way into most production audio and control systems.

9 Cat 6 is backward compatible with Cat 5, such that a Cat 6 cable can work on a Cat 5 device, but a Cat 5 cable will not work on a Cat 6 device. These names are short for "Category 5" or "Category 6" and are just names for evolving standards of wiring for network devices. The standard RJ45 connector is what is also known as an "Ethernet" connector, and until recently was available on just about every laptop and desktop computer— though now many computers require you to use an adapter to change a smaller port (such as USB-C ports) to a network jack.

10 RIP Brooklyn's Galapagos Art Space, with the worst FOH mix position I've ever had to suffer—and in the days before iPad control was a thing!

11 I'm very sorry for this joke that is only barely a joke.

12 Subnetting can be much more complex, and more powerful, than the simple example I have described here— but advanced subnetting is beyond the scope of this book.

CHAPTER 19

Show Control

You are seated in the audience for a grand cirque spectacle. As the show begins, lighting shifts, live musicians start performing a song, a rich field of sound effects plays, video projections hit the walls around the theater, actors announce their presence onstage, acrobats lower into the space on flying trapeze rigs, the stage deck itself transforms from a flat panel of flooring to a range of trap doors and elevated walls (a maze of decking left behind), from within the traps, dancers seem to positively fly from beneath the stage into leaping arcs above the stage floor, but none of them wear any wires. The audience oohs and ahs, and for the typical attendee, the goings-on are nothing short of magical. For the technical designer, however, there is a complex and rigorous system (or, often, there are multiple interlocking systems) that control all of this action.

In a typical conventional stage play, a stage manager issues calls over a headset to light, sound, and stage crew operators to perform their tasks. While a grand spectacle such as the one described earlier certainly still requires the guiding hand of a stage manager, there are some advanced sequences that must be so precisely timed and calibrated that leaving their exact interaction to the vagaries of even the best stage manager's calling is not practical. If the launch platforms that speed dancers from below deck to their aerial arcs do not fire at the exact right moment, a dancer might strike their head on a piece of moving stage that hasn't completed its path. If the music plays too soon, or too late, or if the sound effects don't sync with the actions of the actors, things fall apart. It is here (and beyond the stage, into complex themed installations and the like) where we must engage with the beast known as *show control*.

DEFINITION

In a typical theater sound system, we use sound control subsystems all the time. A cued playback software (like QLab) is such a system. It allows us to sequence recallable actions, to perform them in the order and at the time desired, and to regulate levels, routing, and other parameters key to playback success. In many theatrical

Show Control

A computerized master control system that provides triggering and timing for other subsystems (such as sound, lighting, or scenery automation systems).

environments, QLab is operated by a human, receiving commands over a headset from the stage manager, in a manner that is now decades-old and fairly standardized ("Standby sound 15. . . and . . . go!"), even if the technology for playback is relatively new. This is an audio playback control system that serves us well.[1] However, let's imagine that we need to fire the complex set of cues described in the earlier cirque extravaganza. If we relied on the conventional means of sequencing events, where separate sound, lighting, scenery, and other operators all awaited their specific cue from the SM and then "took the cue" (performed the designated action), as fast as they humanly could upon the verbal "go" command, there would still be a lot of wiggle room for things to get out of sync in a complex scene. In situations where we don't want that wiggle room to exist (or worse, where that wiggle room can endanger performers' or crew members' safety), we often employ show control systems, which can then cue multiple other actions from one master program, such that they always happen the same way, with the same timing, every single time we need that scene to be onstage (or in the theme park ride, or the corporate lobby installation, or what have you).

There are a number of different types of show control systems, and while it is beyond the scope of this book to go into great detail about them, it behooves a contemporary sound system designer to have at least a passing understanding of what the basic standards are, and how sound might need to interact with them.[2]

To put it simply, whether the show controller device operates a program that is self-contained, responding to venue conditions and acting autonomously (as is usually the case with most theme park systems), or the show control system is still human-triggered but serves to calibrate a host of other show subsystems, the general purpose of a show control system is the same: calibrate and align the performance of sound, lighting, video, scenery, and other systems such that complex sequences are repeatable without show-damaging variations.

A typical show control system for a mid-sized production might be configured like so (Figure 19.1):

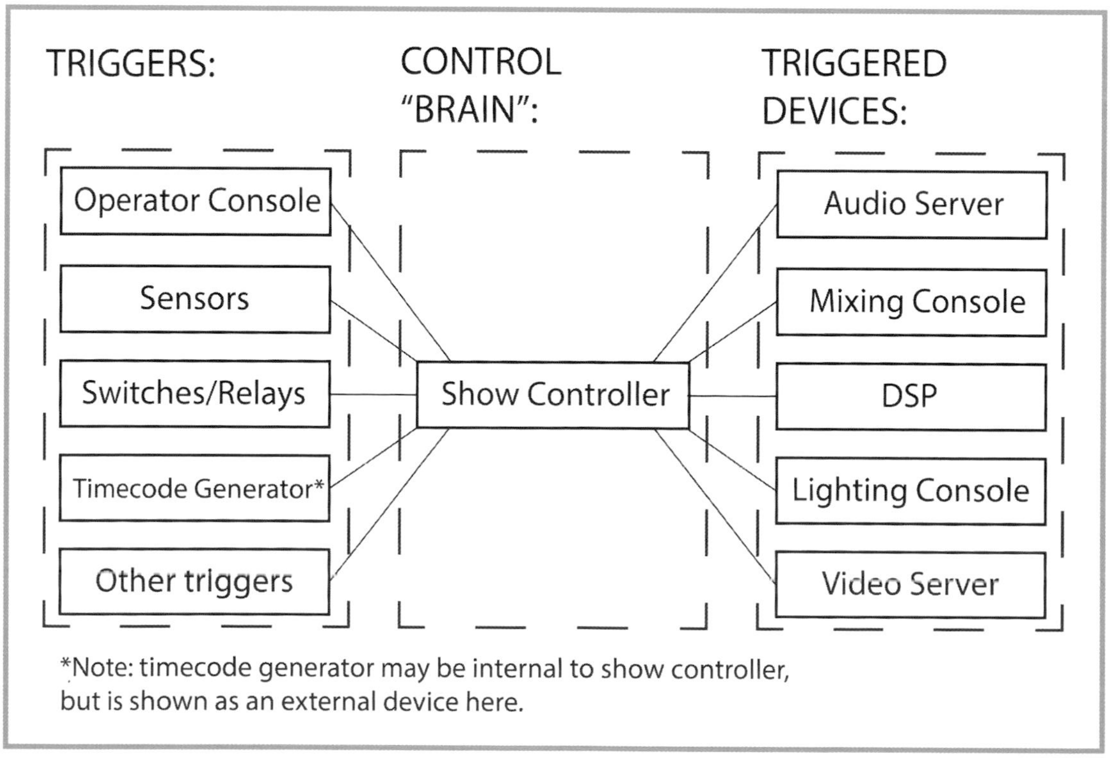

*Note: timecode generator may be internal to show controller, but is shown as an external device here.

FIGURE 19.1
A Typical Mid-sized Show Control System (Conceptual Diagram)

Complex systems scale up from this model, taking the simple signal path of control messages and interweaving it with more devices, possibly redundant controllers, and other outboard hardware that allow the system to handle multifarious configurations.

CLASSIFICATION

There are two fundamental approaches to sequencing events via a show controller:

- Human-operated
- Autonomous

In human-operated control systems, there is, quite evidently, a human operator. In such a system, the operator may await an actor's cue line before pressing the "go" on the complex sequence that triggers the stage automation, the sound, the lighting, etc. While the predetermined sequence may be calibrated to the millisecond, the moment at which that sequence *starts* is determined by the human operator's engagement of the trigger mechanism (whether a literal button on a computer, a virtual button on a touchscreen, or via some other technological control method).

In autonomous systems, there is no human operator. The sequence may be triggered automatically by a performer walking past a sensor embedded in a certain part of the stage, or it may be triggered by a ride vehicle entering a "scene" of a theme park ride, or by the time of day as set in the show's clock, or some other fixed parameter. Once that parameter is met, whatever it might be, the predetermined sequence is fired.

There is a third approach, that is merely a synthesis of the prior two. In such a system, some parts of a program may be human-operator dependent, and others may be completely automated (an example might be a stage show in which the show is designed to only start at a human operator's cue, leaving room for the house to be held if audience members are still lined up outside come curtain time, but wherein the show itself might run in a completely automated fashion once begun, triggering events off a timecode/timeline that is preset).

In the foregoing examples, we've already identified two different means of triggering actions in a show sequence:

- Events
- Time[3]

An event-based trigger might be something like the actor passing a sensor onstage. The sensor itself will send a signal once it is tripped, and that signal tells the show control system to perform the assigned action.

A time-based trigger might be taking cues based on one of two forms of time:

- Wall-clock time: the time of day as set in the show systems. For example, a casino with a fountain that performs an elaborate light and water show every evening at exactly 7:00 pm might take its cue from the wall clock inside the show systems.
- Timecode: a running clock, started at the beginning of a show program, that specifies certain events to take place at given values relative to the start of that timecode. If a show starts at 0 hr, 0 min, 0 sec in timecode, it immediately starts counting up in seconds, mins, and so on. At exactly 1 min 22 sec in, for example, it might trigger a sequence of explosions around the stunt show stage. That means that whenever the clock was started, the first round of explosions will always be 1 min 22 sec later, regardless of the "wall clock" time.[4]

When designing sound systems that must interact with show controllers, a designer must consider what type of program is being demanded, in the foregoing terms, in order to determine if the device(s) they specified will meet the challenge. Show control systems can often send commands in a wide variety of languages, from simple UDP (a basic computer-level command language, highly customizable), to MIDI or MSC (MIDI is Musical Instrument Digital Interface, designed for use controlling synthesizers but implemented in many playback systems, mixing consoles, and other devices as a trigger language; MSC is MIDI Show Control, a specific offshoot of the MIDI language with custom cue forms for lighting, sound, automation, and other device types), and OSC (Open Sound Control, another language developed for advanced communication between sound devices, very flexible and increasingly in use for inter-system communication). These are only a few of the available options. Verifying what control language(s) your show controller will speak is critical before selecting sound equipment that must respond to that controller. In addition, some show systems will be content to fire a command into a sound device and trust that it will work. In other systems, a return message (feedback) is needed, verifying that the sound system received and is executing the cue. Selecting sound devices that perform as desired in these terms can also be critical.

A sound designer must also consider the demands that will be placed on the sound equipment by the show control paradigm. If you are designing a stage show that will perform at most twice in one day, PC or Mac computer-based playback systems might be appropriate. If you are designing an installation for an airport, or a theme park, or another environment where the sound playback must run all day every day for eternity (or at least several years), your laptop-based solution is likely to fail far too soon into the production's run, and you will need to specify a more robust media server.

Show control systems can take on a number of physical forms. From self-contained devices (like a Medialon Showmaster Pro—a rack-mounted brain that is very good at operating autonomous programs all from one central machine) to distributed systems (like an AMX-based platform, which might consist of several small control hardware devices, GPIO widgets, touch panels, and so on). Often, a show control system will land somewhere between the two—with some dedicated hardware running critical functions and some distribution of smaller processing or sensing tasks to other hardware or software. Even autonomous systems normally have at least one operator station, for use in case of emergencies—so that the show can be overridden/canceled in the event that the building needs to be evacuated.[5]

While in the AV world, many sound system designers end up designing control systems as well as sound systems (I certainly do), this book is not a control system design manual. As such, we'll proceed to the specification section to discuss specific languages, connection types, and other such info that is salient to the sound system designer choosing gear to work in a given show control paradigm.[6]

SPECIFICATION

There are not many specs we must consider when evaluating whether our audio systems will work with our show controllers. Here are the four that are most important in this equation.

Control language and connector: what language will the show controller be issuing for its triggers? What connector/ cable will that message be transmitted over? Can your sound device accept and respond to that language?

- UDP/Ethernet—direct network messages, either in a standardized or customizable format, RJ45 connectors wired as Cat 5 or Cat 6
- Serial—either via USB or RS232 or the like
- GPIO (General Purpose In/Out)—typically 2-wire connections carrying simple control messages, often via relays
- SMPTE timecode—setting and sending timecode messages, often over XLR
- MIDI/MSC—either over MIDI cables (three or five-pin DIN connectors) or over Ethernet
- OSC—usually over Ethernet
- DMX—not usually applicable to sound devices, but often specified in control systems for lighting or video, and now available within QLab (hence its mention here)

Latency: how fast can your audio device respond to prompts from the show controller? This spec isn't always listed, and you may need to contact the company directly for this info. This is only relevant if the timing of events in your program is extremely precise.[7]

Feedback: I don't mean that screechy noise when you put the mic too close to the speaker, I mean—can your audio device or system return confirmation that it received a control message and is acting on it? Most advanced DSPs and media servers can be programmed to deliver this kind of message. Even QLab could be configured this way (by nesting audio cues with network cues inside an overall group cue, using the network cues to send a message back to the controller each time audio is fired). It may be challenging, for example, to find a mixing console that performs well in this respect. However, if your show has a mixing console, chances are it also features a mix engineer, so the human operator in that equation can serve as the confirmation/feedback mechanism.

Cue type: Does your audio system work via cue events? Recallable presets/scenes? Does the show controller need to actually tell a fader to move in a mix system because the fader can be remotely triggered but the device has no recallable scenes? Understanding what programming events/mechanisms are native to your sound device is important for making sure your show control system will properly control the device.

CONSIDERATION

What questions must a designer ask (and answer) in order to select sound equipment to work with their show control system?

1. Will the devices receive messages in a language that the controller speaks?
2. Will the devices react to messages in a timely fashion?
3. Will the devices provide feedback if needed?
4. Can the events that need to be triggered/sequenced from the show control perspective be triggered/sequenced within my sound device?

If you are also the show control designer, there are a host of other questions to be asked, but the only one I will list here is the obvious one: does this project require a master show controller, or can decentralized sound, lighting, and other systems suffice?

Show control is becoming increasingly common in advanced media productions. While you may never program the show controller itself, understanding what the show controller may be tasked with doing, what in turn the audio system will be tasked with doing, and how those can interact, is critical to designing contemporary sound systems.

In the next chapter, we will briefly examine some basic trigger types, sensors, and customized parameters. We will examine these in the context of triggering a sound system directly, rather than triggering a show controller which then triggers a sound system—but please understand that anything we discuss as triggering a sound system in the next chapter can very likely also be used to trigger a show control system as a whole.

221

Notes

1 To be fair to QLab, it has grown in complexity in recent versions and is becoming a fairly robust show controller all its own, with options to send network messages, DMX lighting cues, video cues, and so on. But for our purposes at the moment, we refer to it as a sound control system.
2 For a much more detailed look at show control and the evolving standards and techniques, I heartily recommend *Show Networks & Control Systems* by John Huntington (in its second edition as of this writing).
3 I am indebted to John Huntington for his clear explanations of these concepts, from which I am cribbing loosely in this chapter.
4 Timecode is typically some variation of the universal SMPTE timecode—courtesy of the Society of Motion Picture and Television Engineers—which expresses time in HRS:MINS:SECS:FRAMES. MIDI timecode is a common variant, which expresses the same values but adds up to a quarter-frame value beyond frames. Other variations exist, but within all of them, the principle remains that whenever the timecode sequence starts, all events triggered by that timeline will happen at the same relative distance to the starting point, every single time. In some older systems, it was necessary for SMPTE timecode to begin at 1 hr, instead of at 0. This was due to the fact that if you needed to then sequence something just before your starting point (effectively changing the starting point), you would either need to shift every existing event forward in time to accommodate the new event, or you would pass back around what is known as "SMPTE midnight," in which case your first cue isn't at the start of your sequence, but is actually at the end of a 24-hr cycle. In many modern devices, an alteration to the code has been made to differentiate between –5 sec (5 sec before the original zero starting point), and 23 hr, 59 min, 55 sec (which –5 would equate to if you pass SMPTE midnight). In practice, some devices still automatically start all timecode clocks at 1 hr, both as a way of ensuring interoperability with older devices and out of respect to convention and user expectations.
5 Though even in this case, most autonomous systems also have a hard-wired fire-control trigger, such that if a fire emergency/alarm is set off, the show is automatically canceled and/or put into a "safe mode" that allows for

emergency announcements, exit lighting, etc. You don't want to rely on low-skilled operators to determine when there's an emergency if you can have well-designed systems that trigger emergency status instead.

6 And again, if you find yourself needing to actually design a show control system to work with your audio system, and you need guidance, you can't go wrong with John Huntington.

7 That said, this can be relevant. For a particular theme park attraction, I had to work with QSC's Q-SYS engineering department to determine the latency of their media server playback system, available within the Q-SYS cores as an option, so I could be sure that the audio would be as instantaneous upon triggering as needed. Their system exceeded our expectations, and I specified it. It took a few phone calls, but the info was not hard to get from the professionals.

CHAPTER 20

Triggers, Sensors, and Customization

A show control syllogism: all sensors can be used as triggers, but all triggers are not sensors.

DEFINITION

In a production sound system, we often want a particular action to take place upon the prompting of some input value other than one called by a stage manager. In these cases, we need to program our triggers. A trigger for, let's say, the playback of a sound file, could be virtually anything. As discussed in the previous chapter a trigger can be a wall-clock value, a timecode value, or any of a host of other input types. This chapter will address some of the more common variations of those *other* input types from the physical standpoint (devices *used* as triggers) and give some examples of how such inputs might be used to customize the actions of a sound system. Before getting into any detail, I must mention that this is by definition a very limited list. It is beyond the scope of this text to go into great detail about the wide variety of industrial sensors and such devices that are available, and almost any of them could potentially be configured for use in a sound control system. This chapter will address some of the most basic and common of those found in professional production.

> **Trigger**
> A data message containing a value or values that will be associated with an action within a show system.

> **Sensor**
> A device that detects or measures a physical property and records or transmits the resulting value.

CLASSIFICATION

There are a few major types of sensors or input devices that we find in production systems. Some are simple electromechanical devices (like a switch), and some are more complex (like an optical position sensor), but all share the basic trait of sending information to our production sound system to trigger it to change its state. Any of these devices may be used in essentially one of two ways:

1. Momentary—a momentary trigger may be either receiving sensor data in the form of one value at one instant (like a switch that is flipped to a new value) or receiving data continuously (like a position sensor), but in either case, a momentary trigger is only looking for one value among a range, and when that value is reached, the assigned action will be performed within the sound system. This could be as simple as a switch turning from OFF to ON, or as complex as an encoder that transmits over a 0–255 range reaching the value 187 (or any other assigned numeric value). A momentary trigger is not looking for a range of info, only for one piece of information.

2. Continuous—a continuous trigger system will deliver a state of information, either as a fixed position (a switch set to ON that triggers a continuous action until the state is reset to OFF), or as a stream of data (a position sensor that continues to modify a sound parameter with every new position recorded of the object(s) in question). Continuous triggers are providing enough info that the sound system can use their recorded values as input for behaviors that need to affirm the trigger's status at all times during the intended action's operation—for example, a filter cutoff frequency could be mapped to a continuous sensor's values, such that the cutoff is constantly changing, whenever the data coming from the sensor changes.

Switches (Figure 20.1): a simple switch, two-pole, either opens or closes a circuit path. In your home, an electrical switch may be mounted on your wall to turn on your ceiling lights. In a production system, a simple switch can be wired to GPIO ports in a controller to allow it to trigger anything. A set may have a physical switch mounted upstage of a scenic wall, to allow an actor to flip the switch to ON, triggering the "doorbell cue" from the sound system, which only turns off either at the predetermined end of the sound content, or when the actor flips the switch back to OFF, opening the circuit, and triggering the stop cue for the doorbell. In special applications, relays (switches triggered by a pulse of energy) may be found as well.

Encoders (Figure 20.2): often in the form of a physical knob, these sensors read the position of a central knob (or fader/slider) and transmit that value as either a level of voltage or as a numeric (digital) value. Used either for continuous value control (e.g., controlling the level of an audio signal), or for single value triggering (e.g., when the knob's détente reaches 1 o'clock on the dial, a sound is fired, but no action is performed at any other position).

Entry sensor: most commonly, a photoelectric sensor (Figure 20.3) that merely sends one value when its beam of light is not broken, and another value when the light is broken. Often mounted over or adjacent to a doorway, such that when the door is opened, the beam of light is interrupted, and a show system (receiving this changed-state message), is triggered.

Inclinometer: a device that measures the slope of an object with respect to the earth's gravitational force. Can be placed in a scenic, prop, or costume item to then transmit a value to a production system. This could, for example, be a continuous value trigger, assigned to an effect value—say, the cutoff frequency of a low-pass filter—such that when the object is perpendicular to the stage floor, the filter is open, but as the object tilts toward parallel to the floor, the filter cutoff rolls steadily down until it is filtering everything above 50 Hz. Or, this could be used as an absolute value trigger, such that when an object is at any angle other than 82° to the floor, nothing happens, but when the object passes through 82°, a sound file is triggered.

FIGURE 20.1
A Basic Switch—Honeywell 101TL1–2 Toggle
Source: Courtesy Honeywell International, Inc.

Position sensor: these take a wide variety of physical forms, but in general, these are sensors that track objects. They come in two essential formats—absolute (where in general is the object) and relative (displacement—how far is the object from its starting position). Position may be sensed in a linear domain (closer to or farther from the sensor), an angular domain (is the object now on a different axis from the sensor), or in a multi-axis, highly detailed, domain (Figure 20.4). Position sensors may trigger a wide variety of actions, and as with the inclinometer, values utilized as triggers may be a single value or continuous. While, often, these sensors utilize light in order to track object positions, some advanced Radio Frequency (RF) transmitter and receiver systems can be calibrated to determine accurate position (this is part of the principle of advanced actor-tracking systems that retime mics to allow for localization to shift as an actor moves around a space, such as Track the Actor, or StageTracker).

SPL sensor (Figure 20.5): a sound sensor, responding to varying levels of sound pressure. Again, can be used for continuous or single-point triggering. A MIDI Trigger (Figure 20.5) is a variation of this, whereby either a certain SPL level or a certain impact value (such as the physical shock of a drumstick striking) sends a MIDI message to a synthesizer or control system.

FIGURE 20.2
A Rotary Encoder—Honeywell 510E1A48F416PC Encoder
Source: Courtesy Honeywell International, Inc.

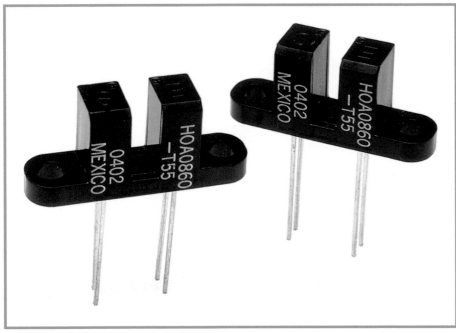

FIGURE 20.3
A Photoelectric (Optical) Sensor—Honeywell HOA0860-T55
Source: Courtesy Honeywell International, Inc.

FIGURE 20.4

A Multi-Axis Position Sensor—Honeywell 6DF-1N6-C2-HWL

Source: Courtesy Honeywell International, Inc.

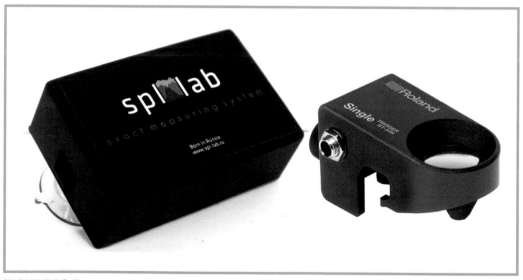

FIGURE 20.5

(Left) An SPL Sensor—SPL Lab Exact Measuring System, (Right) A MIDI Drum Trigger—Roland RT-30I I

Source: (Left) Courtesy SPL Lab; (Right) Courtesy Roland

These, again, are just a *few* of the most common sensors in use in show control systems. However, one need only glance at the Wikipedia page "List of Sensors"[1] to be overwhelmed by the sheer range of options for sensing physical values and transmitting data in response. That data can be flexibly assigned to virtually any parameter in a computerized sound system, and thus control systems can be triggered in thousands of different unique and creative ways. Themed entertainment is making increasingly sophisticated use of such devices, in order to create interactive and "gamified" environments.[2]

SPECIFICATION

There are essentially three values a designer must understand about any sensor when incorporating it into a sound system:

1. What kind of data is delivered: hi/lo switch value? Numeric range or single value? Something else? If it is a numeric value, is it transmitted as digital information (or as an encoded voltage value)? If it is digital, in what format or language is the message written? Can my system read that data directly, or do I need another software/hardware controller to translate the values into a message format my sound equipment can understand?
2. How many wires does it take to deliver that message (and does my sound system have enough GPIO ports to take the message in, or can the signal transmit over Ethernet)?
3. What does the sensor need for power? Many small sensors run on standard 5-V power, but there are variations. Do you need to provide a power source, or do the GPIO ports on your device provide 5-V rails?

CONSIDERATION

What questions must a designer ask (and answer) in order to select sensors or triggers for their sound control purposes?

1. What actions are being sensed? Are they continuous actions or hi/lo, on/off, values?
2. Where are those actions? How close to the action does the sensor need to be in order to properly read the action and deliver useful data? How accurately must the sensor read a given action?
3. How fast does the sensor need to transmit in order to provide usable data?
4. Does my sound system have GPIO or appropriate network connectivity to read data from the sensor, or do I need some kind of logical controller at the front end (e.g., a microcomputer translating GPIO voltage levels into OSC values which it then sends to QLab)?

Depending on the sensing solution, there may be a host of other questions that need asking. As there are many, many types of sensors, it is beyond the scope of this book to inquire after every dimension of the selection of every individual sensor type.

While automated sensors are increasingly found in contemporary production audio systems, the most common trigger is still our good friend the stage manager. However, our stage manager can't do her job without a robust communications infrastructure. While we may not think of it as part of the sound system, com system design is still, almost always, the job of the sound system designer. In the next section, we'll evaluate contemporary production communication systems.

Notes

1 https://en.wikipedia.org/wiki/List_of_sensors accessed 1/30/18.
2 In one real-life case (that will remain nameless due to trade-secret laws), scenic items in a themed environment are covered with capacitive thin film, such that guests to the park can touch different objects and thereby trigger sound, lighting, and animatronic actions within the environment. The guests are not told this is possible, and so the exciting sense of raw discovery when they accidentally come across it is enhanced by the seemingly magical responses of the environment to the guests' actions.

Audio Com Systems

In any given production, or at any given venue, workers need to be able to communicate with one another. During live events and at all sorts of entertainment and business venues, one of the most common solutions to the problem of getting people in touch is to use audio intercom systems, colloquially referred to as "com." Whether it is the retail staff at Old Navy or the stage crew for *Wicked*, countless professions depend on headset-based multi-way audio communication systems.

DEFINITION

Com

Any variety of multi-way direct communication systems, from simple two-way radios to complex, IP-based digital matrix systems, used by production crews to communicate mission-critical information without interruption. Most often a series of audio devices—with hard-wired and/or dedicated radio band communication links—though increasingly tied to video feeds for more detailed communication.[1]

CLASSIFICATION

A simple com system might consist of a single-channel multidirectional communication network, in which a central base station is controlled by a primary operator (e.g., a stage manager), and the distributed team members (e.g., stage crew) each tie into the main station via remote beltpacks, some of which may be wired directly into wall patch points (around a venue) that tie directly to the main station. Others may be communicating via wireless beltpacks, which receive and transmit signals from and to a wireless base station (which itself has to be wired directly to the main station).

In such a system, seen in Figure 21.1, anyone may communicate with anyone else, but everyone else can hear you.

Such a system is very limiting in its operation, of course, because any overlapping conversations essentially turn to cacophony, so this is really only viable for very small operations with very few communicating parties.

In a two-channel system, the controller of the base station has the option to talk to either channel independent of the other. If the crew has two-channel beltpacks, they also may communicate with either channel of the crew, but, typically, as channels expand, production departments are subdivided to their separate sub-channels. In a two-channel system being used for stage production, Channel 1 might belong to the lighting team (the board and spot ops), and Channel 2 might belong to everyone else.[2]

In many small- to medium-sized productions (or even a Broadway show with very little technical cue-calling work to do), the four-channel analog party line com system (Figure 21.2) is very common. In a four-channel system, lighting, sound, and scenic/stage crews of a theatrical production can each have a dedicated communication channel (for communication that doesn't pertain to the other departments), leaving a fourth channel for a video projections crew, a feed to a music conductor, or any other necessary department or departments.

In many contemporary large-scale productions, IP-based digital com systems are very common. From Broadway musicals to the Super Bowl, programmable and flexible digital com systems are capable of much greater communication complexity. In these systems, the operator stations often feature many direct toggles or buttons to which can be assigned banks of different communications endpoints, either singularly or in groups. Say a stage manager needs to communicate directly with several major stars by calling directly to their dressing rooms backstage, these rooms can each be assigned their own control on a panel, and they can also be grouped in any way or combination desired. Each crew can easily have their own channel, and sub-channels become easy as well (e.g., a lighting crew that has a separate communications path to just the spot ops in the gallery vs. talking to the board op or the assistant designer or whomever). These systems are customizable, flexible, and worlds more advanced than any system available in the analog domain. While control panels and matrix software have been available for some time now for digital systems, it is fairly recently that we've begun to see real developments in wired beltpack connections via Ethernet cable (rather than analog XLR via a digital-to-analog converter system). With the use of PoE[3] switches/interfaces, a single Ethernet cable can now deliver multi-way audio to a beltpack and headset, and the equipment is much lighter than traditional analog gear, which crew members tend to appreciate.

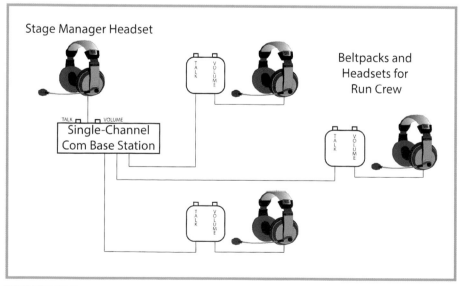

FIGURE 21.1
A Basic Single-Channel Com System

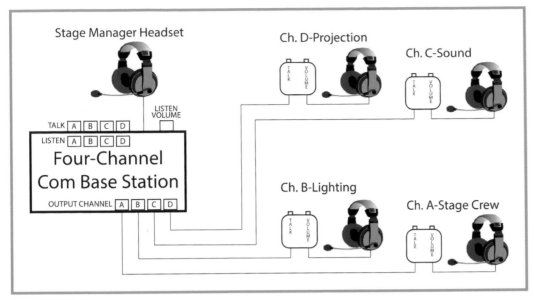

FIGURE 21.2
A Typical Theatrical Four-Channel Com System

As in the lighting world, where old two-scene preset consoles and Altman 6x9s still populate smaller venues, analog com is going to be with us for some time, but the digital era of audio intercom is spreading quickly, due to the enormous advantages in the customization (and thus the rendering more efficient and effective) of your communication paths.

As you can see in in Figure 21.3, such a wide variety of communication options is only feasible with a digital system.

SPECIFICATION

While each production intercom system will share certain basic components (a base station at the center, some form of endpoint communication—beltpacks, wall panels, etc.), there are a host of variations, and the differences between conventional analog party line systems and digital multi-point systems become greater as we drill into the specifics. In this section, we will examine the most common types of equipment found in intercom systems, identify general differences between analog and digital systems, and identify any relevant specs.

- Base stations (Figure 21.4): base stations are the central control panel of the intercom system. In a traditional analog system, there is typically one base station only, used by the stage manager or another production supervisor (though there are remote stations, which essentially copy the features of the main station to a secondary location, for an assistant stage manager, for example). In a digital system, there can easily be multiple base stations (often called *user panels*, or *control panels*) each configured with custom control options and communication routes. In an analog system, the base station also typically serves as the power supply for the system itself. In a digital system, the power supply is typically separate, and the base station serves as a control surface front end for the processors and power housed elsewhere (much like a digital mixing console whose I/O is housed in an outboard stage rack).
- Relevant specs include the following:
  - Channel count—how many distinct communication paths is the station capable of?
    - Analog: one to four channels common
    - Digital: nearly endless (contact specific manufacturer for details)
  - Frequency range
    - Analog: often limited to intelligible speech range, between 500 Hz and 10 kHz, for example
    - Digital: should be full-range 20 Hz–20 kHz

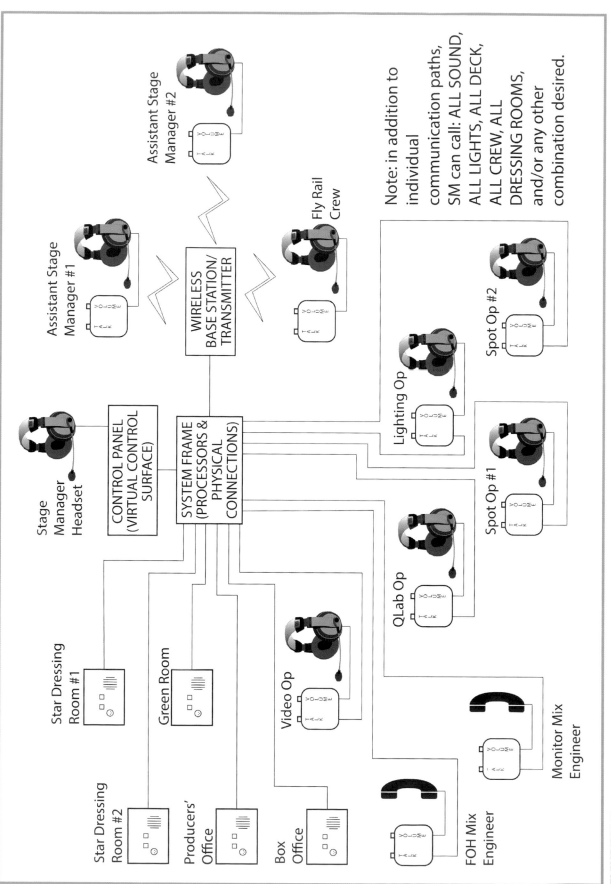

FIGURE 21.3
An Example of a Digital Intercom System

FIGURE 21.4
(Top) Analog Intercom Base Station With Four Channels—Clear-Com MS-704; (Bottom) Digital Base Station With Assignable Communication Routes—Clear-Com V12PD
Source: Both Courtesy Clear-Com

- Talk/listen options—any digital system should be fully customizable, allowing the user to talk and/or listen to any source/path as desired. Some analog systems will offer dedicated talk/listen controls for each channel, some cheaper systems will only allow talk AND listen to be engaged at the same time. Having dedicated controls for each makes a much more versatile system.
- Headset/panel mic—many base stations allow either a headset connection OR a panel mic mounted directly on the base stations, but usually not both at the same time. Make sure your base station offers the option you desire.
- Panel speaker—does your base station offer an in-station speaker? For television com systems, this is often desirable, as the com operator is usually situated in a soundproof room/booth away from the production space. In theater, the stage manager may be situated close enough to the audience or production area that listening to calls over a speaker may be undesirable, or at least unnecessary.
- Program in/out—we often want or need to run a copy of the general production audio through the com system to backstage dressing rooms or other locations. These feeds are referred to as "program" feeds, as they are sending the live program material to production locations. While this can be (and is sometimes) accomplished by sending feeds directly from a mixing console or DSP to backstage speakers, it is also very common to use the intercom system to pass this signal instead. In this case, we're not usually concerned with high-fidelity audio, so much as we are concerned with communicating basic timing information to production personnel. An actor who only has scenes toward the end of an act can, therefore, keep track of where in the play the performers are so that they can ensure that they make it to the stage for their cues. Analog systems, again, sometimes limit the frequency response of the feed, where digital systems are typically

full range (at least in transmission—most wall panels, whether analog or digital, have limited frequency response in their speakers, simply due to the relatively small size of the drivers). Most program inputs can accept line-level feeds from any point in a production system, from board/DSP feeds to a program mic hung in the audience area only for the purpose of communicating program materials (typically this mic has to run through an external preamp which then sends a line-level signal to the com system, though on occasion the com system will allow for a mic to connect directly to a program feed path at mic level—which is more common in digital systems).

- IFB—Interruptible foldback systems allow for a dedicated monitor feed to be sent to talent (program feeds, often used in television, to send everything but the host's voice to a news show host's earpiece, for example). The nature of IFB means that the director/technical director/stage manager can interrupt that program feed with an overriding signal that allows the talent to receive timely messages with no interference from other signals. If you've ever seen newscaster hold his or her hand to an ear and say, "I'm being told we have breaking news now," they are being told that over their earpiece, which is connected to an IFB system. Both analog and digital systems offer IFB, but they are not a feature of every intercom system, so care must be taken to ensure that feature is available in your base station if your production calls for it.
- Button types—this is not typically an option for analog systems, but digital systems offer a range of options for control. Some feature push-buttons that can be toggled through several states, some offer latching switches (like up/down light switches on a wall), some offer rotary switches/encoders. The difference between these is mainly of preferred style of operation—as each generally offers the same basic functionality as the other—different operators will find one or another control type most suited to their particular production.

All of the aforementioned are described as they apply to wired base stations. If your system calls for wireless com, you will also need a wireless base station. Most of the relevant specs noted earlier apply to wireless stations as well, with a couple of added parameters that you will need to consider.

- Wireless transmission frequency—different systems will transmit over different wireless bands. As with planning wireless mic coverage, you will want to analyze available bands in your area to ensure you are not choosing a system that transmits and receives on a channel or channels that are crowded. The FCC's website offers useful planning tools for examining wireless signal band usage in your area.
- Beltpacks per station—each wireless base station will have a maximum number of beltpacks that can connect to it. A 4-channel base station might serve, for example, 25 beltpacks. If you need more than 25, you will thus need a second wireless base station.
- Number of antennae supported—each wireless base station will have a maximum number of transceivers/antennae to which it can connect. The art of designing antenna arrays for wireless com is beyond the scope of this book, but suffice it to say, you will need antennae near any production areas, and you, of course, want to make sure your base station can communicate with all relevant antenna positions.
- There are a number of other specifications you will find in intercom base stations, but for the most part, they are similar to those in other audio systems. If you wish to use a Telex headset with a Clear-Com base station, for example, you should ensure that the pin-out of the wired connection is the same and that the impedance specs match appropriately—in truth, most intercom equipment will either be obviously inter-operable or obviously not. Typically, a system designer will specify all major control equipment within a com system to be manufactured by the same company, from the same product line, as it is an easy way of making certain that the gear will all send the proper level of signals, will appropriately null the talk signals to avoid feedback in operator's own headsets,[4] and will generally function smoothly.
- Beltpacks (Figure 21.5): beltpacks are the devices worn by individual crew members that allow them to connect to the com system.[5] The design of a beltpack is such that it may have one or more channels of access to the system, and while signal levels can be adjusted, the default mode is to constantly listen to the channel or channels to which the pack is assigned, with an optional talk button for responding as needed.
  - Channel Count—most beltpacks offer either 1 or 2 channels of connectivity. In digital systems, we're starting to see the occasional 4-channel beltpack, but these are still rare.
  - Controls—volume controls are common, as are talk buttons, and call buttons (these glow an indicator light for everyone on the selected channel, to alert people that the party pressing the light needs attention—this is useful if/when someone else in the team has silenced the channel and needs to be alerted to turn it back on).[6]

- Connectors—Multi-pin XLR (4, 5, or more) are very common, to support dedicated two-way communication paths, for both headset (4-pin) and system connections (4-pin in single channel beltpacks, 5-pin for dual channel, often with an unused central guide pin). However, as technology evolves, this standard is evolving with it. We are now seeing the first wave of Ethernet-connected beltpacks—while these still use 4-pin XLR to connect to their headsets, they use the ruggedized latching EtherCon connector type to connect back to the digital matrix infrastructure.
- Wall panels (Figure 21.6): wall panels are common in production venues. They are often found in production offices, green rooms, dressing rooms, and the like. These panels typically act as miniature base stations (in every way except providing power to the system). They often have multi-channel communication ability, feature both headset and panel mic connection options, offer a speaker for in-room program or paging feeds, and so on. Specs will typically track along with the aforementioned gear.
- Headsets/handsets (Figure 21.7): connections to a beltpack typically come in one of two basic forms: the headset or the handset. Headsets consist of a

FIGURE 21.5
(Top) An Analog Two-Channel Beltpack—Clear-Com RS-702; (Bottom) A Digital Two-Channel Beltpack—Clear-Com HXII-BP
Source: Both Courtesy Clear-Com

headphone or two, as well as a boom-arm mic whose position can be adjusted. Handsets are analog telephone-style communication devices, with a talk button mounted in the arched piece between earpiece and mouthpiece. Headsets are by far the most commonly used interfaces between beltpack and user, but handsets are very common for sound mix engineers, who don't want to wear the earphones throughout a production, as they need both ears to mix!

235

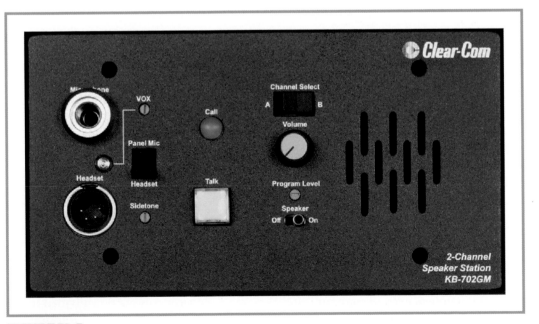

FIGURE 21.6
An Intercom Wall Panel—Clear-Com KB-702GM
Source: Courtesy Clear-Com

FIGURE 21.7
(Left) A Single-Ear Headset—Clear-Com CC-110; (Right) A Sound Operator's Handset—Clear-Com HS6
Source: Both Courtesy Clear-Com

- Headsets come in a variety of forms: single-ear or dual-ear, heavy-duty closed-back to lightweight open-back.[7] Handsets typically come in one format alone. Headsets and handsets should all transmit a relatively wide frequency response, and typically will connect via some variety of 4-pin XLR, though there are a couple of different form factors to this connection (4-pin with a notched central plastic pole, for example). While much of the infrastructure of a com system will typically be designed from components made by the same manufacturer, headsets and handsets are often interchanged between manufacturers—it is common to see a Clear-Com system with Telex headsets, or vice versa.
- Paging stations: paging systems, used to directly call performers and crew over speakers scattered throughout a production area, will be examined more closely in Chapter 22, but for now, suffice it to say that many intercom systems also feature dedicated paging stations. These are microphone stations designed to enable the operator to make announcements over a range of speakers, whether backstage to cast and crew, over the entire extent of an amusement park ride in case of emergency, over an airport terminal to seek a missing passenger before boarding ends, or the like. In theatrical production, paging needs are typically handled via the intercom system itself, but in larger distributed systems (like airports), the paging system is often a subsystem of the DSP that requires its own special attention.
- Video com: see Chapter 23 for more on video com.

CONSIDERATION

What questions must a designer ask (and answer) in order to select an intercom system for their production purposes?

1. How many production crew members need to communicate? How many departments are they divided into? This is the main pair of questions. If you have a run crew of 4 people, and they are easily divided into sound and lighting crews, a two-channel analog party line system will suffice. If you have a run crew of hundreds, divided into more sub-crews than can be easily counted with your fingers (such as the crew for a large television event), a contemporary digital matrix system is essential. In between, there are a range of options that

may suit your needs, and this is one area of production in which compromise is fairly common. A production may ideally call for 6 channels of communication, but if an analog 4-channel system can be made to work, enormous savings can be had using an analog system versus a digital system.

2. Does the production call for wireless com? If crew members need to roam over a large production area, quickly and in the dark, wireless com may be ideal. If crew members stay at their stations on the sides of the stage throughout the entire production, with no need to ever move, wired com will suffice. If you need wireless com, how many antenna locations do you need? Does the base station accommodate this many antennae, or do you need more than one base station?

3. Are there backstage areas that require paging? If so, can wall panels be installed, or would smaller base stations with speaker panels be more ideal (is this an installation system or a touring system)?

4. Is a program feed necessary? If so, what are the destinations? What will be the program source? Is there a suitable feed that can be taken from mixing console or DSP, or do you need a dedicated program mic and preamp?

A robust communication infrastructure is perhaps the least glamorous part of a sound system,[8] but is essential to ensuring that all of the other equipment is operated properly, in time with the other production systems and performers, and safely.

In the next chapter, we will briefly examine dedicated paging systems that are not a part of the general "production com" family of equipment, followed by a quick dip of our toes into the water of video com.[9]

Notes

1 For more on video com, see Chapter 23.

2 On this hypothetical production, the lighting team has *way* more cues than all the other departments, so I've given them their own channel.

3 Power over Ethernet.

4 Nulling is the process of setting just how much of an operator's own voice/mic signal is heard in their own headset. This is important—if we don't hear *any* of our own voice back in our headset, we don't think our mic is working, and we tend to yell or overexaggerate our voice. If we hear *too much* of our own voice, we hear it as a weird delayed after-effect, or as feedback. This is a crucial part of cell phone design, as well, and early cell phones (and those today that don't provide enough suggestion of our own voice in the speaker we're listening to) are the reason that people tend to yell into cell phones, even though their voice is totally audible on the other end of the call.

5 Some wireless systems (particularly in retail) forego the beltpack and instead include rudimentary controls mounted directly on one of the earpieces of the headset. However, for most entertainment production systems, the need to touch one's head is cumbersome and unwieldy, and doesn't allow the flexibility needed by production staff, so beltpacks are the standard interface between system and headset.

6 Again, I will throw lighting under the bus here: this is very common when, for example, lighting designers are talking in detail to their programmers and spot ops, and the SM has decided to stop listening to them until they are ready to move forward. The lighting team may then use the call light to alert the SM to listen to their channel again and thus to communicate that they are ready to move forward with the rehearsal.

7 As with any circumaural (over-ear) headphones, the option of closed-back versus open-back is decided by whether or not the user needs/wants to hear the outside world while using the device (in which case an open-back headset is appropriate, often single-ear), or needs/wants to isolate themselves from it (in which case the closed-back headset is appropriate, sometimes but not always dual-ear).

8 For most designers—however, I know some folks who make their living as dedicated communication system designers/engineers, and they truly love their work. It is only on gigantic productions and production budgets that this role is separate from that of the sound system designer.

9 I know, I know, it's video and not sound ("Booooo!" shouts the peanut gallery). However, again, on all but the largest productions, if video com is needed, it's usually the sound system designer's job to configure it. It won't be a long detour, I promise.

CHAPTER 22

Paging Systems

In a large swath of systems, paging will be handled via the audio com infrastructure. As such, it might seem dubious that paging systems receive attention in their own chapter, but in systems that do not require a typical production com setup, there is still often the need of paging. I can't ever recall seeing an airline ticket counter employee wearing a Clear-Com beltpack and headset, but I couldn't count the number of times I've seen those employees utilize a dedicated paging system to make announcements to their gate, terminal, or the airport at large. This chapter is devoted, however briefly, to sound systems in which paging is a definite requirement but there is no need for traditional headset-based production intercom.

DEFINITION

Paging systems can be found in various settings, from airports and retail establishments to theme parks and military bases. The essence of a paging system is quite simple—an endpoint user activates a microphone, that mic is directly connected to speakers in a venue or venues, and the user makes an announcement for all in the venue to hear. Figure 22.1 shows a basic paging system.

Most dedicated paging systems (i.e., those separate from production-style com systems as identified in the previous chapter) make use of *paging stations*, which are dedicated mics with a range of controls attached. Some are quite simple, consisting simply of a single push-button activator, which will open the channel for the microphone audio to a single location (Figure 22.2). Increasingly, however, paging stations are designed with a range of built-in options, often accessed either through a numeric

Paging System

Any of a series of technological solutions allowing a user to directly address a room or rooms via an audio communication link. Often used to call doctors to a telephone call in a hospital, to alert travelers that a flight is about to close boarding, or the like.

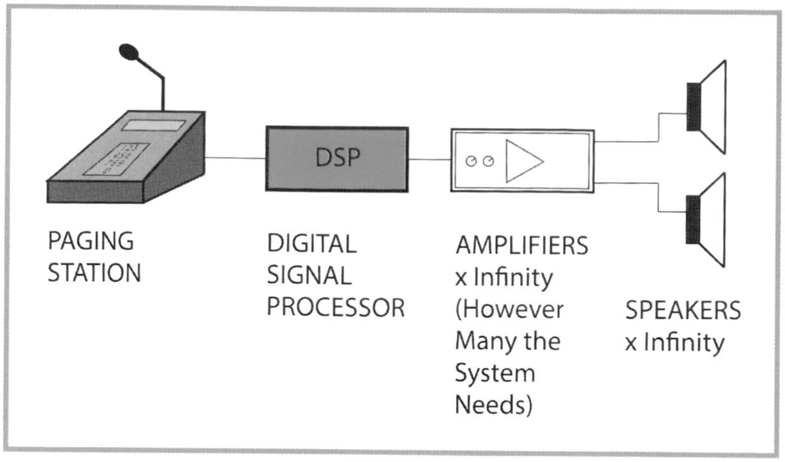

FIGURE 22.1
A Basic Paging System

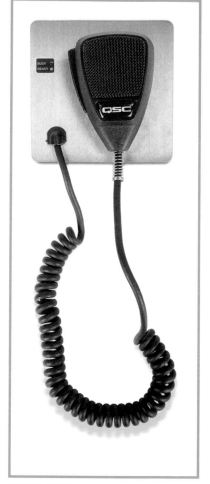

FIGURE 22.2
A Single-Button Paging Station—QSC PS-X
Source: Courtesy QSC, LLC

keypad or a touch-panel display. Such systems offer the ability to lock out communications to any user who does not enter the correct passkey, and to route communications flexibly to a range of locations, either by selection of a numeric destination value or by touching an assigned routing button on a touch-panel control.

Paging systems of this nature are typically concerned with vocal intelligibility in both frequency response and dynamic range control, and are not suited for more high-fidelity audio reproduction.

CLASSIFICATION

Paging systems can be classified into three major categories:

1. Audio intercom-based systems—such as those described in the previous chapter
2. DSP-based systems—these are the most common in contemporary installation work. In these systems, the paging stations and endpoints are routed through a master processing controller (the DSP), which performs all the connection opening/closing of signal paths as demanded by the user. Most major DSP manufacturers also market some variety of paging station(s) that can be specified to work directly and easily within the DSP architecture. Of course, in this paradigm, any microphone whose signal is routed into the DSP can be used as a paging mic. Many small and medium-sized theatrical systems will simply use a switched SM58 for the stage manager's paging station, and all communication lines will be permanently open to backstage, muted by setting the microphone's switch itself to the OFF position. On the flip side, these systems can be enormously complex, with dozens of dedicated paging stations, custom access control for a variety of tiered users, and sophisticated signal routing offering everything from point-to-point communication to blanket area announce features. In these systems, paging is normally only one function of many that the DSP architecture is performing. It may be hosting playback audio for the background of a shopping center, routing production sound to audience-facing speakers, and so on, in addition to its paging duties.
3. Paging-only systems—there are dedicated paging systems, designed only for paging functions, that can be found in corporate offices, warehouses, and the like. These systems still need a central processor, but in this case, the processor is far less configurable than a standard DSP and is simply in charge of routing paging mics to destinations, and nothing else.

In all of these systems, and especially in the last category, telephones may be used as endpoints in the paging system. It is common, for example, to see a checker at a retail establishment pick up a telephone, dial a specific access code, and be connected to the store-wide paging system for an announcement.

SPECIFICATION

When selecting a paging system, it is important to consider some of the following specs:

- Number of destinations/channel count: how many locations can your system route signal to and from?
- Connection types(s): some paging systems run on basic analog audio connections, but increasingly paging systems send audio over IP. If the latter, does your system use a proprietary data format (as many in the DSP-based world do), or will it be inter-operable with endpoints from other manufacturers (as some in the paging-only world do, making use of standard VoIP formats for audio transport).

When selecting paging stations, the following may be important:

- Access control: does your station need user access control, and if so, what type will suit your needs? If you need a range of users to have access via a single station to a variety of different paging endpoints, a station with one single push-button activator won't cut it. On the flip side, if the only paging being done is from one station to one endpoint, you may not need a complex touchscreen-based station.
- Control type: numeric keypad, touchscreen? How large a touchscreen (full-color or single-color)? If keypad, are keys assignable or connected to fixed routes (it is assumed all touchscreens will be assignable)?
- Automatic logoff timeout: some paging systems require users to manually log out after logging in for access. Many, however, provide an automatic timeout after use so that the user will be automatically logged off. Some of these systems allow custom timing of the user logoff. This can be very handy in situations where a user might be interrupted during or directly after use by a customer or other party that is not allowed to use the system.
- Microphone type (Figure 22.3): the three main types of paging station mics are handheld, gooseneck, and telephone handset.

CONSIDERATION

What questions must a designer ask (and answer) in order to select a paging system?

1. Is this system intended to interact with a larger audio system? Even if it doesn't need to *interact*, is there a larger audio system in place, of which the paging system can act as a subsystem (such as a DSP framework)?

FIGURE 22.3
(Left) Paging Station With Handheld (CB-type) Mic—QSC PS-1600G; (Right) Paging Station With Gooseneck Mic—QSC PS-1600H
Source: Both Courtesy QSC, LLC

2. How many users are intended to have access to the system? Do they have differing levels of access depending on their level of authority?
3. How many endpoints will the system have? As a sub-question, how many of those are paging callout stations vs. loudspeaker destinations? Are any of these two-way or multi-way (if so, perhaps a solution from the previous chapter is in order)?
4. What other actions will a user be performing during times at which they need to page? This question is centered around determining what microphone type(s) are needed. A user with both hands free can typically make use of any of the mic types, but a user who will often have one or more hands occupied (a factory foreman, for example) may need a gooseneck so that paging can be done with one or more hands otherwise occupied.

Paging systems as a subset of intercom systems are a distinct audio need in some environments, and it is important that users of paging systems have the access they need, instantly, with the easiest form factor of operations possible.

To complete our look at communication systems, in the next chapter, we take a moment to address video intercom.

CHAPTER 23

Video Com

We Know It's Not Audio, But It's Almost Always Our Job Anyway!

The subject of video communication systems is a large one. In the industrial world, quality-control cameras and video monitors have long been employed along production lines to ensure consistency of performance, and for use in investigations of any errors. This toolset has begun to bleed over into the entertainment world, with everything from theme parks to theatrical productions often requiring some kind of "video com" or at least video monitoring. While large-scale projects may employ a distinct video designer (or at least an AV designer with experience), on small- and medium-scale projects, video monitoring is often the province of the sound system designer.

DEFINITION

In its simplest form, video com can be thought of as an augmentation to audio com systems. In theatrical productions, for example, it is increasingly common to find cameras used to supplement stage manager communications or orchestra conductor activity. A musical may have an offstage choir too large to fit into the downstage pit of a proscenium theater, and so a camera may be mounted to transmit the conductor's cues backstage to the choir who is singing into microphones. This may be a supplement to the audio com system with which the SM has called the choir to places to be ready for their cue. In many theaters, the use of infrared cameras has become very common to allow stage managers to see scene changes being performed in the blackout and to verify that performers or scenic items are in place before calling the beginning of the next scene.

In its most complex form, video com and surveillance camera systems overlap. In a theme park, it is very common for an attraction or a parade route to have cameras mounted showing virtually every location on a public travel path, such that any issue in performance can be identified

Video Com

Any of a host of technological systems that allow for visual communication between locations that are physically remote from one another.

immediately. These systems may not just be used to identify errors with a dance crew on a parade float, however, they may also be used to identify vandals injuring the park itself, and to help security teams apprehend said parties.

Video com is also found at stage doors in many facilities, such that the person or persons requesting stage access can be seen before access is granted, in this case relying on visual identification as an access "code" rather than a keypad or other such.

CLASSIFICATION

These systems can be classified in hundreds of different manners, as the boundaries between com systems and surveillance or production-line monitoring systems are permeable. For our uses, however, we'll limit ourselves to three basic classes of systems:[1]

1. Single source: systems designed to monitor one source of action and transmit that data to one or more points of monitoring. The conductor cam to the backstage choir is a single-source system. Even if there are two choirs, in separate rooms, each monitoring the feed, it is a singular feed of one position in the venue, communicating one set of information to all parties involved. Figure 23.1 shows an example of a single-source system.
2. Multi-source, single-subject: systems designed to deliver complex sets of data about one general location. This might be a set of cameras all positioned in different places around a theater showing the stage manager a host of angles on a scene change. Even if the cameras are showing distinct parts of the stage, the multiple sources are still designed to add up to information about a single subject—whether the scene change is complete, for example. Figure 23.2 shows an example of a multi-source, single-subject system.
3. Multi-source, multi-subject: these systems are used in productions like a theme park parade. There are different cameras on different points in the parade route, and they are being used to monitor parade performers, equipment, and timing, as well as to monitor atmospheric conditions (has it started raining outside?), audience activity (are guests behind the barricades, or is someone trying to climb the float?) and other such. These systems are more like surveillance systems used for entertainment than true video com but are included here as they are in very common use in large-scale entertainment productions. Figure 23.3 shows an example of a multi-source, multi-subject system.

244

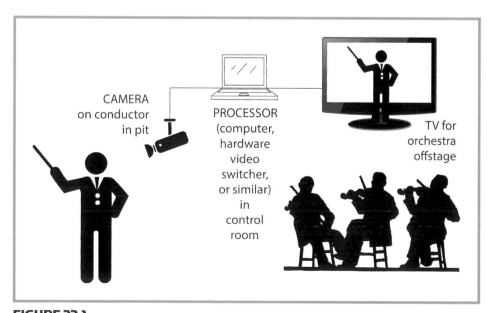

FIGURE 23.1
A Single-source Video Com System

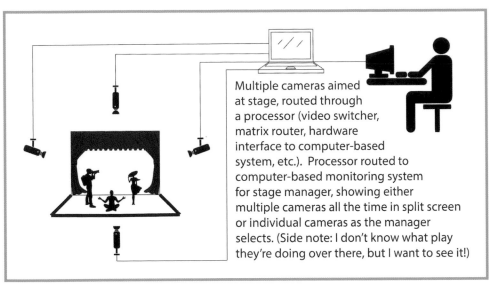

Multiple cameras aimed at stage, routed through a processor (video switcher, matrix router, hardware interface to computer-based system, etc.). Processor routed to computer-based monitoring system for stage manager, showing either multiple cameras all the time in split screen or individual cameras as the manager selects. (Side note: I don't know what play they're doing over there, but I want to see it!)

FIGURE 23.2
A Multi-Source, Single-Subject Video Com System

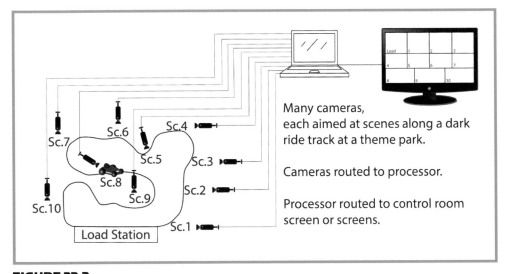

Many cameras, each aimed at scenes along a dark ride track at a theme park.

Cameras routed to processor.

Processor routed to control room screen or screens.

FIGURE 23.3
A Multi-Source, Multi-Subject Video Com System

Message in a Bottle

I think one thing that I've learned through the years: never specify the new-fangled piece of equipment. And, generally, if your client really wants it, advise them why that could be a bad idea, especially with mission-critical things, like a console, DSP, and loudspeakers. I see big mistakes made. And the interesting thing about that is that video evolves, I would say, five times the speed of audio. So while it works one month, the next month there's a new HDMI protocol and everything stops working. So that's a really frustrating thing.—David Kotch

SPECIFICATION

The range of specifications that come into play when selecting video equipment is far too broad for any detailed coverage in this book. As such, I will not be going into detail about refresh and scan rates, video formats, and so on. I address here only the broadest of concerns.

- Resolution: how good is the image you are transmitting? Resolution can include everything from actual pixel count to performance in low-light, and what particulars are important to your design will depend on what is being monitored, and what the ambient light conditions of that monitoring are.
- Color/BW/IR: many camera systems for "confidence" monitoring are still black and white, as color info isn't thought to be as important as basic shapes. Some, as previously mentioned, will be IR cameras (and some perform both regular light and infrared monitoring).
- Analog/digital signal: some systems still transmit an analog video signal; many are now video over IP/Ethernet. What your system will demand depends on what central routing/processing you are using, which will also help define what kind of outputs you will need for video endpoint monitoring.

CONSIDERATION

What questions must a designer ask (and answer) in order to select a video com system for their production purposes?

1. How many cameras and monitors are needed? Does any user need the ability to switch between multiple cameras on a single monitor? Does there need to be a matrixed display showing multiple cameras on one screen simultaneously?
2. What resolution is needed? Do we need to see whether a conductor raises an eyebrow, or will the rise of the baton be enough information?
3. Is low (or no) light monitoring necessary?
4. Is this system small enough that I can configure it without advanced video knowledge, or is this a system that requires high-level video expertise? This is a tough question, as any designer typically wants to be able to solve any problem posed. However, as video systems get more complex, a designer who is really just a sound system specialist may reach a point at which they either need to undergo advanced training in order to properly specify needed video gear, or at which they need to admit they are not specialists in the video field and call for a video designer to be added to the project.

Some sound system designers feel it is unfortunate that video systems are so often placed in their laps. Many (if not most) sound system designers don't innately have any specialty or training in video, and while there are similarities (both systems essentially push around electronic real-time data), the differences are vast. Video systems of the types mentioned earlier could be (and undoubtedly are) the subject of whole books of their own. They are really only included here to address the most basic and common needs that are sometimes demanded of sound system designers.

OK! We must have been through all possible sound system equipment types, right? We've got to be ready to move on to how we configure and use these systems . . . right? *You know how this works by now.* Not quite. We have one more gear-focused chapter before we get into Part II. That chapter is on the stuff that connects all our pieces together. CABLES!

Note

1 Please note that, unique in this book, the terms used to classify systems here are my own and are not (as the majority of terms used elsewhere in the book are) industry-standard classifications. While the architecture of these systems would be recognized by any video system designer with enough description, the very nature of video monitoring systems is so complex and multifaceted that I have yet to experience a good standard set of terminology for the variable styles of monitoring system in wide use.

Cable Types and Their Uses

How important is cabling to the performance of a system?[1] Cabling is simultaneously completely critical (since without proper cabling no signal travels anywhere through our system) and also not terribly important (since any decent cable of the proper length and type will work about as well as any *other* decent cable of proper length and type). Sound cables are often referred to by their connector type, their length, and/or the signal type they are carrying. While we won't treat length in any detail here,[2] we will address the other categories.

DEFINITION

Cable

A linear run of conductive material that carries either an analog or a digital signal from one device to another. Most commonly, cables consist of insulated copper wires in some configuration, though fiber-optic cables have also become more common—as such the cables are either conducting electricity or light.

CLASSIFICATION

Sound cables carry one of two basic types of signals: analog or digital. Analog audio signal is a continuous signal, and digital signal is a discretized signal (made of individual bits of data). Within analog cables, there are two major subdivisions: unbalanced and balanced.

Unbalanced cables are two-wire cables, one carrying signal and the other serving as the electrical reference and return path for backflow current.[3] Such signal cables are much more susceptible to noise. An RCA cable (Figure 24.1) is an example of an unbalanced cable with typically very small wire diameter and little insulating material, thus making it very susceptible to interference.

As such, these cables tend to be most useful at a length of 25' or less, as any longer runs are increasingly vulnerable to picking up electrical interference from other cables and devices in the area, and thus suffering from hum or buzz in the signal.

In Figure 24.2, you can see how the wires inside an unbalanced cable function.

Note the numbering. Wiring diagrams are always listed by pin numbers, and those numbers must remain consistent across all devices and cables. The "ground" wire is always pin 1. The "hot" lead is always pin 2.[4]

By contrast, a balanced cable (Figure 24.3) adds a third wire, which is ALSO a signal-carrying cable, with its signal inverted in polarity. The transmitting device contains a differential amplifier that allows for the pin 3 signal to be

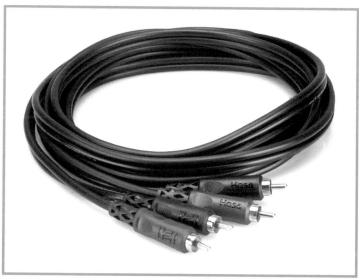

FIGURE 24.1
RCA Cables
Source: Courtesy Hosa Technology

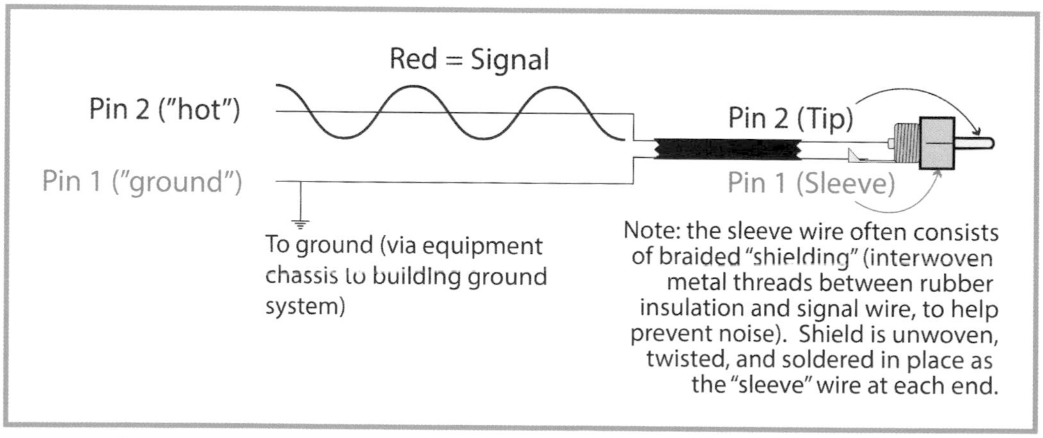

Red = Signal

Pin 2 ("hot")   Pin 2 (Tip)

Pin 1 ("ground")   Pin 1 (Sleeve)

To ground (via equipment chassis to building ground system)

Note: the sleeve wire often consists of braided "shielding" (interwoven metal threads between rubber insulation and signal wire, to help prevent noise). Shield is unwoven, twisted, and soldered in place as the "sleeve" wire at each end.

FIGURE 24.2
Signal Connections in an Unbalanced Cable

flipped. Now, if we have a signal on pin 2, and an inverted signal on pin 3, one would think that at the receiving end where the signals are summed, the signals would cancel out and we would hear nothing, as that's how signals that are 180° out of polarity work. However, at the receiving end, there is ANOTHER amplifier (both differential and summing), which flips pin 3 again, and *then* sums the signals from pin 2 and 3. Other than the 6 dB of gain you get from summing two voltages together, what is the advantage of this? Isn't one signal strong enough?

This is where we introduce our friend Common Mode Rejection (CMR). The principle of CMR is this: imagine that our balanced audio cable is running from a microphone back to a mixing console. Along the way, it crosses paths with a high-voltage moving light mounted on the stage floor. This moving light (and all of its assorted computer parts, motors, etc.) is giving off an EMF field, and by running our signal cable near it, we may pick up that EMF and find it induced as buzz or hum in our signal.

In Figure 24.4, we can see that when noise is induced in a balanced audio line, it is induced in the same polarity on both of the signal cables.

This noise is now in polarity agreement between pins 2 and 3. However, as we see in Figure 24.5, when we invert pin 3 at the receiving end, the noise is now 180° out of polarity with itself, thereby canceling out. We have *rejected* the *common mode* (noise) in our line. Thus, a balanced cable is not only vastly more noise resistant but *also* gives us 6 dB of gain at the summation point. Therefore, a signal that might have had a 25' maximum run as an unbalanced cable now has a typical maximum run of 1,000'. Nice!

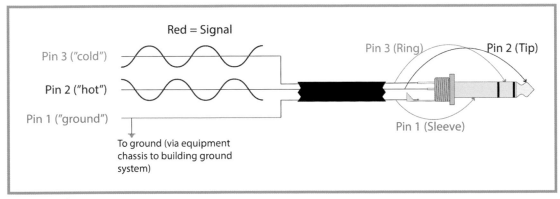

FIGURE 24.3
Signal Connections in a Balanced Cable

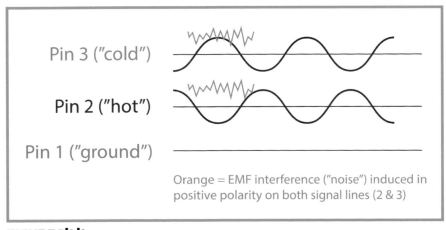

FIGURE 24.4
Balanced Signal Cable With Noise Induced

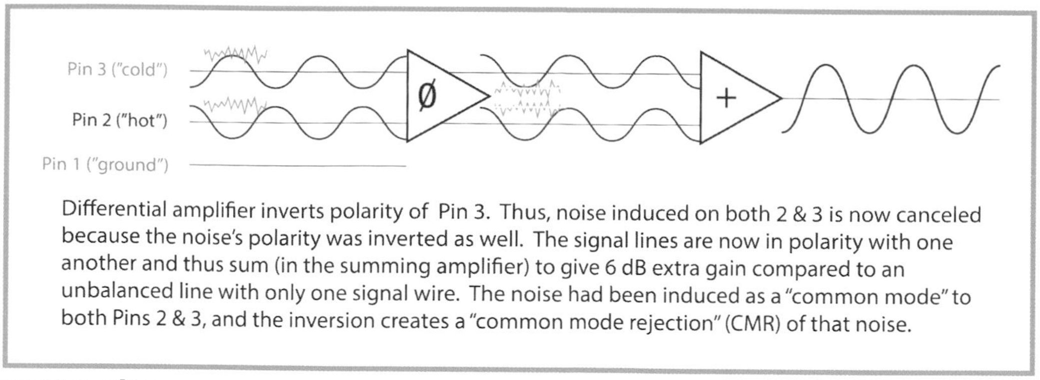

Pin 3 ("cold")

Pin 2 ("hot")

Pin 1 ("ground")

Differential amplifier inverts polarity of Pin 3. Thus, noise induced on both 2 & 3 is now canceled because the noise's polarity was inverted as well. The signal lines are now in polarity with one another and thus sum (in the summing amplifier) to give 6 dB extra gain compared to an unbalanced line with only one signal wire. The noise had been induced as a "common mode" to both Pins 2 & 3, and the inversion creates a "common mode rejection" (CMR) of that noise.

FIGURE 24.5
Common Mode Rejection in Action

Passive speakers use unbalanced connections, though, and the signal runs are usually well over 25'. How do they prevent the noise? Well,

a. They don't always; they are more susceptible to noise than balanced signals run to active speakers (another advantage to active speakers); and

b. They reject much interference by virtue of their thick wire gauge and heavy insulation. Speaker levels are high-intensity signals, and any high-intensity signal will be less susceptible to interference than a low-intensity signal.[5] The high-power signal requires a thicker cable so as not to melt the wire with the heat of the intense signal.[6] The thicker cable requires heavier and thicker insulation, which in turn helps reduce interference as well.

Audio signal cables are often referred to by gender. A plug-end (the prong that plugs into a socket—called the *jack* or *receptacle*) is known as the MALE end, while the jack end is the FEMALE end.[7]

In the following section, we will identify the major audio signal cable types and their uses by category, beginning with analog and moving into digital.

- Analog cable types:
  - Unbalanced:
    - RCA: named for Radio Corporation of America, aka "phono plug" as these are commonly found on "phonograph" (record) players. These cables are generally found on record players, CD players, and other consumer-level audio equipment (though use of RCA has been on the decline since the rise of the "mini" / 3.5 mm / 1/8" connector type). Generally, RCA cables are found in pairs, one each for left and right of a stereo signal. See Figure 24.1 for a picture of RCA cables. Pin 2, the signal cable, terminates at the center pin. Pin 1, the shield, terminates at the metal ring around the central pin (as shown in Figure 24.2).
    - TS (Figure 24.6): "tip sleeve." This is typically a single cable of ¼" diameter, used to carry a monophonic signal. It is named TS because the two wires in the cable are terminated at the tip and

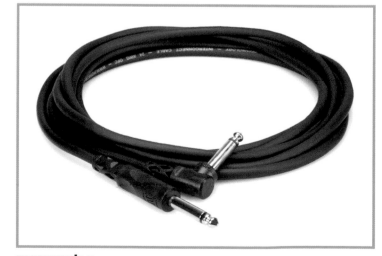

FIGURE 24.6
TS Cable
Source: Courtesy Hosa Technology

sleeve of the plug, respectively, separated by a plastic insulating band. These cables are also known as "instrument cables" as they are the most common cable for connecting electrified instruments to their onstage amplifiers and/or DI boxes. These are also colloquially known as "phone plugs" a nomenclature that goes back to their original use as patch cables for telephone switchboards. Tip is always pin 2, sleeve is always pin 1.

- 3.5 mm: this cable type goes by many names. The truest name is 3.5 mm, as that is the diameter of the connector. It also goes by "mini," "mini phone plug," "1/8"," and so on. This is typically a single plug that carries both the left and right signals of a stereo pair. This cable is actually a pair of TS signal leads, with each signal lead acting as its own pin 2, left at tip, and right at ring, with pin 1 terminating at the sleeve. These cables are found on cell phones, laptops, and all manner of other devices for sending analog stereo signals into a system. When interfacing a device like a phone with a mixing console, we will often find a cable that is 3.5 mm at one end and splitting to two ¼" or XLR connectors at the other end for ease of manipulating left and right independently at the console (Figure 24.7).
- NL2/NL4/NL8: aka "SpeakOn." These are speaker cables. An NL2 contains two wires, where an NL4 contains four (for bi-amping, typically) and an NL8 contains eight. In an NL2 cable, the wires are terminated as + and—signals that correspond to the + and—terminals on power amplifiers. NL2 and NL4 is a distinction in wiring, as both use the same connector (Figure 24.8). NL8 are often used for more complex multi-amped speakers. The NL8 connector looks identical to NL2/NL4, except that it's larger. NL connectors are twist-locking connectors that make sturdy connections that are unlikely to be accidentally kicked out by an errant stagehand.
- Banana plugs (Figure 24.9): these are typically only found on power amplifiers, and even then, most commonly on older models. They are so named because they sort of, vaguely, look like bananas. They slip out of their jacks quite easily and are typically avoided in professional systems wherever possible.
- Insert cables (Figure 24.10): these are special cables designed to both send and return a signal via a single connector. Used with mixing consoles that have analog insert points, these allow outboard gear to be connected to a single channel of audio—very common for outboard EQs, compressors, and gates. In an insert cable, the single point that connects to the console appears to be a TRS connector (see next in balanced) but is in fact a pair of TS runs in one connector (much like our 3.5 mm headphone cable). At the other end, there will be an unbalanced pair of signal cables, one to send to an outboard device's inputs, and the other to return from the output of the device back to the console. The pair ends are usually TS, but can also be RCA, or (rarely) unbalanced XLR that only makes use of two pins. In such a cable (also known as a Y cable), the tip of the single-point connector is the send, and the ring is the return, with each sharing the sleeve as the electrical reference.[8]

251

FIGURE 24.7
3.5 mm to dual-TS Cable
Source: Courtesy Hosa Technology

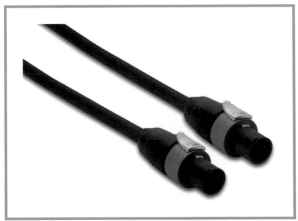

FIGURE 24.8
NL2/NL4 Cable
Source: Courtesy Hosa Technology

FIGURE 24.9
Banana plug
Source: Courtesy Hosa Technology

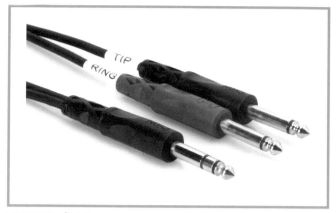

FIGURE 24.10
Insert Cable
Source: Courtesy Hosa Technology

FIGURE 24.11
XLR Cable
Source: Courtesy Hosa Technology

FIGURE 24.12
TRS Cable
Source: Courtesy Hosa Technology

252

- Balanced:
  - XLR (Figure 24.11): there are many forms of XLR, but in audio, the most common is the standard three-pin XLR carrying balanced audio. This is often referred to as a "mic cable" as a vast majority of mics are terminated in this connector type. The connector was originally invented by James Cannon, and so in some circles is still referred to colloquially as a "Cannon Cable."[9] Cannon's original connector was the Cannon X, with subsequent models adding the well-known locking mechanism ("L") and surrounding the female end with rubber ("R"), hence XLR cable. XLR can also be found in four-pin fashion, used in com systems, and five-pin, used for other specific disciplines (such as com systems, and carrying DMX—a standard lighting control protocol). In addition, XLR is also a digital cable type, carrying either S/PIDF or AES/EBU signal types, which we will discuss later. In an XLR cable in the United States, pin 2 is the "hot" (normal polarity) lead, and pin 3 is inverted. In Europe and much of the rest of the world, these are switched, which mostly becomes relevant when changing XLR to another connector type like TRS, where you need to know which lead connects to the hot/tip connection point. XLR cables are often commonly used to carry balanced line-level analog signals.
  - TRS (Figure 24.12): tip ring sleeve. This is the balanced version of the TS cable. It features two plastic bands instead of one, as there is one extra lead to isolate. In a TRS cable (used commonly for line-level interconnections), the tip is the pin 2 "hot" lead, ring is the pin 3 inverted ("cold") lead, and sleeve is pin 1.
  - TT (Figure 24.13): "tiny telephone" cables. These are patch cables used in studio patchbays, typically balanced, for routing signals through permanently rack-mounted effects. These are wired just like TRS ¼" cables and are often color-coded to make signal runs easier to trace in a large patch bay.

- Digital cable types:
  - XLR: the common three-pin is often used to carry one of two digital audio types. AES/EBU (Audio Engineering Society/European Broadcast Union) is a two-channel digital audio format, and S/PDIF (Sony/Phillips Digital Interface) is another two-channel format written in a different code language. AES (or AES3 as it is sometimes called) is thought of as the "professional" standard where S/PDIF (also sometimes transmitted via RCA) is the "consumer" standard.
  - BNC (Figure 24.14): Bayonet Neill-Concelman Connector, so named because of its twist-locking "bayonet" style connector.[10] BNC is a coaxial cable, which is to say, a single wire in the center of the cable, surrounded by insulation, which is itself surrounded by shielding wire, which is then surrounded by more insulation. BNC cables are used to carry a variety of digital audio signal types, with a variety of channels. Care must be taken to use the proper BNC for digital audio, as BNC is also used for wireless microphone antennae, but the cables used for each task are of different native impedance. The antenna BNC is generally 50-Ohm, whereas the digital audio connection calls for 75-Ohm. BNC is also used to carry Word Clock signals in many systems.
  - Optical (Figure 24.15): aka "ADAT," "TOSLINK," "Lightpipe," etc. Optical cables are fiber-optic lines that carry the digital audio information encoded as flashes of light. Optical cables have the advantage of being very resistant to EMF interference, but also have the disadvantage of being somewhat fragile to bends and sold with non-locking connector ends that are easily dislodged.[11] The connectors for ADAT and TOSLINK cables are often the same, but an ADAT signal cannot be read by a TOSLINK device, as these are different encoding languages. Optical cables may carry MADI signal as well (Multichannel Audio Digital Interface), but MADI can also be transported over network cables or other multi-pin form factors.
  - D-Sub-miniature: "D-Sub" connectors are consolidated single connectors carrying many signal leads. The DB25 (Figure 24.16) is commonly found on audio gear, and it can be used to carry analog or digital audio (there are also D-Sub connectors with different pin counts, but 25 is one of the most common).

FIGURE 24.13
TT Cables
Source: Courtesy Hosa Technology

FIGURE 24.14
BNC Cable
Source: Courtesy Hosa Technology

FIGURE 24.15
Optical Cable
Source: Courtesy Hosa Technology

FIGURE 24.16
DB25 Cable
Source: Courtesy Hosa Technology

254

FIGURE 24.17
Network Cable
Source: Courtesy Hosa Technology

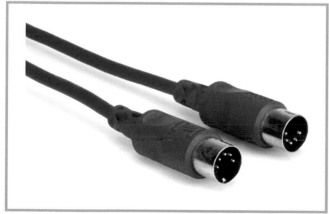

FIGURE 24.18
MIDI Cable
Source: Courtesy Hosa Technology

- Network (Figure 24.17): aka "Ethernet," "Cat5/6," "RJ45." The standard computer network cable has increasingly become the universal transport cable for digital audio. It can carry many channels, in many protocols, over long distances with little or no interference. The connector itself is the RJ45 ("Registered Jack") connector, and while there are different pinouts for different protocols, each newer version is backward compatible—hence any Cat 6 cables can also carry Cat 5 signals, but not the reverse.
- Non-signal:
  - MIDI (Figure 24.18): Musical Instrument Digital Interface, a cable type that originated with the need to send synthesizer control info from a keyboard controller to a rack-mounted synthesizer device and has expanded in use to send control messages to all types of devices (lighting and sound consoles, computer interfaces, etc.). MIDI messages are also frequently sent over network cables.

SPECIFICATION

There are not a ton of specifications to consider when selecting cable, here are the most important.

- Analog vs. digital
- Connector type
- Length
- AWG (American Wire Gauge)
- Termination/pin-out (most relevant for network cables, or when attempting to interface between US gear and gear from overseas)
- Impedance (for digital audio cables)

Beyond these specs, as with any gear, listen to colleagues and professionals on recommendations for manufacturers. Most reputable manufacturers will sell their items for approximately the same prices. Anything drastically expensive[12], or drastically cheap, is probably to be avoided.[13] For any who are inclined to detailed manual labor requiring good hand-eye coordination, there is a vast savings to be had in buying spools of unterminated wire and connectors and building your own finished cables.

CONSIDERATION

What questions must a designer ask (and answer) in order to select cables for their system?

1. What cables does my equipment require?
2. Do I need to specify cable models, or can the system integrator do this?[14]
3. Do I need spares?[15]

Message in a Bottle

Raccoons and sound equipment do not mix well. Shortly after undergrad, the summer I spent doing outdoor theater, at the end of the season, they tear down and put in storage the sound equipment, however, they leave the cabling runs up. So when they put their house system back in (*they* put it back in, they had to move these big honking speakers), it was my job to get everything up and working. And, it wasn't working. And I was real young, and I was like, "Oh geez, am I doing something stupid?" Like, I read the manual for the console cover to cover and was like, "Nope, that's still right." Had to make sure everything was plugged in, and it was. Finally, I was like, "Well, I've just gotta trace the cables." And I'm not even really sure, nobody's really sure where the cable runs are—I had to trace them. Finally, I found one of the cable runs, running across under the stage, which was this crawlspace that was about 4' tall, that was dirt and mud, and discovered that raccoons had chewed through the insulation of the speaker runs under the stage. They hadn't chewed the copper, they'd just chewed the outside, the insulation. But then all the copper had rusted! So great, so I trim the part they'd eaten away, and soldered it back together, did an ugly solder job and put it back together, and still the cable wasn't working. Apparently, they had chewed off the insulation in the fall, and then it sat open for months and months, so there was this strip of about a foot of insulation that was chewed off, but the inside was rusted for about 12'. It took a while to figure out. But, finally, once I'd trimmed out, like, 12' of cable, it was fine, it worked.—Elizabeth Atkinson

OK! We've done it! We made it through Part I, and we should now have a basic sense of all the pieces that go into a contemporary sound system. Now, while this is not an integrator's manual, we will take a short detour in Part II through the setup, installation, and testing process. It behooves any designer to understand these processes, and though we won't go into great detail, it will be a useful detour before Part III, wherein we examine the process of system design, step by step.

Notes

1 Hint: not as important as speakers!
2 What is the proper cable length for a job? The shortest length that can effectively perform its task. In the case of installation wiring, this is often a very precise number, in the case of concert setup, it is typical to leave a coiled foot or two (called a "service loop") at each mic position in case they need to be adjusted once the performers are in place.
3 Current that is returned along a signal path due to resistance at the terminated destination end. A speaker is a good example of a device that delivers high rates of backflow current. At a certain frequency, the speaker is most electrically resistant—this is its so-called impedance peak. At that frequency, it becomes very difficult to induce the driver to move in a controlled fashion, and much of the power that is sent to it is reflected back by that resistance along the signal cable path.
4 In the United States. In Europe and elsewhere, the hot lead is pin 2 in unbalanced cables, but pin 3 in balanced cables.
5 Witness the difference between your local college radio station and the mega-size station owned by Clear Channel. The Clear Channel station has a much more powerful transmitter, and thus much less ambient interference than the college radio signal, whose low-power transmitter sometimes fails to send the station into dorm rooms on the same campus.
6 In the US, cable thickness is measured in AWG (American Wire Gauge), a numeric range that is somewhat odd. In AWG, the smaller the diameter of the conductors, the *higher* the number in AWG. A 24 AWG cable is much thinner than a 10 AWG cable. RCA cable, for example, is often between 18 AWG and 24 AWG, whereas production speaker cables are often 10–14 AWG. There are calculations that can be done for signal loss, as a cable is

still a resistor, and length and gauge change the resistance of the run; however, for all but the most high-fidelity (and permanently installed) systems, this is not something we need to pay too much attention to.

7   This does, of course, portray an old-school, penetrative, cis-gendered sex paradigm, but this isn't a book about politics, so for now just know that the pointy end that goes into a hole is the male end and the hole that receives it is the female end.

8   A brief note on Y-cables versus "splitters": an insert cable is a Y cable, as each side performs a different function from the other. This is true for a 3.5-mm plug that splits to each side of a pair of headphones, a Y-cable each carrying distinct info in unbalanced format. A splitter, on the other hand, takes a full signal and sends it in its entirety to both sides of the split. A headphone splitter, for example, takes Left and Right from the playback device and copies both to each jack on the splitter, allowing two sets of headphones to hear the same material at the same time.

9   I have heard this terminology used most often in TV/video crews, but it has fallen increasingly out of fashion.

10   BNC is sometimes erroneously referred to as a "British Naval Connector." This is a so-called "backronym" in which a name has been associated with an acronym after its creation.

11   Increasingly, large-scale audio systems use fiber-optic data lines to send audio over large distances (particularly in permanent installation systems), converting the signal back to more common analog or digital signal/connector types at each end. These fiber systems are not classed as "optical" audio, even though they are truly sending audio via pulses of light. This distinction is made because the connections at the sending device and receiving device are seldom optical, but get converted to optical data by a transcoding device at the head of the fiber run. Professional fiber systems of this type often feature sturdier locking connectors.

12   Don't fall for the "gold-plated conductor" scam that inflates audiophile cable in price by orders of magnitude. Yes, gold is a *very marginally* better conductor than copper, but I defy even the most seasoned professional to actually hear the difference between a gold cable and a copper cable. This is a marketing gimmick, aimed at rich old guys who have enormous expensive home theaters, and it has no place in the professional world.

13   With the exception of purchasing in bulk. If you buy a ton of cables all at once (for example, when outfitting a new venue or production company), you will often be able to see a huge price break because of a large-volume order. If you're buying single cables, though, and one is *drastically* cheaper than most of the others, it is *likely* (though not always) of inferior manufacturing quality.

14   I've done plenty of designs where the ultimate installer/integrator will choose all the brands/models of cable, and I merely specify connector and signal types. Conversely, I've also done plenty of designs where I had to specify every last detail, down to which model of Belden network cable was appropriate for the installation. It is beyond the scope of this book to examine the differences between, say, plenum-rated install cable versus outdoor-use cable, but suffice it to say that the better you understand every part of the equipment that goes into your systems, the more insurance you have that your system will be installed as intended.

15   For a concert system, I typically want to have two mic cables on hand for every one needed, so each has a spare. For a permanent installation, I typically only specify the cable needed for the install, with very little to spare—only enough to allow for errors in termination on site.

PART II

How Do You Set Up, Test, and Calibrate a Basic Sound System?

How do you assemble a sound system? How do you test your gear and troubleshoot problems? Beginning designers often have to do this work themselves and all designers need to understand how their system fits together. This section examines the installation process.

CHAPTER **25**

Signal Flow, From Installation to Verification

If you are a professional sound system designer, the likelihood that you will be personally responsible for installing your own system is very slim. However, for students, early career designers, and anyone working on shoestring budgets for any reason, system installation and testing are definitely a part of the system designer's world. It behooves a system designer working at any level of the profession to have a solid understanding of signal flow and gain structure. Without a strong grasp of how audio flows through a system (and how to fix when it doesn't flow, but should), a system designer is at the mercy of whatever installation or integration team has been hired for the job. Trying to design a sound system without this knowledge is like trying to design a set without understanding how a drill gun works—a bad idea, that is likely to fail, or at very least waste everyone's time and money.

In this section, we will examine the nuts and bolts of signal flow and gain staging, system testing and troubleshooting. In the following section, we'll address system calibration and optimization. After these two sections, we will proceed (at last!) to the process of system design.

Signal Flow
The order in which an audio signal travels through a sound system, from input to output.

In an analog system, audio signals flow in fairly predictable directions, but in a digital system, the opportunities to lose a signal along the way due to a routing error grow exponentially. Let's start with a fairly simple analog system.

In Figure 25.1, you will see a basic analog concert sound system, drawn as a signal flow block diagram.

260

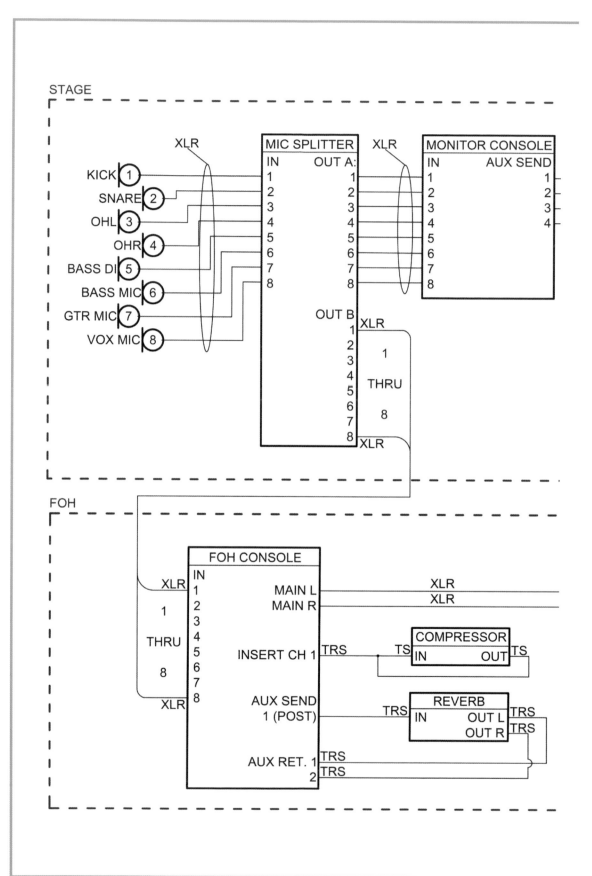

FIGURE 25.1
Analog Concert Sound System With FOH and Monitor Consoles

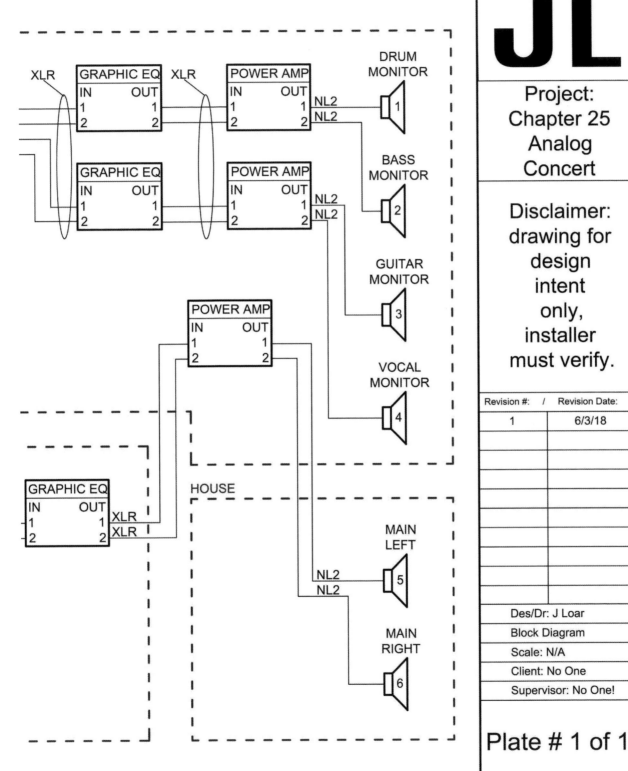

This kind of drawing[1] represents the sound devices in a system and all of the signals that travel between components. As you can see, each signal cable is labeled by cable type, and each device is listed by device type. Let's trace the flow of signal through this system, beginning with inputs, and identifying every signal path (and signal direction) along the way.

Let's begin with the mics. Microphones are placed onstage to capture the sound of the instruments, and that signal travels in one direction to an outboard microphone splitter. This splitter is in place so that each mic can send two copies of its signal, one to the FOH console, and the other to the Monitor console. However, if we just used a splitter cable, each set of preamps would pull on the signal, causing problems for the operators, and noise in the signal line. So, these mic signals travel through an *isolating transformer*. This allows the signal to be split and for each console to set preamp gain without interfering with the other. When the mic signal reaches the consoles, we perform our first *gain* of the signal and begin the long process of *gain staging*.

> ### Gain Staging
> Setting the electrical level of a signal at every stage of its flow such that it is traveling at a *nominal level* through the system. This is the level at which the strongest possible signal level is reached before hitting any overload distortion.

At each stage of a sound system (analog or digital), opportunities arise to raise the gain level of a signal. In the case of microphone-level signals traveling to preamps, the pres raise the gain of those signals to *line level*, such that they are maximizing the signal to noise ratios in their channels. The importance of proper gain staging can hardly be overestimated. If you have not set proper gain at your microphone preamps, but then attempt to get a good strong level at the *next* major stage of gain (typically the channel fader), you have now relied upon that fader (and all the circuitry of the effects along the channel strip path) to generate level that these devices should not have to provide. As a result, along with raising the level of your microphone, you will also be raising the level of the inevitable self-noise generated by any audio circuitry. As a result, your signal is much more likely to contain noise, hum, and hiss by the time it hits the speakers.

Now, once that microphone signal has reached nominal line level at the preamps, it travels through the channel strip. If your channel strip has equalizers, each of those has the opportunity to boost the gain (at a certain range of frequencies). If your channel strip features a compressor, there again is another opportunity to raise the gain. Once the signal has traveled through those effects (whether they are used or not), it travels to the channel fader and assigned outputs. In the FOH console, this is typically the next stop on the way to the main sound system, wherein the monitor console, this is typically on the way to the stage monitor speakers. The channel fader itself provides another opportunity for gain, or for reduction. If the fader is set to 0/U, otherwise known as *unity gain*, then the fader is neither boosting nor cutting the signal, merely passing it through. From the channel fader, we enter the assigned outputs, found in the master section. This is another opportunity for gain or reduction. A master fader in a properly calibrated system should be able to be set at unity and stay there, leaving the vagaries of momentary signal intensity to the channel faders and the mix engineer's decisions. The console, of course, is the second major stage in our analog system—the processing stage.

In an analog concert system (and in some digital systems, though less often these days) it would be common to find some outboard signal processors as well. Whether these are outboard EQs, compressors, or reverbs, the signal paths to those devices are worth understanding.

Effects that are intended to replace a channel's signal with an entirely processed version are patched as *insert* effects.

> ### Insert
> A signal path whereby a signal is sent and returned to and from an effect device, from the same point in the channel's signal flow.

As we can see in Figure 25.1, the compressor is attached via an insert cable, which sends a raw signal to the processor and returns processed signal to the console, all at the same point. Also in Figure 25.1, we can see a reverb effect, which (in typical fashion) is patched via a post-fader aux send. This returns reverberant signal to two separate return channels on the console, allowing A) stereo effects, which give reverb more sense of spatialization, and B), the potential for the reverberant signal to be changed in level, to be EQed, or processed itself in a variety of other ways.

Once our signal travels out of the console and outboard processors, it travels through system processors. In a digital system, or a hybrid system, this would likely be a DSP. In our analog system, it might just be some old-fashioned graphic equalizers, which provide yet another stage of potential gain/reduction. From there, we leave the processing stage and travel to the amplifiers.

The amplifiers, of course, raise line-level signals to speaker level, which is potentially massive gain. The amplifiers send a signal to our speakers, and thus, we have output!

At every point of this signal chain, the potential exists to both weaken a signal and increase the noise floor, or to over-strengthen the signal and reach distortion. The art of gain staging is the art of checking and confirming levels (both with meters and audibly) at every gain position, such that we continue to achieve strong nominal operating levels without veering too quiet or too loud.

INSTALLATION

When installing a sound system, the order of events will vary on every job. If you're sharing resources or space with other production or design teams (and you almost always will be), you may not have the aerial personnel lift when you'd most like to install speakers, so you begin with the FOH equipment (for example). However, the order of *verification* of system function should always follow the signal flow.

If you know that the mics are properly sending signal to the preamps, that the pres are properly sending signal through the channel strip, that the channel fader is properly sending signal to the master section, that the master section is properly sending signal to the system processors, that the system processors are sending proper signal to the amps, and yet still you don't have sound in the speakers, you know that the signal is being lost somewhere between amp and speaker. Whenever you are working with a system and not getting sound where you expect, begin with the start of the signal flow, and trace.

Now, in a digital system, signal paths are on the one hand simplified (Dante sends many channels over single cables) and on the other hand more complex (as there are more destinations routed from the DSP). Figure 25.2 shows a digital concert system with expanded functionality (compared with Figure 25.1).

To begin with, your system may or may not feature the aforementioned analog transformer-based splitter. In many digital concert systems, a new paradigm has taken over, whereby one console sets the gain, and the other console simply takes that gained signal as is. In this configuration, the secondary console[2] will generally have a control called "digital trim," which does not affect the gain setting but does allow the mix engineer to reduce the incoming level should it be necessary in order to provide enough range on the fader for the mix operation. In major professional systems, the analog transformer-split is often still preferred, as each mix operator can then set their own gain structure as desired. When a show is being recorded for archival or later release purposes, we often find a three-way transformer-split.

Beyond this seemingly prosaic issue, we find that digital systems adhere to the same principles as analog systems, but with added complexity. Certainly, as we see in Figure 25.2, a digital system may have multifarious destinations that, combined, result in a much more complex signal flow, with many more "points of failure" (opportunities for a signal to be mis-patched, improperly gained, for a wire to break, or the like). However, we also add to this the vagaries of network traffic routing and clocking, both of which may impede a signal that otherwise *seems* to be routed appropriately.

At every stage of a digital system, verification must include not just cables, signal routing assignments, and gain structure, but IP addresses and clock settings. A signal could be routed from the FOH console to the DSP, but if the signal is digital, and the FOH console and DSP are set to differing sample rates, the signal will not reach its

263

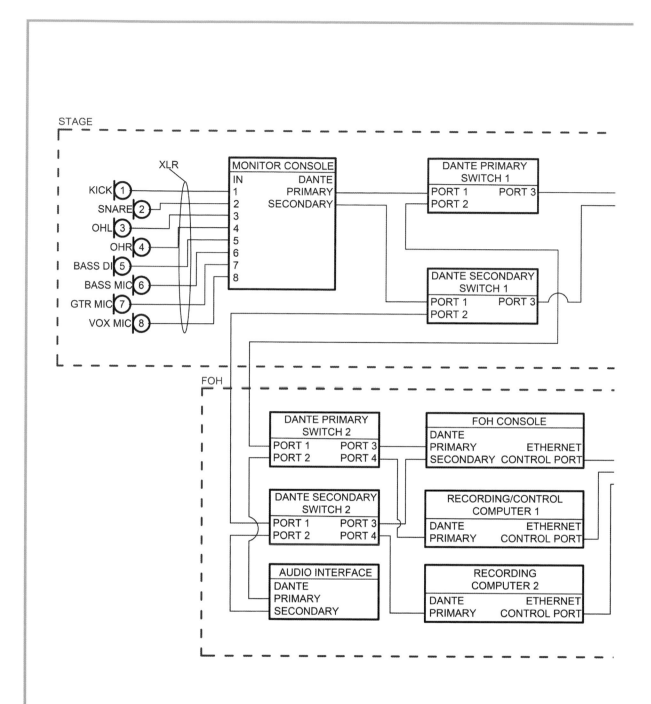

264

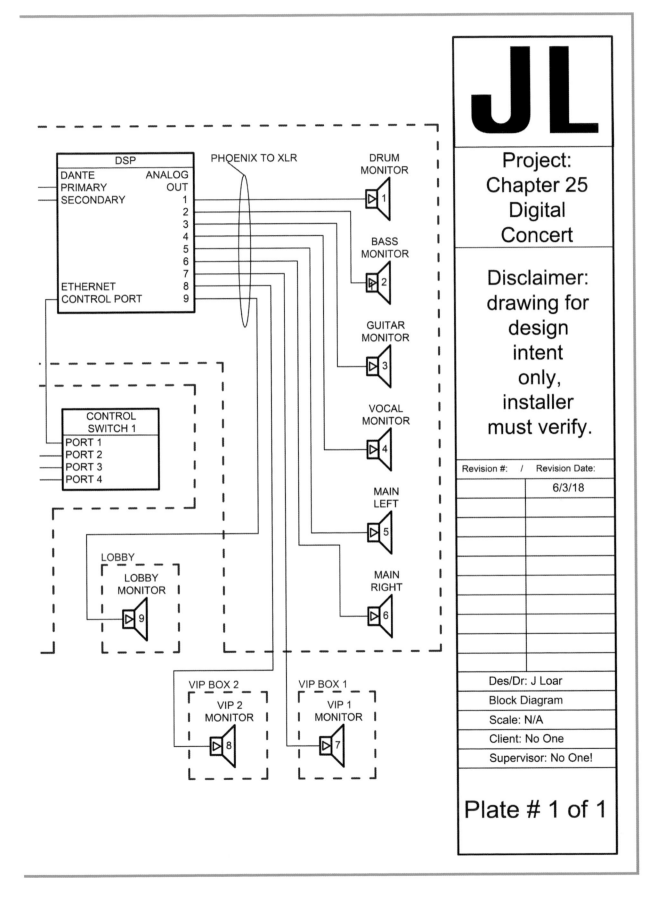

destination.[3] If your Dante patch on the FOH console is set to one NetID, and the DSP to another, the Dante controller software won't see both of them, and patching between them will be prevented.

Regardless of whether your system is analog, digital, or some combination between the two, installation must be a meticulous and well-documented process. Typically, an installation crew will be working from a series of documents[4] that detail not only every single piece of gear that will make up the system but also each interconnection, both physical and digital. For many systems, this documentation will get as detailed as showing each cable, by type, gender, and length. As the system is installed, a crew will check off each piece of equipment and signal connection that is put in place, so as to ensure that work is not duplicated. The job of organizing the effort and keeping this checklist is the job of the A1, or Audio Lead. As cables are installed, care is taken not only to make certain that slack is properly coiled and secured where appropriate, but also that each end of every single cable in the system is labeled (ideally with a label maker, for clarity) describing the destination of said cable at its other end. Thus, when a signal fails to connect, the cable itself can be checked first. If the cable is properly labeled, and properly connected at both ends of its run, then a replacement cable can be tried without worrying that the patching itself is the problem.[5]

Once all equipment has been installed, connected, and powered, verification can begin. Verification of a system is merely the process of running a signal along every planned signal path and confirming that said signal arrives at the destination at the appropriate strength and without unexpected noise, interference, or other problems. This is done with every component of the system (again, with a checklist), until everything is working. When testing mic runs, it is conventional practice to set an average gain, such that signal runs can be confirmed—understanding that, especially in a concert system, the levels will likely change once the band (and not just the A2) is on stage.

Once the system has been verified as functional, testing (the process of running show-like levels through the system) and troubleshooting begin.[6]

Notes

1 Which will be covered in detail in Part IV of this book.
2 It can be FOH or Monitors, depending on how you've designed your set up.
3 Or it WILL reach its destination, but it will do so in jagged, broken up form, exhibiting a terrible kind of digital blurring in the time domain.
4 Again, see Part IV for more detail.
5 More on troubleshooting in the following chapter.
6 Note that while I am laying this out in a linear order, this won't always be the case. On concert touring systems, for example, troubleshooting might begin during verification, as soon as a problem is discovered. Especially on shows with larger crews, multiple stages of system work may overlap, so long as they are being organized by the A1.

CHAPTER 26

Testing and Troubleshooting

What is the difference between *system verification* and *system testing*? Well, depending on where you sit, there may not be one, but I have separated these here because it allows us to illustrate an important distinction in thought process (if not always in action). What we refer to here as verification is the process of simply confirming that every component of the system in question is operational and passing the signal where it is supposed to go. What we refer to here as testing is the process of running a verified system at levels akin to full operational level. An example allows us to clarify this distinction: a *verified* concert system is up and running, and while there may be issues to resolve, most if not all of the signal paths are passing audio; a *tested* concert system is one in which we not only have signals passing, but signals are appropriately gained, speakers are achieving intended SPL without distortion, network is stable, show presets are saved to consoles and DSPs (in a digital or hybrid system), etc. In practical reality, verification and testing may be but levels of detail in one continuous process. For our discussion here, once we've verified all of our connections, we begin the process of testing.

TESTING

Before installation, every piece of system equipment should have been verified to be in working condition. In the process of loadin and installation, however, innumerable things can happen that may compromise the performance of a piece of equipment. Cables may be crimped, breaking signal leads, connectors on hardware may be snapped, someone may have dropped the condenser mic and shattered the ceramic. Often, the initial verification process turns up these problems, but it is by running the system at intended levels, ideally over a bit of time, that subtle problems can be discovered. Running speakers one by one during initial verification, for example, may seem to proceed without error. However, as we combine speakers into systems and subsystems, we may, for example, discover that speakers are out of polarity with one another. This is often recognized by virtue of a certain "hollowness" of the sound, but this hollowness doesn't always reveal itself until the out of polarity speaker is run in combination with a speaker in correct polarity.

Tales From the Field

I did a production of *George Lucas' Super-live Adventure* in Japan. Francois Bergeron and I did this show and everything was—it was the beginning of digital mixing and playback, everything was bleeding edge. We were in the middle of a show, and the generator providing power for the show blew up. Apparently—I never saw this—but the motor blew up and landed on a bunch of cars parked next to it. When it blew up, it blew every MIDI chip in our system. So everything came to a standstill and stopped. They were able to get the lights back within probably minutes by swapping generators, but no audio at all. One would think—you're in Japan, it should be easy to get MIDI chips, but not instantly because it took a day or so to repair this—there were no MIDI chips available! We had to have the components shipped from America, even though they were made in Japan, shipped from America to put them in the system. Meanwhile, what we would do is mix the show entirely manually for the rest of the run. The whole show was timecode driven and synced to video, to lighting, to projection, to flying automation, everything. So Francois and I, we'd just sleep under the consoles just fixing it as components came in.—Jonathan Deans

Verification may take place as components are installed. A mic connected to a console may (on a rushed loadin call) be checked as soon as it is connected (though general practice would call for all the mics to be set up and patched before testing any of them). Testing, on the other hand, is generally undertaken after the full system is installed, as a precursor step to calibration and optimization. As issues or problems are uncovered, one of two things will happen:

1. The issue(s) will be corrected immediately; or
2. A list will be made of existing issues, which will be addressed as a group by the crew.

We may find the former case when we are installing a system that needs to be fully operational later that day—this is often the case for small and mid-sized outdoor festivals. We may find the latter case on larger productions, permanent installations, and the like.

A good system test will incorporate all anticipated system functions, sending a signal along all possible routes, testing com, and so on.

TROUBLESHOOTING

It is very rare that a system of any complexity is installed with no errors or issues to be found. In the rare case that a system passes all testing phases without issue—celebrate! It is cause for cracking a cold beverage of your choosing. In all other cases, it's time to start solving problems.

If the testing process has been done methodically, in signal flow order, problem spots should be self-evident. Wherever a problem arises, there is a very basic process to undertake in order that said problem can be fixed expeditiously.

1. Locate the problem—is the signal failing to reach a destination? Is it reaching the destination but clouded with noise? Something else?
2. Trace the signal backward from the problem spot along the signal path—while verification and testing are usually done from point of signal origin to final destination, troubleshooting often functions in reverse. If an anticipated signal fails to turn up at its destination, follow the signal back to the last juncture (patch point, gain stage, etc.). If the signal is clear and strong at the previous juncture, then you know the problem is somewhere in the last leg of the signal path. If you don't have signal at the previous juncture, trace backward again, and again, until you locate the signal. Once you've located the signal, you can then begin to identify the issue preventing the signal from passing to its intended destination.
3. Check the obvious—I have witnessed countless young designers, when faced with signal problems in their systems, refuse to check something so prosaic as a cable, because they feel that whatever is preventing the signal from arriving at its destination must be some complex, computer-related setting.

Tales From the Field

I'm called in to do a system that's one of these super-complicated systems with all sorts of crossovers, matrix delay, and these kinds of things. And I go in there—they've been having intelligibility problems for 7 years now; they've been struggling. This is the kind of job that people would call me in to do. It's like, "Well, you know, nobody else will sort it out. Call in the Fixer." It's like Harvey Keitel from *Pulp Fiction*—the Cleaner. I'm the Cleaner; that's me. So I come in there, and it's a three-way speaker, and I say, "OK, first thing we're going to do is listen to just the woofer and measure it. I'm going to put a mic right up close. Now just the midrange. Now just the horn." And I look, and I go, "You know there's something really strange here, because there's more low end in the midrange, and there's more mids in the low range." And that's when you see *that look* come over their faces. I said, "It looks to me like your low-frequency amplifier is driving the mid speaker, and your mid amplifier is driving the low speaker. And they swap it and kaboom. OK, so this is in all of their speakers now. It's the same model speaker all over the room. I'm booked for 3 days and 30 min into the job, and you know your check is going to clear. Everything else you do is just details. You just found a major screw up that they've been scratching their heads about for 7 years.—Bob McCarthy

Message in a Bottle

There were many waterloos in my day. And it's hard for me to remember some of the exact—like who the client was, or who the band was—but I can tell you that when I had a major design crash or fail of some sort, that has always been something that is *so* valuable. I spend all the time that I can to find out what it was, and that's the beauty of the FFT analyzer is that it reveals the failure in a way that you don't get if you're walking around with a real-time analyzer, or just guesswork. So you can really analyze, "Oh, now that's what we missed there." There are many great moments like that where tuning fails or design fails.—Bob McCarthy

When solving system problems, it is good to rule out the most obvious potential causes first. A (not-at-all comprehensive) list of factors to check when the signal isn't arriving as intended:

- Is the cable plugged into the right devices/channels?
- Is the cable good or bad?
- Are all the devices in question powered on?
- Is the channel/signal path unmuted?
- Is the fader turned up?
- Is there a master control (master bus fader, VCA, group fader, etc.) that is turned down?
- Is the channel assigned to the master output path (or another appropriate output path)?
- In a digital system:
  - Is the physical input patched internally to the channel in question?
  - Is the clocking set properly?
    - Sample rates are identical in all connected devices
    - Only one device is set to master, the others are set to slave

In a majority of situations, the three steps noted earlier will solve the problem.

THE WEIRD ONES

Have you ever taken your car to a mechanic, and tried to describe a strange noise or behavior that the car has been making for weeks, only to have it perform completely normally in the garage? You swear over and over that this problem is real and concerning, but for the life of you, the car just won't exhibit the issue while it's in the shop; 20

min after you drive away, the noise returns, and 30 min later you're stranded by the roadside due to *whatever it was* malfunctioning.[1]

This same phenomenon can arise in any technical system or equipment, including our sound systems. When testing a system, especially at the end of a long workday (though it can happen anytime), the odd phantom problem can arise and drive us crazy. I recall mixing monitors for a large Latin jazz ensemble in Harlem one night, in a system run through a transformer-isolating splitter. Throughout the course of the performance, channel after channel from the splitter to my console seemed to drop out, but only intermittently. I would hastily re-patch a mic to a different splitter input, only to have it drop out again. I would change mic cables, and re-patch again, and still the issue persisted. This system had tested perfectly well prior to and during soundcheck, but in the moment of truth, my signals were flitting away like ghosts. In the moment, I felt patently insane, because I had done everything I could possibly do in the moment, and still spent the night chasing signals. As it happened, the transformer-isolating splitter itself was malfunctioning and intermittently short circuiting wiring internally, causing signals to fail and find a quicker route to earth than through my mixing console. The wiring was loose, and as the band (which was quite loud) shook the stage (which was not vibration-damped), shorts were connecting and disconnecting throughout the performance.

My point here is that in the situation where you are faced with seemingly unsolvable problems, all you can do is follow every logical step until you've exhausted your options. In some situations, you will happen upon a solution that may not have occurred to you,[2] or maybe you will simply discover that a function that a given piece of gear claims to be able to perform robustly is in fact not possible with that piece of gear.[3]

At any rate, the weird, confusing, intermittent problem is the hardest to solve. If you've got plenty of loadin and testing time, hopefully, these issues manifest during this phase so that you can identify and ferret out such anomalies before you need to commission the system for full operation. If you're on a tour, and your system loads in and out every single night, these problems can be quite frustrating, as there is rarely enough time or space to set up gear such that you can replicate these conditions outside of show mode (you'd better hope you've got a good repair station in the back of the bus).

Regardless of the difficulty of your production situation, your duty as a system designer (and/or engineer) is to check off all the logical steps in the problem-solving process. It can help to do so literally, making a list of everything you've tried so that (again) you can avoid duplication of efforts.

Ideally, your system issues will be easily solved, and you'll be on your way to tuning. In the worst case, you may need to replace cables and hardware, rebuild software patches, and the like. At any rate, your system should be fully tested and operating without problems before you turn to the calibration and optimization process.

Notes

1 *Whatever it was* is the technical term, right? Can you tell I'm not much of a car guy?

2 When I was working primarily as an FOH mix engineer, I had worked primarily on digital consoles that offered direct outs for recording from each individual channel. On most of the consoles I had experienced, direct outs could be patched from a variety of signal path locations, and there was generally a trim knob for each to ensure that the level going to Pro Tools wouldn't overload and distort. One day, I had the misfortune of using an FOH console that included a patch page, a level control, *and* a mute, in the direct out signal path, with no two of the three settings visible on the same screen at the same time. I had never experienced a console that offered a direct out mute control (and I'm still not entirely sure why that would be a valuable or useful tool, though that's another issue), and as I was struggling to get my recording rig running before downbeat, I couldn't figure out why no signal was arriving at Pro Tools. It took some fast exploration through console menus that I didn't think would be relevant to the issue before I found the mutes, and then I had to go channel by channel unmuting every direct out and re-saving my show presets. This was not something I would have expected in my experience to that date (though it was forever burned in my brain as a possibility). Sometimes the solution is *simple*, but not *obvious*.

3 I was beta-testing a piece of hardware that suggested it would work better when connected to an AVB switch but would work (in reduced channel count) via the AVB outs of a MacBook Pro (via the network cable port) sans switch. After much testing, back and forth with the manufacturer about error-ridden directions in the manual, and a lot of head-scratching, both my team and the manufacturer realized the essential truth that the equipment *did*, *in fact*, require an AVB switch in order to perform properly. This was a pre-public test, so it is somewhat expected that things won't perform to spec. In another notable case, I was interviewed and written up in a trade publication about my use of a brand of plug-in effects in a live mixing environment. At the time of the interview, the plug-ins did not work in the console I was being interviewed about. The manufacturer insisted that the plug-ins *would work very soon* in the console, with just one more update, so I (at the insistence of a publicist) conducted the interview as if I used those plug-ins every day in my work. As it happens, the plug-ins were not made to work in that console until *several years after that article appeared*. The article highlighting me and my use of that gear still ran and was available for nearly a decade afterward online.

Measurement Tools and Systems

As recently as the 1970s, sound system calibration was a process of routing each output path through graphic equalizers, turning on microphones, and then riding each EQ control up until there was feedback, then pulling each control down again until the feedback went away. This was called "ringing the system," and it was a blunt tool at best. As microprocessors got smaller, faster, and cheaper, techniques for the measurement of sound became available outside the laboratory, in the field of battle (production). By combining the essential technique of the FFT (Fast Fourier Transform—an algorithm that allows a signal to be measured in real time and converted from time information to frequency information) with increasingly affordable computing devices, engineers and designers were able to measure the real-world performance of their sound systems accurately. Figure 27.1 shows the the first major FFT system for sound system calibration, Meyer Sound's SIM1.

Suddenly, engineers could see frequency response in real time, could analyze time arrivals via impulse response measurements and phase comparisons, and could begin to bring sound systems out from the Neanderthal age ("make it fucking loud!") to the present ("make it loud, but also clean and clear and articulate!"). The importance of this transition can't be understated in terms of the advancement of sound system design as an art.

In this section, we'll first address the common tools used for system measurement and calibration, then the process of optimization, followed by some brief (and far from comprehensive) discussion of acoustic treatments.

In Figure 27.2, we can see a sound system measurement setup.

FIGURE 27.1

Meyer Sound's SIM1 (Source Independent Measurement) System—the First Major FFT-Based Sound Measurement Computer System for Sound System Calibration

Source: Courtesy Meyer Sound

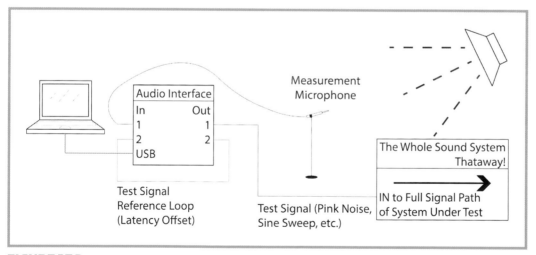

FIGURE 27.2

A Schematic of a Typical Sound Measurement System

The setup pictured is a simple one, consisting of one measurement microphone, a computer running measurement software, a hardware interface, and cables connecting the test equipment to the system (and to itself).[1] Let's examine each component of this system.

- Measurement software:[2] A sound system measurement software must be, at minimum, capable of the following:
  - Generating standard test signals: typically, sound systems are tested with a combination of pink noise (a kind of staticky-sounding test signal that features randomized sound at equal energy per octave), and sine/tone sweeps (called pink sweeps in SMAART) that excite the entire range of human hearing in a calibrated sequence

- Sending test signals via hardware output to a sound system under test
- Taking in the signal of *at least* one measurement microphone (and often multiple microphones)
- Feeding a reference/timing offset signal to itself
- Performing FFT/IFT (Fast Fourier Transform and Inverse Fourier Transform) conversions on time-based data (sound), to allow for frequency- and time-domain representation of data
- Displaying frequency/spectrum data, generating impulse responses, and ideally displaying phase, magnitude, and coherence data (more on these in the next chapter)

- Measurement microphone(s): microphone(s) for measuring sound systems must have an exceptionally flat frequency response over the entirety of the human hearing spectrum. Physically, this means they must be the following:
  - Condenser mics—no other microphone type is electrically sensitive enough to do this work
  - Very small diaphragm—microphones of this type have a diaphragm smaller than the wavelength of the highest measured frequency, such that its natural resonance is well above the human hearing range and is thus uncolored in response
  - Flat frequency response—across the whole auditory range
  - Omnidirectional—to avoid proximity effect and other directional anomalies in measurement
  - Very low self-noise

- Audio interface: typically, a measurement system consists of the computer running software, connecting physically to the sound system via a hardware audio interface.[3] This interface, at minimum, requires the following:
  - At least one microphone preamp/input (up to however many mics you plan to use/your measurement software can handle at once)
  - At least one line-level input (for the loopback)
  - At least two line-level outputs (one to the system, one to itself)
  - Stable latency[4]

- SPL calibrator (optional): an SPL calibrator is a device that is fitted over the capsule of a measurement microphone and produces a test signal at a standard value (typically ~94 dB SPL, aka 1 Pa of pressure). By using a calibrator, you can then set your test system to register that value as 94 dB SPL, and thereby take an accurate measure of SPL during the rest of your measurement process. While this is not necessary for all system optimization tasks, it can be helpful if part of your system spec demands a minimum SPL at certain locations (or if you are matching SPL coverage around a venue with disparate surround effects speakers, for example, to create an even and diffuse coverage field).

- Cables: your cables must be free of noise and interference.

In a simple two-channel measurement, we are comparing the reference signal to the measured signal run through our system, time-aligning them (so that the time difference between reference and measurement doesn't erroneously show up as frequency anomalies on our screen), and then performing EQ, delay timing, and the like, based on what we find. This process involves measuring frequency response (as magnitude, phase, and coherence

275

Message in a Bottle

The thing about sound that people really have to think about is that, of course, it's invisible, everybody knows that. But it's also virtually—you're going to laugh at me but—it's virtually inaudible. And what I mean by that is that you can only hear it where you're standing. It's very different than lighting. Everybody knows when the lighting spot operator falls down and misses, you know, the spots fall off of the actor. But nobody knows about the sound except for right where they are. And so we're all in our little cocoon, all in our little private spaceships. So we have to try to communicate about this invisible thing that *might* be different. Just because you're standing where you are, and I'm standing where I am. This is not a swimming pool, and you're not measuring chlorine content. You don't put the sound in, wait 4 hrs for it to evenly spread throughout the room, and then put a dipstick in. Sound does not spread by osmosis.—Bob McCarthy

data) and impulse responses (for timing data). In the next chapter, we discuss the basic system measurement and calibration process.

Note that while the terms and graphics we use are specific to Rational Acoustics' SMAART, they apply to most professional measurement software, and the principles are the same across the board (physics doesn't change, no matter what tool we use to measure it!).

Notes

1 Most measurement systems require one channel of loop, sending the test signal out of the interface and back into itself so that the measurement computer can account for the latency of the interface itself. More on this later.
2 In this section, we'll be using images from Rational Acoustics' SMAART software, but there are many alternate programs available.
3 Recent developments have allowed some measurement systems to send all signals digitally (via Dante, for example), with the exception of the measurement mic(s) signals. Any system or process description in this and the following chapter is applicable to a Dante system as well, but we will stick with a hardware interface in our discussions here as it makes the signal flow concepts easier to visualize.
4 One of the reasons that Dante interfacing between a measurement system and a sound system has only recently become viable is that until recently, many computers running the Dante Virtual Soundcard interface software were not stable in their latencies. Since we count on the interface having a stable latency in order to time-align our reference signal with our measured acoustic signal, this caused problems in measurement. Recently, this has improved.

CHAPTER 28

Measurement Practices, System Calibration, and Optimization

In a perfect world, sound system designers would mount their speakers, those speakers would deliver frequency-flat, time-coherent signals at equal level across the entire audience area right off the bat, and we'd all be at the bar in time for happy hour.[1] In our real world, there are a host of steps between the testing phase and operational readiness. In between, we must measure the performance of our system in the installed environment and adjust signals accordingly.

Speaker systems strive for flat frequency and good time-alignment for a number of reasons. Flat frequency helps prevent feedback in systems running open microphone channels[2] and is (even in effects speakers) useful in that it allows content creators to ensure that what they produced in their (hopefully, flat response) studios sounds as anticipated in their live systems. Time-alignment ensures that a large audience can share a sense that the sound is coming from the stage and can experience sound as continuous and (relatively) uncolored wherever they may be in an audience (particularly important at standing room shows, where the audience will move about frequently and at will).

In this chapter, we will begin with a very simple example: a single-point-source speaker delivering dialog to an audience in a theater. From there, we'll proceed to another level of complexity: two point-source speakers delivering dialog to an elongated audience, with the first acting as "main" and the second acting as "delay" speakers, respectively. Following that, we'll briefly address some of the considerations that come into play when tuning speaker arrays. Along the way, we'll address key techniques and terminology associated with the system optimization process. As I have warned multiple times throughout this text already, I will warn again: this is far from a comprehensive reckoning of all the considerations that come into play when calibrating complex systems.[3]

By examining these three simple setups, we will encounter principles that can multiply and combine to form the basis of practice for any large or complex sound system. While the variables shift and expand with each additional speaker, the principles remain the same.

> **System Optimization**
>
> The process of examining speaker system performance and adjusting frequency, time, and level delivery to achieve the intended system goals.

What is the goal of a professional sound system? Many designers will say that the goal is always to deliver flat (reference-quality) frequency response, amplifying signals that are time-coherent from point of origin (e.g., a stage) to the audience position, at equal signal level to all audience members. This, in many cases, is a reasonable (and these days, *mostly* achievable) goal. However, this definition does not suit every system. In the case, for example, of a large concert venue, this definition of a system design goal is logical and common. In theatrical sound design, however, there are often multiple subsystems at work (e.g., one for music, one for effects, one for dialog), and different designers will have different goals for these systems—one may want dialog to be delivered in the aforementioned "reference-quality" fashion, but another may feel it is desirable for the dialog to fall in level a bit (not sacrificing intelligibility) by the back of the house, to give it a sense of naturalism and pull the listener back to the living people onstage. Effects systems often cover audiences in a wildly uneven fashion, delivering localized sounds that are intended to be heard by all but not at consistent levels. A system's goals should be determined prior to design, and the measurement, calibration, and optimization process should always keep those goals in focus— and ideally help the system to better achieve those goals along the way.

To begin, let's consider the simplest possible sound reinforcement system: a single speaker, amplifying an actor's voice, in a proscenium theater (Figure 28.1).

The actor in Figure 28.1 is wearing a wireless microphone, and her voice is being sent through the mixing console at FOH, from there to a DSP, then to an amplifier, and to the speaker. Before we've done anything to measure or optimize the system, the sound is undergoing several changes from its acoustic form to its amplified form:

- Each stage of the signal path is imparting frequency characteristics. Even without setting any EQ in the console or DSP, the microphone and speaker themselves will effectively act as bandpass filters, limiting the spectral range being transmitted.
- Time is slightly blurred, as each stage of the system will imprint the sound with a small bit of latency.[4]
- The speaker couples acoustically with the room, which itself imparts more frequency character as well as time interference in the form of reflections from the walls, ceiling, and floor.[5]

Now, if you trust some audiophile blogs, you might think that the sound is already "ruined" by the time it gets to our single speaker.[6] From a strict measurement standpoint, this is—to an extent—true. However, all of us as listeners are accustomed to listening to sound in far-from-pristine environments, and with all the coloration that mics and speakers can impart. Our goal, then, for such a simple reinforcement system, is to deliver a signal that is as representative of the input source as possible.

First, we must measure our system's performance in its neutral state.

In Figure 28.2, we can see the axial coverage of our loudspeaker in side elevation view.

Tales From the Field

I got into designing sound systems because I'd become deeply involved in the measurement of sound systems, that's sort of a reverse of what you might think, but basically I'd become involved in the earliest measurements of taking the modern FFT analyzer out into the field, and that led me how to learn how to tune other people's designs, and then as time went on, I was able to see what worked in this person's design and what worked in that person's design, and what didn't work in this person's design and then built up a library of design techniques with the ability to visualize the parts of a design and how to make a design that I knew that I could fully implement all the way to the tuning process.—Bob McCarthy

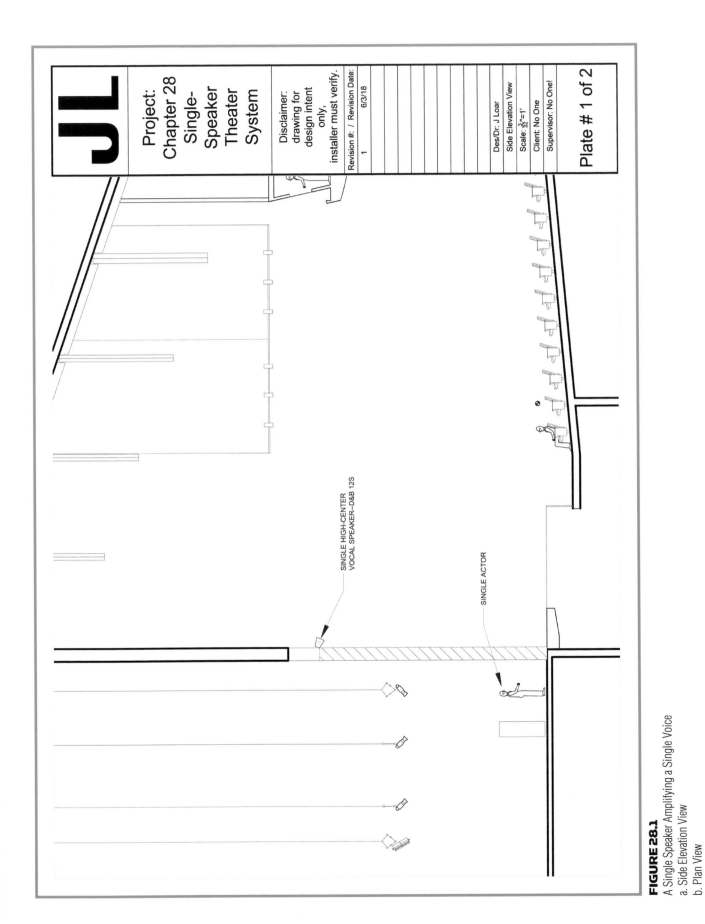

Project: Chapter 28 Single-Speaker Theater System

Disclaimer: drawing for design intent only, installer must verify.

Revision #: / Revision Date:	
1	6/3/18

Des/Dr: J Loar

Side Elevation View

Scale: $\frac{3}{32}$"=1'

Client: No One

Supervisor: No One!

Plate # 1 of 2

SINGLE HIGH-CENTER
VOCAL SPEAKER–D&B 12S

SINGLE ACTOR

FIGURE 28.1
A Single Speaker Amplifying a Single Voice
a. Side Elevation View
b. Plan View

280

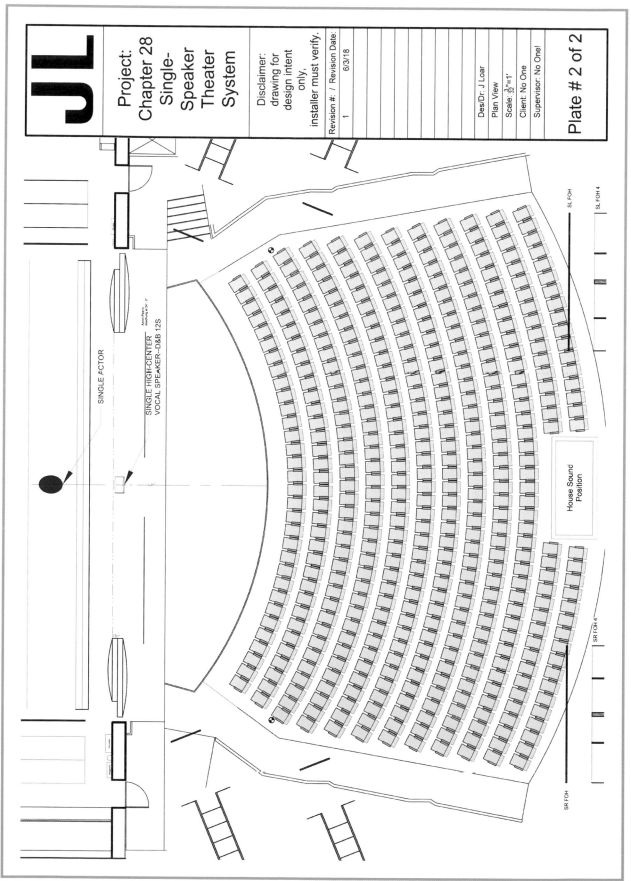

FIGURE 28.1

Continued

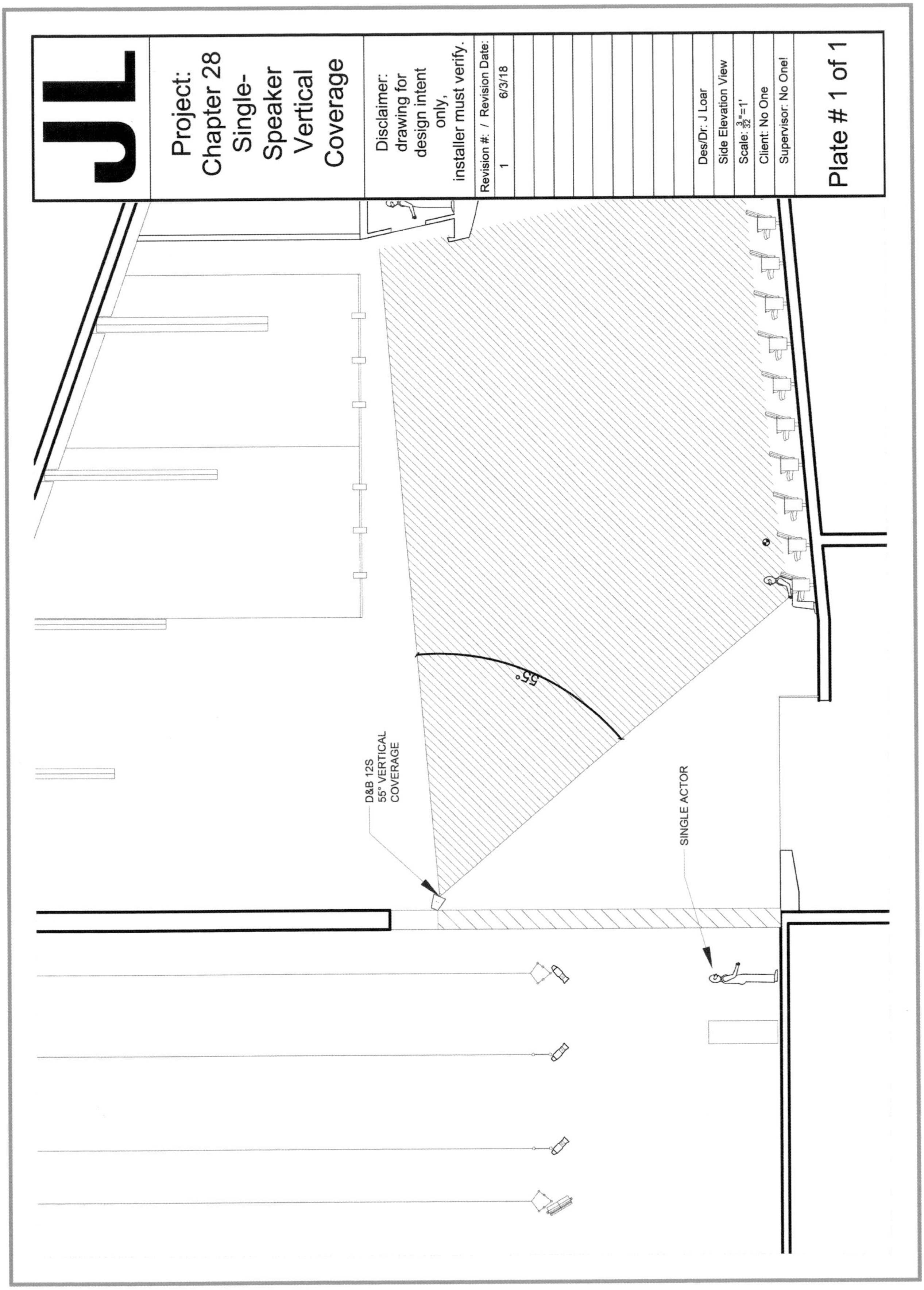

FIGURE 28.2
Axial Coverage of Single Loudspeaker Over Audience Section

As we begin to measure a speaker's performance in a venue, we must always keep in mind that a speaker is generally providing sound to the audience members across a range of seats or audience areas. While it would be tempting to measure the speaker's performance at a single, on-axis position, any EQ changes we made to its performance would be relevant only for listeners at that single spot. Any measurement process, even of a single speaker, must take measurements from different spots around the coverage area.

A measurement system generally sends a calibrated test signal to our speaker(s), such that the software we are using can easily identify differences between the signal as transmitted, and the signal as received in our measurement mic(s). As detailed in the previous chapter, our software will send a signal (such as pink noise) to two places at once:

- Out to the system, via the signal path most similar to what we are amplifying (in this case, through the console, DSP, amp, and speaker—just like the path of the actor's mic).
- To itself. This loopback signal exits the audio interface in our measurement rig, and directly enters the interface via another channel. The purpose of this loopback signal is to ensure that the information we're seeing in our software is time-coherent.

Any audio interface (including a virtual interface like Dante Virtual Soundcard) will have some innate latency in its signal path. By sending a looped signal out via the interface and back directly in, and comparing that signal to the signal sent out to the system and back into the interface via the mic allows us to calculate the delay between the two and adjust the software accordingly. Prior to any delay measurement, the loopback signal should generally be arriving sooner than the measured signal. There are various processes for finding delay time, but all of them involve some form of running a transfer function measurement. By examining the reference signal and comparing it to the measured, the software can find the time difference, and set a delay on the loopback signal, such that we are comparing apples to apples in our measurements.

> **Transfer Function**
>
> A mathematical function for relating the input of a system to the output of a system.

In SMAART®, there is an easy way to tell if we've set our delay times properly, and this is the Phase readout. Phase traces can tell us two key pieces of information: the relative *time relationship* between two signals, and the relative *phase* between the two. NOW WAIT JUST A GOLDURNED MINUTE! Isn't phase and time the same thing?

No. Not quite. Phase, counted by degrees on the wheel of compression and rarefaction, tells us about how complete a given wave is in a particular cycle path, but we top out at 360°—one cycle. After the cycle, we go back to 0° and start again. So, if we examine the phase relationship between signals, we can examine offset from 0° to 360° but for a given frequency, that's all we've got. Now, if we remember way back to when we calculated wavelength, we established that there is a given space each frequency takes in order to complete one cycle. Corresponding to that is the concept of *period*, which is the given *time* it takes a frequency to complete one cycle. The equation for this is

$$\frac{1}{Frequency} = Period.$$

So, if we have a 100 Hz signal, it takes 1/100th of a second to complete its cycle (10 ms), but a 1,000 Hz signal takes 1/1,000th of a second to complete (1 ms). Therefore, it takes half a millisecond for 1,000 Hz to reach 180° on the phase cycle, but it takes 100 Hz 5 ms to do the same thing. If we think about two copies of a single complex signal (containing, at minimum, some components of both 100 Hz and 1 kHz), offset from one another by a quarter of a millisecond, 1 kHz will be 90° out of phase with itself between the two signals, but 100 Hz will only be 9° out of phase with itself. Therefore, signals can be out of phase but "in time." By "in time", here, we don't necessarily mean exactly time-aligned (as in arriving simultaneously), we mean offset in time to the same relative amount, consistent across frequencies. Signals can also be out of time (not in a consistent time relationship with one another), but *in phase* at a particular frequency.

When we are looking at a single signal in SMAART's® phase window, we are looking at a line that represents a comparison between the measured signal and the reference signal. If we see a flat horizontal line, our signals are in perfect time-alignment. If the line is sloping downward (reading left to right), the measurement signal is arriving later than the reference signal (as it would before setting a delay in the SMAART® system). If it somehow slopes upwards, the measured signal is arriving *before* the reference signal, which is uncommon unless you've set something up incorrectly.

In Figure 28.3, we see two different phase windows from SMAART®.

Each is comparing two stored "traces" or measurements of signals (either two different positions of the same mic on the same speaker, or two speakers, or . . . these two could be anything, but the important feature is that they are measuring the phase response of the same signal in different settings). On the left, we see signals that are completely out of time with one another, but in phase at one (marked) spot. If these were two speakers we were attempting to time-align, we would have failed! On the right, we see two signals that are in time, as well as (for the band from 100 Hz to 200 Hz) in phase. These are very well-aligned signals! Note that the lines disappear at the edge of the plot and reappear "wrapped" around the top as if they had continued around a cylinder. This is a standard method of showing phase as it passes the edges of a given plot.

Now, even if we're not comparing two speakers (in our system shown in Figures 28.1 and 28.2, we've still only got one speaker), we need to measure it in many locations. If I'm going to EQ the performance of the speaker for the entire hall, but I only measure it in the front row, I can get it sounding *awesome* for those patrons, but it might be at the expense of the patrons in all of the rest of the audience. So, when measuring a speaker's performance we typically measure at least a few locations within its coverage area. I will often measure a bit at the edges of coverage, and then several positions within the main, central, on-axis area. Store all of those traces, and look at them overlaid (or even, using the tools of the software, average them mathematically). Then, EQ changes can be made based on what will improve speaker delivery for the WHOLE coverage section of the audience, not just for the expensive seats.

This is an important point—when measuring a speaker, as we move the mic so we can examine different positions within a speaker's coverage area, we must re-calibrate the delay time, as any change in mic position will mean a propagation time difference from speaker to microphone. In years of tuning systems, this is the mistake I see people make most frequently, and young designers sit, scratching their heads, wondering what acoustic reflections

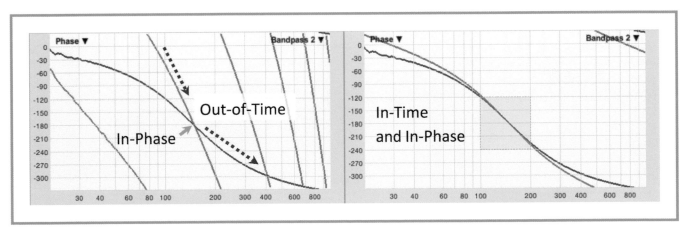

FIGURE 28.3
(Left) Signals Out of Time, but in Phase (At One Spot); (Right) Signals in Time and (for a Narrow Frequency Band) in Phase
Source: SMAART® Screen Captures Courtesy Rational Acoustics LLC

or anomalies are causing the phase response to disintegrate, when all they need to do is re-measure the delay time between reference and measured signals.

> ### Impulse Response
>
> The reaction of a dynamic system to a change in state. In audio, impulse responses are used in a variety of fashions: from exciting a sound system in order to measure time relationships (between a measurement mic and a loopback signal, between two speakers that carry the same material, etc.) to creating reverberant recordings used in convolution reverb effects to capture the signature time decay of a space.[7]

Let's take our first system measurement. We place the mic somewhere in the coverage area of the speaker—taking care to note exactly where the mic is (what seat #, what height, any info possible). We send a test signal, and make sure that the loopback input and the mic input both have strong signal level without overload, and that the signal level for each is as near to the other as possible (as this makes the best test). We run our impulse response measurement to set our delay time (or, if we're using the latest SMAART®, we can use the simple *Delay Finder* tool), and then we run a test signal of our choosing through our speaker. As we run that signal,[8] we examine a pair of windows: phase and magnitude.

If our phase measurement shows that we are in phase (or very near to it) across the frequency range of the speaker, we can generally proceed to examining magnitude. If we are not in *perfect* alignment (perfect alignment is rare in acoustic spaces), we need to consider trends. If we're in good alignment till the very HF range, but there are many reflective and/or diffusive surfaces near the speaker or the mic, we can assume that the phase/time issues we see are the result of the HF reflected content reaching us. As always, we must consider the case (and we'll talk a bit more about how acoustics impact this work in Chapter 29). In SMAART®, there is also a third value, called coherence, that is overlaid on the magnitude window.

Magnitude is a measure of amplitude relative to a +/− 15 dB display. If your signals are properly gained, the baseline coverage of the speaker should hover around 0 on the display, and divergence above/below the zero line will demonstrate (via the X-axis) frequency areas in which signal is not "flat" in frequency response.

As we can see in Figure 28.4, our speaker is not flat in performance.

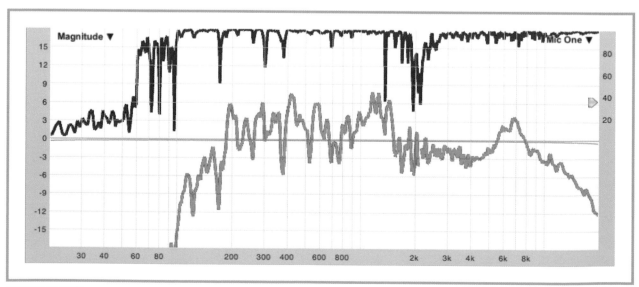

FIGURE 28.4

A Magnitude Measurement, Pre-EQ

Source: SMAART® Screen Captures Courtesy Rational Acoustics LLC

This is to be expected (with any speaker, in any space, before calibration). We can also see the coherence trace overlaid (with scale on the right-hand Y-axis). Coherence is a value measured on a scale of 0–100, that serves to tell us how confident the software itself is that what we see in the magnitude trace is actually accurate. Coherence is calculated by taking the average of the measurement and reference signals and dividing by the power of the reference signal. The more coherent the measurement (the less clouded by reflections or other erroneous data points), the closer to a flat line at 100 the coherence trace will be. Outside of the range of the speaker's frequency performance, we expect the coherence (and phase, for that matter) to disintegrate, but we want as coherent a measure as possible within the frequency range.[9]

So! We've measured our speaker! Time to EQ, right?

Ha, ha, ha . . . NO.

As mentioned earlier, we must measure across our whole coverage area, because otherwise we are EQing perfectly for the measured location, but likely ruining frequency coverage elsewhere. First, we store the measurement data we've captured in our first test—this is referred to as *capturing the trace* (the trace being the lines of data in our measurement views). When we capture the trace, we will name it something that tells us exactly what we've measured—usually some combination of measurement position and info about the system status, e.g., "Seat 4G, Pre-EQ, 4.2' High." Now we will move our microphone to a new location and measure in the same way as before (set the new delay time, run the test signal, examine the phase trace to ensure alignment between loopback and measurement, and then examine magnitude response).

We repeat this process for a range of points across our coverage area. Depending on the size of the coverage zone, it is common to measure some of the corners of the on-axis area as well as at least one center measurement.[10] Capture traces for each measurement, and then select them all to view in the magnitude window at once.

As we can see in Figure 28.5, the measurements diverge from one position to the next, but we can see some trends.

In each measurement, we're seeing a bit of a lump around 2 kHz, which is a perfect target for some gentle, wide, parametric EQ. Now, because we've meticulously recorded the positions for each measurement, we can adjust our EQ (on the output path through the DSP, not on the input signal in the console), and then re-measure each position with the new settings. If we've done enough reduction in the EQ, it will be revealed in the new measurement sequence. If we've done too much (or too little) that will be revealed as well.

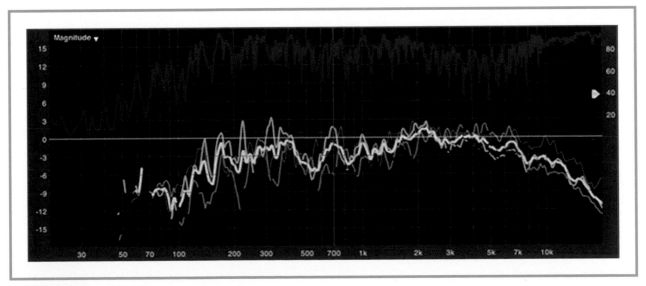

FIGURE 28.5
A Series of Measurements Over One Speaker's Coverage Area
Source: SMAART® Screen Captures Courtesy Rational Acoustics LLC

When we're EQing a system, it is important to understand the following:

- By adding a filter, we've induced some phase shift. Thus, we should expect our phase trace to be worse at the frequency center we've just adjusted. This is one of the many reasons we try to use as little EQ as possible and to use broad, gentle filters to correct overall trends rather than intense surgical notches (which introduce much more drastic phase shift).
- It is best, always, to cut rather than boost. If a speaker's magnitude is measuring too low in a specific frequency range, it is usually worth boosting the measurement mic signal until the overall level is at or above 0, and then use EQ to reduce those lumps that rise above. Any EQ will introduce phase changes, but reduction via EQ will typically be less artifact-prone than boosting with EQ.

In advanced system tuning, engineers will often make use of a tool known as an "all-pass filter." At first glance, this seems to be a contradiction in terms. If the job of a filter is to allow some content to pass and other content to be, well, *filtered out*, the idea of an "all-pass" seems like putting clear lenses into sunglasses—like it misses the point. However, an all-pass filter actually allows the user to change the phase of certain frequencies, thus correcting issues on a micro level without the blunt instrument of EQ or re-timing the entire signal. In complex installation systems, this technique has become increasingly common, but for a system as simple as our single-speaker example noted earlier, it is unlikely to be necessary.

Once we've EQed our system so that it is providing a relatively even response over its frequency range, we have finished the first stage of optimization. Next, we might consider time-alignment.

"Wait!" I hear you saying. "There's only one speaker! I thought time-alignment is something we do between speakers when we have delay speakers for a large audience!" Well, yes, but sometimes (especially in live theater) we will time delay a main vocal speaker back to the location of the actors themselves. This can help preserve the illusion that sound is coming from the performer, rather than the speaker, even as we are amplifying the voice.

286

As we can see in Figure 28.6, without delay, the signal from the speaker reaches the front row of the audience faster than the sound from the upstage live actor.

This will cause the audience to localize the sound to the speaker, not the performer. By using the speed of sound, we can calculate the time difference in propagation between the live voice and the amplified voice. Once we have done so, we can set a delay on the speaker output, such that we eliminate that time difference. Thus the sound travels from the actor, as it passes the speaker, sound fires from the speaker, and together they travel (in phase) to the audience. However, even in this simple time operation, there is a complication. By calculating based on one actor position, we are time-aligning the speaker to that spot, and if the actor moves up or downstage, that alignment will change. Notice that the downstage actor's voice will reach the front row *before* the speaker's signal. This is getting complicated. Unless we have a lot of money for sophisticated real-time actor-tracking software (which re-times the individual actors' signals based on position onstage—this software exists and is becoming quite common on Broadway, but is also terribly expensive, in part due to the need of radio trackers on each performer), we need to align our timing based on an average of performer positions. If we align to the downstage-most edge, as soon as the actor moves upstage, we are back to localizing to the loudspeaker. If we align to the far upstage edge, as soon as the actor moves downstage, the speaker will be behind the live sound. The latter is preferable to the former, though both are time-incoherent and will lead to some comb filtering. If the reinforcement levels are subtle, allowing the

Message in a Bottle

Learning a dual-channel FFT measurement system, whether that's SMAART or SIM or whatever, is really, really, really important. The other thing I think that's important to the tuning is time-alignment—really understanding time and phase alignment in the system is super important. Delineating the difference between time-alignment and a frequency response issue is really important. Obviously fixing time-alignment in the physical space is almost always better than the digital space if you can. But you can't always do that.—David Kotch

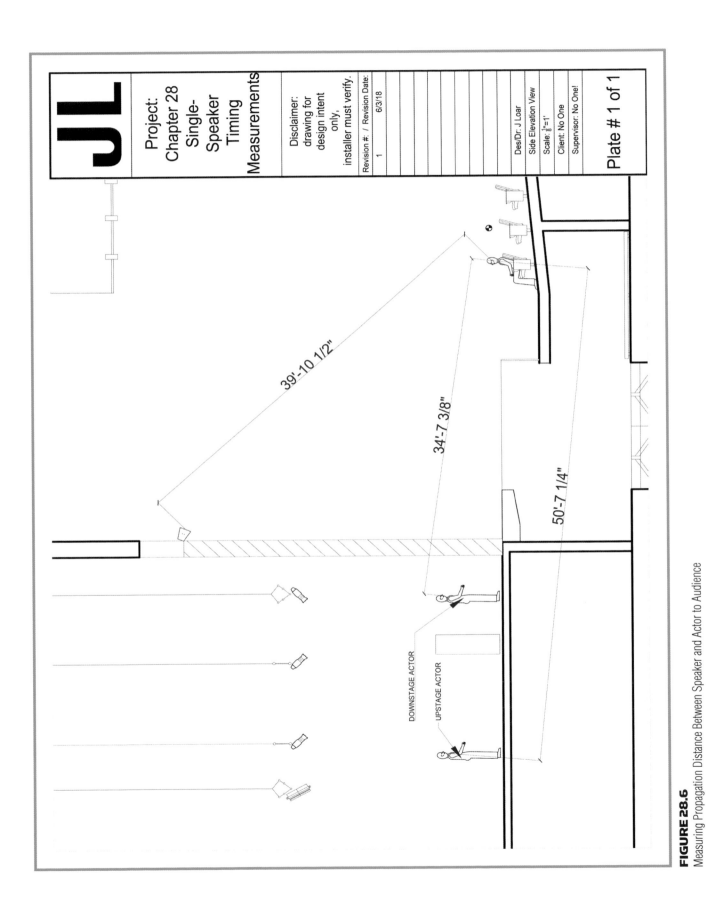

FIGURE 28.6
Measuring Propagation Distance Between Speaker and Actor to Audience

speaker signal to arrive just behind the live sound will be acceptable, so long as the signal is not more than about 30 ms behind.[11] On a very deep stage (where the difference between the upstage edge and the speaker placement will naturally be longer than 30 ms because of physical propagation time), the alignment will usually be set based on an averaged up/mid-stage position.[12]

And, of course, all of the aforementioned is considering time-alignment for *one* audience position. We also need to average the audience positions, just as we do the actor positions, when time-aligning so that we create the best experience for the most people possible.

At any rate, now that we've time-aligned our speaker we're ready to set a speaker-protecting limiter. Limiters are set in any sound system to prevent speaker overload, which evinces itself in two primary forms:

- Thermal overload—related to continuous power over the speaker's handling limit, which sounds like an oversaturated signal, with blurred and poor transient response, and can, if left unabated, melt the voice coil
- Peak overload—related to the XMAX, or maximum excursion that can be handled by the drivers and their suspensions (spiders and surrounds); this causes audible break up and if left unabated leads to blown speakers (tears in the driver mechanism)

There are a multitude of approaches for setting a system limiter. If you are working on a permanently installed system, especially one that is trying to fully maximize the SPL delivery of every speaker, you can calculate voltage values for maximum excursion of a speaker's drivers at different frequencies and set a complex path of limiters each tied to a separate frequency range of the driving signal, which is split into multiple bandlimited paths. This rather elaborate process can squeeze the true maximum performance out of a speaker (especially at the LF end, where driver excursion is most at risk of overload, since excursion increases as frequency decreases). However, for most sound system applications, this kind of Byzantine complexity is unnecessary. If you have selected loudspeakers that comfortably deliver enough SPL to meet your design goals, the process of setting speaker protection limiters can be considerably simpler.

Generally, a loudspeaker will provide a specification that identifies maximum power handling. Often, there are some variety of two such specs: peak and continuous.[13] A peak value is an instantaneous maximum that the device can handle prior to distortion. The RMS value is approximately 0.707 of the peak value for a given test signal (though in actual program content, RMS may measure lower or higher than this value, depending on such musical considerations as compression). For the purposes of setting a basic speaker protection limiter, we are primarily concerned with peak values—though again, in complex systems, some engineers will deploy slow-attack limiters triggered by continuous values above the RMS power handling limit in addition to the peak limiters.

A peak limiter is deployed simply to prevent the speaker from being blown. Let us imagine a speaker whose maximum peak power handling is listed at 600 W into 8 Ohms. If we remember our Ohm's law equations, we will recall that

$$V = \sqrt{WZ},$$

where V = Voltage, W = Wattage, and Z = Impedance.

Impedance, of course, is a varying value—a typical 8 Ohm speaker, when measured with a multimeter, will read around 6Ω with a steady-state test signal, but the value can range from as low as 2Ω to as high as 20Ω or 30Ω, depending on frequency. We will begin with our 8Ω nominal value. At 8Ω, the maximum peak voltage the speaker can handle is 69.28 V^{peak}. If we examine our possible range of values, at 2Ω, the maximum voltage is 34.64 V^{peak}. At 30Ω, our maximum voltage is 134.16 V^{peak}. If we desire a perfectly safe limiting setup, one that is *guaranteed* to never allow peak distortion, we can set our limiter based on the lowest value (34.64 V^{peak}). However, reputable manufacturers' specifications are listed at nominal values because in general, it is safe to operate as if those values are stable. As such, if we set our limiter based on the 8Ω value, we will (with rare exception) be able to avoid peak overload—and we will have a stronger usable level than if we set based on our very safest plan.

In order to set our limiter, we must use a multimeter/voltmeter.[14] We will disconnect the amplifier from the loudspeaker and connect our meter to the output terminals of the amp. We will run test signal (usually pink noise) through the amp—if the amp has a level control, we will begin with it all the way down (if it does not have a level control, we'll adjust the signal level within the DSP). As we run signal level up, we will watch the voltmeter. We will run the signal until we are 2 dB or so *above* the voltage handling we calculated for our speaker (in this case, 69.28, so we run our signal to 71.28 V). We then engage the limiter in our DSP signal path and reduce the threshold such that the limiter is reducing the level measured on our meter back to the safe operating level (just below the max we can handle). We have now set a speaker protection limiter. Our threshold is set to engage the limiter once we've exceeded the maximum voltage level the speaker can handle. If our limiter has a ratio control (many do not), the ratio should be set to *at least* 20:1 (and as high as possible beyond that). If we're living on the edge, we'd set the value based on our absolute max usable voltage, but that's not generally advisable, as it leaves you far less protected from peak damage.

Now that we've set a limiter, our single-speaker system is optimized! Congratulations! Now let's make it more complex![15]

In Figure 28.7, we see that our simple one-speaker system has now expanded.

The audience section has become deep enough that we need to supplement the primary speaker with a second one, out over the audience. This is a very common situation. Whether an outdoor concert with a large audience or a theater needing under-balcony speakers to help reinforce vocals, we often have to send the same signal to speakers that are displaced over a significant distance. When doing so, we need to consider a few factors, both in terms of placement and in terms of timing.

When placing what is generally known as a "delay" speaker, we must first understand why we would choose to place it in any given position. The need for supplementary coverage is driven by a few factors:

1. Level: in an outdoor setting, a typical reason for placing a delay speaker in a given location is that the main speaker(s)' coverage declines (as any sound level does) over distance and is approaching a level too low for concert reinforcement. Depending on whether your system is designed to maintain a constant level for all audience members, or to decline steadily from the stage to the back of the venue, your delay speaker may be closer or farther from the mains. Regardless, there is often a distance at outdoor shows where supplemental speakers are required in order to avoid an unacceptable drop in level.[16]
2. Frequency shadow: in many theaters, the audience area consists both of floor-level seating and balcony seating. For those audience members below the balcony, high-frequency content from flown main loudspeakers is shadowed (absorbed by the people and furniture atop the balcony). Thus, we often place supplemental speakers under the balcony to restore those missing frequencies for the audience seated at the back of the floor section (as in Figure 28.7).
3. Critical distance: in a long indoor venue, it is possible to reach the acoustic point known as critical distance, which is the spot at which 50% of the sound perceived by the listener is reflected, rather than direct, sound.

289

Tales From the Field

Your task is intelligibility, and so there you've got to deal with the interaction with a live source on stage, but then the interaction of multiple speakers and imaging and timing. Intelligibility is all about clean, coherent radiation of energy. So you've got 15 subsystems trying to pretend like they're one big speaker system, timing becomes a massive issue. Broadway sound design guys like Tony Meola and Andrew Bruce, and you know any number of those guys that were part of that—these were guys who adopted the SIM analyzer as part of their tools. Guys like Jonathan [Deans] with LCS [the precursor to D-Mitri], building the LCS systems for all his Cirque du Soleil stuff. These guys were the ones that really pushed the boundaries of alignment technology because they were trying to do two things: they were trying to cover everybody evenly, which meant a distributed sound system; they were trying to support intelligibility, which meant timing and transition and imaging. And they wanted something that, in a lot of cases, didn't sound like a sound system, that sounded very accurate and natural.—Jamie Anderson

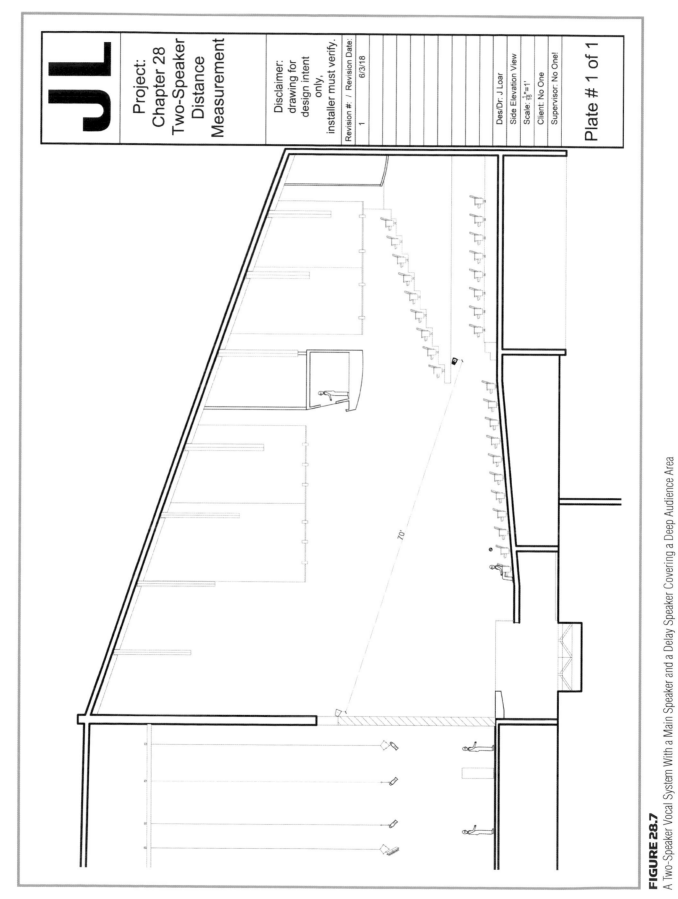

FIGURE 28.7
A Two-Speaker Vocal System With a Main Speaker and a Delay Speaker Covering a Deep Audience Area

In certain circumstances, this is appropriate (such as listening to a choir in a cathedral), but in others (a large seminar like a TED Talk for example) it becomes necessary to supplement the main speakers' direct sound by virtue of delays in order to keep the direct to reverberant ratio weighted toward direct sound (often with the goal of improving speech intelligibility).

Once you are working with a system with a delay speaker (or 10), time-alignment becomes very important. Looking once again at Figure 28.7 we see that the delay speaker in this case is about 70′ into the audience from the main. Remembering that sound travels at roughly 1,130′/second, 70′ takes sound 61.9 ms to travel from the main to the delay speaker. Now, if we just fired the sound live through both main and delay speakers simultaneously, the sound from the main would reach the sound from the delay outside the Haas range (the range within which the two signals would seem to combine and increase in perceived "size"), which means that for the audience closest to the delay speaker, there would be first the sound of the "delay" speaker, then a quick echo from the main speaker, resulting in a blurring of the signal. Therefore, we need to delay the signal sent to the delay speaker (hence the name "delay speaker") by the same amount it will take sound to travel from the main to the delay. So why don't we just delay by 61.9 ms and be done with it?

Figure 28.8 shows the system from Figure 28.7 but with the addition of coverage cones.

When time-aligning two overlapping speakers carrying the same signal, it is often good practice to measure the distance from the main speaker to the first point at which the delay speaker covers the audience rather than the physical speaker position. By choosing the first point of coverage of the delay speaker, you are also choosing the first spot where the two speakers' coverage overlap. Whenever you are timing speakers that overlap in coverage, the first point of overlap is the best spot for time-alignment, as it is the point at which signals have the chance to sum to add level (by being in time with one another), or to comb-filter (which produces gain and reduction in level, as well as an inarticulate swirling of sound that blurs transients and results in loss of clarity).

By measuring the time arrival of a swept impulse response, one can determine the delay between the main signal's arrival and the down-audience supplemental signal, we can properly set the delay so that sound travels coherently from the stage to the audience with as little temporal ripple as possible. Of course, once we've combined speakers, we must consider frequency response again. The frequency response in the summation/overlap range is likely to be different than in either speaker's isolated coverage area. Do we flatten response in the zone of overlap, in the isolation zone, or neither (creating a sort of average between the two)? There is no one correct answer to this question, but what it (as always) reveals is that any sound system build is a matter of compromise. There is no such thing (and never will be such thing) as a "perfect" sound system, with all sources in ideal time-alignment and delivering flat frequency response.

Next, let's change to a different venue, and examine a typical "J-shaped" speaker array[17] and a bit of how calibration changes with this kind of source.

The array shown in Figure 28.9 is a 12-box array.

The array is level at the top, and the splay between box 1 (top) and box 2 (next down) is 1°. The splay between box 2/3 is 1.5°. The splay between 3/4, 4/5, 5/6, and 6/7 is 2°. Between 7/8 is 3°. Between 8/9 is 4°. Between 9/10 is 6°. Between 10/11 and 11/12 is 7°. There is a total splay of 37.5°. This J-shape is a fairly common curvature, particularly with sloped audience sections, as it allows on-axis coverage for the first several boxes without undue splay and makes use of increased splay to help cover audience on the flat floor.

When tuning line arrays, we must first understand something about how array speakers are designed to work. By virtue of their configuration, the typical line-array box offers very wide horizontal coverage by very narrow vertical coverage. Of course, as we've previously discussed, those axial coverage features apply to high (and some mid) frequencies—those transmitted by the tweeters (usually horn-loaded). As we progress into low-mid and low frequencies, the coverage pattern will (as with any speaker) become blobbier and less bound by the pattern control evinced at the high-end of the spectrum. When planning a line array's coverage, then, we must set the angles of coverage such that the high frequencies of each box are assigned to cover a dedicated range of audience seats. In this way, it is similar to positioning point-source speakers, except each point-source is physically connected in a line. We

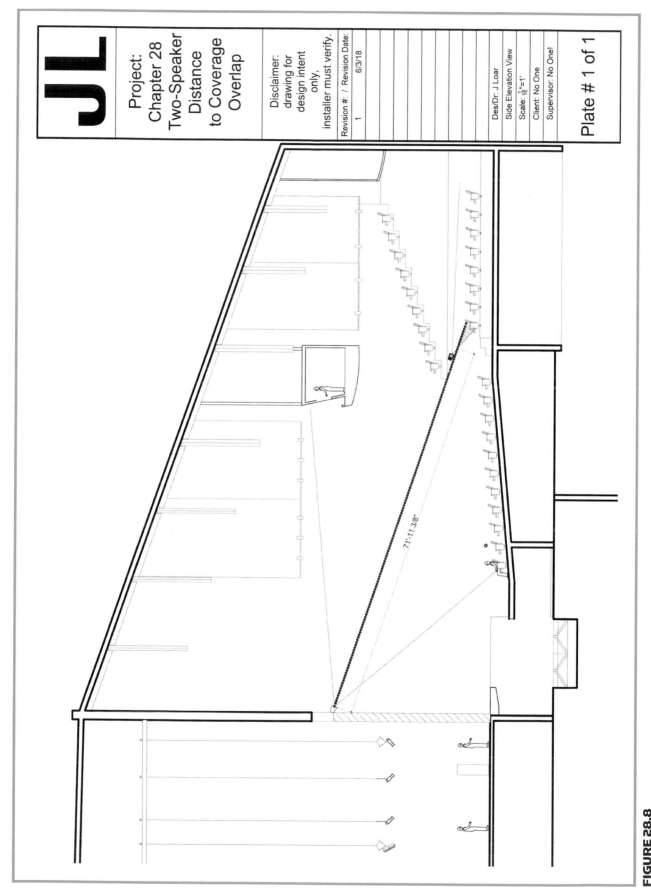

FIGURE 28.8
The System From Figure 28.7 With Coverage Angles Shown

must also then assume that LF content will sum quite a bit and recognize that when measuring an untuned array playing all at once, we should expect to see a disproportionate level of LF in our initial measurements.

Now, when measuring and calibrating a vertical array, we often begin with verifying the coverage of the topmost box, playing (at first) in isolation, with the other speakers in the array muted. Why should this be? A couple of reasons:

1. If we are intending for the array to send even level to all audience members within its coverage area, the limit will be set by the topmost box, as it is traveling the farthest to reach its audience area (seeing the most level decline over distance).
2. By making sure that the aim of the topmost box is correct (e.g., not tilted too low, or aimed too high), we can be certain that any adjustments to the splay of the subsequent boxes will be accurate and can remain as set during verification/measurement.

Once our topmost box is set, we need to verify that the splay angles we chose in our speaker modeling software adhere to the acoustic realities we find in the space. If we have the time, a simple way to do this is to use our measurement mic(s) to verify evenness of HF coverage across the vertical range intended to be covered by each speaker in isolation, and verifying that the edges of that HF coverage fall at the edges of the next speaker's coverage (without unsatisfying dips between from splay that is too wide, or unsatisfying lumps of summation, from splay that is too narrow). Each speaker in an array should be granted "custody"[18] of its own audience area, providing clean and clear HF/mid content to dedicated spots in the audience. Once we've verified that each speaker is splayed properly, and we've covered the whole range of the audience properly, we can begin to tune.

We will often break a vertical array into sections, treating each like a sub-speaker within the main array. The reasons for this are that on the one hand, a 20-box array (not an uncommon sight in concert work) would take a very long time to tune if we had to individually tune every single box, and on the other hand, it is frequently the case that several boxes in an array are positioned to cover a collective area of sub-sections that each feature very similar acoustic properties. In our sample drawing (Figure 28.9), we can see that the top four boxes are set to cover an upper sloping audience area for which the acoustic properties should be very similar. There are no strange outcroppings or balcony rails to cause reflections in one area but not in another. There is a compound slope that changes the acoustic angle of incidence for the next audience section. This middling slope is covered by the next three boxes. Beyond the coverage of those middle three boxes, the audience flattens out, and we curve our array accordingly to cover those zones.[19]

Based on the slopes of our audience, the box count in this array, and our need to simplify tuning, we might at least break the array into groups of two speakers to tune together. The on-axis coverage points break down well that way. This is still six items to align (though that is approaching reasonable). We *might* be able to bump some of these up to groups of three, but we'll need to be careful to think about time-alignment. Over a large audience section, we need to ensure that early arriving signals from lower array elements don't overlap too much with boxes above them in the HF range, otherwise, we risk the audience hearing multiple arrivals and blurring more than necessary. At any rate, let's assume we're tuning in groups of two, and walk through the process. The topmost pair will be Group A, the next down is B, down to F at the bottom.

We will first tune Group A as if it is a single-point-source speaker (which in effect it is). We can then tune Group B. As with our point-source combinations, we then investigate the zone of transition between Groups to evaluate whether time-alignment is needed, or whether level/frequency change is needed. It is common practice to gradually reduce the signal level of arrays as we move down the line of speakers, as we want the audience to experience consistent level, and for each Group the propagation distance to the audience is smaller (and thus level has had less time to decrease before reaching our audience). In our example, this is a standing room event, where patrons are likely to be filling all of the spaces in the coverage areas (there are no fixed seats/aisles) during the performance, so we need to make sure that the zone of transition is handled smoothly. Ideally, the array covers the audience such that as propagation distance between speaker and audience increases, so too does audience distance from the stage increase proportionally, such that no time delay is needed.

At any rate, as we tune Group A, down through F, we can get a good level in each, and decently flat frequency response. In the transition zones, we must be most careful with the pileup of LF content, as the less directional

293

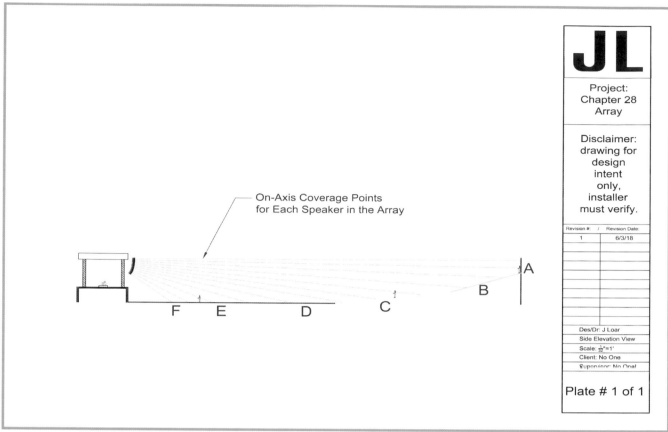

FIGURE 28.9
d&b J-Array Covering a Gradually Sloped Audience

Message in a Bottle

Always keep your eyes open; keep your ears open. Don't let your ego stop you from doing the right thing. I think a lot of times we get our ego invested in an outcome, and it turns out that really what you are proving is that there is an error in your design that could be improved. It's not all good or all bad. But keep your objectivity as much as you can so that you're open to what other people are telling you and what your instruments are telling you. In the end, you've got to use all of your senses, your ears, your eyes, your brain, your experience—what makes sense. If it *seems* too good to be true, that's not a good outcome either, 'cause it *isn't*. So I always tell people, if you find magic in the system, that's not good, because it's going to go black on you at some point. Stick with the non-magic solutions. The other thing I would say is that the KISS approach is always a good one, that's Keep It Simple, Stupid. You want to make sure that you don't try to overdo things. It's a big piece of my tuning. People will look at my tunings and go, "Wow. You didn't put very many filters in, and they're all quite broad and really simple," and it's like, yeah, that's because this is going to make this thing so that it's just smooth over the room and continuous over the room. But I'm not trying to micromanage something so that if there's a change in temperature of two degrees everything goes out the window. We're just trying to clean up the current canvas, but we're not going to bleach it out. You can reach a point of diminishing returns. If I can't get it done in five filters, then I stop right there. Look for something else.—Bob McCarthy

coverage at those frequencies tends to lead to lumpy coverage with extra lows. Conversely, we must also be careful with our EQ shading. A line array, even one subdivided into segments like this, ultimately works on the array principle, which means that especially at low frequencies, the whole array to an extent is operating as one large speaker. The longer the array, the more LF tends to turn directional, beaming forward rather than spreading outward (due to the combined surface area of the drivers acting as one larger driver . . . 12 boxes with 8″ woofers will, in an array with *no* splay at all, act as a speaker with 96″ driver, which can send LF content much more directionally than any single 8″ driver can). The beaming of such LF content can tend to have a lower vertical center than the HF content, so by EQing out some of the LF in lower boxes, we can steer the lobe upward (and the lower audience will still be getting plenty of LF). However, we can also destabilize the array beam and end up with less *predictable* coverage in the LF region—to nodes of hot and cold level across our audience. This can be a slow process when you are just learning (as can any tuning and speaker combination, for that matter), but if you are careful, methodical, and take good notes, you can achieve very good results. Generally, it is advantageous to use multiple mics spaced over the whole coverage range of the array, such that multiple positions can be averaged without having to move one mic around.[20]

At any rate, while calibrating a large array (or a system with multiple large arrays) gets more complex than a point-source system due to the multiple sources and the summation differences across frequency, by breaking each array down into manageable coverage areas and subsystems, we can approach the system tuning just the same way as we would with any point-source system—just with more steps.

If we always keep our minds focused on getting individual coverage areas calibrated, and then carefully ensuring that areas of transition between coverage areas do not evince unacceptable nonlinearities, then we can assemble even the largest system.

To once again quote Bob McCarthy,[21] if we keep in mind that any speaker system can be calibrated by remembering that "A + B = A"—which is to say that a first speaker subsystem, whether a part of an array or an individual point-source, gets combined with a second subsystem to form a new, larger, working subsystem—we can slowly build up any combination of speakers to a coherent coverage plan. If A + B = new A, then New A + New B = A the Third, and so on, until the whole sum of subsystems is a coherent overall system.

To summarize a topic that many find overly complex:

- We verify that each speaker is putting sound out (in correct polarity) to the area it is intended to cover.
- We examine frequency response and work to achieve a flat response across the zone of coverage (an *averaged* flat coverage, not *very flat* in one spot but poor in others).

Message in a Bottle

There's a lot of people that use analyzers to confuse themselves. I think the whole point of it is that it's just a tool that you're using to make your decisions. People will confuse an analyzer with a methodology. You don't SMAART a system, that's stupid. You Josh or Jamie or Bob a system.—Jamie Anderson

---

- We combine the first speaker with the next carrying the same content[22] and ensure that they are time-aligned if necessary and that in combination the frequency response does not suffer unduly.
- We rinse and repeat till our system is tuned!

Once we've optimized the system, we're done, right? Ready to use it?

Most of the time. In large installations, or productions with long runs, we sometimes find in tuning that there are problems we cannot solve with speaker settings (reverb time that is too long, for example), and we need to make adjustments to the actual acoustic performance of the space. This is not always a sound system designer's job (in fact, many times it is the job of trained acousticians). However, it is worth a bit of examination of common acoustical problems in venues, and typical solutions, if only so that we are familiar with the terms when we are on a job working side by side with an acoustics team. In a large installation, it would be expected that acoustic treatment would be modeled just as would the speaker's performance, and ideally any needed acoustic treatment would be done prior to the designing of a sound system for the space. However, there are plenty of times when a production is being mounted in an unconventional venue (such as the Armory, in NYC, which is a literal former armory but often hosts avant-garde music and theater performances). In these cases, we may need to recommend acoustic treatments that become evidently needed during our speaker modeling process, or (unfortunately) later, once we've mounted our system and found anomalies we neither expected (due to inaccurate modeling) nor can control with speakers alone.

Notes

1 Not every sound system designer drinks. Just *most* of us.

2 Feedback is of course still possible, it is just less likely when there are no particularly egregious nodes of frequency boost in the speakers' coverage.

3 The user guide for SMAART® itself is over 200 pages, and quite useful not just for operating SMAART®, but for understanding the calibration process. Bob McCarthy's (quite excellent) book is a weighty tome largely dedicated to this very topic.

4 Even analog gear delays a signal a very tiny bit, but the moment we enter the digital signal chain, there is a latency from the A/D and D/A process, however negligible, that changes the phase relationship of the amplified signal to the live sound.

5 Assuming, of course, that your venue is not an anechoic chamber.

6 Stop it. With rare exception, stop trusting audiophile blogs. They are often as filled with misinformation as useful information. That said, *Sound on Sound, Tape Op, Stereophile,* and a few select others, are trustworthy.

7 The first impulse response measurements involved either popping a balloon or firing a starter pistol in a space and recording information about the decay of that "perfect" impulse. Today, we usually use calibrated test signals (often a sweep), to accomplish the same result but with more precision and accuracy.

8 Which is usually a calibrated noise signal, but can be any signal, actually, so long as the copy is perfectly provided both to loopback and speaker under test.

9 For that matter, if we're measuring a speaker at close range, the phase response should be generally linear with two major exceptions: outside the frequency range of the speaker (where we expect no phase alignment) and at the crossover points between drivers. Crossovers are filters, and any filter (of any kind) creates phase shifts—it is the inevitable physics of filter design. Thus, a speaker with tweeter covering 2 kHz–20 kHz, and a woofer from 100 Hz to 2 kHz, will exhibit non-linear phase response below 100 Hz and around 2 kHz.

10 Keeping in mind that the "on-axis" coverage is only relevant in mid and high frequencies that perform directionally, and that if we need info about potential LF spill back onto the stage—for feedback consideration—we may need to measure outside the on-axis area for a full picture of coverage. For now, we'll stick to on-axis measurements for simplicity's sake.

11 Within 30 ms, we're within the Haas/precedence range, where a pair of signals will be mentally combined to seem like reflections of a single source. Beyond 30 ms or so, the sound is perceived as a distinct echo, which is what we always want to avoid.

12 In a vocal *replacement* system, it is rare that we would worry about delaying this single speaker at all. In a rock concert system, we would *never* worry about delaying this speaker. This is really chiefly a theatrical concern.

13 It is worth noting that continuous measurements are not all created equal. "Continuous" is a catch-all term that may be referring to AES power handling tests (pink noise with a 6 dB crest factor—ratio of peak to RMS value—with drivers loaded in specific conditions), it may refer to "program" continuous, which is usually a dynamic musical signal with intense and mild passages, or some other value. It may be measured over 2 hr, or over 8 hr, or more. A good speaker manufacturer will at least detail what their measurements mean, and how they were taken, and some manufacturers will actually provide several versions of this spec: an AES, a program, and so on.

14 It is important to note that your multimeter must have a readable voltage range higher than what you are trying to measure and set. I once tried to set a 40-V limiter with a meter that went to OL (over limit) reading right at 40 V. This proved impossible, as the test signal (noise) fluctuated enough that it kept blinking right out of readable range.

15 You knew that was coming, right?

16 "Unacceptable" drop would have to be defined by project. For one system, a dip of 3 dB is too much, for others, 10 dB might be the rubric. It depends on your system design goals.

17 In loudspeaker texts, it is technically improper to call a J-shaped array a "line array" because a true line array is a straight line of speakers with no splay or curvature to the box positions relative to one another. However, note that in the real world of installation, these are still called line arrays by *almost* everyone.

18 To use Bob McCarthy's term. I am indebted to Bob's explanations of tuning arrays for the material in the following section of this chapter. And in general, as Bob is the acknowledged master of array tuning, and has taught me a lot through his work.

19 In many such curved arrays, it is common to have boxes of identical vertical coverage in all the top positions—in this case the first six boxes—and boxes with wider vertical coverage in the lower positions. For this example, however, we've chosen boxes with identical vertical coverage all the way down for simplicity's sake. It does not drastically change the calibration and optimization process to add wider coverage boxes at the bottom, it just changes our calculations of the "center" of each zone.

20 Most contemporary measurement systems, SMAART® included, allow for multi-mic measurements. You are still measuring each mic as a paired source with the reference signal, and each time you must reset the delay between mic and reference loopback, but by having multiple mics you can save quite a bit of time in a large venue.

21 Speaking at the AES National Convention at the Javits Center in NYC, in 2017.

22 Always remember this. Speakers that carry different content 100% of the time do not get tuned as part of the same operating system, even if they are controlled via the same console and DSP. If you have vocal speakers that never carry instrumental music and instrumental music speakers that never carry vocals, each of those two sets gets optimized independently of the other. There is no need to time-align a vocal delay to a main music speaker, and vice versa, in this scenario. There is definitely no need to calibrate sound effects speakers as though they are part of a vocal delivery system—and often, sound effect speakers work with their own rules . . . a sound effect speaker only needs to be time-aligned if another speaker carries the same sound effect(s) at the same time (and even then, maybe playback is aligned at the cue level, rather than at the system level).

CHAPTER 29

Acoustic Treatments and Other Solutions

The topic of acoustics is what a former professor of mine once referred to as a "master discipline," which is to say a deep and involved topic that is difficult to master, and whose masters tend to be seasoned experts who possess a wealth of experience to fill in and compliment (and in some cases contradict) the academic writing on the topic. It is *well* beyond the scope of this book to address acoustics in any satisfying depth as a subject unto itself. As such, the goal of the following chapter is to simply present the basics of how sound can interact with surfaces and some of the most common treatments employed to alter these interactions. It is not usually a sound system designer's job to recommend acoustic treatments (though I do know a few acousticians who are also system designers, and their work in systems is improved by the depth of their understanding of acoustics). Instead, it is often the job of the sound system designer to express their needs and goals to an acoustician, who then will recommend a course of action in response.[1] If the sound system designer understands the basics of how sound can interact with an environment acoustically, they are better able to understand how an acoustician's choices can/should influence their system choices, and in certain cases can spot when an acoustician is working at cross-purposes to the sound system designer's intent.

BASIC INTERACTIONS

If we remember all the way back to Chapter 3, we will recall the five basic forms of interaction that sound can have when it encounters an object in its path:[2]

1. Reflection
2. Diffusion
3. Diffraction
4. Refraction
5. Absorption

However, now that we're considering venues, we're going to consider one further variation of these, which we will term RESONANCE. In a resonant chamber, frequencies can excite an entire enclosed volume of air (such as a room itself) as if that volume were a musical instrument. We will begin with resonance, because this is one of the major issues we face in enclosed spaces.

A major problem we face with enclosed venues is what is variously referred to as *standing waves*[3] or *room modes*. These are resonances that build up between two parallel surfaces that resonate at frequencies whose wavelengths are either the same distance as the space between the surfaces, or half or one-quarter that distance, or two times or four times that distance.

In Figure 29.1, we see a room with parallel walls.

The distance between the left and right walls is precisely 15′. By using our wavelength to frequency formula, we can calculate that the frequency whose wavelength is 15′ is 75.33. . . Hz. This means that when we excite the room with 75.3. . . Hz, that frequency will "stand" easily between those walls and excite the room at that frequency beyond the initial impulse we played. Now, if the walls are less than four times the wavelength in their linear dimensions (or less than 60′ in each direction, which these are), this is not a true reflection (even though the waves stand between the parallel surfaces), but a resonant mode in an enclosed cavity. For every doubling of frequency (octave up), the wavelength is half the size. At 150.6. . . Hz, the wavelength will be 7.5′, which can complete two cycles between the walls and still stand. The chances of it destabilizing faster are greater at one-half wavelength, and as we proceed up to 301.3. . . Hz,[4] the wavelength of 3.75′ can again stand between those surfaces, but we're getting less stable as the frequency rises (due to more likelihood of other acoustic phenomenon interfering based on the size of obstacles). At 3.75′, four times the wavelength is of course 15′, so the walls (which are greater than 15 x 15) can directly and truly reflect 301.3. . . Hz, which is part of the destabilization of the likelihood of standing waves at that frequency (as the waves will tend to reflect at an angle of incidence, and if the angle of incidence is anything less or greater than perpendicular, standing won't happen as easily).

300

In Figure 29.2, we can see that if we aim a speaker directly at the wall and play 301.2 Hz, it will beam directly to the opposite wall.

However, because the wall is large enough to provide a true reflection of this frequency (and any higher), as soon as the source is not 90° to the wall, neither is the reflection. *True* reflection being an equal angle of incidence and an angle of response, we start to get more directional as the frequency rises, and reflections are less likely to form perfect standing waves.

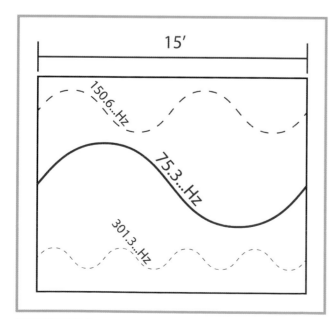

FIGURE 29.1
Room With Parallel Walls

If we double the original wavelength, to 30′, the frequency is 37.66. . . Hz. A 37.6 Hz signal can fold back on itself and form a standing wave that is a one-half cycle delayed from the original signal, but again, the farther we deviate from the center wavelength, the less likely a true standing wave is to occur; 37.6 Hz is more likely to just diffract and spread outside the room.

So, why does a frequency of wavelength equal to the distance between walls resonate the space if the walls are less than the size needed to reflect the frequency? Remember that *true* reflection and *general* reflection are two different things. When we speak technically about reflection, we are talking about the equivalent angles of travel. However, when items reflect at different angles (which we technically call diffusion), standing waves are fairly easy to form. We also see evidence of diffraction, in that a room resonating at 75 Hz will probably transmit that 75 Hz outside the room (unless the room is acoustically isolated, which is a whole separate topic and discipline).

These room modes we've been discussing so far are what are known as *axial* room modes, modes along the axes of the room. There can also be modes that arise *tangentially* or *obliquely*, which are combinations of angular travel bouncing off of more than two obstacles. These can cause acoustic issues in spaces, but between the fact that they are less prevalent (typically) than axial modes in the listening experience, and the fact that calculating their values is more complex, we will leave them aside for now, only saying that there are other resonances in a space than simply axial modes.

So! How do we deal with standing waves and modes? There are two basic tactics: placing diffusers or absorbers.

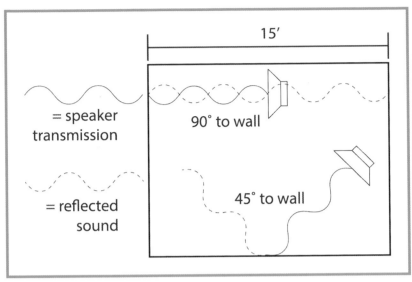

FIGURE 29.2
301.2 Hz Angles

In Figure 29.3, we can see a quadratic diffuser and two different models of absorber.

Diffusers are used where HF content has built up, either because of room modes or because of specific angular reflections (such as the wide coverage of nearfield monitors splashing against side walls in a small recording studio). These diffusers take HF content (with its small wavelengths) and scatter it in a wide range of directions, producing a diffuse sound field that has less focused intensity at the high-frequency problem areas (where HF has "beamed" to a collected spot). It is common to find these on walls and ceilings in studio spaces. We do see variations on this technique in live venues, but they are seldom shaped like "diffusers," more often we see architectural and decorative elements that are serving the dual purpose as both interior decoration and sound scattering device. Diffusers are typically made from hard materials, such as plastic, or hardwoods with a finish.

On the flip side, absorbers are sometimes placed to convert sound energy to heat energy via friction. Absorbers must be highly porous. The absorbers shown in Figure 29.3 are smallish, and thus focused on HF energy, though

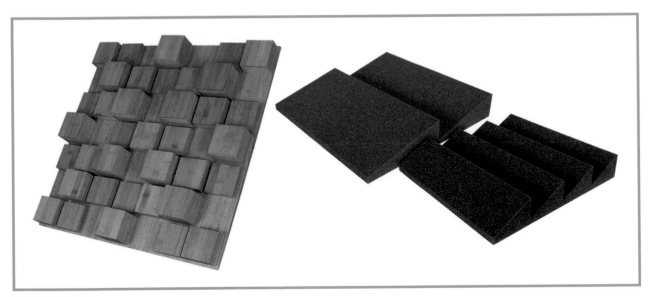

FIGURE 29.3
(Left) Quadratic Diffuser (Auralex Sustain Metro); (Right) Absorbers (Auralex DST 112 and 114)
Source: All Courtesy Auralex

there is another, larger type of absorber, known as a "bass trap." Bass traps are LF-focused absorptive systems that are either simple absorbers of large size, or resonant chambers designed to resonate at the frequencies needing to be controlled, but also to surround those chambers with heavy absorption to isolate that bass resonance from the rest of the room. It is beyond the scope of this chapter to go into detail about bass trap design, but these are commonly found, again, in studios, and less frequently in live venues.

Diffraction (the travel of LF content around barriers) is a huge issue in the acoustics of a space, one that is very difficult to control. High-end recording studios usually "float the room," which is to say build a room within a room, completely isolated vibrationally by neoprene supports, and with cavitatious bass traps hidden behind the interior walls. This is seldom a practical solution for, say, a live venue (though it is not unheard of in high-end venues, especially those dedicated to music with wide dynamic range, such as chamber or orchestral music).

Refraction (the changing of the angle of sound's travel when passing through media of different densities) is seldom a concern in venue and system acoustics, *unless* you are designing a system for a swimming pool/bar in a Vegas casino. Then, you will need to think about the change in speed of sound through water vs. air, and how that might impact design choices and multi-speaker summation.[5]

FIGURES OF IMPORTANCE

There are several acoustic measurements that we want to consider and may need to be aware of when designing a sound system for a space. Here are a few of the most common (in alphabetical order) and some basic considerations for the system designer:

- Clarity/C80: the ratio of early sound (that arriving at a listener before 80 ms have passed from the initial impulse) and late energy (after 80 ms), expressed in dB. The higher the number, the more clarity can be expected. A concert hall, where long reverberance is often desirable, often aims for a C80 of between −4 and +1 dB, whereas a nightclub would want a much higher C80.
- Noise: we often take one-third-octave-band measurements of the noise floor of a given space, so we can be aware of what HVAC or other hums and sounds we need to amplify atop (or treat acoustically).
- RT60: Reverb Time defined as the time it takes for a sonic impulse in a space to decay by 60 dB. The longer the reverb time, the more "wet" the space sounds. A cathedral typically has a very high RT60 number. A bodega typically has a very low RT60. In spaces with high RT60, we often need more speakers spaced through the audience area in order to increase the ratio of direct sound to reverberant sound, for clarity.
- STI: Speech Transmission Index, a measurement of speech clarity over a space. It is reported as a value between 0 and 1, where 1 is ideal clarity of speech, and 0 is very poor. In any space where speech needs to be able to be transmitted and understood by all in the audience, a value of 0.6 or higher is desirable.

As you may have noticed, most of the earlier measurements have to do with time response and clarity of signal. If you are working in an established venue, and you don't have the opportunity to take real acoustic measurements of the space, you can certainly estimate with tools like AFMG EASE® and other such software. A knowledge of the acoustic properties of a venue is extremely helpful when selecting speakers for that venue. Without understanding how live or dead the room is, what the time decay is, and other such factors, we may select less-than-ideal speakers and place them poorly. If I am selecting L/R mains for a live venue, and I know that the side walls are dense absorbers, I know I can splash a bit on those walls with my coverage without ruining the sound for those seated in the potential reflection zone. However, if those side walls are highly reflective, dense, hard, nonporous surfaces, I know I will want to select L/R speakers with narrower horizontal coverage and try my best to aim them away from the walls, such that the audience is primarily hearing direct sound and not comb-filtered reflections summing with my direct sound.

This summary of some (very basic) acoustic terms and concepts is woefully brief and underpowered, but it is time to move on. For more on the field of acoustics, see the "Bibliography and Further Reading" section at the end of this book.

Message in a Bottle

[On working with variable acoustics systems, such as Meyer Sound's Constellation] I have had a couple of opportunities to work with those sorts of systems. The outcomes have always been a little mixed. Sometimes really exciting, sometimes just OK. You just can't control them in ways that I find are really effective. If you're trying to *create* reverberant characteristics, those systems work, I think, really, really beautifully. I find the Constellation works best when you're trying to create a kind of environment and less effectively when you're trying to do interesting or even subtle movement gestures. I don't find it as effective for that, though you can do it. You can maneuver those systems, but I find the time it takes to do it is not worth the energy—I don't find the outcome compared to the time in is necessarily a winner. But if you've got a good operator/programmer, that is really valuable.—David Budries

Now, we move into Part III, where we begin to examine the process of designing our sound system in detail.

Notes

1 Or vice versa. I have certainly worked on projects where the acoustics team was first hired and described their intended goals for a space, which I then worked to assist in accomplishing with my system design choices. But since this is a system design textbook, and not an acoustics textbook, we'll pretend for a moment that we're always in charge of this conversation.
2 See Chapter 3 for the basic definitions of these terms, which we will not restate here.
3 Standing waves are sometimes useful. For example, the resonant body of an acoustic guitar is designed to make use of standing waves as part of its acoustic signature/timbre.
4 Yes, 2 x 150.6 is 301.2, not 301.3, but I am simplifying to the tens place numbers calculated based on the wavelength size and a speed of sound estimated at 1,130 ft./second.
5 If you've never been to Vegas, this is a real thing, and it's very strange.

What Type of Sound System Do You Need?

What are the special questions that need to be asked in each discipline? What are the differences in expected process? What are the schedules and special terms? This part examines different sound system types by discipline.

Understanding the Source Material

The process of designing a sound system can be conducted in many different ways, with many different subsequent orders of events. While this chapter and those following it represent one way of organizing the tasks associated with just about every sound system design, keep in mind that this is but one version of how the process might be conducted. We begin here with understanding the source material being amplified, but I have worked on productions where we knew the venue, the budget, and many other factors before the source had even been truly detailed.[1] At any rate, each of these chapters presupposes that you have been brought into a process well underway, such that, for example, the source material being amplified is known in advance. Feel free to depart from the perceived order of events laid out here, depending on the needs of your project and information available. While the individual steps in Section 11 are each essential to a system design process, they can be conducted in a wide variety of orders.

So! Let's assume you have been hired to design a sound system. Congratulations! It's always nice to have work. Ideally, one of the first steps you will take is to gain a full understanding of the source material to be amplified. If you understand the source material clearly, it can inform every subsequent design choice. For the purposes of our discussion here, there are two main categories of source material:

1. Production—any project which is centered around designing a system to amplify one event/installation, with every design choice suited to delivering the content associated with that particular production. This could be anything from a concert tour by a headlining artist to a theatrical production to a theme park ride. All of these are "single" productions, in the sense that you are not creating a flexible system designed to handle a range of different events and content, but creating a system catered to the needs of that particular project alone.

2. Repertory—while I borrow this term from theater, I use it here to mean any system designed to accommodate a range of events/projects, whose exact parameters are unlikely to be fully known at the outset of the design process. In this category I count everything from a conference room in a large office complex—which,

yes, will mostly host meetings, but may need flexibility in terms of what configurations the room will see, how many participants are involved, etc.—to concert halls, theaters, and other such spaces in which many different projects will take place. These systems need to be designed to accommodate a range of activities, not just one project.

Within each of the above categories, there is, of course, an endless variety. There are as many different types of source content as there are imaginations to dream them up, and each project (or venue hosting projects) will demand different things of its system designers, but we can, at least, identify some of the major subcategories of source content, and some key points we will need to understand before designing a system to suit them. Note that some of these subcategories will overlap between the two types. A single concert or tour by a single artist might fall into the "production" category, while designing for a concert festival or a concert venue will fall into the "repertory" category. In Section 12, we examine the types of sound systems in detail, but here we strive to identify the initial questions a designer needs to ask and answer about the source content in order to gain enough understanding to develop a plan of attack for the system.

PRODUCTION SYSTEMS

We can generally break production systems into five major subtypes:

1. Concerts
2. Theater
3. Corporate events (aka "special events")
4. Themed entertainment
5. Art installations

Within each subtype we can establish further subcategorization:

- Concerts:
  - Small scale—defined here by audience size, not band size, as a 3-piece band might play to 60,000 people, demanding a large system, and a 20-piece band might play for an audience of 200, demanding a much smaller system. For our purposes here, we will define a small-scale concert as one presented to 1,000 people or less.
  - Large scale—audience of 1,000 or greater. While there is definitely a difference between a system designed for 1,000 people versus one designed for 60,000 people, the difference is less one of strategy than of scale, whereas a system for 1,000 or less may approach the design from a different strategic plan.
- Theater:
  - Straight plays—non-musical theatrical works (which may contain music cues, but no outright *musical numbers*)
  - Musicals
- Corporate events (aka "special events"):
  - Single-room
  - Multi-room—many corporate events will spread over a large space such that there are multiple parallel happenings, for example in a convention setting. This is still a single overall production event, but there may be diverse system needs across the different quarters of the event.
- Themed entertainment:
  - Rides—these tend to be designed once and fixed for some time thereafter.
  - Shows/parades—events that may change during different seasons in a park, but whose design needs are consistent so long as those particular "shows" are running.
- Art installations:
  - Site specific—works that are intrinsically tied to the location in which the work is being mounted, and that would have to be entirely redesigned if moved to another space.
  - Self-contained—works that would be consistently presented no matter what venue the work is situated within, designed to suit the needs of the production, and to be installed and re-installed identically wherever the production is mounted.

REPERTORY SYSTEMS

We can generally break repertory systems into six major subtypes:

1. Performing arts venues
2. Houses of worship
3. Themed environments
4. Media production facilities (e.g., studios)
5. Commercial environments (e.g., retail spaces)
6. Corporate Environments (e.g., office spaces)

Once again, within each subtype we can establish further subcategorization:

- Performing arts venues:
  - Concert halls
    * Classical/orchestral
    * Jazz
    * Rock/pop/hip-hop/electronic/etc.
    * Multi-genre
- Houses of worship:
  - Standard
  - Megachurch—defined by the size of the audience and the scale of multimedia production involved
- Themed environments:
  - Theme parks—open (or "parkwide") areas within a park that must accommodate a range of content
  - Hotels—particularly those associated with theme parks
  - Cruise ships
  - Casinos
- Media production facilities:
  - Music studios
    * Recording
    * Mixing
    * Mastering[2]
  - Television studios/soundstages (here I am counting studios generally as post-production facilities, vs. soundstages for active filming)
  - Film studios/soundstages
  - Video game studios (including virtual reality/augmented reality)
- Commercial environments:
  - Retail stores (including grocery)
  - Shopping malls
    * Indoor
    * Outdoor
- Corporate environments:
  - Office buildings
    * Offices
    * Conference rooms
    * Lobbies
  - Warehouses/production facilities—assembly lines and other such, as distinct from *media* production facilities

The first question any system designer must ask about the source content being amplified is which of the earlier subcategories the work falls into. If you know you are designing a straight play, you can eliminate many of the concerns you would have when designing a system for an automotive company's factory.

Once you know which of the earlier subcategories your work will fall into, you can begin to drill farther down into the understanding of the particulars of your project. Let's examine, one by one, the subcategories noted earlier, and list some of the pertinent questions you must answer in order to gain understanding of the content in question.

PRODUCTION SYSTEMS

CONCERTS (SMALL SCALE):

- What needs to be amplified? In many concert settings, an audience of less than 1,000 may or may not need amplification. An orchestral group in a lovely concert hall with good acoustics may need no amplification at all. A rock band whose guitarists insist on bringing their full Marshall stack and turning it up to 11 for a tour of small clubs will not need that amp stack to be mic-ed. A folk group playing in a coffeehouse might need mics and speakers for the vocals but not for their instruments.
- What needs mics vs. DIs? Does that ukulele have a pickup? Or do we need a mic in front of it?
- How loud does the source material *want* to be? A heavy metal band will expect much higher SPL (even for a small audience) than will a folk duo. This will define a lot about how you plan speaker coverage, amp power, etc.
- How many channels do I need on my console? How many mics and DIs, playback channels, etc.?
- Will there be monitor mixes? How many? This determines either how many aux sends are needed on the FOH console, and/or what specs are needed for a dedicated monitor console (as well as, of course, speaker and amp considerations onstage). Will any of the monitors be IEMs (in-ear monitors), instead of speakers?
- Do I need a full-range system? Again, a folk duo probably doesn't need dual-18' subwoofers, but the trap stars certainly will.
- Will this production be mounted indoors or outdoors? Weather-resistant tents over the FOH position, weather-sealed speaker enclosures, and the like come into play when working outdoors.

CONCERTS (LARGE SCALE):

- Ask all of the same questions as noted earlier for the small-scale concert, with the exception of "What needs to be amplified?" By the time we get to large-scale systems, with rare exception it can generally be assumed that everything needs to be amplified (even an orchestra, if the LSO is playing with a pop artist at the O_2 Arena, they will need to be mic-ed, and classical purists be damned).

THEATER (STRAIGHT PLAYS):

- Do any actors need to be mic-ed? In smaller venues, or for productions striving for a natural feel, actors are not always mic-ed.
- Playback:
  - What kind of playback content will be used during the production? Sound effects? Music? Do these effects need to seem to be real and be noticed by the characters onstage (diegetic sound), or are they more for the audience's enjoyment and understanding, and the characters don't notice or attend to them (non-diegetic sound)?
  - Does the playback need to come from a particular place or places onstage, or to surround the audience, or both?
  - What is the frequency range of the anticipated content? What kind of show is this? A realistic show about the Iraq war probably needs to deliver a lot more SPL and frequency range than a bedroom drama about a couple's relationship disintegrating (though this will, of course, depend on *how* that couple's relationship is shown to fall apart). Mood, content, and desired effect on the audience should all be established at this stage of the production process.[3]

THEATER (MUSICALS):

- Ask the aforementioned questions for straight plays, and also
- Is there a live orchestra/band, or is the music canned (playback from recordings)?
- If there are live musicians, where are they situated?
- Do the musicians need amplification?

CORPORATE EVENTS:

- Is the event single-room or multi-room?
- Does the event feature an amplified panel of presenters? If so, how many presenters at one time?
- Do any of the presenters need audio playback connection for their presentations?
  - Sound from PowerPoint or other such software (presenter controls playback)
  - Sound from offstage playback (offstage operator controls)

Tales From the Field

I was doing a production of *The Piano Lesson* by August Wilson. So the set that our designer did had a big stair unit that sort of went all the way up from downstairs, from upstage left it rose to upstage right. And part of the visit of Sutter's ghost at the end of the play is, as Sutter's ghost was sort of haunting that house—that sequence—Boy Willie ran upstairs to get something, and when he ran upstairs, Sutter's ghost was up there and sort of frightened the crap out of him and caused Boy Willie to sort of tumble down the stairs. Well, we get into tech, and I've been in rehearsal with everybody, so we all knew the beats of the visit and sort of rhythmically how everything was gonna go, but we hadn't spatialized it out yet because we were just in rehearsal. Well, I put a big speaker at the top of the stairs—hidden, you know, way up out of head height so no one is going to hit their head. We get to that part of the play, Boy Willie runs upstairs, and as soon as he gets up there, we take that cue and it's this huge loud *guh guh guh guh guh* coming out of it, out of a[n EAW] JF200 that happened to be 6" away from the actor's head. So we get through the sequence, then he comes out to me—very politely, but very firmly—he said, "We've got to do something about that speaker." So that was, for me, a lesson—I did not recognize that I needed to include the actors in that particular conversation, because we rarely have to include the actors in those conversations. Lots of your positioning is really, it's a conversation with the lighting designer, and maybe with scenery; this situation—I just missed the boat and should have involved the actors as well.—Vincent Olivieri

Message in a Bottle

For producers, and general managers, and production managers, and technical directors—I would say that even though sound is entirely untouchable and entirely invisible, the quality of the sound that your audience hears is so critical to the pleasure and their perception of how good the show is. Far too often, an audience will drive to a musical in their SUV with 5.1 surround sound playing the Broadway cast recording, and then they'll watch the show, and then they'll go home and watch the review of the show on their Dolby 5.1 system connected to their TV. Sound quality is way different now than it was when I was growing up, and I thought opera all sounded terrible because my grandmother's record player only put out 300 Hz to 12 kHz. Our audiences are—maybe not consciously, but—they're way savvier about what good sound is than they used to be. And so it's not something that can be relegated to the "good enough." It needs to be as good, it needs to be as expertly done, as your set and your costumes. Not more. Not any of that. But it does need that degree of quality prioritization. And that can get lost.—Robert Kaplowitz

- Does the event require teleconferencing (either traditional or via Skype or the like)?
- If the event is multi-room:
  - Does the event require announcement capabilities separate from presentation systems? Often conventions will have a few stages for specific presentations, as well as a convention-wide announcement system for safety messages and other such emergency communications.
  - How many rooms? Do all of them need amplified sound in some fashion?

THEMED ENTERTAINMENT:
- Rides:
  - What type of ride is this?
    * Dark ride—slow ride around a track in an enclosed building, think *Haunted Mansion* at Disney
    * Roller coaster
    * Water ride—water rides usually include lots of splashing, so equipment used in water rides (even if they are enclosed in a building) needs to be moisture-sealed

- Is there themed content in the queue, the ride itself, or both?
  * Roller coasters, for example, often have heavily themed queues demanding sound reinforcement, but little content once the ride starts
- Is there both onboard (in ride vehicle) and offboard (outside the vehicle) content? Just one or the other?
- Does the ride vehicle or rider control any of the content by their interactions (whether by direct guest interaction within a vehicle or environment or being triggered by a moving ride vehicle)? Or is the content presented passively?
  - Shows/parades:
  - Shows—ask all the same questions that would be asked for concert and theatrical productions (starting with: what kind of show is this?).
  - Parades:
    * Is there both onboard (in parade vehicle) and offboard (outside the vehicle, on the parade route) content? Just one or the other?
    * Does the parade vehicle or performer control any of the content by their interactions? Or is the content presented passively?
    * Do any of the parade performers need to be mic-ed?
    * What is the parade route? Does the parade need to interact with the parkwide systems, or will parade content be amplified exclusively via its own dedicated systems?

ART INSTALLATIONS:
- Site-specific:
  * What is the site? What factors about this site are unique such that they need to be incorporated in our understanding of the content?
  * What content will be amplified? Music? Effects? Voices? Experimental noises?
  * How is content triggered? Is it delivered passively or are guests interacting with triggers?
- Self-contained: the same questions as for a site-specific piece, minus the unique factors about the site that impact content.

As we can see, production systems are all asking variations on the same questions. Do we need mics? Are we playing effects or music? What type of effects or music? How loud and full range does it need to be? How many channels? Where does sound need to come from? Etc.

Now, when designing for repertory systems, we must ask questions not only of the content of one show or event but also of the *range of events* that a given venue will produce/host.[4]

REPERTORY SYSTEMS

PERFORMING ARTS VENUES:
- What type(s) of arts will be presented in the venue? A space that hopes to host both orchestras and major rock bands (like the Hollywood Bowl) is going to have different requirements for its sound system than a venue that only hosts improv troupes. Once you have identified what kinds of events will be hosted in the venue, you must ask the following:
  * What is the possible range of needs expected based on that type of events? Again, a venue producing loud rock bands every night probably needs a host of standard rock and roll instrument mics that a venue hosting only string quartets and chamber ensembles will not.
- How many channels of I/O (in what formats) will this content demand? Does the type of event require separate FOH and monitor consoles or will one console suffice? Et cetera.
- How fast do changeovers between events need to happen? A venue that produces only one event per night with days in between individual productions will have different needs than one that produces an 8:00 pm show of one type followed by an 11:00 pm show of entirely different nature. Changeovers will not only impact staffing but equipment choices—long changeover times may mean re-aiming speakers and re-tuning the venue, short changeover times may mean your speaker plot must work for all potential events, as there will be no time for adjustments between productions.

HOUSES OF WORSHIP (HOW):

- Is amplification needed? Many small churches believe they need amplification when in fact a little acoustic treatment will solve their intelligibility issues without any sound reinforcement at all.
- How many sources (and of what type) need amplification?
  * Small HoW often need a couple of wireless mics, and maybe the ability to play stereo audio back over a small system.
  * Large and megachurches often feature multimedia productions as complex as some large concerts or Broadway shows, with dozens of wireless mics, full bands, video with audio sync, interactive triggering from the performers, and so on.
- Any of the questions from the show categories noted earlier apply here as well if the church presents elaborate enough events.
- Are there special color needs for visible installed sound equipment? Many HoW want their speakers to be white or some color other than black (in fact, most speaker manufacturers offer custom paint jobs, and besides themed entertainment, HoW are the primary driver of this market).

THEMED ENVIRONMENTS:

- Theme parks:
  * The parkwide spaces (or spaces that are not part of an attraction, a retail space, or a restaurant) often need sound systems as well. What is the content being carried? If the system in question is just for emergency announcements, a full-range frequency spectrum is not needed. If the system is for emergency announcements, but also for holiday music all day from October through January,[5] then full range might be in order.
  * Hotels: increasingly, hotels connected with theme parks are whole themed environments themselves, with full décor, soundscapes, and interactive exhibits. All of the questions for any themed attraction would be applicable here, in addition to the needs that arise from normal hotel operations (paging of guests, lobby music for ambience, etc.). Even many hotels not connected with theme parks now feature lobbies with music playback, emergency announcement systems and the like.
- Cruise ships: floating theme parks, with all the of the associated areas of focus: attractions, shows, hotels, etc. All of the questions for those disciplines must be asked, in addition to which the designer must ask whether the system being specified will be able to run for a long time at sea with little need for maintenance (particularly maintenance that requires new parts, as getting parts out to sea is difficult, and cruise ships often require that their systems be operable for many years without updates). Any outdoor equipment must not only be weather resistant but also *salt* resistant, which is another order of protected. There is some overlap between equipment used on cruise ships and equipment straight-up used on Naval battleships.
- Casinos: an environment that is designed to make you forget about the passage of time while you spend copious amounts of money is *going* to use sound as part of its design scheme.
  * What content is being reinforced? Music? Effects? Paging?
  * What mood is the content trying to achieve?[6]

MEDIA PRODUCTION FACILITIES:

- These kinds of venues play host to a very wide range of content, and the key difference between these and other venues is that where other venues may incorporate recording of events in an auxiliary way, recording is central to the media production facility's mission. One central question for any production facility design will be how many sub-spaces are contained therein, and thus how many different content streams all need to be able to be used simultaneously. A studio with one control room and one live room for music recording is very different than a post-production campus with multiple control rooms, editing bays, mixing/composing suites, central media storage servers, and the like.

- Music studios:
  * What kind of music will be produced here? The loudspeaker choices for a studio focused on rap music may be different than the choices for a studio focused on orchestral film scores.[7]
  * Will production be done using analog media or digital, or via some combination thereof?
  * How many channels of signal need to be available (both for capture in recording and for playback)?
- Television studios:
  * What kind of show(s) will be produced here? The studio for a greenscreen talking-head show[8] will be much more minimally appointed than the studio for a large multi-camera sitcom.

* Who needs to hear? Besides the control room, do speakers need to cover a live studio audience (unlikely for a news broadcast, likely for a comedy show)?
* What kind of sync is necessary between sound and picture?
* Will facility-mounted, boom-mic arms be needed? These are common to facilities producing talk shows, for example, where an audience will be seated on the floor and a balcony or loft with deck-mounted telescoping boom arms will allow for overhead mic coverage without obstructing audience views.

- Film studios/soundstages:
    * Is there any need for a reinforcement system of any kind? If you are producing music videos, you need to be able to play the songs in question so that the performers can sync their movements to the music. If you are just shooting narrative dramas, will there be any need for a permanent system of this kind at all, or can such a system be rented when needed?
    * Film production audio (on-set recording systems) are not typically specified by system designers for the studio facility itself . . . but often such facilities will keep one field set of equipment on hand—a boom mic/arm, a field recorder, a few wireless lavs.
    * Emergency announcement system?
- Video game studios:
    * The range of content and styles of production in the video game, virtual reality, and augmented reality worlds is vast. These facilities often feature several dedicated smaller content creation/mix/programming suites, but increasingly may also feature spaces in which to demo material in full surround as it is developed. The signature question to be asked is: what styles of content will be delivered, and what monitoring setups are needed in the facility in order to do quality-control checks before the product is compiled?

COMMERCIAL ENVIRONMENTS:

- Retail:
    * What content will be amplified? Mere announcements? Muzak (of some fashion)? Will there be special events? If a system is announce-only, sound quality and frequency response can be much more limited than in a system that is playing music and/or hosting special events. A Disney store probably demands higher quality from its audio systems than a local supermarket with one branch.
- Shopping malls: while shopping malls are technically retail environments, I have separated them here to address the audio that is "mall-wide" and not in each individual store.
    * Indoor: what content is being amplified? Same questions as for a retail environment, but with the added consideration of facility-wide announcements, zones of coverage (Do we need to be able to address the food court separately from the whole mall?), and so on.
    * Outdoor: same as indoor, but with weatherproofing considered.

CORPORATE ENVIRONMENTS:

- Offices:
    * Do individual offices need to be able to teleconference with other spaces either within the building or outside the building? Is teleconferencing true telephony (over terrestrial phone lines) or via Skype or another "virtualized" IP-based platform? Does such conferencing need to interact with overall room sound systems? Is there a need of a central announcement system?
- Conference rooms:
    * How many people can gather in this space, and do we need to be able to amplify all of them for large meetings? Will individual mics be necessary, or will table-mounted boundary mics suffice? What are the teleconferencing and presentation system needs?
- Lobbies: these spaces are becoming increasingly complex showpieces for large companies to show off, featuring video walls, sound content, interactivity, and more. As such, they can often be treated as themed installations, with all the same questions of how the system is engaged, what content is being transmitted, and what is needed for emergency announce, etc.
- Warehouses/production facilities:
    * Does this system require AV production monitoring in real time? Does that monitoring also need to be recorded in case an accident happens and recordings need to be reviewed to evaluate the accident and how to avoid it in future?

* How loud is the production equipment, and how loud must the delivery system for emergency announcements be in order to supersede the production facility's self-noise?

As you can tell, some of the questions for "repertory" systems verge into the topic of the next chapter ("Understanding the Venue") and the following chapter ("Understanding the Audience"). Again, while I've broken these concepts down into component parts in order to allow us to examine them in an orderly fashion, in real-world design projects, many of these considerations are part and parcel of one another. Whether or not the announcement system works in a given space will depend a lot on the size of the space, the acoustics, the audience therein, the activity levels therein, and other factors that all become interdependent. However, by establishing (in this example) that we need an emergency announcement system, we know something already about our design: it must prioritize speech intelligibility at all listening locations and must reach SPL such that it will be understood at any position in the space where we might find a worker/guest/customer.

Once we understand our source content, we can begin to examine the space in which we will be delivering that content in more detail.

Onward!

Notes

1 Many theme park projects work this way, where content is being developed as the ride itself is being designed, and systems start by creating an environment that will support all *potential* content, and choices are refined along the way as content becomes clearer. In some cases, the deadline for the system is well earlier than the deadline for content, and so—to an extent—system ends up determining what content can be produced/supported.
2 Many facilities will offer more than one sub-discipline within music production. For example, many recording studios are also mix studios.
3 This is not a book about narrative design, but when designing a system for any narrative project, a basic understanding of narrative arc and terminology can only help you. I require all my student sound system designers to also take a course in narrative design (and vice versa) such that they are all speaking the same language when they engage in work across the technological and creative divides.
4 Note that for traditional repertory theater productions, wherein several shows will share a theater, rotating in and out of the venue night after night, the general planning questions align with the production system questions, but treat all productions sharing the venue as one overarching production. The design of true "rep systems" in this way must account for the needs of each show. This is also true of most concert festivals, where the equipment and systems selected must suit the full range of shows to be presented.
5 I swear, soon we'll be hearing Christmas music in stores right after 4th of July.
6 This question could be asked for any theater design, for any theme park design, for any design of creative content at all, really. I list it here, however, because more than most environments, casinos, retail spaces, and the like are *very* concerned with controlling the mood of the guests, as mood control typically corresponds to better sales. A casino may want the guest to feel energized in certain areas of the floor and relaxed in others. It may want to make the guest feel glamourous. Energized content may need a good transient response from the delivery system, relaxed content might prioritize a smooth, even response with the blurring of time cues, glamor may demand intense surround sound such that a guest feels totally immersed at all times. These are not necessarily mutually exclusive demands, either. I can't tell you how many design projects I've worked on where a client says two seemingly contradictory things and feels they must both be achieved. "I want the guest to feel really pumped up but also totally at ease and like they could stay seated forever." It is your job as the designer to figure out how to create a sonic space that convincingly achieves both those goals (along with the content designer, of course).
7 They also might not differ at all. I often advocate for creating production facility systems that strive to be flexible enough to take any client of any genre down the road (space allowing). That said, this is not everyone's design philosophy.
8 Like *The Joel McHale Show* (Netflix's already-canceled version of his old show *The Soup*).

CHAPTER 31

Understanding the Venue

Whether you are a designing a system for Madison Square Garden or the park down the street from your house, a solid understanding of the venue is key to any successful system design. In this chapter, we will examine the fundamental questions a designer must ask and answer about the venue they are working in before specifying a system for that venue.

QUESTIONS

- Outdoor or Indoor?

  This is a huge question. Systems specified for outdoor work must either be rated to prevent moisture and particulate matter (like sand or dirt) from getting inside of it or be set up in such a way that it can quickly be sealed from the weather that may arise. Speakers are often available with environmental ratings,[1] but, for example, mixing consoles usually are not so rated. It is assumed, when designing a system for an outdoor concert as an example, that the console will be situated under a waterproof tent, and that in the event of seriously inclement weather, the waterproof road case will be close enough at hand that the console can quickly be protected. Waterproof tarps are also a common tool for outdoor production—in a situation where dead-case land is too far from FOH to quickly case up a console, it can at least be quickly wrapped in a waterproof tarp until the case can be retrieved and that will, hopefully, buy enough time to prevent destruction of the valuable electronics.

    In addition to weather considerations, the other main distinction between an outdoor event and an indoor event is in the acoustic response. An indoor venue by its nature will have more reverberance (unless it is deadened with acoustic treatment). The very fact of having a ceiling over a venue creates both sound reflections and often allows for more variety in rigging points; outdoor events by nature do not have ceilings and thus do not have that added reverberance to contend with, and also require ground-mounted structures (like truss towers) to create rigging points anywhere other than the stage.

- I Iow are the acoustics?

Is the space very live? Very dead? What are the acoustic properties of the space? Understanding the acoustics of the space in advance of selecting speakers can be very advantageous. As mentioned in earlier chapters, a venue with a very long reverb time may pose challenges for the delivery of speech intelligibility, necessitating more speakers for direct coverage. A venue with resonant cavities may need acoustic treatment in those cavities before those resonances can be overcome (no speaker choice will prevent a cavity from resonating on its own). Again, the better you understand acoustic principles, the more you can take them into account as you design systems, and the clearer your delivery will be.

- What is the RT60 (reverb time)?
- Are there uncontrolled room modes/resonances that need to be addressed?
- What is the noise floor in the venue (and does it stay stable or change—an HVAC system might measure at one loudness in mild temperatures when it is not working hard, and quite another when going full-bore in more extreme temperatures)?
- What are the relevant surfaces made of? If there are side walls/ceiling, are they hard and reflective? Soft and absorptive? Diffuse? Some combination?

- What style of venue is it?

This seems obvious, but it is likely that the same speaker plot would not work for both a proscenium theater presenting a rock concert and an outdoor Shakespeare in the round production.

- Theaters (Figure 31.1):

  * Proscenium/end-stage: audience on one side, facing the stage, usually through a portal (proscenium arch) opening
    - Balcony: some proscenium venues feature balcony sections, others do not. The presence of a balcony section often necessitates over- and under-balcony fills.
    - Mezzanine: some venues may feature "mezzanine" sections, which is to say a mid-level balcony between the orchestra (or floor) section and the true upper balcony.
    - Box seats: usually on the sides of a venue (at least for an end-stage venue), close to the stage, and often requiring direct coverage fills for wealthy customers. Some venues (like the Hollywood Bowl) offer a box-seated section down on the floor, where other venues would instead sell expensive orchestra-level seats. These boxes (unlike the side-clinging boxes at many opera houses) are not isolated acoustically from the main orchestra section, and as such do not require special speaker coverage separate from the main audience area.
  * Alley: audience on two sides, facing one another, with performers in the middle
  * Corner: two audience sections, at 90° to one another (or thereabouts) with performers in the wedge space between and in front of the audience
  * Thrust: audience on three sides
  * Round: audience on four sides (some "round" stages are actually circular, but many are square or rectangular—they are still called "round" because the audience is situated around the stage)

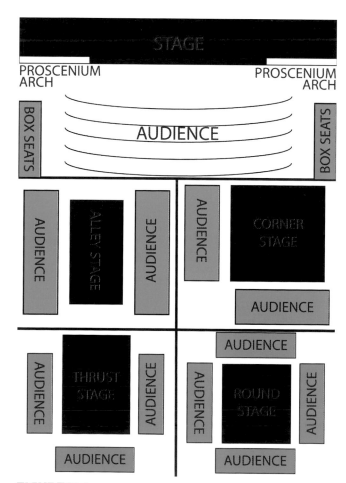

FIGURE 31.1
Common Theater Layouts

Message in a Bottle

For me, it's always about the venue. I'm always trying to assess the venue and thinking "how do I make this particular production work in this venue." Once I understand the venue, it's what is the play, what are the needs of the play, how much of it is reinforcement, how much of it is sound effects, how much of it is realistic, how much of it is abstract, how much of it is immersive? Once I understand what the balance of those different components are then I design a system that addresses those needs. So if it's primarily reinforcement with a small amount of sound effects that's one thing, if it's a completely immersive system where I can't really define through the text or the rehearsal process where sounds will be coming from, I'm going to specify a pretty sophisticated multi-channel loudspeaker system with a tremendous amount of capability, not knowing exactly how I might utilize that capability, but knowing that I need that flexibility in order to survive that particular show. A multi-stage system typically for me will include a presentational frame, front fill, high-center positions—sometimes split, sometimes more center-cluster-like (depending again on the needs of the production), the development of lateral fills, the development of rear fills, ceiling fills—and depending on the size of the house how many ceiling fills might be necessary, I'm gonna determine that, and then if possible if there are plenums and things like that underneath the house then we'll actually use some floor fills. That is then repeated onstage. So I'll think about lateral fills onstage, rear fills onstage, ceiling fills onstage, monitoring, anything necessary like that. That will be the template I use to try to design a system.—David Budries

* Concert halls: much the same as theaters, though less frequently featuring a proscenium arch, and often featuring rounded walls/audience sections. Usually featuring higher ceilings than a typical theater (to increase acoustic volume, reverb time).
* Nightclubs:
  * End-Stage
  * Round
  * Three hundred and sixty degrees: in these venues (typically set up as dance clubs), a DJ will perform in a booth either at one end of the venue or in the center (sometimes in a corner), but the audience isn't (at least by default) facing the DJ as a performer. The sound system is typically configured to completely surround the dance floor, with all speakers aiming toward the center, to create an immersive pulsing dance experience. Aiming speakers directly at the center from opposing sides of a venue can cause phase problems, but if the club is well-enough attended, some of these problems are literally absorbed by the dancing patrons before speakers from one side can sum with speakers from the opposite side.
  * Multi-room: while this does not speak to the configuration of any single space within a club, it is worth noting that many large nightclubs actually consist of several different sub-venues inside, isolated acoustically from one another, which will often feature distinct performers in each space simultaneously.
* Convention centers: these venues are known for featuring, on the one hand, large cavernous spaces with little definition, which can be used for trade show exhibits and the like, and side rooms that are configured more like corporate conference rooms (though often with worse acoustics, as the divisions between these rooms are sometimes just temporary borders extended for the duration of the event). Large trade show floors are often acoustic nightmares, designed like warehouse space with hard ceilings and little to no acoustic treatment. Reverb time is often high, and speech intelligibility is poor by default. Add in thousands of attendees all making noise as they explore, and it can be very difficult to hear in these environments. Distributed speaker systems are usually in order if there is to be any convention-wide announce or broadcast necessary.
* Retail stores: retail stores vary considerably from one to the other in size and scope. There is a big difference between a Super Walmart and Jim's Local Corner Store. Chances are if you are actually being paid to design a sound system for retail, it veers closer to Walmart than the local bodega, at least in terms of style

of installation. Most retail spaces these days are essentially modified warehouse/shed spaces, with metal ceilings and hard surfaces everywhere. That said, there are some distinct variations within the field worth identifying.

- Big-box: like the aforementioned Walmart, big-box stores are typically large spaces with little or no consideration for acoustics in their design. Often, these spaces forego any drop ceilings, so ductwork, conduit, and all other such things are exposed above the shoppers' heads. This can make speaker mounting challenging, and often speakers end up being hanging pendant speakers, or small wide-coverage speakers mounted to steel poles facing straight down.
- Department stores: while traditional department stores are not as common as they were in the 1980s height of the shopping mall craze, they still exist all across America and overseas. The distinction between a big-box store and a traditional department store is mainly one of architectural and decorative style. Often, department stores are multi-story spaces, subdivided into distinct shopping zones. Each zone may have some decorative uniqueness, and typically these spaces favor the dropped ceiling, finished look that suggests a more luxurious environment. As such, the acoustics are seldom as poor as in big-box stores (though they are still often less than ideal). Like convention centers, or multi-room nightclubs, these spaces are often best designed as a series of contiguous spaces each functioning as its own sub-zone. That said, there is still often a need for a store-wide announce/emergency function in sound systems.
- Supermarkets: similar to big-box stores in most ways.
- Restaurants: the variety of designs seen in restaurants is staggering. From small bistros catering to no more than 25 patrons at a time to vast Vegas cafeterias catering to hundreds, there is seemingly no limit to how a restaurant can be configured. Increasingly, restaurateurs are investing in multimedia systems to provide a more intense and flashier environment to lure in patrons. Some restaurants are offering film screenings as part of a dining experience. The salient point is that each restaurant will necessitate individual evaluations—there are no generic rules that apply here, other than that acoustic treatment is increasingly important to create spaces in which guests can converse freely and not feel overwhelmed by a cascade of noise from their fellow guests (and from the multimedia playback, which may be both a draw and a deterrent if it's too overwhelming to guests).
- Bars: see restaurants. Often (not always) music is louder in bars than in restaurants, but otherwise, everything said earlier about restaurants applies here.
- High-end boutiques: the wealthier the clientele, the more expensive the items sold, the more likely that acoustics will be considered when designing a retail space. In the case of many contemporary high-end boutiques, the acoustic environment is quite lovely—just live enough to feel open and not oppressive, just dead enough to prevent being overwhelmed by reverberation. Any sound system will prioritize clear, direct sound, at audible but not overwhelming level.
- Shopping malls: indoor or outdoor, malls are still a fixture of life in the 21st century (even though the advent of online shopping has caused a wide decrease in mall attendance). Indoor malls tend to have long reverb times, hard surfaces, and lots of reflection. Outdoor malls (often outlet malls) don't suffer the same reverb time, but still feature lots of hard surfaces (concrete paved walkways, glass-fronted shops).
- Et cetera: there are infinite varieties of retail spaces beyond what is listed here, and it is beyond the scope of this book to list them all. When working on a retail space, the main guidelines are still rooted in content—are they playing music, just using the system for emergency announcements, both? Something else?
- * Corporate offices: as anyone who has worked in an office building can attest, while there are variations (typically found in large shared spaces such as lobbies, meeting rooms, etc.), the general style of office buildings is fairly consistent: large open floors with cubicles (or true open-plan office space, with workstations and no dividers), ringed by hallways and larger offices for management. Silicon Valley office designers have been pushing the envelope (the current Apple headquarters looks more like a building from a science-fiction movie than a typical corporate office tower). Each part of an office building may have some need for sound systems, but in a majority of office spaces, the need is limited to emergency announcements. Corporate lobbies and futuristic headquarters, however, often have design goals as unique as a theme park, and so must be evaluated individually.

* Warehouses:
  * Warehouses used *as* warehouses: large, cavernous spaces, with high noise floor from the equipment involved in inventory management.
  * Warehouses used as art spaces: more and more often in our first-world "post-industrial" era, warehouses are often used for art and performance installations, as they are large raw spaces that can be customized as the creators desire. The acoustic properties of a warehouse will remain: long reverb times, hard surfaces, etc. Some artists consciously make use of these properties in their creations, while some are simply seeking the space, and want to counteract the natural washiness of the sound in such spaces.
* Theme parks: custom-designed spaces, to be evaluated individually, often featuring large distributed sound systems.
* Hotels: we are principally concerned, as sound system designers, with high-end hotels and their lobbies, where increasingly you will find DJs, video/sound installations, and the like. Again, these spaces vary widely in shape, size, and character.
* Something not mentioned here: there are many different types of venues that fall into either overlapping categories from the aforementioned or are unique and unclassifiable—meaning they must be individually assessed.

■ What is the electrical power situation?
Sound equipment often requires quite a bit of electrical current. Amplifiers particularly draw significant current, and for large and loud installations, the evaluation of circuit/amperage needs is important. In addition, sound systems should never share circuits with lighting systems. Does the venue have dedicated circuits for sound? Will you need to bring in a three-phase power distribution setup to turn high-voltage rails to lower-voltage usable service?[2] If we have dedicated circuits, how many? What amperage is served over each? What geographic region am I in, and what does that mean about service voltage, typical electrical connectors, and so on?

■ Where can equipment safely be mounted?
Before specifying equipment, it's good to know where it can be placed safely. If you are specifying a 700lb speaker array for a point that can only support 500lbs, you'd better rethink your design. If you specify an FOH console that is 60″ wide by 40″ deep, and your FOH position can only hold a console that is 45″ wide by 28″ deep, you'd better rethink your design. There are many situations in which a design will be adjusted simply because the ideal sonic plan won't fit in the room or can't be safely operated or supported.

■ What is the HVAC/temperature situation?
Large equipment racks (for amps, computers, etc.) often give off quite a bit of heat. Where is the equipment to be stored, and is that space climate-controlled? Gear should neither get too hot nor too cold in the course of its operation, as either extreme can damage equipment. If the production is outdoors, what is the climate? If it is too cold, heated enclosures (or just space heaters, if the event is temporary) may need to be specified to keep computers and other such gear from freezing up (literally). If it is too hot, rack fans at very least must be specified, and in certain situations, air-conditioned equipment racks may need to be specified.

■ How large is the venue?
We'll address audience size in relation to the design process in the next chapter, and we've already mentioned acoustics earlier, but literal dimensions of the space and air volume (for an enclosed venue) will have an impact on speaker selection. This leads to questions about audience location as well—again, see the following chapter.

■ Are there any neighbors we need to be concerned about?
This might seem like a joke, but it is a surprisingly important concern. I worked for a time at one NYC nightclub that shared a building with some upscale luxury condos. Most of the performance space was well isolated acoustically, but the service elevator shaft that landed next to the main stage ran vertically the entire height of the building. It had not been properly isolated, and so on busy Saturday nights, the upscale residents of the aforementioned condos continually called the police to complain about the noise. This is relevant in a number of settings. Outdoor concert venues often have noise curfews due to agreements made with residents of nearby homes and shows that go past curfew will incur large fines (as will shows that exceed a certain max SPL). Knowing the constraints of an environment before designing a system to be used within that environment is key.

■ What is the protocol for gaining access to the venue?
Any time you are designing for a preexisting venue, it's a good idea to know how to gain access to the venue, even if that's just information you pass along to your installation team. While you may never need to initiate

access on your own, it is a good piece of information to have, and worth pursuing. If you need (as we often do) to acquire noise measurements of the space before completing our design, this access question becomes very relevant.

- *Insert any question that seems relevant here*
  As I often tell my students, the stupidest question is the one you didn't ask. Now, in the professional world, there are times when it is clear you should know something (and you don't) and it may be worth your time to try to do some research and learn what you need to know before asking people on the team and revealing your ignorance.[3] That said, asking a question is a million times better than remaining ignorant and wasting your own and others' time by designing something incorrectly because you didn't ask the question.

There are a ton of details that you might want to know about a venue, and what is most relevant will change depending on the type of project you're doing, what type of content will be amplified within that venue, and so on. Learn as much as you can about the sound source, and the venue the source will be presented in, and then (as we examine in the next chapter) begin to assess the needs of the particular audience in detail.

Notes

1 See Chapter 16 for more on the IP rating system for environmental factors.
2 Such as the Motion Labs portable distros that are common throughout the industry. Three-phase power is usually high-voltage service, in three "hot" legs, each referencing a common neutral and ground. This service is usually terminated in large camlocks and must be stepped down and distributed for use by sound equipment (though large industrial equipment sometimes uses three-phase directly without the step-down transformation).
3 This has happened to me a lot working on themed and architectural projects, where terms standard to those industries were bandied about, and everyone around the table understood them except for me. If I could spend a few mins on Google and learn the answer ("Hey! RCP means Reflected Ceiling Plan!"), I would do so and save face. If my quick research didn't turn up an answer, I'd ask, though—because it's better to reveal ignorance in one moment and have the correct info on which to base subsequent decisions than to skip an important detail and fall too far down a rabbit hole of design mistakes because you didn't understand what someone was saying.

CHAPTER 32

Understanding the Audience

As most stand-up comedians will attest: crowd work can be tricky. Trying to gauge the mood and desires of an audience is something that—to some degree—everyone involved in entertainment does. This is true of sound system designers, except that we're not usually insulting someone's tacky leisure suit.[1]

Understanding our audience means understanding their relationship to the source content and venue in a way that allows us to serve their needs without getting in the way. When designing a system, here are some of the questions we must ask (and answer) as we plan.

QUESTIONS

- What is the relationship of the audience to the venue/event?
  If we're designing a concert sound system, we understand innately that the audience is made of up concertgoers, who are at the event (at least in part) to listen to and enjoy the performance of music. If we're designing a system for a hotel lobby, we need to understand that the audience (the hotel guests) have not come to this lobby to witness a performance (even if there *is* a DJ over in the corner). Very likely, they are in the lobby either checking in, checking out, or passing through to/from their rooms. As such, the system must be different from our concert. If a concert features directional sound, perceived as originating with a performer on stage, loud enough to be central to the audience's experience at the event, a hotel lobby typically features diffuse distributed sound, and comparatively low-SPL levels to allow for conversation. These are but two examples, of course, but this question is always the first we must ask of our audience—why are they here, listening to our sound system, and what do they expect from that experience?
- What SPL does the audience need/expect?
  In a heavy metal concert setting, the audience is likely expecting quite high SPL—sometimes as high as 120ish. If a jazz concert reached 120 dB SPL, most audience members would flee, holding their fingers in their

FIGURE 32.1

These Two Groups Are Not Seeing the Same Kind of Show

Source: Left: Public Domain Photo—no photographer credited, via www.pxhere.com; Right: Public Domain Photo by Meline Gharibyan, via www.flickr.com; both images Creative Commons 2.0 License: https://creativecommons.org/licenses/by/2.0/. Per license requirements: photo on left was flipped horizontally, photo on right was flipped horizontally and cropped

ears. Setting reasonable SPL targets is very important to the success of a system design. If we're designing an emergency announcement system, what is the ambient noise floor of the space? Different in a library than in a factory, right? We typically want to get at least 12–20 SPL above the ambient noise floor for emergency announcements, and/or reach at least 85 SPL in otherwise quiet environments. One of the first specs a system designer will assess when planning a system is how loud it needs to be. If you know what SPL your audience wants and/or needs, you already know something about what speakers will or won't work for your job.

- How large is the audience?
  This will impact everything about your design. If you are trying to cover 10 people in a theme park ride's car in one particular action scene of a dark ride, the speakers you can use to accomplish this are very different from the speakers you will need if you are amplifying a pop concert for an audience of 25,000. Human beings are absorbers, soaking up HF content, so the larger the audience, the more you will need extra speakers to connect HF direct sound to the listeners (hence the use of line arrays for large shows, which are designed to project HF content to discrete audience areas on axis to each individual box in the array). Power, coverage angles, speaker count, all will be (at least in part) determined by audience size.
- Where is the audience?
  If this is a show, what configuration is the venue? If this is a themed install, a hotel, an office, or any other space, where are the guests/customers, and how do they use/move through the space? What is the distance

Message in a Bottle

It's not about how much 4 kHz hits somebody, it's the quality of it. The classic *I burned your food, but I gave you extra* doesn't help.—Jamie Anderson

> **Message in a Bottle**
>
> When you're doing a sound design, a system design (or both, hopefully) you should imagine yourself sitting in every seat, in every section, split the whole theater up—as you're looking at the drawings. If you're in the theater, sit in every seat, and say "what am I hearing? How am I hearing it? Why am I hearing it? And is it good?"—Jonathan Deans

between audience and speaker(s) (or what should it be, if you can determine it)? The closer a speaker is to an audience member, the lower SPL it can produce to still reach a desired listening target. The farther a speaker is away, the louder it must be overall,[2] in order to reach the same SPL at the listener. Does the audience stay in one place (as in a seated theater) or move around (as at a concert festival outdoors)? Is audience density fixed (say, in a venue with fixed seats or occupancy limits) or down to how tightly the audience crowds toward the front of the stage when the headliner comes on?

- What are the projected demographics of the audience?
  This is an oft-overlooked factor of sound system planning but is still a very important one. If your audience is made up of senior citizens,[3] you can expect that some of your audience will need hearing aids (which may tie into to your assisted listening plan), and in general, you will want to ensure that speech intelligibility is prioritized in your system concept. If your audience is made up of young punk rockers,[4] you won't be as worried about speech intelligibility, but you will need to ensure that your system is loud and intense, and punches the audience in the chest. Height can be a factor as well—designing systems for spaces in which an audience will be standing, we need to take into account the average audience height (as well as the general boundaries of the range). The average height of the audience for a children's theme park ride will be different than the average height of the punk show (probably). What seems loud to one group can either be timid or face-melting to another.
- How much is the audience paying for the experience?
  This is, admittedly, a cynical question—but an important one. The level of refinement expected by an audience who spent $6 to see a show is different from that expected by an audience who spent $600. While we never want to shortchange our audience, no matter what they paid or didn't pay, the reality is that we as designers will spend more money on experiences that cost more for the audience to attend (not least of which because our budget tends to be higher).

The better we understand our audience—size, shape, and intended experience—the better we can deliver that experience to them.

Next, we must address a very important factor that no one likes to talk about: *money.*

Notes

1. Not as part of a show, anyway. . .
2. Remember the inverse square law!
3. As it always was for a friend of mine who used to work on the annual AARP conferences.
4. Like at *insert name of trendy underground nightclub that will already have closed by the time this book is published* here.

CHAPTER 33

Understanding the Budget and Available Resources

You knew this was coming.

The whole time, we've been talking about gear, dreaming of expensive consoles and network-enabled microphone receivers and DSPs, and inevitably, there must come a reckoning. We must, at some stage, understand what exactly things cost, versus how much money we have available to spend, and hope that our plans fit within range. Of course, if we're experienced at this, we can probably identify *before* we've spent time dreaming up our system whether or not we'll have the money to accomplish our goals. However, especially when you're designing systems for the first time(s), it can be challenging to get straight answers about what things will cost. Most real topline professional audio gear is rented and sold not by Sweetwater or Guitar Center,[1] but by "vendors"—pro audio companies like PRG, Masque Sound, 8th Day, and the like. You can't buy a Meyer speaker from Sweetwater, so you have to buy it from an *integrator* (a term used for vendors who do more than just sell, they also typically install and commission systems).[2] Thus, when you are doing research and building your budget, it is pretty common that you can't just Google the items in question and get realistic prices. You actually have to *talk to people*. Depending on whether you approach sound work from the arts or from the engineering side of things, this may cause you more or less anxiety, but is a totally necessary part of the design process.

However, before you even get to pricing out your ideal system, we need to consider some things:

1. Is this system going to be used in a temporary condition (e.g., for a production with a limited run)? Or is it being specified for "permanent" installation (e.g., for a theme park ride)?
2. Does the venue or producing organization have preexisting equipment stock that is available to you? If so, do they expect you to use that stock?
3. Are you provided with a fixed budget, above which you cannot rise? Or are you expected to prepare a budget based on your system needs which then needs to be approved by the producer(s)?

First, we must understand our production. If we're specifying a system for a 6-week regional theater run, that's a very different prospect than creating a system for an installation meant to last for 10 years. For starters, with rare exception, a temporary project will use existing stock, will rent equipment, or some combination thereof. While you will likely be buying items considered "expendables" (mic elements, batteries, mic dressing, etc.), you probably aren't buying a $100,000 console for a 6-week production. You might well be renting that same console, though, if it has features you absolutely need.

Now, if you are working with an existing venue or production company, said company may well have their own equipment stock. Or, in certain cases, you may be brought in to design a system in a situation where the vendor has already been contracted, and you are thus limited to using gear from that vendor's stock.[3] Sometimes, it can feel annoying to have your hands tied by existing stock, but whenever stock is available, it is our job to be reasonable and *at least* consider it. You may find yourself in a situation where the existing stock will not do the job needed, and so you have to persuade the producers that what you need (framed as what your sound system should deliver to the client and audience) is important enough to the success of the project that it is worth spending money and acquiring outside gear, rather than using what's already available. It is certainly *possible* to convince a producer that your show can be eminently more attractive and exciting to the audience if they just spend some extra money—but this can be a difficult conversation, as sound is rather obviously invisible, and thus tough to market. It is usually not worth making political enemies over the choice of a few speakers, as this business is small, and you will run across the same people over and over again. However, if the existing stock has been evaluated and found wanting it is your job to let the producers know that the system won't live up to expectations, and in what ways it will fail if you must use existing stock. Never promise the moon when all you've got is a cardboard cutout of the moon—nothing sours producers on a designer faster than being told they will receive A+ work and being delivered a big fat D-. If your existing stock can't get you past D-, you have to be clear with your team about that fact. If getting to A+ isn't possible within the budget, maybe you can compromise and find yourself at least with C+ work, which is still an improvement—and which may ultimately be better for the production (and your reputation) than insisting on A+- level systems, if it means you won't be bankrupting the other production departments and irritating producers who then may not hire you for their next job.

In the event that you have been provided with a fixed budget—a fairly common event, as it happens—you will need to know how to estimate your costs. A fixed budget typically arrives in one of two forms:

1. A minuscule budget, perhaps only enough to cover expendables (and sometimes, not even that much).
2. A budget roughly adequate to the production needs, though often just a *bit* lower than you might hope.

There is a very rare third case, the fixed budget that is larger than you could possibly need, but since this is essentially the mythical unicorn of budgets, I will only say that in these rare situations, you have to judge whether you will be better off delivering novelty and excitement from your system choices at every turn, or in just saving the producers money which can then be allocated to other production departments.[4]

At any rate, in the case of #1, how you spend your pittance is an important issue.[5] If you've got a production with 12 wireless mics for actors, and you've got a production budget of $100, it seems likely that you will need to spend much if not all of that money on mic supplies—expendable items like batteries, skin tape, floral wire, and the like (and you'd better source them as inexpensively as possible!). If your theater has all that stuff already (in which case, bully for you!), maybe there is a particular effect in the show that requires a specialty item to be rented. Chances are, for $100, you'll be lucky to rent *anything* for the run of an actual production—but it depends on your needs.[6] If your company is so strapped for financing that your production budget is this low, you should be aware that they likely have little or no margin of error for all of their production budgets—so you need to spend the tiny amount of money you've been allocated judiciously, and take care to *not* go over budget.

At the professional level, however, it is fairly common to run into a fixed budget that is more moderate in size. Depending on the scale of the production, this number can be anywhere from several hundred to several thousand dollars.[7] In the case of the midrange budget allocation, you will need to examine priorities. Often, this situation arises with companies that own some, but not all, of the equipment that they will need to accomplish the production in question. Imagine you are working for a theater. That theater owns a mixing console that, while perhaps missing one major feature you wanted, will suffice for your upcoming production. However, the set designer has reconfigured the space and the audience for this show and the speakers that the venue usually uses for their vocal reinforcement are not the correct shape and coverage pattern for your new audience areas. Rather than rent the console that performs that one missing feature, you will likely start to consider renting better vocal reinforcement speakers (since after all, speakers are the most important part!). How do you go about figuring out how much something costs to rent? We must begin to develop our network of dealers.[8]

First, you must identify who rents what gear. Many AV houses will carry different brands of equipment, but some do not—some will only feature speakers or consoles by one manufacturer. You can contact AV rental vendors directly to ask if they stock the equipment you desire, and if they do, you can move into requesting a quote. If they do not stock what you are seeking, you can either:

- Write that vendor off your list
  or
- Ask them to propose substitutions

Vendors are used to proposing substitutions, and if they are at all worth their salt, they will propose equipment with roughly equivalent functionality (e.g., speakers with the same frequency responses, SPL levels, and coverage angles you requested, ideally in the same or very similar form factors; consoles with the same or greater channel and routing path counts, etc.).

When preparing a list of equipment for a rental quote, you must be thorough. Every piece of equipment you need must be on that list (and in correct form, spelled properly, with model numbers where appropriate). This means not just the make and model of speakers you want, but the model of the fly hardware, the types and lengths of cable (if you don't already have enough), the adapter that you're forgetting you need because that output of your console is only TRS and the input of that rack effect is only XLR, and so on. A full list, with quantities of each item, should be prepared before you begin contacting any vendors, because as soon as you need quotes, you need that list ready to email. It is wise to budget extras of things like cables. While a rental shop that sends you a broken cable isn't doing its due diligence (and can be expected to replace that cable), you often won't have time to wait for their second cable to be delivered. If you have rented a few extras, you will also be in much better shape when the drummer accidentally kicks the connector to his drum subwoofer monitor and breaks it.[9]

If your production is in a region with more than one AV provider, it always pays to get more than one quote. Sometimes, you will find the same gear renting for wildly different prices from different vendors. Sometimes, one vendor will point out a particular accessory you will need with an item on your rental list that you hadn't anticipated, while another will leave that detail out. Sometimes, a vendor doesn't have the gear you wanted but will rent you something nicer than what you specified because it's just been sitting on the shelf in the shop collecting dust, and they would like to make some money on it—any money at all.

Even if the substitution offered is very similar (neither nicer nor more limited than what you requested)—and especially when the gear they are offering is of slightly lower quality than what you've requested—if there is no one else clamoring for that particular equipment at that particular moment, substitutions can be quite favorable to you, as you can often negotiate a lower price. That said, if there *are* other parties who are interested in that gear, getting a discount is less possible.

329

Tales From the Field

I really had to struggle with budget-based compromises on *Fela*, too. We had a system designed—and this is the politics of it—we had a system designed, and we had one shop that was ready to provide everything, and then another shop came in and with relationships and whatever, came in under, enough under that it made a sizable difference to the GM. And then that second shop did a lot of, "well, we can give you *this*." It was a lot of the back-end substitutions, the loudspeakers mainly stayed the same, but amps changed. And so I was powering KF650s with three different manufacturers of amp, for high, mid, and low. There were all these weird phase problems in the box. That means we end up having this manufacturer's amp actually sent to us with inverted poles and, you know, it was terrible. It was very frustrating. And so that, I would say, that's where the failure point has been for me.—Robert Kaplowitz

Rental prices are extremely flexible. They will depend on the demand for the gear at a particular shop in a particular market at a particular time. Some vendors, for example, may cater heavily to the summer concert circuit, meaning they might have equipment gathering dust on their shelves during winter, at which point they may cut you a much better deal for renting it. Inversely, a vendor that caters heavily to the seasonal indoor theater companies of the area may not have as much work during summer, when said companies may be taking an annual break. At any time, a shop that has gear unlikely to be rented is going to be a target for negotiation.

How do you know if a shop has gear just sitting around? After all, isn't it in the vendors' best interest to convince you that their gear is sought after 24/7/365? There are three primary ways:

1. Know the market: if you are dealing, for example, with Masque Sound in New Jersey, you should be aware that they are one of three main production sound companies that serve Broadway. Broadway theaters run year-round, so when might you catch a discount from them? When one of the Broadway shows they were contracted to closes early. By networking with fellow professionals, and reading trade publications, you can learn which shops are providing audio gear for which shows. If one of the shows on that company's roster gets terrible reviews and closes quickly,[10] the company now has gear on its hands that it had planned to have out of the shop for weeks if not months. Even if they get paid a major cancellation fee, that gear is now not actively generating income, and a shop would always rather generate *any* income on equipment than *none*— especially in a geographic region (like the NYC metro area) where real estate for storage is not cheap. If your run is going to be soon enough, and short enough that it is unlikely another large Broadway show is going to come in and sweep the equipment up, you can often get rentals for much cheaper than you'd expect.[11]
2. Ask: sometimes, vendors will be honest with you—especially if you bring them a lot of business over the years. I've designed systems where I met low budget targets by calling vendors and just asking what they've got on the shelf that hasn't gone out in a while and making the best system I could out of the leavings of other productions.
3. Make an actual friend in the shop: someone who ideally isn't in the sales department (and thus has no incentive to get extra money out of you), someone who just works with the gear. They can often get you inside information about what has been sitting unused for a long time.

Aside from having gear that is collecting dust, how else can you get a discounted rate from rental vendors? Bidding wars. Many producing institutions (especially universities) have company policies in place demanding that any expenditure over a certain dollar amount be competitively bidded out. Some such institutions then have a subsequent rule stating that whichever company (providing appropriate services) comes in with the lowest bid numerically must be awarded that bid. This is a double-edged sword. On the one hand, such policies often make it difficult or impossible to quantify such intangible values as a company's reputation for being pleasant or difficult to work with, or the maintained quality of the equipment stock (Broadway shows are much less rough on consoles than outdoor concert festivals where beer is easily spilled into fader banks). That said, such bidding policies can allow you to conduct low-key bidding wars between vendors, driving prices down. I have been, as a renter of equipment for sound systems, in the position of playing three vendors off of one another, such that their services ended up thousands of dollars lower than the initial offers. Every time one came back with a lower price, I would go to the other two and simply say, "I'd love to work with you, but my institution has a policy that we must accept the lowest bid. I have a vendor who has come back lower than you. If you can get lower than their number, you're back in the running, but I can't accept your current bid." If the vendors have gear that they want to get off the shelves, and if they have enough profit margin to make the reduction in price feasible, then you can often get a much nicer system for quite a low price. This tactic is limited by several factors:

1. There has to be a fixed timeframe, and this has to be a real project (not just a speculative estimate). You have to have a deadline for final selection; otherwise, the vendors will feel you are just driving them into the ground and may never even lay money on the table with any of them.
2. You have to be incredibly polite the whole time. No one will be driven into a bidding war by an asshole unless the amount of money you are waving around (*after* vendor discounts) is still mountainous. Even then, you'll get better deals if you're nice to everyone.
3. It helps if you have a preexisting (and positive) relationship with the vendors. Or, barring that, if you've at least got a reputable producing entity behind you. If you're an unknown, working with an unknown

company, the likelihood that vendors will play the bidding game with you is much lower than if you are a known entity, and/or affiliated with a known entity.

Sometimes, you can also negotiate a lower price based on the financial status of your organization. If you work for a university that has recently suffered budget cuts—especially if those budget cuts made the news—you can ask the vendor point blank if there's any way they can lower the price, since your institution has been hit hard. Even without an organizational crisis, sometimes, you can even get a discount just by straight-up asking for one. I have honestly secured lower prices from vendors sometimes simply by saying "we just don't have quite enough to reach your number here, but we really need this equipment. Is there any way we can get that price down just a little farther?"[12]

Once you have your bid or bids, you can compare it to your fixed budget. If the rental fits within your allocation, great! Make sure you still have enough money for expendables, and you're set! If the rental does NOT fit within your budget, you have to make choices. Do you try to negotiate a lower price? Do you cut gear?

A lot of this will come down to *how far* over your target the estimate is. If you're $50 off, then by all means negotiate (unless the whole rental was $100. . . I'm still talking about more sizable budgets here). If you're $5000 off of a $30,000 budget, you may need to compromise what you're renting (but if you're $5000 off on a $500,000 budget, negotiate!). If you're within striking distance but you can't seem to negotiate a lower price, consider whether the equipment causing the overage is important enough to go back to the producers seeking the extra money you need, or whether it's time to cut something from the design.

Finally, if you are designing any kind of large-budget production, the likelihood is that the producers don't necessarily understand sound itself well enough to specify a budget target in advance.[13] In this case, you will likely be tasked with designing a system to meet the needs of the production, and then estimating what that system will cost. As part of that estimate, you will need to justify every major choice in the design—usually in terms that non-technical people can understand—so as to validate the expense. You will want to secure quotes and ensure that the quotes do not expire before you will be ready to get the go-ahead from the producers.[14] You will also need to ensure that a quote holds the gear for you, like a stock option. Usually, whoever requests a quote on a set of gear basically "reserves" it, meaning you have the right to pull the trigger first regardless of whether someone else requests the

Message in a Bottle

Budgets will scale that—to say that "well, this much you can do, this much you *can't* do." So then obviously, the needs of the production are moderated, modified, by the budgeting process. If you're a good negotiator, you can educate (in a sense) your producer about a particular need or goal and in that, hopefully, get them to put more resources into completing a more ideal picture for the particular production. I always let the production be the "thing that is being served," and not one ego or the sense of toys to play with. If you want to serve the production, here are the things that I think are really essential, here's what your budget allows—do we have any wiggle room? Or do we need to back off? And if we need to back off, then the director needs to know that this objective is not gonna be met, that objective is not gonna be met, but we can get really close with a lot of the other objectives. I try to make it very, very practical and very much about serving the production. People are respectful of having that information. Some of them—a smaller number now than ever—are impatient with that and just say, "Just do it, and tell me what it's gonna cost." There are a few people who just don't want to know what the issues are, but the best production supervisors really are curious. They really want to know what is going to be the best way, and how am I going to make an investment that will service this theater and the productions we do in the long run. And I can tell them about things that they can buy, they can own, and I can tell them about things they shouldn't bother buying because the technology changes too rapidly and they should rent. There's always a dialog about the scale, scope, and needs based on the production, for the venue, and that's really where it all comes down. It's about language, it's about intention, it's about budget negotiation, and about the reality of what you can and cannot do.—David Budries

same gear later. However, in some cases, even having the first quote will not reserve the gear—if you are renting 2 speakers, and a concert tour suddenly calls that wants to rent 50 of the shop's speakers, including your two, chances are the shop will prioritize their order over yours, even if you were quoted first—they should, however, be decent enough to warn you in advance when you are in this situation or make an offer of a discount for alternate items or on a later project if they have to cancel your quote.

Working on purchase projects (for long-term installation) is not actually so different from preparing rental budgets. If the equipment is being sold exclusively by vendors who are AV companies (e.g., Meyer or d&b speakers, which are only sold via authorized vendors), you will still contact them with a bid list, and request a quote. You can still negotiate discounts, and the more you are purchasing, the cheaper you can get things per unit. However, when you prepare a speculative budget for purchased items, you will usually want to observe two major rules:

1. If the items are vendor-provided, and you get multiple quotes, include the quote in the middle of your range in the total, acknowledge the higher quote, and save the lower quote so that you can come in under your estimate with the producers after the first round of submissions (during the value-engineering process).
2. If you are specifying off-the-shelf items, specify the MSRP.[15] If you can even find a price list for items that are only sold through AV companies—so long as it is an official price list from the manufacturer, you want to list these numbers. They are ALWAYS higher than what will actually be paid, and professional practice dictates that you want to provide an estimate that will produce the designed system NO MATTER WHAT. Then, if you come in under budget at the actual purchase, which you have set yourself up to do, you will be viewed as a responsible leader. Also, if the discounted vendor's prices shoot up between the time of your budget estimate and the time of purchase (which can certainly happen on years'-long jobs, particularly if the initial quote has expired because your producers have pushed back their "pull the trigger" date), you will still come in much closer to your estimate (which was based on higher quotes) than if you had listed the lowest available price.

Generally, don't list the lowest available price when you are preparing a speculative budget. Protect yourself. It is bad for your reputation if you submit low budgets but always exceed them in operation. Far better is to operate in the reverse—budget middling to high, and always deliver for a lower price. Producers love this.

Once you've prepared your speculative budget, prepare for the "dreaded VE"—Value Engineering. This process goes by other names: cost/benefit analysis, slash and burn, etc., but no matter what you call it, the process of going through your design budget and seeing where you may have to cut things can be painful. Of course, we want to defend our choices as best as possible—and if your producers are really interested in sound and can be made to understand what all that money is really gaining them, you are more likely to be afforded the money you requested. If your producers don't understand sound, or if you are bad at explaining sound to them, you may be faced with killing some of your darlings.

This is the life of the designer. We don't get to be precious about our work. While we all strive for the highest-quality sound delivery systems possible on every project, we must be ready to toss ideas out and make sacrifices to specific areas in service of the good of the whole project. Again, if you lose functionality, be very clear with your producers (and directors, if applicable) about what they are losing, but always be prepared to compromise. This is a team sport. Sometimes, you won't be working for a company with large budgets, and sometimes, you will have to sacrifice something you think is critical to the production. This can be rough, but it is our job to make the best of the situation we are in.

That said, when you design a system for a company with high budgets, it is often wise to include some equipment in your initial estimate that you don't absolutely need. This shouldn't be equipment that will just sit in cases in the venue, you should have a good use for it planned should the expenditure be approved, but producers often want to feel like they are keeping costs down in every department and entering Round 1 of VE with some gear you can safely cut is a useful and common tactic.[16]

There are some situations, however, where no compromises can be made. For example, if you are designing a theme park—there will inevitably be life-safety systems involved. Emergency announcement players, battery system backups, speakers that will remain operable in extreme environmental conditions—whenever you are dealing with life-safety designs, you must deliver a system that will be reliable and useful 100% of the time, with no exceptions. This is often not an inexpensive proposition, but when safety is involved, our duty is to protect our clients and audiences 100% of the time. This also applies to any equipment that is flown. Speakers mounted overhead can easily kill people if they fall. Safety in rigging is very important and can cost money. Safety is the one area on which we must never cheap out. If doctors take the Hippocratic oath, I propose the following Sound System Designer's Oath:

> *I swear to do everything in my power to make my systems safe for users, guests, and clients. I swear to acknowledge the limits of my abilities—to request that rigging plans be evaluated by a qualified rigger if I am not one. I swear to never choose to compromise safety for the sake of money. I swear that if a Producer makes a decision that will compromise the safety of my system, I will register my objection,[17] expressing the safety risks clearly.*

This sounds easy. However, when the choice is between a very expensive life-safety system that you need, and the five super-awesome toys and tricks you *wanted*, even the most responsible designer can be tempted. *After all, what are the chances of something going wrong? We could have so much more cool sound!* This is the false idol of system design. We must never sacrifice safety to flash. Where safety systems are required, they must always be designed and utilized properly.[18]

Finally, know that whatever the budget of your production, there may be obstacles to getting what you want. Jonathan Deans related a story[19] about having to convince Steve Wynn personally that a Meyer Constellation system was necessary in order to improve the acoustics of a Cirque du Soleil show that was playing in one of his casinos. The production venue had poor acoustics and a ton of absorptive material in the form of deep carpeting and lush seating, which made the space feel dead and thus had a chilling effect on audience reactions that was undesirable. In order to fix this, Jonathan wanted a Constellation system—a virtual acoustics platform that can convincingly and drastically change the acoustics in a room by virtue of arrays of a ton of speakers and mics mounted all around the space, and very sophisticated real-time digital modeling. Jonathan Deans is one of the preeminent sound designers in theatrical history, and Cirque is a commercial juggernaut, running wildly profitable productions around the world. You would think that when he says he needs something in order to get a certain job done, people would automatically believe him. But Constellations don't come cheap. Jonathan had to get a Constellation system on loan and installed as a test in the venue, and walk through it with Wynn to convince him. Once Wynn heard what Constellation could do—it is truly quite astonishing how successful it is at changing the sound of a room—he was convinced. But even Jonathan Deans still has to present a case for his expenditures to the producers.

Managing resources and funding is critical to a successful system design and implementation. Designers whose shows make the most of small budgets, whose work comes in under budget, and who understand what they can accomplish within a certain price range are designers who work again and again. When you first begin designing systems, you won't know what things cost. Internet research is a great starting point, and just calling vendors is another. Don't request a formal quote from a vendor unless you actually have funds and intend to rent/purchase. If you're just getting an estimated number for a speculative project, tell them that, and they can usually give you a rough estimate faster than a true quote anyway.[20] As you design more systems, you will get a sense of what things cost, and your estimating time will shrink from project to project. If I could provide you with a "real" range of prices for certain types of gear, I would, but such a thing really doesn't exist. It varies so much depending on where the project is, how much overall it will be renting or buying, who you are, who your producers are, and so on, that to give you a false range for "speaker rentals" or "console purchase" would be misleading (and even if it were accurate, it would be out of date in an hour).

Budgeting can be intimidating for young designers, because you may be spending other people's money for the first time. Do your research, find the best prices possible, and create the most useful and interesting work you can with the resources available.

OK! Whew! Everyone's always glad when we're done talking about money, right?

Now let's talk about something more fun: choosing speakers!!!

333

Notes

1 Though, to be fair, I have bought a ton of mid-level and studio gear through Sweetwater. Thanks, Jeffrey Green!
2 "Commission" is used here in the sense of commissioning a new ship, which is to say, bringing it up to speed for use by the owners/operators, and kicking off its term of service.
3 I designed a concert system in upstate New York, where I was brought in by the design firm, who had been hired by promoters working in an existing venue. Neither the promoter nor the venue had gear stock, but the promoter had a longstanding relationship with a sound system provider, so we were limited to using that provider's stock. This constrained our choices in many ways, but we were still able to carve a useful and listenable system out of the project (despite the gear not being my first choice, in any context).
4 I'm looking at you, scenery and costume departments!
5 Pro Tip: don't refer to it as a "pittance" in front of the producers.
6 I worked on a production that featured a Bob Barker-style long thin microphone being pulled out of a lunch bag that a character was carrying around. It had to work, wirelessly. We found the mic at someone's local studio in town and borrowed it, and we rented the plug-end wireless transmitter and receiver for not much more than $100 to cover tech and a 6-week run. More on the flexibility of rental rates later in this chapter.
7 Usually, if you get beyond the "several thousand" range, you aren't given a fixed budget number, you're preparing a system and laying out what it would cost, and then seeing if the producers bite. More on this later in the chapter.
8 Note that much of what I am about to discuss regarding rental houses works best in metropolitan areas, where there will usually be at least 2 or 3 (if not several) AV rental shops from which to source quotes and eventually equipment. In smaller markets, where there may be only one such rental vendor within a reasonable geographic distance, the negotiation tactics I discuss next will likely not work. When there is no competition in a market, you are at the mercy of the single vendor and must tread lightly. However, in most places where a sound system designer would be paid for their work, several rental sources will be available.
9 Again, I am a drummer. I blame only myself!
10 Some notorious flops have closed after less than a week's performances!
11 I once was able to rent eight Meyer UPAs for 7 weeks for $1,000 because of this situation. Three months after that production closed, I went to the same house to rent the same speakers for another show for the same amount of time, and the quote was $7,500 because they were in demand again.
12 Admittedly, this particular tactic only worked because of my particular personality, and because I was asking this of vendors to whom I brought thousands of dollars of business every year. That said, I've found that so long as you keep things pleasant, it doesn't hurt to ask.
13 With apologies to any technically knowledgeable producers out there. I've worked with a few of you, who gave me targets for large projects that turned out to be very close to reality. However, the majority of you didn't pretend to know sound well enough to estimate, so you left it to me. Thanks!
14 For those who have never been through this process, quotes always have an expiration date. Before that date, the quote must be honored by the company (with rare exception), after that date, all bets are off.
15 Manufacturer's Suggested Retail Price.
16 Don't make it *too* obvious, though. And always make sure you are specifying gear that you still think will benefit the production.
17 *. . . in writing, for my legal protection. . .*
18 Which often means redundant systems so that if one fails the other kicks in. More about redundant systems in various chapters of Part IV.
19 At the USITT Conference, 2018, in Fort Lauderdale, while chatting on a panel with the legendary Abe Jacobs.
20 Some vendors will be quite helpful in this situation, others won't give you the time of day.

CHAPTER 34

Selecting Loudspeakers and Positions

Loudspeaker Modeling

If loudspeakers are the most important part of a given sound system (they are!), they should probably be selected and placed early in the design process (they should!). So how do you select speakers? What factors are important, and how do we use specs, speaker modeling, and other tools to understand how our system of coverage will work before we actually select and install equipment?

If we've already been through the steps laid out in Chapters 30–33, then we should have a pretty strong idea of what jobs we need our speaker system to perform. Let's break those concepts out and examine how each specifically impacts what speakers we select.

1. THE SOURCE MATERIAL

Analyzing the source material will tell us a ton about what kind of a system we need to deliver. First, by establishing what type(s) of content will be delivered by our speakers, we can identify several key performance factors:

- Frequency response: do your speakers need to be full range for music delivery? Are they delivering human voices, in which case they need to be mostly full range but may sacrifice some at the LF end? Are they delivering sound effects—if so, *what* sound effects? The frequency range required for the sound of frying eggs is substantially more limited than the frequency range required for the sound of a busy heavy manufacturing plant. Speakers are not all designed to do the same jobs. While it can be liberating (on creative projects) to specify full-range speakers wherever feasible to allow for any content to be played from each location at a whim, this is not typically how systems are configured. Speakers will be, at least in part, selected on the basis of what frequency response they can deliver, and whether that frequency response suits the intended source content.
- Sound pressure level: how loud does your content need to be? This is based on not only what type of content you are delivering (voices, music, other sounds), but also on the type of project. A bedroom drama will

typically reach much lower peak SPL levels than a Metallica concert. While content is the first determiner of SPL, this will also be determined in large part by the size of your venue. A small black-box theater is likely to need lower SPL in general than a giant concert stadium—but that will be determined by the content first. A small black box that must convincingly replicate a war zone or an earthquake may actually reach a higher peak SPL than a stadium featuring a very quiet folk music act. If you are an experienced listener and sound person, determining target SPL can be easy. However, when you are first learning, it can be challenging to know how loud something needs to be. How loud is too loud? There are a few things you can do to help figure out how loud your program needs to be:

- Measure sounds with a meter and keep a log. Hardware SPL meters will be more accurate (generally) than meters on your smartphone, but even a smartphone meter can give you a good guide to how loud various experiences are. If you are designing a production, the best guide may be to attend a similar production and surreptitiously collect data. Collect data wherever and whenever possible—when you are listening to music at home or in the car or at work, when you are at a coffee shop that is loud (or quiet!), or when you are watching movies. Collect SPL measurements in a wide variety of places and situations until you feel comfortable that you know how loud different situations are (or should be).

- Follow some very basic rules:
  * Any system that communicates life-safety (e.g., evacuation) information needs to rise above the noise floor of the environment by at least 12 and ideally 20 dB (up to but not exceeding the 120 dB SPL "threshold of pain"). If the safety system is playing in a generally quiet environment (or if, as in some themed installations, triggering of the life-safety system will override all other playing sounds), we must deliver at least 85 dB SPL for life-safety announcements (which provides 20 dB of headroom over the average conversational level).
  * Unamplified spoken conversation (between non-actors) tends to hover around 65 dB SPL. If you are designing a system for, say, a hotel lobby, you need sound to be loud enough to be audible, but not so loud that guests can't comfortably talk with the registration desk. Anywhere that the content is background music to be played under civilian (non-performer) guests, use 65 as a guide. We may want to be able to deliver 85 but run the system much quieter on a day-to-day basis.
  * For vocal delivery systems, a baseline of 85SPL will "feel amplified" in most quiet environments. We always want to aim for at least 12 dB of vocal delivery headroom available above the noise floor of the environment or other program material—with the exception of very high-power concert systems, in which vocal delivery will be matched to the high SPLs of the music delivery.
  * Amplified concert systems in the United States are often designed to deliver 120 dB SPL throughout the venue, with a plan of operating on average 20 dB below that. In Europe, noise laws are stricter, with some countries holding max SPL to 100 or below, and on average operating quieter than equivalent American systems.
  * Observe local noise ordinances—some nations require events and systems to operate below a max SPL cap, and many outdoor venues have noise controls in place set by the local municipality.

- Think about what kind of experience the audience needs. Do the seats need to rattle, or does the audience just need to hear a speech clearly?

- Localization: the source material will help to determine where sound needs to be localized. Do we need to source actor voices to a stage? Do we need sound effects to emerge from hidden parts of a themed environment? Or are we delivering retail Muzak, where we just need even coverage across the entire space, localization aside?

2. THE VENUE

Already, some of our plans based on source material have been bleeding over into the venue itself. This is because one production mounted in a small black box may make very different demands than the otherwise same production re-mounted in a large proscenium theater. The same band playing in a grungy nightclub will need a different system than when they graduate to playing arenas. The venue will determine a lot about what shape our speaker coverage will take.

Tales From the Field

With *Spiderman*, I did a surround immersive system that had speakers up in the dome, in the roof, and all around, and the idea was that . . . there was Bono and Edge music, and originally the music was really good, really good storytelling, just excellent. What happened was that as the show went on, the music was transformed into something else, not due to Bono and Edge, but due to other areas, it became something that they didn't write. So that being said—I wanted to be able, at times, for the spectacular Marvel events, to be able to immerse the audience with sound, with music, with effects, to go with the visual actions onstage—which were very vivid, very crazy Julie Taymor stuff and really fantastic images and ideas. When it came down to the Peter Parker moments—regular human beings—to go back to a regular standard theater system, as if it was *Oklahoma* or something—just very focused on the individual actors. But what I did was—I had this great thing where I could track the actors, follow them around, and place the time delay for their voices where it needed to be. I also knew it wasn't really going to work in that way because if you're going to do a rock show, then go into dialog, the dialog is not going to be natural and *Oklahoma*-like, it's going to be very well reinforced, and so the voices are going to come out of speakers. So here's this product where I thought, "Oh, wouldn't it be cool when Spiderman flies around the room"—and there were lead guitars that played the same theme while he flies around the room—"that the lead guitar from the proscenium mix got pulled out and tracked around and followed Spiderman around the room, flying in right behind him, as he flew around, whatever he did." I tried it, and it worked great, and we thought, "Wow! This is fabulous, really cool!" The music was perfect for this, the actor would go around, and the tracking worked fabulous, so cool. So, Julie Taymor came in that section of the show was performed, and she said, "Wait, wait, wait, wait! Why is the guitar following around the room?" Well, it's going around where Spiderman goes. She said, "But why?" And I said, "Well, it's because it's his theme, and it's tracking around him, and it's really exciting." And she said, "No. No. Put it out of the speakers in the front. Don't have a guitar following him around." So it went horribly wrong because I'd spent all of this time and all this effort, plus, there wasn't much money being spent on it because the tracking system was on loan for me to try things, so it wasn't a massive cost, except for my time. Where it went horribly wrong in that case was that it was just completely killed by the director, because the director had no interest in doing anything really sonic—actually with the show at all.—Jonathan Deans

First, we have to know (and work from) a correct set of technical drawings of our venue. If we're working in a raw space, or on an architectural/themed project, these will just be drawings of the venue itself, ideally with any available mounting/rigging points visible (or at least noted). If we're working on a stage production, we'll need accurate drawings of the set as well as the venue itself, and any/all audience areas.[1] In order to plan loudspeaker coverage, we need to assess where the audience and/or guests will be, what condition they will be in (seated in fixed seats, standing, moving, etc.), and what experience we are trying to give them. We also need to assess locations where speakers can reasonably be mounted. I have worked in many venues where the acoustically ideal positions for loudspeakers were simply impossible, due to lack of rigging points, and lack of budget or workable plan to install temporary rigging points. The venue will impact many features of our speaker system design, beginning with the following:

- Coverage angles: a speaker's coverage angles are very important to how successfully it can communicate the desired info to the desired audience. Consider Figure 34.1.

  In this drawing, we see a small proscenium theater. Let's imagine, first, that we're trying to create a speaker system for vocal reinforcement of the performers on stage. What are our typical options for a venue of this configuration? We can place mains at the left and right extremes, either ground-stacked in front of the arch, or flown above the same spot (Figure 34.2).

  This poses localization challenges, as any sound loud enough to be amplified several rows back risks drawing the listener in the front few rows to the speaker, rather than the actor who is center stage. Why doesn't this work, when in everyday life we listen to LR stereo speakers all the time and feel that there are sounds localizing

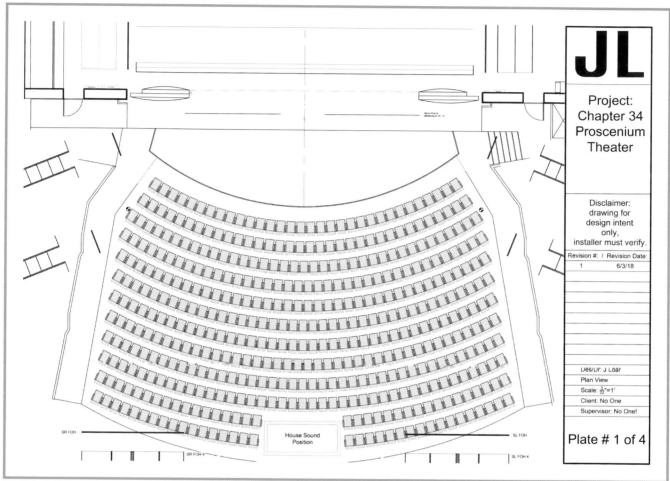

FIGURE 34.1
A Small Proscenium Theater

to the center between them? In everyday life, in order to image the "center" of a L/R pair, we must be seated between the two speakers, basically equidistant from each. The more off-center we get as listeners, the more one speaker or the other will dominate the sonic image, and thus pull us to localize to that speaker. In the theater in Figure 34.2, the audience seating area is wider than the proscenium arch, so there is a fair chunk of the audience that isn't even *between* the two-speaker positions, not to mention central enough to actually hear a center image. Thus, we might want speakers to be centered on the stage. How can we do that, though, when we can't place speakers on the deck between actors and audience (no director would allow for it)? The typical solution is to create one high-center position and to use front fills mounted to the lip of the stage—as in Figure 34.3.

In this scenario, front fills will generally be covering the front few rows of the audience with the high-center speaker (or cluster, or array) covering the rest. The venue as drawn is quite wide. If we are planning a high-center speaker position, probably its horizontal coverage should be quite wide. How wide? Let's draw some angles and find out! (Figure 34.4)

There are some angles we can rule out right away. For example, I have drawn a 45° horizontal coverage cone. This is substantially narrower than our audience and will not deliver enough width to be useful in the high-center position. We want our high center to cover the entire width of the audience. Next, I've drawn a 90° cone. This is closer, but still shaves off seats on the sides. Looking at this position, we need a speaker (or speakers) whose coverage is *at least* 110° wide (even at 110°, there are 1.5 seats on each side that don't get direct on-axis sound, so 120° would be even better, but given the nature of a reverberant space, those couple of seats would

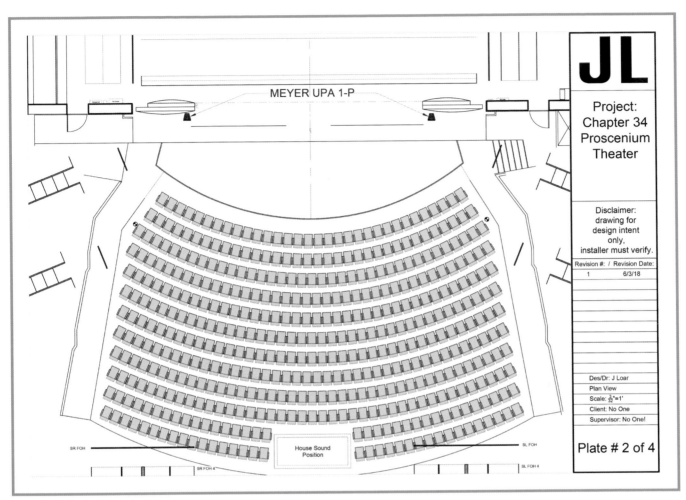

FIGURE 34.2
The Same Theater, With Left/Right Main Speakers

probably have *acceptable*, if not great, coverage from a 110° box). We can accomplish this a couple of different ways: we could use two point-source boxes, each with 55° horizontal coverage, aimed out from a mutual center point as a horizontal array (assuming their vertical coverage also covers the depth of the audience)—though this will create an LF pileup dead center; we could select "line-array" boxes with wide coverage (110° or greater), and stack as many as are required to cover the throw depth of the audience; or we could find the rare (but real!) point-source single box that covers 110° wide—always making sure that none of our solutions drop physically low enough to impede the sightlines of the back-of-the-house seats.

This process, of placing speakers and drawing rough coverage angles, is very often an early step in the speaker planning process. Once we know where we can place speakers, we can gauge how we need them to cover the audience. Any speaker delivering vocal reinforcement or music is likely to be a system that wants to cover the audience or guest areas more or less equally, such that the content is communicated with equal clarity and strength to everyone. Of course, if we remember our physics, we know that coverage angles listed in a spec sheet are targets reached only at certain frequencies, and the actual coverage of the speaker will vary by frequency. This is why it becomes very useful to use speaker prediction/modeling software, which we'll address later in this chapter.

At any rate, the practice of drawing coverage cones from your speaker mounting positions can at least guide you toward selecting speakers that will cover your target areas appropriately, and thus give you suggestions as to which models to skip entirely versus those that might be worth running in speaker prediction software.

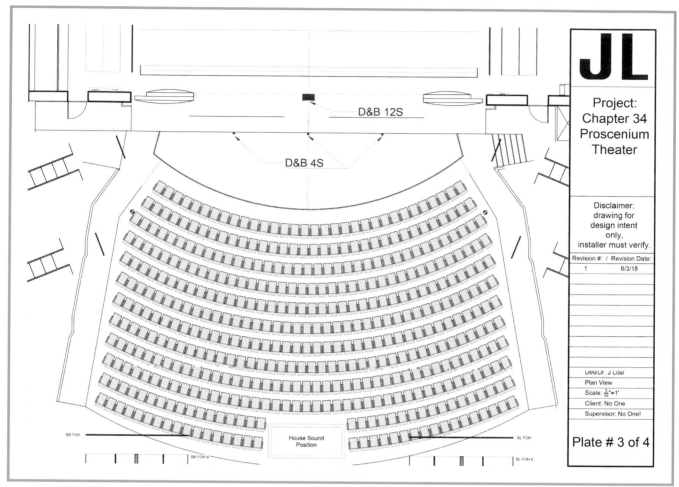

FIGURE 34.3
The Same Theater, With High-Center Cluster and Front Fills

- Form factor: specifically, size, weight, and hanging hardware (if the speaker is flown). Will your chosen speakers fit in the positions you have available? This should be a pretty straightforward process, but don't forget that you're usually not just placing speakers in a space, but also sharing that space with other designers. On productions, this often means lighting, video, and scenic automation/flying designers are also placing stuff around the venue. For architectural and installed designs, these may come into play, but you also may be fighting for space with HVAC ductwork, plumbing, physical support beams for the structure, and so on. While a good production or project manager will overlay different departments' drawings in order to identify spaces being requested by more than one design team, a good designer is ALSO requesting drawings from the other departments and checking this themselves. Sometimes, you want a speaker in a given position, and you can

Message in a Bottle

I think you need to be comfortable with your tools whether that's ArrayCalc or MAPP Online or, frankly, pen and paper. You know, particularly if you're doing point-source loudspeakers, you can figure a lot out with a protractor and a scale rule to make sure you're getting the coverage that you need. I think you need to be comfortable drafting, because as a sound system designer, one of the things that you're going to have to do is reconcile what you want to do with the way the physical universe works, and you can't put lights and speakers in the same place. Physics doesn't allow it.—Vincent Olivieri

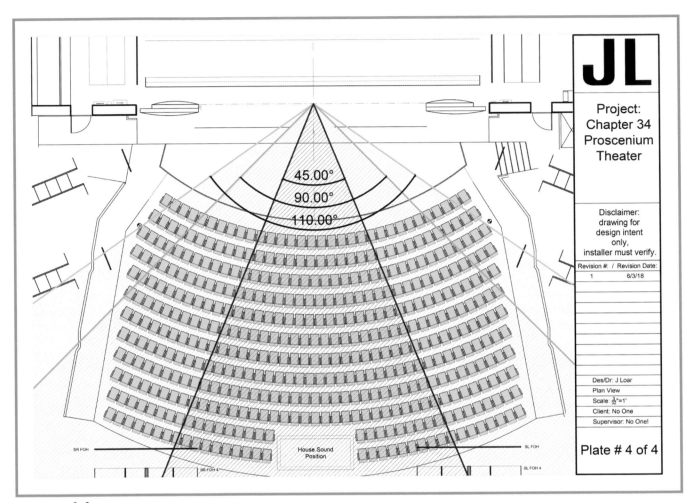

FIGURE 34.4
Different Horizontal Coverage Angles From the High-Center Position

make a persuasive argument that it's more important for your speaker to be there than someone else's lighting instrument or HVAC duct, but sometimes, you will be forced to move from your intended positions, in which case you need to then reexamine your predictions.

- Position: the venue often dictates where speakers can and cannot be placed. If you're working in a concert venue and there are weight-rated rigging points for the L/R mains but none above the high-center position, you may well be limited to L/R mains. However, in some venues and for some projects you may need to specify the creation of new positions. This might be as simple as adding a new crosspipe into an already existing Schedule 40 pipe grid in a black-box theater, or this may be as complex as installing brand new rated anchor points above a drop ceiling in an architectural environment.

3. THE AUDIENCE

From our audience information, we can glean what kind of delivery will be expected. If we are designing a sound system for an EDM nightclub in Vegas, we can pretty well guess that we'll be expected to deliver a 360° coverage system producing very high SPL across the full audible frequency spectrum. If we're designing a system for teleconferencing in a corporate meeting room, we can guess that the SPL will be a *lot* lower and that we don't really need to worry if we're able to accurately reproduce 27 Hz.[2] Along with the content and the venue, knowing about the audience will help selection of speakers quite a bit. Where is the audience? How large is it? Does it remain seated in fixed seats, or does it move? Is it a mosh pit at an Avenged Sevenfold show, or is it a group of corporate lawyers listening to a talk? Audience info should impact the following:

Message in a Bottle

When you realize that a sound system has been designed with care and thought, and, hopefully, supported by an appropriate budget, the fact that it is like a finely tuned race car—that you've got this thing at your control that's actually an extension of your imagination, this gigantic pile of wires and magnets and all this stuff—when you can use that effectively, that becomes, in my mind, that's exciting. You're adding some storytelling in a most subtle but effective way. To me, there is a kind of orgasm of great sound design that you can get to this point where the system behaves the way you want it to and is responsive in the way you need it to be, and as a result, the experience of the audience is incredibly successful and well-appreciated. Also, finding simple solutions to complex problems. It doesn't necessarily require a ton of money, just the right piece, the right loudspeaker in the right location—those simple solutions are very rewarding as well.—David Budries

- SPL
- Coverage angles
- Delay system need/placement—are we OK with sound emanating from the stage and getting somewhat quieter toward the back of house? Do we need to place delayed speakers through our venue to supplement the direct sound from the stage speakers and thus to ensure clarity?

4. THE BUDGET AND AVAILABLE RESOURCES

Budget will, sadly, determine a lot about our speaker system. You can design the loveliest d&b system with main arrays at the stage and surrounds all around the walls of the venue, but if your budget is $2,000, there's no way to even rent that system, not to mention buy it. Save yourself unnecessary work, and design to your budget, unless—and this does sometimes happen—you need to make a point to a producer that the budget you've been afforded will not produce the desired result. At times like that, it can be helpful to model several parallel designs, to demonstrate how the system will perform within your budget, and how a budget increase of X amount could improve the system performance.

Anyway, once you have a good understanding of how the source material, venue, audience, and budget will impact your potential speaker selections, it is time to begin speaker prediction. First question: *what is speaker prediction?*

Speaker Prediction

Also known as speaker modeling, speaker prediction is the process of drawing the dimensions of a venue into a dedicated prediction software, placing speakers (real models performing with real measured data), and running predictive models to examine how coverage area and SPL will function across different frequencies. By engaging in speaker prediction, we may select the best available speakers for our jobs, place them in the most acoustically optimal locations, and save the labor it would take otherwise to determine these factors by trial and error.

Speaker prediction, as such, has become an essential part of most professional system designers' work. Most speaker manufacturers make a proprietary version of speaker prediction software that is available for free via their websites. Meyer has MAPP XT, d&b makes ArrayCalc, and others offer their variations. Each of these has some advantages (the first of which being that they are typically free to download and use). However, most of the free software also comes with some big limitations:

- They don't usually model in 3D, sticking only to 2D planar images.
- They are proprietary, meaning they can only model one manufacturer's devices, and not hybrid systems (which are very common).
- They only cover some models (JBL, for example, offers array calculators for its vertical line-array speakers, but no free tool to model hundreds of its other units).

The industry-standard software for speaker system modeling is AFMG's EASE® (Enhanced Acoustic Simulator for Engineers). This software allows you to build a 3-dimensional venue, assign real acoustic textures to all venue surfaces that mimic the real-world performance of walls, curtains, floors, ceilings, etc., and to place speakers of almost any manufacturer in a design together. EASE® doesn't have the cleanest graphical interface, but they remain the industry standard not only because a properly configured EASE® model is reliably matched to real-world results, but also because they have managed to convince just about every professional loudspeaker manufacturer in the world to measure their speakers in the EASE® data format, such that a majority of professional loudspeakers are available to model (including a ton of vintage and no-longer-made units). For the remainder of this chapter, we will be discussing modeling and prediction with the aid of terms and techniques (and graphics) from EASE® but know that most of the general principles apply to other modeling software as well.[3]

A BIT OF TECHNICAL BACKGROUND ON HOW SPEAKER PREDICTION WORKS (DON'T WORRY, WE WON'T GET TOO INTENSE):

The basic premise of speaker prediction software is that someone has measured the real-life performance of a given speaker model and recorded this data in detail. This data includes polar pattern measurements both vertically and horizontally of the speaker's performance across the majority (if not the entirety[4]) of the audible frequency range. It also, for any line-array-style speaker, will include the details of that particular array system—so if you select a 2-box array vs. a 20-box array, it will allow you to configure that array in any fashion that would be physically allowed by the hardware. This includes splay angles between boxes (so you will be limited to splay settings that can actually be achieved with the hardware),[5] gain shading, and other such options.

From the database of speaker information, particular models will be loaded into your venue, and from there you can select a wide variety of settings, from time-alignments/delays to aiming of coverage, from intensity of amp settings, to any desired EQ filters.[6]

Message in a Bottle

My approach is like this: I look at a room and every seat could be members of a family. And I look to see who's going to establish custody for every member of the family. Somebody's got to cover that seat and that row, and either he's getting covered by somebody or he's in the transition of coverage—he's in the custody exchange between somebody and somebody else. So, I will look at that and it starts with "what areas can I cover with the main system, and what areas then get parceled out to get covered by fill systems such as either side fills or center fills or front fills or under balcony fills or over balcony fills or rear fills or etc.?" So I basically partition things out, and you know there are some times when it's very straightforward and very few fills are needed, and other times it's the House of Many Fills because of the shape of the room—or the positions that we're allowed to use for the mains. You can take the same room and if we could put the mains in the correct place we could make the sound design really simple, but because we're forced off into some corners, now we need a center fill, or because we're too high, now we need under-balcony fills even though the balcony isn't that deep, but there's no line of sight anymore because they pushed our speakers so high . . . these kinds of things. So for me, I try to cover as much as I reasonably can with mains, but you can reach a point where if you're trying too hard, you're going to end up stretching the mains out, pushing them into side walls and these kinds of things. And there comes a point where you're just better off to parse or partition out the system and work it out like that. That's kind of my flow—I look at it like a big quilt, and I stitch together big pieces of fabric to cover the main areas and then smaller patches to connect the rest, and then I plot out where my transition is between the main and the front fills. That's the third row probably—or where is my transition from the mains to the side fills? All of these things are looked at. It's essentially how I'm going to thread the whole together in the tuning process. So whenever I design a system, I can tell you this, I never add a speaker into a design without seeing exactly how it's going to fit into the tuning. What it's going to be timed to, where it's going to set its relative level, where I'm going to time it for the transition to the other systems—all of these things, they're all in the equation of the design.—Bob McCarthy

The software will perform some variety of modeling—the simplest variety of modeling is statistical, making use of averages of the measured speaker performance and predictions about the performance of a venue based on the room's physical volume (for example); the more advanced variety of modeling is based on what is known as "ray-tracing" technology. In this paradigm, a venue is assigned lifelike acoustic textures, and each speaker has a given number of "rays" assigned to fire from it, within its polar coverage. You can set not only how many rays per speaker, but what "order" of ray-tracing to perform, which is to say, how many reflections you calculate.[7]

From there, the software generates plots that can not only measure SPL (both direct and total/reverberant), across frequency, but can also produce models of speech intelligibility indices, and many other common acoustic measurements.

The results come out as color bloom plots across the space, something like Figure 34.5.

That's all well and good, but the process of setting up a speaker prediction (and getting useful results) can be daunting for many designers. Here, I present a basic order of events for the selection of speakers, and the process of configuring and running speaker predictions. Again, I use EASE®, but these steps can apply to most any prediction software (with the exception of the steps in which we configure the acoustics of the venue, which are not available in most freeware prediction programs). Once we've laid out the steps, we'll go into some detail with each one.

HOW TO SELECT AND MODEL LOUDSPEAKERS:

1. Begin by using basic coverage angle drawings overlaid on existing drawings of your venue, to determine some rough ideas about speaker models that may work for your system. Make initial selections based on coverage angles, SPL production, frequency response, and form factor.
2. Examine whether any of your systems need modeling. Vocal and music systems generally benefit from modeling, however, depending on the size and scope, there are times when sound effects delivery systems *can* benefit from modeling, and times when it is pointless to model them (more on this in the detailed sections that follow).
3. Build a venue (or at least outline dimensions of the audience area) in your modeling and prediction software—taking care to include important details, and also (as we'll discuss next) to leave out details that will only confuse the modeling process.
4. Assign acoustic textures to the venue that approximate the real-world textures of the venue's surfaces.

344

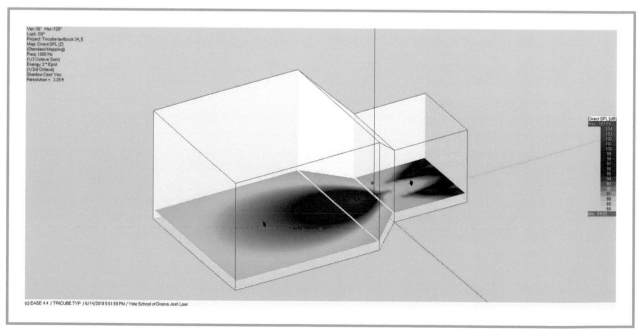

FIGURE 34.5
A Sample EASE® Plot—Speaker Coverage on an Audience Area (Statistical Prediction Without Active Venue Acoustics)

5. Establish audience areas in your model (in EASE®, assign them as "Audience Areas" so they will model with the acoustic properties of a bunch of people, rather than as a wall or another surface).

6. Place your selected speaker models into the venue. Ensure that you label speakers properly. You may be modeling both music and vocal subsystems for a stage musical, but you only want to run models of subsystems that carry coherent content—meaning, don't model all the vocal and music speakers together, model all the vocal speakers as one pass, and all the music speakers as a separate pass. It is easier to do this quickly if you've labeled the speakers in your model with names that allow you to easily remember what job each speaker is supposed to do. Configure your speakers—setting position and aiming, power settings, time delays, etc.

7. Run statistical models to see if basic direct sound coverage is about what you expected. Refine position and aiming of speakers in statistical modeling mode until they are basically as desired.[8]

8. Run ray-tracing models with real acoustic textures to see how the venue will impact the coverage.

9. Refine speaker models/positions/aiming as needed based on the results of ray-trace modeling.

10. Produce a design that features good, even coverage (as good and even as possible within budget or the gear stock available). Collect all the data about mounting positions and angles and other settings from your model, and then use it to install your system!

Let's examine each step in more detail. We will begin with our hypothetical proscenium theater from earlier in the chapter. First, let's assume that we're planning to create three separate subsystems in the space, one for vocal reinforcement, one for stereo music playback (canned, not live) and one for sound effects. Our hypothetical stage production is of *A Midsummer Night's Dream*, and we will assume that the vocal reinforcement system is working to reinforce (but not *replace*) the voices of the actors; that music playback is for interstitial music and occasional underscoring, but never needs to rise to a very high level (max 90 SPL, perhaps); and that the sound effects system needs to produce enveloping ambience of forest, and a few sounds localized to the stage, including Bottom's transformation to an ass, various other fairy magic, and some offstage hunting party noises.[9] In the following figure, I have borrowed a scenic design from my colleague M. C. Friedrich for this play.[10] Figure 34.6 shows a scenic design for our hypothetical production.

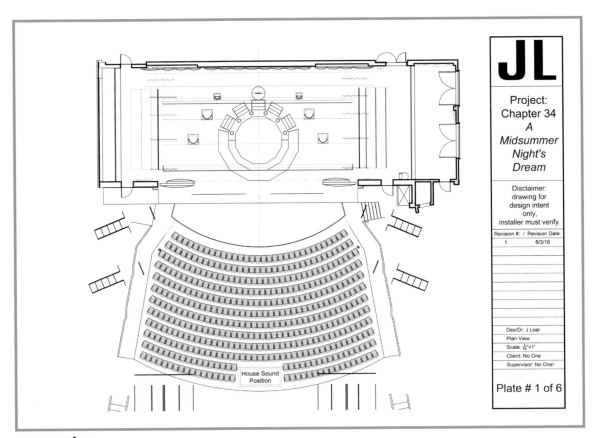

FIGURE 34.6
A Small Proscenium Theater, With Set for *A Midsummer Night's Dream*
Source: Set Designed/Drawn by M. C. Friedrich, Courtesy M. C. Friedrich

> **Message in a Bottle**
>
> The goal of your speaker system is to give everybody the same show. So you're trying to cover everybody evenly. Whether you're flying a big point-source array of MSL3s or KF850s, or you're flying a line array, you're trying to address those guys in a way to try and produce a more equal level and get better pattern control. What you're trying to do is put up a speaker that just naturally has the coverage pattern that matches your venue. And so manufacturers give you some tools to help set up that array in a way that creates that consistency so that your top boxes have one degree between them, whereas as you go through it, there are two degrees between the next guys and three or five [between the next], and you're bending that array to create a speaker whose pattern is asymmetrical. The truth of the matter is that, in general, you can get the high frequency consistent everywhere. The whole idea is to treat it like it's one big speaker. So you're going to take a measurement point or a couple measurement points. I like to take multiple measurement points just to kind of randomize and average out the position and the position-dependent anomalies to just focus on the array response. But then you end up treating it like a speaker, and then slowly add together your speaker systems to make one big speaker—or actually it's the left/right system, two big speakers, or let's generate three big speakers—and the person that's using it doesn't have to worry about outfills, side fills, front fills—the system design takes care of that.—Jamie Anderson

So, let's go through the process of selecting and modeling a speaker system for this imaginary production!

Step 1: to plan the speakers for our production, we must examine the three areas of content separately. Because this imaginary production is of a straight play, we will assume that vocal clarity (dialog) is of paramount importance. We will assume that since music is only interstitial and possibly light underscoring, it is not terribly important. We will assume that SFX, while important, are minimal and thus can be addressed as such. First, let's make a vocal reinforcement plan.

As discussed earlier in the chapter, a common proscenium theater tactic for vocals is to create a high-center cluster, paired with front fills on the fascia of the stage, with tops of boxes flush with stage deck level so they don't obstruct anyone's view. Beginning with our high-center cluster, we will first determine what kind of coverage we need in the horizontal plane. Looking back at Figure 34.4, if we are to cover the whole width of the audience, we will need a speaker or speakers that have at least 110° of horizontal coverage (as we stated earlier), and ideally 115° or 120°. Now, let's flip the room on its end and examine vertical coverage.

In Figure 34.7, I've drawn angles from our high-center position to cover the entire audience.

We see that we would need coverage of roughly 55° vertically to cover the audience. Now, we could use point-source high-center boxes here, but this is a tricky pair of axial dimensions to match (120° x 55°). Earlier, we looked at a d&b point-source box (the 12S) that can cover 110° x 55°, but since we're designing in a vacuum, let's see if we can match our coverage even more ideally, and not leave out those seats at the end of the aisles. We can assume that the bottom 10° or so of coverage in the vertical can be removed and assigned to be primarily handled by the front fill system. This leaves us 120° x 45° of coverage, which is still a challenging dimension to find in a single-point-source enclosure. We could mount a pair of point-source boxes, each 60° wide by 45° vertical in coverage, but looking at d&b's website quickly,[11] I can identify boxes with 80° x 50° coverage, or 90° x 40°, or 75° x 45°, but using a pair of any of these would mean either creating a wide zone of overlap that falls right in the middle of our audience, creating phasing/comb-filtering issues that are unpleasant to listen to, or focusing the speakers out wider and splashing all over the venue side walls, giving us a ton of reflected sound. Because the proscenium arch is tall enough (enough space between rigging point and the opening of the arch), what we can do instead is try a small vertical array. Looking at the options available from d&b, I find that they offer a few boxes with 120° wide coverage. The smallest of these boxes is the Y12, which is 10" tall at the front. If we mount an array of 4 or possibly 5 Y12 boxes, it will hang less than 50" below the rigging point (exact height TBD pending splay of boxes), and according to a rudimentary sightline study (Figure 34.8), will leave the set visible to all audience members.

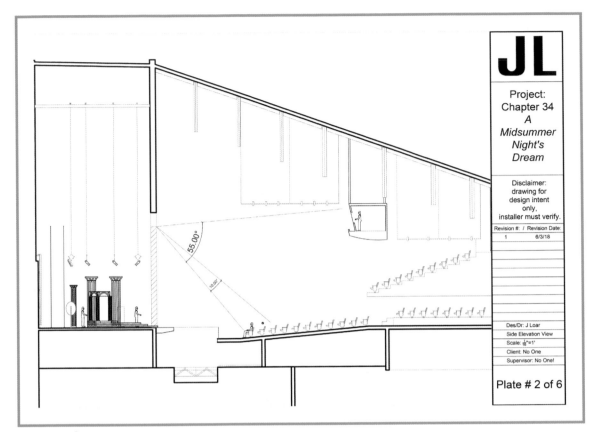

FIGURE 34.7
High-Center Position Vertical Coverage Angle

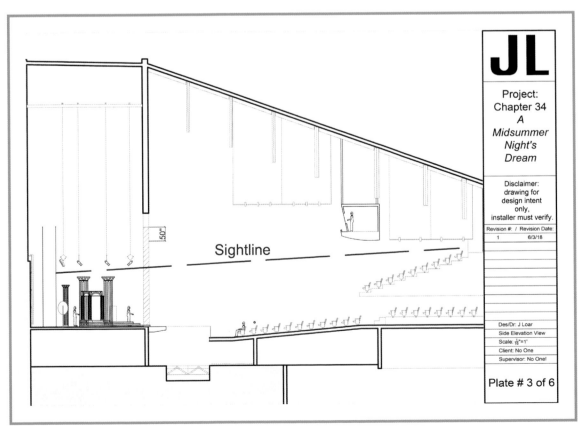

FIGURE 34.8
Initial Sightline Study, High-Center Array

Now, for this array, we'll start with a four-box configuration, to see if we can accomplish enough vertical coverage with that many boxes. Line-array speakers often do not list vertical coverage, since it varies depending on how many boxes are in an array, and on the curvature of the array itself. If we find we're easily accomplishing the desired coverage with four boxes, we might test with three to see if we can get away with one less. If we are finding patchy coverage or not a broad enough vertical pattern, we'll start by adding one more box.

We also need front fills. Given the width of this stage, and our desire to be able to localize actors to the center, we'll want 1 center fill, for sure, and at least one to either side, for a minimum of three front fills (depending on coverage). Selecting front fill speakers for vocal reinforcement can be challenging. Our first consideration must always be size and ease of mounting—a giant box with great pattern control won't look good in front of a stage set and will generally be rejected by producers (electrified concert front fills often just sit on the stage deck itself, and as such tend to be larger). If we want a small, unobtrusive box, d&b's 4S (Figure 34.9) is a prime example of this form factor.[12]

FIGURE 34.9
d&b audiotechnik 4S
Source: Courtesy d&b audiotechnik

However, care must be taken to examine how far downstage actors will be playing with mics, and how loud they will need to be. The 4S is a 100° conical box. As a result, the potential for splashback on stage is real. If your performers don't play too close to the downstage edge and are able projectors of their natural voices, the feedback potential can be managed. Audience members are absorbers, so we have to assume attenuation of the vocal frequencies by the time we reach the outside edges of each front fill speaker's horizontal coverage cone. We must place enough front fills to cover the front row (Figure 34.10), and so that we can keep the overall level of the front fills as a group very low.[13]

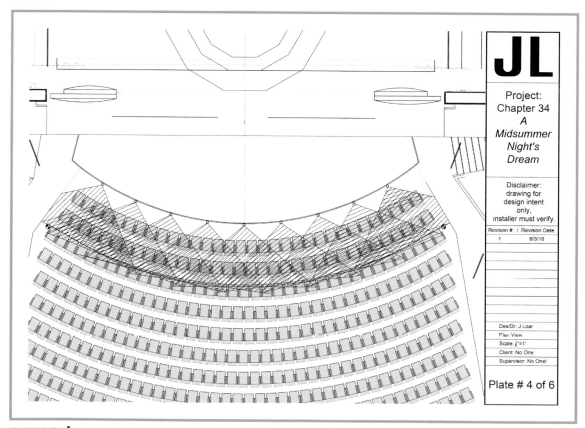

FIGURE 34.10
Front Fill Horizontal Coverage Angles

Now, we've got a rough vocal system. Let's select a music system. We'll be playing stereo music, and we're not concerned with localization. For this particular project, we will try using a standard L/R mains design. Since music plays a secondary role in this production, we're going to test Meyer UPJs, along with USW-1P subs. This will give us flat coverage loud enough for our needs, down to about 35 Hz (with some less-than-flat coverage below that). The USW sub can be mounted to the side, or directly above, the UPJs. Horizontal coverage of the UPJ is 80°. In most theaters (and even more concerts), we don't typically run "true" stereo. If we run a system where the left main speaker only covers the audience on the left of the house, and vice versa for the right, a keyboard part panned all the way left won't be heard by audience members on the right of the venue. Thus, it is fairly common, knowing that left and right signals won't typically be identical, to specify wide-coverage speakers and aim them with a fair amount of overlap (cross-firing), such that they create some sense of L/R difference, but allow all of the audience to hear all of the musical parts (Figure 34.11).

Now, as to our sound effects system. In order to create a show with fully immersive audio, we need the audience to feel like they are *in* the environment, so we are going to create an elaborate surround system, with speakers ringing the walls of the venue, speakers flown overhead facing down, and even speakers aimed at the walls and ceiling (to use the surfaces as added "diaphragms" that spread sound and discourage localization). In the spirit of attempting a mixed-manufacturer system, we're going to use L-Acoustics boxes for effects. We'll ring the venue with 100° conical X8s, that cover down to 60 Hz. We'll create a matrix of flown 5 XTs, also 100° conical, though only covering down to 95 Hz. These will allow bird flight and other movements above the audience. Since this theater has a plenum (a space below the audience and/or stage), we will place two SB28 subwoofers, to rattle and thump from below.

So! Now that we've selected our initial speakers, we can move onto the next steps.

2) Do our systems need modeling? Vocals—yes. We'll be time-aligning between flown and deck-mounted speakers, and we need to check feedback potential onstage in the vocal range. Music systems—sure! We can generally tell that our system will deliver what we want, but it'll help us to select the best rigging options and aiming angles. SFX

349

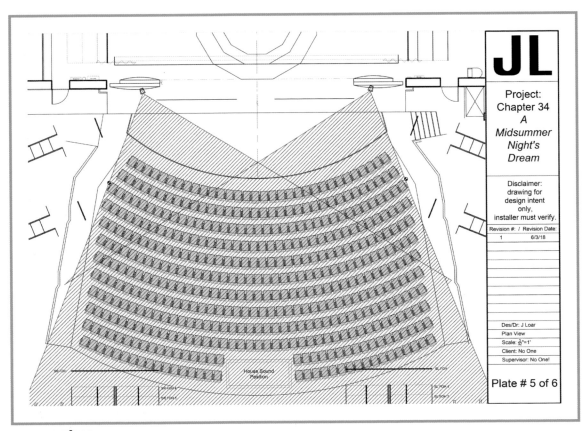

FIGURE 34.11
Music System Horizontal Coverage

systems? Not this time! We are placing plenty of speakers in a large diffuse system, and content will move through them spatially rather than playing universally through them. We can tell already that they will do the job we need.

3. BUILD A VENUE IN A MODELING SOFTWARE

EASE® provides the user with a selection of stock venues to start with, which can be modified in terms of features and dimensions. Because we are working with a standard proscenium space, we can begin with EASE's ® "Theatre" prototype and modify it. It defaults to giving us a balcony, which we need.[14] It is beyond the scope of this book to go into how EASE® works in any detail but suffice it to say that if you read their user guide,[15] it will explain in detail how to work with and configure your venue. Of major importance here, for our consideration, are three points:

- Internal dimensions of the venue must be accurate.
- A closed room must be set up as "closed" in the modeling software (otherwise it will perform modeling as if the venue is open—like a stadium with no roof—and won't be accurate to the performance of a closed indoor theater).
- An EASE® (or other software) venue model needs to include some important details and leave out some details that will confuse the software. If, for example, our theater had pillars in the audience areas, we'd definitely want to include those in the model so that we could calculate how the acoustic shadows would be cast. If we have doors that stay open during performances (as is the case in many nightclubs), we'd want to model those openings. However, if our doors stay closed, we can generally leave the doors off the model entirely (unless their acoustic texture in real life is *so different* from the surrounding walls that they will drastically change the reverberant sound fields around them). Never include details like molding or wainscoting—these kinds of details throw a model off. Modeling software, as a rule, does not know how to prioritize the reflections from one surface over another, so a 1″ strip of molding will take as much time and processing power to model as a 20′ wall, but its real-world acoustic impact will be minuscule. Go for the broad strokes, and make those as accurate as is reasonably possible, but leave out frill and decoration unless it has very specific and prominent acoustic properties (e.g., if your walls are lined with acoustically calculated quadratic diffusers, you probably need to include some version of those; if they just feature a few framed pictures of various donors, then by all means *leave those off*).

Once we've built our venue (Figure 34.12), and checked the physical space for errors, we assign acoustic textures.

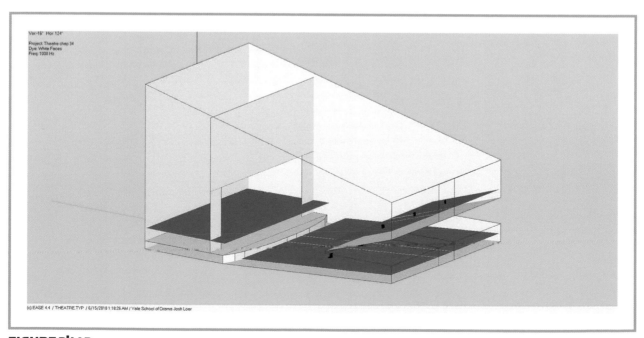

FIGURE 34.12
An EASE® Model of our Venue, With Audience Areas (in Green)

4. ACOUSTIC TEXTURES

In EASE®, we can (and should!) assign acoustic textures to all of our surfaces. One of the most common mistakes that people make when first using EASE® is to leave all of their walls and other surfaces untextured. What is the problem with this? Not only is the venue not going to contribute to the model's performance, but it will actively work against the accuracy of your predictions if you do not assign acoustic textures. Why is this so? The default acoustic surface in EASE®—what is assigned to every surface until you specify something else—is called ABSORBER. This is an "acoustically perfect absorber" (something that does not exist in the real world), that absorbs all frequencies that strike it. Thus, not only will unassigned surfaces not *contribute* to the accuracy of the models you run, they will actively *injure* said models, by acting more absorbently than any real-world surface. So, we need to assign textures.

This sounds easier than it is in practice. First, we must assess the actual surfaces in our venue. If this is an existing venue, ideally, we can pay a visit and take some acoustic measurements, or at least note what kinds of surfaces exist in the space. If you are not able to visit (perhaps you are designing a show in another state and won't get to the venue until tech week), you can ask venue staff to provide as much information as possible. Once you have a decent idea of what real-world textures you are dealing with, it is time to hunt through the EASE® materials database. This is a challenging task: the comprehensive list of materials with their acoustic properties is only available by running a report via the Import/Export widget within the software. The report that is run is formatted in a difficult-to-read list (Figure 34.13) and takes some time to search through. It is, however, well worth the effort.

Once you have done this, it's time to select our audience areas.

5. AUDIENCE AREAS

In EASE, audience areas are a distinct type of object. They appear similar to faces (though by default they will be outlined in green in our working window (and green solid in our modeling window); faces will be shown as yellow, white, blue, or orange in the working window (they can be shown as white or other colors as you

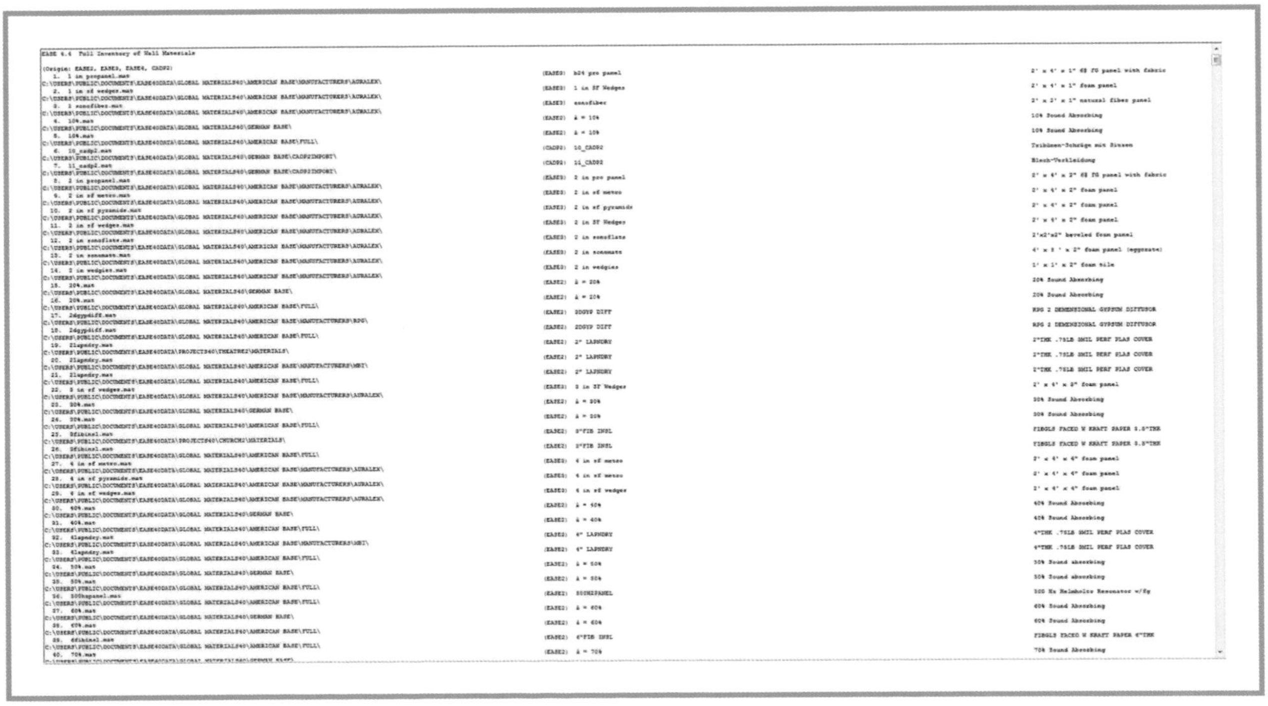

351

FIGURE 34.13
Part of the EASE® Acoustic Textures Inventory (Note How Small the Scroll Bar Is on the Right of the Screen . . . the List Is Much Longer Than What Is Seen Here)

prefer in the modeling window). Audience areas will be modeled as spaces containing an average pack of human beings (with the concomitant absorption and transmission properties). Anywhere our audiences will appear, we need an audience area. We may run models of the space without audience areas turned on (so we can assess the venue itself) and then run models with the audience area activated. In general, though, modeling on the audience areas is the single most important function of EASE®.

Take care to ensure that your audience areas are established at the proper height. By default, the audience areas load in an assumed average seated height, but this only works if your audience *will* be seated. Because we've based our theater on the available template, almost all the audience areas were already placed for us, though I have added one above the stage for standing height of actors, to estimate feedback from front fills.

6. PLACE YOUR SPEAKERS!

Now, we place our speakers. We need to insert loudspeaker objects, assign the proper models, and set all the properties. For most passive speakers, you can set amplifier power in the properties menu; for most active speakers, EASE® will assume you are running at full power. We then will check our positions to ensure our speakers are positioned as they would be flown (or otherwise mounted) in the venue (Figure 34.14). We can turn on a feature that draws a line from the center of the speaker to its on-axis points through the venue, which allows us to set the aim easily. We will also set any time delays we anticipate needing. In our vocal system, for example, the front fills are closer to the audience than the high-center cluster. We will examine the point in space where the coverage for the high-center cluster takes over from the front fills (the spot at which the high center will be louder than the front fills). We will then measure the distance from that spot back to the high-center cluster and to the front fills. If the high center will be farther from that spot than the front fills, we may think about delaying the front fills a bit so that the sound travels in linear order from the stage to the speakers to the audience, without a double-image.[16] Once we have positioned the speakers we are going to test (and labeled them!), it's time for our first model!

352

7. STATISTICAL MODELING

The first rule of modeling speaker systems is that we only model coherent subsystems together. There is NO NEED to ever run all of your speakers at the same time unless they are all legitimately carrying the same signal

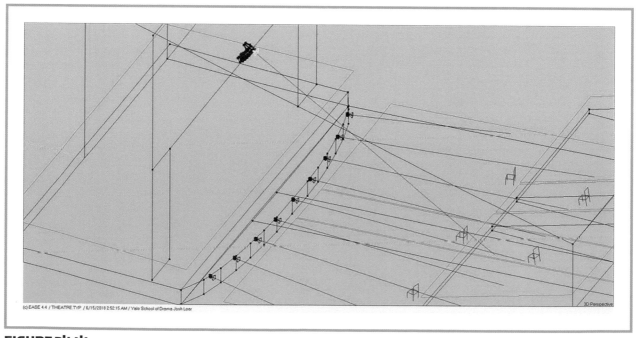

FIGURE 34.14

Positioning Speakers in EASE®

at the same time. Thus, for our example system, we will run our vocal system as one model, and our music system as another.

In EASE, there are the two types of modeling: standard, and standard with reflections. Standard is where we begin, as the models run much faster, and can give us a basic understanding of our coverage so we can refine choices of speaker models and angles before doing the high level (and time-intensive) modeling with reflections.

When we enter the mapping window and select standard modeling, we have a range of options before us. We can select how many speakers (and which ones) to model, which audience areas, faces, and so on. You can run models that will make available all frequencies, or just one; you can model direct SPL, or every acoustic measurement the software is capable of; you can elect to run a model with one-third octave noise bursts, or broadband noise, or with what AFMG suggests is most accurate to our hearing when "walking" the space—three octaves. Again, it is beyond the scope of this book to go into great detail about how and why you would select these various options, but to start, let's assume we're going to use the default settings, model with our vocal speakers, and run the comprehensive set of models on our audience areas and stage surface.

Figure 34.15 is the map of 1 kHz direct SPL.

In running models, we find that the four-box array was too narrow in vertical coverage, so we upped to a five-box array for this image. It appears there is some splash onstage, as we anticipated, both from those 4Ses and from the array itself. Let's take a look at a couple more frequencies (Figure 34.16).

The pattern gets a bit more controlled the higher in frequency we go, but we still have some splashback. Looking at our array, we are covering fairly evenly through the orchestra section (though we would definitely need under- and over-balcony fills). Just for the heck of it, let's replace our high center array with a single-point-source box that covers most of our angular needs—the d&b 12S, with a pattern of 110° x 55°.

As we can see (Figure 34.17), this is a far inferior solution.

The box, though "covering" the angles we need, evinces a strong hotspot with rapid falloff. I modeled this in several positions, with several angles, and Figure 34.17 was as good as it got (other times, there was a gap between front fills and the hotspot, or the hotspot crept onto the stage too much). While there is also a hotspot in our array model, we can shade boxes to avoid this, or know that we can simply run everything at around 90–95 dB SPL, which is available basically across the whole orchestra section, without worry. This

353

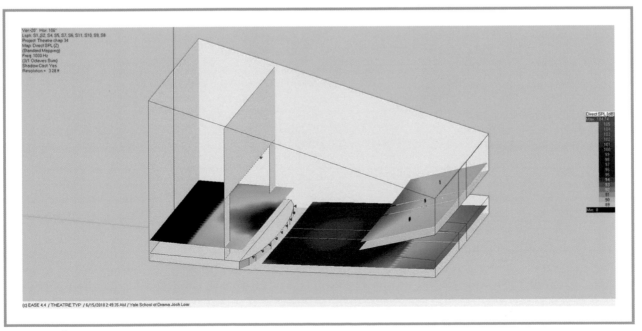

FIGURE 34.15
Vocal System Model #1, 1,000 Hz, Direct SPL
Source: Image From EASE®

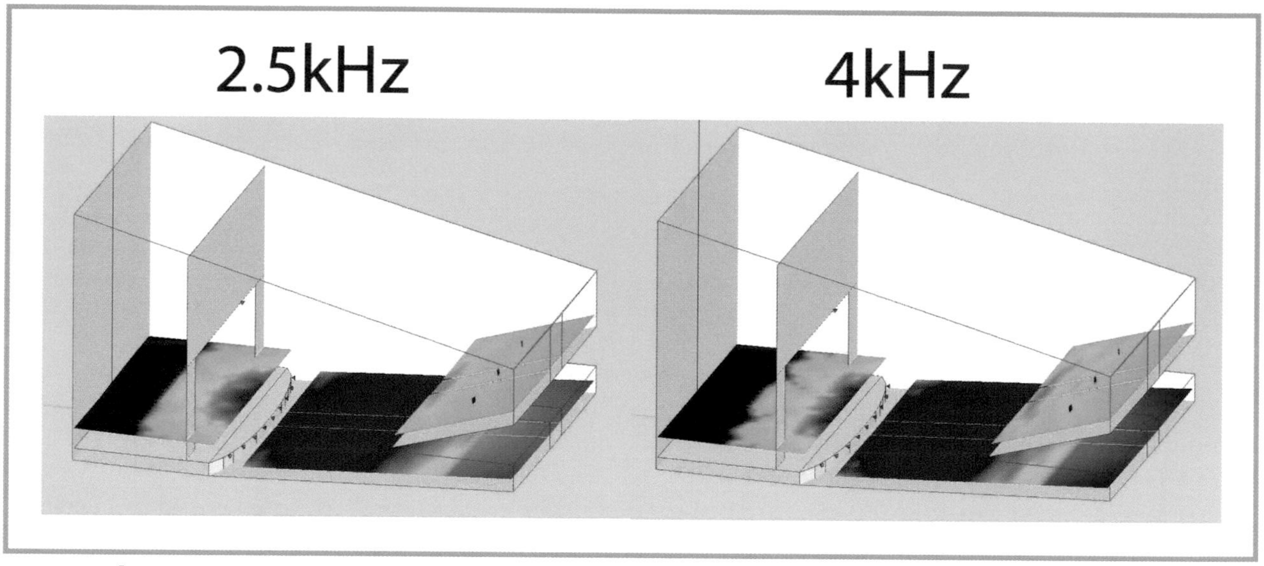

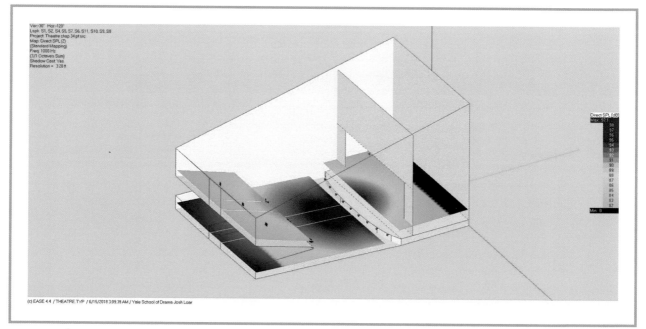

FIGURE 34.16
Vocal System Model #1 (Left) 2,500 Hz; (Right) 4,000 Hz, Direct SPL
Source: Image From EASE®

FIGURE 34.17
Vocal System Model #2, 1,000 Hz, Direct SPL
Source: Image From EASE®

is the value of modeling: without running models, in a vacuum, I might have thought that the 12S would be a perfect solution for this position, but as we model it, we can see how uneven the coverage is, just from the statistical model! The array is more complex and will take more time to refine and get right, but once it is configured properly, it will clearly work better than the single-point-source option. Poor option avoided, thanks to speaker prediction!

At any rate, once we've run the vocal system up to our satisfaction, we would also then model the music system to examine its performance. Since I'd rather this book doesn't become several *thousand* pages long, we'll skip that for now—but know that it's just the same process again.

It is *well* beyond the scope of this book to go into how to read models and adjust the position of speakers to achieve the best coverage.[17] Suffice it to say that we will continue running the statistical models and making small adjustments until our coverage looks about like we'd hope in terms of level, evenness, and frequency response. A system that seeks even response for the whole audience should see no more than +/−1.5 dB of variation across the coverage area. This isn't always possible, but this is often our goal.

8. RUN THE RAY-TRACING!

Once your basic choices are set, run the detailed models. These will reveal properties of the space that may or may not cause issues. You may find a strong reflection off a hard-back wall causing a strange summation or cancellation band in your audience. You may find more attenuation of HF signal due to curtains hung around the venue. There is no end to what you may find, and you will only find it in the modeling and prediction stage if you've built an accurate model and run it with intelligent consideration of the settings. It takes time to learn how to configure a successful model, and how to read and interpret the results. However, don't let inexperience stop you. The only way to get better is to try. If you work in a theater (or attend school where there is a theater), model a single speaker in that theater, and then mount that speaker in the theater and test your model for accuracy. See how closely your model matches what you actually find in the venue. If you can get one speaker working properly, you can add a second, and a third, and so on.

9. REFINE

Improve your aim, speaker model selection, and other settings based on what the model reveals. While taking care not to exceed production deadlines, run as many models as necessary to get your speakers properly configured.

10. HARVEST THE DATA!

Once you've got a successful model system (or set of subsystems!), carefully collect all the data—speaker models, positions, angles of aim and splay, and so on. You can do this in EASE® by running a report, or you can manually copy down the data. Whatever you prefer. Use the info and turn it into a key part of your loadin paperwork.

An important note: while we've only examined the house system in this chapter, any onstage monitor speakers would go through the same process. Many early designers model their house system and forget to model their monitor system. This often leads to monitor systems that are under- or overpowered for the needs of the show, feature too many or too few monitors onstage, and so on. The monitor system is a speaker subsystem like any other and should be treated accordingly.

EASE® and other modeling software is not just useful for your own design work but is increasingly a critical tool for pitching concepts to clients. The ability to produce high-contrast multi-color images representing sound coverage is a way for our notoriously invisible medium to become visual. Clients (the people whose money we want) don't often have a sound background. Telling a client that sound will be 100 SPL is not as powerful as showing a smooth and even color map of that SPL across multiple frequencies. As the budgets for your projects rise, it becomes ever more critical to impress upon producers the necessity of our technology. One of the signature tools in our arsenal is prediction software.

Once we've selected our speakers, planned our coverage, and verified that we are allowed to mount speakers where we have planned, it is time to flesh out the rest of our system!

Notes

1 This is true of theme park designs as well, with the difference that the set is often not in a state to be drawn and delivered to you yet while you are designing your system.
2 Unless your conference rooms are *way* more interesting than the ones I've been in.
3 Students often worry about the use of EASE®, not only because it seems (and can be) quite complex, but be-

cause the software is not inexpensive to buy. However, AFMG has done a wonderful thing and made it so that one purchase of EASE® is a lifetime license to any and all upgrades. Thus, while it is an expensive software to acquire up front, you can continue to use it forever, so long as you keep importing the latest speaker models. Be aware that while much audio software is Mac-only, EASE® is PC-only, and unless your machine is super-powerful, it doesn't do very well working in virtual desktops like Parallels or VMWare. Nevertheless, it is the industry standard.

4 EASE® does not actually model the entire audible spectrum. It models from 100 Hz to 10 kHz. It assumes that anything below 100 Hz is going to be *so* dependent on room acoustics, on sub-array configurations (which are typically spaced arrays, and thus challenging to model accurately), and so on, that any models it produces at those frequencies won't be terribly useful. Since EASE® was first released, mathematical modeling has only gotten more and more impressive, so I wouldn't be surprised if an upcoming update allows a more robust bass-end modeling, but for the time being it is not available. Beyond 10 kHz, it is assumed that these frequencies (while desirable for a sense of air or openness in your system, especially around voices and trebly musical instruments), carry no particularly critical *information* and as such can be left out. It is also assumed that because the wavelengths are so small at that frequency range that a model (which has to, by design, make some generalizations about architectural details—about which we will discuss more later) won't be accurate enough at this HF end to be worth modeling.

5 A common error among new designers is to model an "array" by entering several of the same single boxes and placing them on top of one another in the model. Speaker models should always configure an array in the sub-menus of the array speaker itself, so they truly perform as a combined unit like you would find in the real world.

6 Different prediction software will offer different options. Everything I mention here is available in EASE®.

7 A first-order ray-tracing model calculates one reflection after sound leaves the speaker, a third-order will trace three reflections, say from speaker, to back wall, to floor, to side wall. The greater the number of rays and the higher order of the model, the more acoustically accurate the model will be—but also the greater processing power and time required for that model. It is common practice to run initial tests of speaker models using the statistical modeling tool, and once you've got a plan mostly worked out, then run ray-tracing models only on speaker setups you are mostly confident about. This will be discussed more as we lay out a systematic plan of attack for predicting speaker performance later in this chapter.

8 At each modeling step, you need to check to ensure that what the model returns matches roughly what you'd expect based on your knowledge of the speakers in question and physics itself. Sometimes a model will reveal an acoustic phenomenon you weren't expecting, but one that makes perfect sense if you think about it in detail. However, sometimes (especially when you are just learning prediction software) you can make a simple mistake and the model will return wildly incorrect results. You need to be aware of how a speaker *should* perform, roughly, so you can identify when the model is causing a problem that wouldn't actually happen acoustically. One common version of this problem in EASE® is that by default, wall surfaces are "single-sided" meaning they are only acoustically active on one side of the face object (typically the one facing into the venue). It is very easy to accidentally reverse the faces, and then run a model where no sound shows on the walls at all, because the face is aimed the wrong direction, and the wall of the interior of your venue has no acoustic properties at all. In this particular case, all one needs to do is select the wall in question and run the "invert" command in EASE®, but this *kind* of error is common in young designers' first models (and I have seen this particular error in young designers' models more times than I can count).

9 If you don't know *A Midsummer Night's Dream*, go read it! Or at least read the SparkNotes or something, so you know what I'm talking about here.

10 Thanks a million, M. C.!

11 We'll go d&b for the vocal system, and change manufacturers for the other subsystems.

12 Another is Meyer's MM-4XP, which is slightly smaller even than the 4S, but when designing subsystems, it is easier to keep a coherent "voicing" when keeping speakers all from the same manufacturer (and/or specific product line). That said, there are plenty of times when the demands of form factor or coverage or SPL will mean that you end up with a hybrid system, consisting of multiple manufacturer's products. In our sample system here, we will be keeping coherent manufacturers within each modeled subsystem type, but we'll be mixing manufacturers across the whole system.

13   I have in some unconventional situations used speakers with very wide horizontal coverage and very narrow vertical coverage as front fills with success. Consider the JBL CBT-50LA1. This is a weird, contractor-grade speaker, that most people would never think of for a front fill, however, with a suitably high stage and actors playing very far downstage (and speaking quietly), these have been successful in several theatrical productions as front fills. Typically, we want to use a speaker with less beaming, and with less of a visible/sizable enclosure, but as James Brown always said: "you've got to consider the case!".

14   You will note that my coverage plan is primarily focused on the orchestra section of seats, not on the balcony. Let's imagine that the balcony exists, but is almost never sold due to low attendance numbers. I have built the EASE® model to include the balcony area, as the resonant space is relevant to our coverage, but I am mostly concerned (for the purposes of this chapter) with orchestra-level coverage. If this were to proceed into a full design model we would need to pay more attention to the balcony area.

15   Which they refer to as a "tutorial."

16   This kind of time-alignment is very common for "delay" speakers such as under-balcony fills. When time-aligning between front fills and high centers, there are a number of challenges to consider. If we delay the front fills too much, the sound will arrive after the actors' natural sound, which will be distracting. If we don't time-align the speakers at all, we may have a double-strike from the early arrival of the front fills followed closely by the arrival of the center array. This will also depend on level. If by the time an audience member hears the high-center array clearly it is louder than the front fills by at least 10 dB, then the front fills will just form part of the acoustic environment (so long as the arrivals from both speakers are within the roughly 30 ms Haas capture range, beyond which we hear distinct echoes). This is a tricky proposition, and modeling can help us work out the best solution, but a model is never going to tell us an answer to a question we don't ask, so we need to be aware of this challenge going in. Additionally, if we allow the front fills to arrive just a bit before the high center (but again within the Haas range), we can help draw the sonic image down to the stage instead of up in the air).

17   Bob McCarthy's book is the bible on this subject. Seek it, learn it, build statues of Bob in your theaters (only kidding about one of those).

CHAPTER 35

Selecting the Remaining Equipment and Plan for Operation

It is reductive to assume that all the considerations necessary for selecting all production equipment aside from speakers can truly be addressed in one chapter. In fact, most of Part I is focused on this question.[1] That said, in our seven-step process, selecting remaining gear, and planning for the operation of said gear, is critical to our success. What follows in this chapter will be broken into two basic sections. The first will address in simplified form some of the considerations we need to examine when selecting the remaining production gear for our system. The second will examine in a bit more detail the process of planning the operation of your system, including staffing.

SELECTING THE REMAINING GEAR

Think back to the primordial days of Chapters 1 and 2, and the four standard analog and four additional digital elements that make up a sound system. When trying to wrestle the giant pile of potential gear floating in our brains down into manageable chunks, it is helpful to work through our needs in categories based on these elements. We will begin with the analog elements, because, as we know, all sound systems will feature some version of them. We'll then move onto the digital elements, because as we also know, *just about* every sound system these days is digital at some point or another.

Just to refresh our memories, the analog elements are as follows:

- Inputs
- Processing
- Amplification
- Outputs

Since we've already specified our speaker needs in the previous step of our process, our outputs are covered. Since we know what kind of outputs we will be using, selecting amplification becomes rather simple. Any speaker that needs to carry its own channel of content needs its own amplifier channel. Any self-powered speaker is already taken care of. Any passive speaker will need an amp channel that is rated for the impedance of the speaker and that ideally will provide more power than the max power rating of the speaker.[2] If we're designing a 70V system, we will need fewer amps than if we're designing a high-fidelity concert or theatrical system, but the selection of amplification should be simple.

From there, we move to examining our inputs. What are the input sources? Do we have playback? If so, how many channels, and of what type (are we playing music from a cell phone via 1/8″ connector? Or are we playing 48 channels of sound FX via QLab? Something else?)? Do we have mics? What type? How many? If we are a playback-only show, can the playback system connect directly to DSP (as in a QLab over Dante Virtual Soundcard setup)? Or does it need to run through a console so that an operator has tactile control of the signals?

If we have mics, we'll definitely need a console (at least one, possibly two if we have a show with complex monitor needs, or at least three if we are also sending a recording and/or broadcast split to a production truck or booth). How many input channels of what type do we need? How many output paths of what types do we need? What kind of automated features do we need (scene recall, digital scribble strip labeling, etc.)? What kind of effects does the console feature, and do we need to supplement it with any third-party plug-ins or outboard effects?

Now, what were those 4 digital elements again?

- Converters
- Advanced Processing
- Control
- Clocking

Beginning with converters, if you are using QLab but sending analog, you might need a computer audio interface to convert digital to analog. How many channels do you need to send? What kind of computer connection do you need? What sample rates do you need to support? A digital console will generally have converters already built in, as will DSPs.

Advanced processing is generally performed in our DSP How many ins/outs do you need? What kind of sample rates/bit depths do you need? What signal types does the DSP need to handle? What kind of computer is required in order to program/configure the DSP?

Control is key! How many of your digital devices can be controlled remotely? Consoles, wireless mic receivers, DSPs, playback computers, and more are network-enabled, and when you have the ability to control devices remotely, it is almost always desirable to include that feature in your system.

Clocking might be simple—one Dante sender to one Dante receiver will self-clock and be easily managed. A giant distributed system, however, might require an external master clock, and the concomitant cabling.

We also mustn't forget com![3]

Since we spent so much of the book already[4] examining each of these gear selections in detail and discussing how to plan for their selection, it would be foolhardy to try to recapitulate the entirety of that work here. Suffice it to say that I place the selection of all other production gear AFTER the selection of speakers because the speakers are the most important part of your system.[5] Of course, it is entirely possible that these steps will happen in reverse order, or even interlock and happen simultaneously—I break them out as discrete steps in an ordered process here for clarity of mental organization, not because production always works this way. I have certainly designed systems where I knew the intricacies of all the inputs before I had selected a single speaker, but that didn't displace how important selection of the speakers was to the success of those systems.

PLAN FOR OPERATION

We now move into a fresh topic. How will our system be operated? What staff is needed to ensure its smooth operation (and what training will that staff need)? How long will staff have to be on the job—does the mix operator need to be at a soundcheck and show? Does the installed system require someone who understands basic operation on site 24 hr a day, or does it only require human touch under special conditions? The system designer will specify

> **Message in a Bottle**
>
> Don't bite off more than you can chew in terms of things you don't know, particularly when you're early in your career. But, you know, hopefully, most of the time we're challenging ourselves and we're learning new things and we're trying out this new piece of gear or a new theory or something. But when you're early in your career, there's *so much* that you don't know. And when you get your first couple of gigs where you've got—not carte blanche, but you've got the ability to design your system—I've seen people . . . I've seen my students do it fairly frequently, where they want to add *all* the bells and whistles. And they don't necessarily have the ability to think clearly about time management for budgeting that in. And so they end up getting in the weeds, and they're missing the big picture because they're so focused on the little things.—Vincent Olivieri

needed staffing as part of the design process, so we must examine some basic types of systems and the common staffing needs for those systems' operation. Keep in mind that this is not a comprehensive list, but a general guide—for every show that typically needs one sound operator, there is a similar show that needs three crew members, and yet another that can have a combined lighting/sound operator with no need for dedicated sound staff. Each production or installation must be examined in terms of what the system needs to accomplish, and what human interaction is required in order to smoothly accomplish those goals. It is the goal of the system designer to design a system that operates with the smallest crew possible, both for fiscal and efficiency reasons, however, a system designer who underestimates their crew need has made a grave error—and that error will come back to bite them when the client can't operate their system as desired with the staff as specified. No client likes to have costs *increase* after the fact, and when in doubt, it is generally better to plan for slightly more personnel than you believe you may need. Staff can always be cut if you find the system can be run with a smaller number than you originally planned, and if the budget has to change in one direction after system commissioning, it should always be to get smaller (if you want clients to keep hiring you, anyway).

As we examine different types of systems, we will first list typical staff positions found for each, and then we will ask the following questions:

1. Are there parts of this system that need human artistry in operation? These would include mix engineers for creative productions (music, theater, etc.), or the like.
2. Are there parts of this system that require physical human intervention? Such as wireless microphone crew to mount mics on actors for a theatrical production.
3. Are there parts of the system that need to be monitored for safety? Theme parks are key areas where this might be the case.
4. Are there parts of the system that might benefit from human operation, but could be automated if need be? These are tricky situations, but some exist—for example, in a stage show with very simple cues, can the cues be programmed to be fired by the lighting console, saving money on a dedicated playback operator? Or will the audio cues need to be fired at times in contradiction to the lighting cues, such that having one operator perform both is impractical?

While standard crew positions for a given type of system's operation are worth knowing about, every designer should always ask these questions before turning the stock position list into their real operation plan.

SYSTEM TYPES

- Live events:
  - Concerts
    - Standard crew positions:
      1. FOH mix engineer/A1
      2. Monitor engineer/A2

* Occasional crew positions:
  1. Additional A2s/mic technicians
  2. System engineer (manages connectivity of systems and ensures fluid operation during setup and event run)
  3. Recording engineer
* Standard questions:
  1. Human artistry? Definitely. Placement of instrument mics and mixing of both FOH and monitor feeds require significant human input in order to be done effectively.
  2. Physical intervention? Definitely. Again, placement and maintenance of mics, management of any wireless mics, physical control of mixing consoles.
  3. Safety monitoring? Somewhat. It is the responsibility of the FOH (and system) engineers to ensure that safe SPL levels are not exceeded. In a touring system, the system engineer may also be responsible for speaker fly systems and rigging, which is definitely an area of safety concern. The monitor engineer will have slightly less strict overall SPL concerns (some bands like stage volume well louder than any audience would tolerate), but if the band is using in-ear monitors, safety is of paramount importance.
  4. Possibility of automation? Not really. Real concert work is human from start to finish.
* Time considerations:
  1. Soundchecks—which artists get a soundcheck? Are these full-length soundchecks including the setting of monitor mixes, the playing of individual instruments as well as a group song, etc.? Or are these line checks, where the band gets onstage, quickly verifies each needed line and mix is working, and then either immediately starts their performance, or has the settings saved in consoles and clears the stage for the next act? Or some combination of these? Full soundchecks may take anywhere from 30 min to several hours depending on the act, the complexity of the setup, and how many other artists are sharing the bill.
  2. Loadin/tuning—if you are on tour, you may well be loading in your sound system at a new venue each night, and thus will need to quickly tune the system each night as well. Time for this will vary considerably depending on the complexity of the system and venue schedules. If you are not on tour, but doing shows in an existing system, this time is reduced substantially, if not eliminated.
  3. Stage setup—how long do you need in order to literally set up mics, stands, monitor speakers, and cables? Will increasing your crew count shorten that time, or (as is sometimes the case with inexperienced crew or very small spaces) increase it due to logjam onstage?
  4. Show itself—however long the show is, your run crew must be there!
  5. Strike—how long will be needed both to strike the stage setup, and (if on tour) the system itself?
* Theater
  * Standard crew positions:
    1. FOH engineer/A1 (if microphones or other live sources are in use)
    2. Playback operator
    3. RF crew/A2s—wrangling wireless mics, in-ear monitors, and the performers associated with them
  * Occasional crew positions:
    1. System engineer
    2. Secondary FOH engineer (on some large musicals, though this is falling out of fashion in the age of digital consoles)
  * Standard questions:
    1. Human artistry? This depends on the type and complexity of the show. On a small straight play, with only playback cues and no live mics, the artistry element to the actual running of the show is very small if not nonexistent. In fact, for very small shows, sometimes the sound cues are fired via MIDI trigger from the lighting console, by the lighting operator, and thus there is no dedicated sound crew. On a large musical, with both live musicians and singing performers, and substantial effects playback, significant artistry may be required.
    2. Physical intervention? On any show with wireless mics (or mics of any sort)—yes. On a playback-only show, it really depends. Each night at the start of the call, the playback operator (and mix engineers if the show has them) will run a speaker test. On a show where the lighting op runs the limited sound cues, that person still needs to be able to understand sound well enough to try to troubleshoot a problem if a speaker doesn't fire as needed during pre-show test.

3. Safety monitoring? Like any live event (concerts included) there may be need of a God mic to make emergency announcements if the venue needs to be evacuated. Theater shows are not typically as loud as concerts, so SPL monitoring is not an issue in 90% of cases (though in that other 10%, it can be vital). Again, if any performers are using in-ear monitors, strict care must be taken to not overload the signal and damage the performers' hearing.

4. Possibility of automation? Increasing every day. Playback is already run typically via systems like QLab, and the goal of most such systems is to make the running of the show as automated as possible, while still leaving flexibility for actor timing. Consoles with programmable scenes have become standard across productions of all types. Show control systems feeding timecode or other triggers to the sound system have become quite common. Some shows—like theme park "theatrical" shows—often run everything on a stock timeline, such that once timecode is started at the beginning of the performance, everything else is entirely automated. That said, there still generally needs to be someone on site with some knowledge of the system and operations in case of error or emergency.

* Time considerations:
1. Nightly speaker check—just long enough to run a quick coherent sound (usually a recording of someone saying, "Speaker 1," "Speaker 2," etc.) into each speaker in order. This must be done early enough before a performance to allow for problem-solving time if a speaker does not fire.
2. Mic prep—if there are mics on actors, they must be rigged and placed each night, and hidden in the costumes. Depending on the costumes and cast size, this can be quite a lengthy process or quite a short process.
3. Mic check—each actor (or musician, in the case of a pit) testing their mics each night before the performance. Not as long as a full-length concert soundcheck, but still needs enough time to get through all sources and have problem-solving time if something isn't working.
4. Show itself
5. Mic takedown—getting mics back from actors and stored properly prior to the next night's show

• Conventions and other multi-room live events

* Standard crew positions:
1. System engineer—for any multi-room live event, a system engineer is typically on-site to ensure that all parallel subsystems continue operating as intended.
2. Mix engineer(s)—if there are convention stages, each stage will likely have its own FOH engineer, and sometimes its own monitor engineer, depending on the complexity of the presentation on each stage. In addition, there may be a mix engineer for the convention-wide system (for announcements and the like), or this may be handled as a simple routed patch by the system engineer.

* Occasional crew positions:
1. RF crew/A2s
2. Assorted assistants and runners (sometimes a communication link between two rooms breaks down, and it is helpful to have some runners on site to help not only repair the link but also to send messages from room to room in case of emergency).

* Standard questions:
1. Human artistry? Unlikely. Occasionally, conventions will feature musical performances, in which case that portion of the convention is to be treated like any other concert, but usually conventions are more cut and dried. There may be need for some human operation to manage large flows of signals, but "artistry" is a stretch.
2. Physical intervention? Very possible. Conventions often need RF mics placed on guest speakers, and the like.
3. Safety monitoring? Inasmuch as there will be a need for robust emergency announcement systems, yes. Otherwise, not generally.
4. Possibility of automation? Yes. Much of the work of distributed convention-style systems can already be handled by the more sophisticated automixing and routing products, and while they still require system engineer(s) to monitor their operation overall, there is a lot about a convention system that doesn't require human intervention if designed properly.

* Time considerations:
1. Installation—convention centers might have a baseline announcement system, but your design may need to tie into or expand upon that. Installation of large distributed systems takes more time (and more crew) than for a single-room concert.

2. Event run—how many days? Do you need multiple shifts of crew members because days are quite long?

3. Strike—again, more time than a single-room event.

■ Permanent installations: permanent installations are a different equation, as they usually will be designed to host a range of events (in the case of performance spaces) or to run for a very long time either semi- or completely autonomously. For some of these, like performance venues, we'll identify *likely* staff, knowing that each event itself will have to evaluate its individual needs. For some, like theme park attractions, we'll address what staffing is likely to be needed, though this staff may not be a dedicated *audio crew*, but rather an *operations staff* that monitors audio systems as part of the overall management of the attraction. Time considerations for permanent installations vary wildly from project to project, depending on the scope and scale of the installation. We will address likely installation timelines for permanent systems in Section 12. As far as the events held in venues with permanent systems, they would need to address the same time considerations listed earlier for live events. As such, while we will continue addressing our four standard questions for each type of system that follows, we will leave aside time considerations for now.

• Performing arts venues and houses of worship
  * Likely staff:
    1. Mix engineer(s)
    2. RF/A2s
    3. System engineer(s)
    4. Possible recording engineer(s)
  * Standard questions:
    1. Human artistry? Yes, generally.
    2. Physical intervention? Frequently.
    3. Safety monitoring? Same for any live event.
    4. Possibility of automation? Some in multimedia HoW (House of Worship) shows that may repeat several times per day identically (like theater), otherwise, not much.

• Retail environments, airports, and other public facilities
  * Likely staff:
    1. System engineer(s)—for large public facilities (airports are a good example) it is typical that there will be at least one person on staff who is familiar enough with the system to do basic troubleshooting and maintenance. It is unlikely, however, that this person is very advanced, as integration companies are often brought back in when systems like this need substantial adjustment or maintenance. It is very unlikely that standard retail establishments (grocery and clothing stores, for example) would have anyone dedicated to the audio system on staff, or even anyone with more knowledge than how to operate the user panels. Otherwise, it is unlikely that there will be any audio staff in these environments.
  * Standard questions:
    1. Human artistry? No.
    2. Physical intervention? Highly unlikely.
    3. Safety monitoring? Of the spaces themselves, yes, of the operation of the system, probably not.
    4. Possibility of automation? Very likely. Any sound operations beyond a cashier needing to make an announcement over the PA are likely to be automated with so-called sunrise/sunset features, such that the automated background music (BGM) is set to start automatically based on time of day just before the store opens, and to end just after store closing. These systems try to automate everything possible.

• Theme parks, cruise ships, casinos, and other themed environments
  * Likely staff: This will vary based on what each of these venues offers. For example, a theme park that only features rides but no concert stages will feature fewer audio staff members than one that features performances in addition to rides. For any live performance elements, see the aforementioned live event considerations, as they again repeat here.
    1. Live event staff—see the aforementioned.
    2. Ride operators—for theme park rides/attractions[6] there will typically be at least two operators on even the simplest attractions: one at the guest loading/unloading area to help guests get situated in vehicles, and another in a control booth. It is the latter who will typically monitor and/or interact with audio systems. For very complex attractions, there will be several other operations staff, and systems for these attractions may allow for muting or paging to individual ride "scenes" in case of need/emergency.

3. Venue or park operations staff—equivalent to ride operations staff, but for non-ride attractions or environments. These might be the staff running a largely automated water/projection/light show in a theme park, or the staff monitoring the overall system health of always-running projections and sound systems on a cruise ship. The larger and more elaborate the venue/environment, the more likely that there will be operations staff whose purview is specifically sound systems (or at least AV systems).

* Standard questions:
  1. Human artistry? Depends on the environment. Anything with a live show—sure, though live events in themed environments tend to try to automate as much as is conceivably possible for repeatability (and so they can pay the staff less because they aren't as highly skilled). For rides? Highly unlikely. For continuous content streaming (like video/sound installations lining the walls of a cruise ship passageway)—no.
  2. Physical intervention? Again, live shows—possible. Anything else? Ideally no. There should be no need of a human touching or moving any piece of equipment in a themed installation.
  3. Safety monitoring? Constantly. This is the principal job of ride and venue operations staff in these environments. Themed environments are inherently dangerous because guests can easily get distracted and overwhelmed,[7] and thus find themselves in need of intervention. While ride systems are designed with sensors to help alert operations staff to any deviations from the norm, operations staff is still needed to monitor CCTV cameras and intervene in attraction performance if and when necessary. Of course, sometimes a ride will break down with riders on the track, too, in which case operators are definitely needed to intervene with systems, overriding FX and BGM to make emergency announcements and help guide guests to safety. Even when there are no rides involved, any themed environment is still a lot to take in for some guests, and safety monitoring is one of the key operational jobs in such venues.
  4. Possibility of automation? *Hell, yes.* As much as possible. A well-designed theme park ride or environment should be almost, if not entirely, automated, with the exception of emergency intervention systems.

- Recording/mixing/mastering studios for music, film, etc.
  * Likely staff: too many to list here. Recording, mixing, and mastering engineers. Producers, technicians, etc.
  * Standard questions:
    1. Human artistry? Of course.
    2. Physical intervention? *Mais oui.*
    3. Safety monitoring? Soundstages for filming and other such production—sure. Music and post-production audio studios—unlikely.
    4. Possibility of automation? File backups are often automated in such facilities. Otherwise, generally no.
- Corporate offices, classrooms, and other architectural environments
  * Likely staff:
    1. System engineer: college campuses, large corporate facilities with lots of AV, and other such environments may well hire a system engineer (or several, depending on the size/scale of the enterprise), smaller environments are unlikely to do so.
    2. Operators: any such venue that regularly changes states (e.g., an office building whose conference rooms regularly need to be reconfigured for events of different sizes, different numbers of presenters, and different teleconferencing options) may hire dedicated operators.
  * Standard questions:
    1. Human artistry? Very little to none.
    2. Physical intervention? Some, for larger organizations that reconfigure spaces.
    3. Safety monitoring? Not likely, though AV systems will be designed to interrupt when building fire systems trigger.
    4. Possibility of automation? High. Typically, as many parts of these systems that can be automated will be automated.
- Creative systems—multimedia art galleries, new media, new uses
  * Likely staff: this varies so wildly depending on what the actual installations are doing. There may be operations or system staff involved, or these may be designed to have zero human intervention. It is difficult to put a finger on this in any definitive way because the creative content (and thus the system configuration) will cover a wide range of possibilities.

* Standard questions:
  1. Human artistry? Sometimes.
  2. Physical intervention? Sometimes.
  3. Safety monitoring? Sometimes.
  4. Possibility of automation? High.

The details of staffing a system's operation will often be very determined by the system designer (for theater, concerts, art installations, etc.) and sometimes will not be determined by the system designer at all (theme park attractions, hotels, etc.); that's tough in the latter case because the system designer still may be consulted regarding what they anticipate the staffing needs will be. Regardless, it behooves the system designer to always have a clear understanding of the operational needs of what they are designing. If you design a system but have no idea how many staff (and what kind of staff) will be needed to make sure it runs as intended, you are essentially rolling the dice with whether your system will *ever* run as intended—and thus rolling the dice with whether or not your client will be happy with your work. I have definitely designed systems that required too many operators for the client's wishes, and when they made clear to me that they would not hire as many staffers as needed, I simplified the design so that it could be operated by the appropriate number of people.

Now, assuming you've selected speakers, you've selected the rest of your gear, and you've figured out how your system will be staffed, it's time for the final revisions and refinements. Some of this is a good and desirable process—the process of thinking through your work, eliminating anything unnecessary, adding anything critical you may have forgotten, and being as clear-eyed as possible about how your system will work. Some of this might be painful—matching our desires up with budget restrictions, whether because our numbers ran over the target, or because the client comes back with a lower number than they'd initially said. Whatever the reason, we must remain diligent through this final design stage, or our designs will suffer for it.

Notes

366

1 If you are a reader who skipped ahead to Part III but still have some questions about how to select a given piece of gear, going back and reading Part I will help!
2 For more on why I suggest the amp should be overpowered compared to the speaker, see Chapter 15.
3 See Section 7.
4 Again, see Part I.
5 For any installation or live event system, that is. If we're talking about studios, often the recording devices are the most important part, followed *very closely* by the speakers, but since this book is mostly focused on live and installation systems, I will keep blowing the tune of speakers being the most important.
6 Every theme park ride is an attraction, but not every attraction is a ride. An attraction, in the parlance of the industry, may be a character meet and greet, a parade, or really, any themed "event" or space of any kind within the overall park environment.
7 Or just stupid, as in the case of the idiots who get out of ride vehicles and try to climb up to shake animatronic Jack Sparrow's hand and get very wet, or worse, stand up during a roller coaster run and end up seriously hurt.

CHAPTER 36

Refining Your Choices

You've designed your perfect system. You've specified your gear, planned your needed crew, and prepared detailed technical drawings. Then, someone who has never touched a mixing console in their entire life comes and tells you that you can't have one-third of what you think you need, because they can't or won't pay for it. ARGH!

Thus we enter the refinement stage of design, which can be (and often is) the most frustrating part of any design process. In an ideal world, any changes to the system design after you've made your proposal are changes that you are making based on the evolving creative needs of the project—you've designed a theatrical sound system counting on playback music, and a live band gets added, so you need to add a bunch of gear to properly handle that band— but in the real world, this is only one of the reasons you may need to refine your choices. The principal reason we end up changing our designs after first submission is because the budget changes. The second most common reason is that the project's goals change. The third is because we find a more elegant and efficient solution to something we've designed and want to change it.

Whatever the reason, when you are forced to change a design, it can mean a lot of work. It is challenging enough to get a design completed, with all details accounted for, but to then change even small details can be quite a headache, as you must track each change through every piece of what can be a quite large document set.[1] Let's examine the three causes of design refinements and changes, starting with the least common, and moving to the most common.

FINDING A BETTER SOLUTION

Whenever you are working on a design, the goal should be to create a system with the most efficient, elegant,[2] and simple technological solutions possible for your given set of challenges. Sometimes, creating such a system will be easy. If you're designing a system for punk shows at a small bar, the needs of the system are straightforward, and while there is still seemingly an infinity of gear available from which to select, the ultimate system itself will most likely take on a conventional shape (FOH console, speakers, amps, mics, maybe playback, hopefully stage monitors, and ideally a stage monitor console). However, when designing complex systems, and especially systems with interactivity built into the premise, we are often faced with challenges for which no stock "off-the-shelf" product will solve our problem. In these cases, we find ourselves at our most creative, making use of all manner of electronic components in order to achieve our goals.

Imagine that we are designing a fantasy walk-thru environment for a theme park. The creative team has expressed the desire to allow guests to trigger sounds interactively and has shown us some initial concept designs of the scenic environment. It is decorated to look like an exaggerated jungle scene, with outsize rainbow-colored plants and glowing magical bugs and such everywhere. As such, this is not a mechanical environment, so the likelihood is that we're not going to want to put any visible switches or buttons in place because we will want our interactivity to be triggered in a somewhat "organic" fashion. Immediately, we might think about motion sensors, light sensors, proximity sensors—all kinds of tools we could use to trigger content from hidden devices. However, the producer says they want to have objects that guests can touch that will trigger sounds, but that those objects must just look like plants and bugs. How do we do this? Well, first we might think—"ok, if we have, say, giant banana trees, maybe the bananas themselves could be connected to actuators of some kind, and if you pull the banana, it triggers a relay in the stem." This could work!

This becomes the basis for our initial design (Figure 36.1).

However, as the design process continues, we talk with the creative team more, we think about the intended target audience for this environment. It's probably intended to be an environment to appeal to small children. The actuator devices we've designed may not be easy for small children to operate. In addition, even if it's NOT difficult for children to operate, it's a mechanical part, the other end of the child guest spectrum might be *rough* with the device, and the actuator/relay setup has a high chance of breaking. Designing a system that needs frequent maintenance is never ideal, especially when designing for an environment that is supposed to remain a stable installation for several years. After doing more research we discover transparent capacitive film—a thin membrane that can be affixed to any surface that acts basically like the surface of an iPad. If we shape the capacitive film around each banana, we can create interactive surfaces that trigger whatever we desire (and the effect is no longer limited to one special banana). It's simple, invisible, and most importantly, creates a "magical" interaction. Ta-da!

This is, of course, an esoteric example,[3] but a more prosaic version of this same process happens frequently. Maybe you are designing a system that requires a wireless radio onstage. The theater doesn't own any in-ear monitors, so you are forced to initially try a Bluetooth speaker (the budget is low). Bluetooth is a notoriously terrible solution onstage, it is a weak signal, with a short travel distance, and is highly unreliable, with devices un-pairing frequently

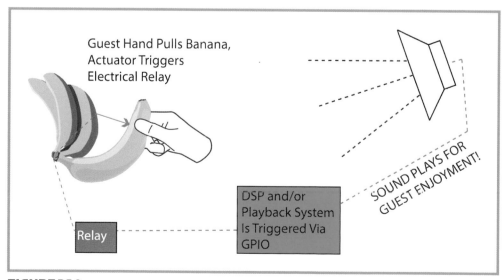

FIGURE 36.1

Banana Relay Actuator (Concept Diagram)

> **Message in a Bottle**
>
> The range of information that they [young designers preparing to design a system] could know is crazy and impossible to know as a young designer. What do they *need* to know? They need to know what they want. There is tons of information that I could say, "Oh, y'know, you need to know how to delay a system," but you know what, you *don't* really need to know how to do that. If you realize you need to delay a system, but you don't know how to do it, that's not all that difficult to learn really quickly. You need to know what you want, and you need to know how to communicate it. You don't need to read Bob McCarthy's book, but you do need to learn some of the foundations. You need to learn, like, the inverse square law, that sort of thing. It's OK if you forget it, you're going to remember the principles. To have some of that basic foundational knowledge, like, technical knowledge, in your system—y'know, in your body—not to necessarily have the calculus at the tip of your brain, but it's rumbling around in there somewhere.—Elizabeth Atkinson

during production. However, perhaps you speak to a local concert venue and can ascertain that they have an IEM they don't use regularly and arrange to borrow it or rent it for very little money. The battery-powered amp to go with it is very inexpensive (or the Bluetooth speaker is active), and you've found a better solution. Maybe you specify one speaker for a task, and just find a more ideal model for your coverage goals later on. We are constantly seeking the best technological solutions, and while it is always our goal to nail that on the first try, it's not always going to be possible—keep yourself open to new ideas as you refine the design.

PROJECT GOALS CHANGE

This is the easiest situation in which we refine our design goals. When a project's goals change, we refine (or completely change) our design to suit. Perhaps you are designing a theatrical system that called for a special speaker for a doorbell cue, and the director cuts the doorbell, preferring that an actor knocks instead. Perhaps you are designing a concert system, and the tour adds an opening act that requires some extra microphones.[4] Whatever the reason, when a project's goals change, we adjust. This is our job.[5]

BUDGET CHANGES

Dum dum DUUUUMMMMMM!

This is the most common, and most challenging, reason that we end up refining our design choices. It can be enormously frustrating to create a system that perfectly suits our needs, and then be faced with reducing functionality, or selecting devices that sound inferior to our original plans because the funds aren't available. That said, this is such a common part of our business, it behooves us to understand some of the reasons why a budget will change, and to anticipate this and plan strategies for the eventuality.

First, why do budgets change?

- Funding available to the producers has changed. Perhaps an investor has backed out. Perhaps another department on the project is running into overages, but the producers have decided the other department's overages are more important to support than our full budget.
- Funding promised by the producers was not secured in advance. This is an unfortunate circumstance, but sometimes you prepare a budget based on the understanding of available funds that the producers have imparted to you, only to find that such funding never existed, and was only speculative. While this doesn't speak well of the producers in question, this circumstance is still more common than we'd hope.
- Our design proposal goes beyond what the producers see as necessary. Sometimes, we manage to deliver core functionality at a lower price than the producers had allocated. Even if we found good, interesting, inventive uses for the remainder of our initial budget, the producers don't see that stuff as necessary and remove the funds from our line.

- The producers are not sold on our concept. This is a mismatch of our pitch/presentation with the producers' desires/expectations. Every designer will have their own style(s) of pitching ideas, and they depend heavily on your individual personality, the personalities of the producers, the project, etc. I can no more tell you how to get in good graces with a *particular* producer than I can tell whether the orange you ate for lunch today was tasty. That said, treat people with respect and be prepared (do your homework). You'll be ahead of the game if you can keep these thoughts in mind.
- The producers simply want to save money, regardless of the implications. This may be the most common situation—where producers would love to have all the functionality you've described as part of your system, but at only 2/3rd of the cost. Producers often just ask questions like, "Can we bring that number down at all?" without context.

So how do we prepare for this? There are a few things we can do before the budget reduction process[6] undermines the intended functionality of our systems.

- Pad the initial budget. To some (mostly producers), "padding" the budget (or increasing it above what is genuinely needed) is a bad word. While we never want to acknowledge that we've turned in a budget with ANY padding in it, it is also widespread professional practice to do so. If we're specifying MSRP purchase gear, it's hard to inflate individual costs, because those prices are generally widely available. In that case, we may specify a few pieces of equipment that are a bit flossier than what we truly need for baseline functionality or add some extra pieces beyond what is *absolutely* required for the operations. Nothing we specify should be useless, and we can't go hog-wild with this tactic, but it is a way to have some things to toss overboard if the ship starts to take on water (so to speak). On a rental system, this is usually done by adding gear to the system, as rental systems are usually priced *en masse*, rather than by line-item. Again, keep in mind with this tactic that you may be expected to defend all your choices, including any "extras," and if you defend successfully, you may be afforded the funds to acquire your "extras," so you need to be planning to use that gear if you get it. Thus, rather than inventing some wild unconventional technique that doesn't fit the show, sometimes, we just specify extra redundant gear, in case of failure.[7]
- Verbal persuasion. A lot our work, even as highly technical people, is really interpersonal work. Persuading someone to spend money on your ideas is a large part of our job, and the better you get at this, the more successful you will be at preserving your budgets. Depending on the producers you are working for, you may have to adjust your style. If you are working for producers on a major hip-hop tour, you may be talking to people who themselves have been a part of hip-hop culture for some time. They may favor conversational persuasion to written persuasion (though getting any decisions that you reach together in writing is still a good idea). They may favor casual language to formal. When working on a multi-million dollar architectural/themed project, chances are that formal language is a bit more important, that written communication is more important, and so on. These are BROAD generalizations. Each person is different and learning the communication styles of your teammates will go a long way to convincing them that whatever you present is worthy of funding. Some of this may involve so-called code switching, wherein speaking with one part of a team is done via one communication style and speaking to another part of the same team is done via a wholly

Message in a Bottle

Patience with directors and clients [is really important]—I think one of the really hard things is managing expectations that oftentimes are budgetary constraints. You know, you start with the best system in the world, and as it evolves, obviously there are budgetary cuts, and by the end of it, you end up with not exactly what you expected and sometimes have a disgruntled client 'cause you went from, for example, L-Acoustics, back to . . . down to . . . whatever, Behringer and others. And now they're disappointed because it supposedly doesn't sound good. So I think you have to manage expectations and have patience through the whole process. I also think clearly documenting things like conversations in email is really important. It's to just make sure everyone's on the same page through the entire process.—David Kotch

alternate style. The better you get at understanding communication styles, the more successful your work will be (anywhere in the entertainment business).

- Being aware of how your system can be delivered at lower than MSRP (or lower than initial rental price). Sometimes, you bid a system, and submit the middle or most expensive version as a rough estimate to the producers (stating that there are more quotes in the offing that have yet to arrive). If you know already that you can get your system for a lower cost than what you've stated to start, you can then undercut yourself when budgets start to reduce.

Now, sometimes, we negotiate, we pad, we barter, we cajole, and nothing works. Our budget is now smaller than our initial target, and there's nothing we can do except work out what to cut and how to save money. This process hurts but is still often necessary. In order to survive it, we must keep a few things in mind:

1. Maximize impact. If you've specified a very expensive console with beautiful advanced features but can save many thousands of dollars by changing to a much simpler console that will still handle 90% of the original's function, and you assess that the lost 10% won't tank the project, that's probably a better place to start cutting than reducing mic rentals or cables. That said, if the production won't work with a simpler console, you may have to thin-slice a ton of other gear categories to get where you need to be. Find the item or items that can have the most impact on budget reduction with the least impact on system performance and cut there first whenever possible.

2. Be clear about what functionality your system loses by cutting any given piece of gear. If you have to lose functionality, be clear and up front with the creative and producing team about that loss and what it means for their project. Don't ever conceal the loss of function from your team—they need to not only be aware of what a reduced budget will lose them in terms of the end result but also need to adjust their expectations so they are not planning on your system delivering something it can no longer accomplish.

3. Set priorities. If you must lose function, and there's simply no getting it back because there just isn't the money to do what you want, make sure you've set yourself clear priorities about which system functions can be lost easiest. That doesn't mean it won't hurt (either you or the system), but it does mean that you've assessed the *core functionality* of the system versus the less essential bits and structured your financial world accordingly. If I'm doing a musical in a very small theater, and I wanted to mic the whole orchestra for clarity of balance, but I also know the orchestra will be audible without mics, I'll cut those before I cut actor mics, because understanding lyrics and dialog OVER the orchestra is more essential to the success of the production.

We always hate to cut things, but in any creative collaboration, you must be ready to compromise. As I've said before, we must be ready to "kill our darlings" for the success of the greater project.[8]

Now that we've gone through the stock steps of designing any system, let's drill down and examine some of the specifics of different types of productions and installations, and what those systems (and processes) will call for as distinct from others. On to Section 12!

Message in a Bottle

The best advice that has been given to me before has been "it's good enough." And it's not saying . . . you're not quitting. You're just saying, "OK, we're ready to go on to the rest of this process." Some people think that they can solve it all beforehand and that's B.S. I think the funniest thing is, one of the number one complaints when you hire Bob McCarthy—who is the master—"We hired him because he's the master." And then he gets done and somebody's like, "But it doesn't sound good!" That's not the point. What he was here to do is to make it all sound *the same*. You still have a lot of decisions to make about what is good, he's giving you a linear system. He's giving you a blank canvas to work on. But still, there are just so many other elements that go together in the show to make it work.—Jamie Anderson

Notes

1 For more on documentation, see Part IV.

2 When I say "elegant," what I mean is streamlined, pleasant to operate, stable.

3 Not *directly* from my life, but pretty close!

4 Though, to be fair, sometimes opening acts on a tour are not afforded much. They may *want* more mics, but they may not be afforded more mics, depending on the budget of the tour and how important an artist the opener is or is not relative to the headliner.

5 There's not really much more to say about this, except that I want to remind up-and-coming designers that this is our job. Young designers get attached to their work, and sometimes are resistant to change. Our job is to serve the project, as a collaborator, not to serve ourselves. If you keep that in mind and remain flexible, not only will this career ultimately be easier, but you will be better liked by clients.

6 Often referred to as "value engineering" on architectural or long-term installation projects.

7 This is challenging when designing a touring system. You are seldom going to get away with specifying a whole backup console (unless it's a GIANT tour), as it will take up too much space on the truck—but specifying extra wireless mics, extra DIs, extra cable, extra outboard effects, and so on, are the kind of things that may help here.

8 I originally attributed this quote to William Faulkner, but in doing some digging, it seems rather uncertain who actually originated the phrase, so I present it here without attribution. Nevertheless, the sentiment is an important one—don't get so attached to your work that you can't change it when necessary.

Concerts, Small Scale

In this section, we will examine a variety of common sound system types. For each system type, we'll address the following:

- Defining the system type as stated in the chapter title.
- What versions of the eight standard analog and digital elements might we expect to find in such a system?
- What unique functions does this system need to perform, and how do we evaluate the production needs?
- What unique challenges do typical venues for this type of system present?
- What unique terms are used in the design process for this type of system?
- What unique gear types are commonly used in this type of system?
- What are the typical process timelines unique to this type of system?
- How do our designs need to be presented for this type of system, and to whom?
- What questions does a designer need to ask (and answer) in order to undertake a design of this kind?

Along the way, we'll show examples of system block diagrams and floorplans that represent typical cases of each system type.[1]

DEFINITION

What do we mean by a "small scale" concert system? Such systems are often conversationally defined by the size of the venues in which they are installed. A tiny dive bar that holds an audience of 50 people is definitely a "small" venue, where a stadium housing a crowd of 30,000 is clearly "large." However, to be more precise, we will here define "small" concert systems as systems designed for live music[2] reproduction that do not feature dedicated monitor mixing consoles, any provision for live recording, or any delay speakers.[3]

STANDARD ELEMENTS

In examining the eight standard elements from Chapters 1 and 2 and determining what versions of those elements we may expect to fund for a given system type, we will also find that some of these elements change drastically from one type of system to another (inputs, for example, will differ quite a bit from a large concert to a theme park ride), but some (such as amplification) will vary less from project to project. I have listed the elements here in the order in which they appear in the earlier chapters, which is to say the four analog elements first, followed by the four digital elements. This is not attempting to preserve any sense of signal flow order, as the digital elements typically interlace with the analog elements (advanced processing, which is often accomplished by use of a DSP, for example, usually happens between the first processing stage and the amplification stage). For more information on the signal flow order of such devices, see the chapters dedicated to those devices earlier in this book, and the sample block diagrams in this section's chapters.

- Inputs: microphones on performers and their instruments, DI boxes, talkback mic at FOH for announcements, two-channel playback for interstitial music
- Processing: FOH mixing console, possibly some outboard compressors or gates (if the console isn't digital and doesn't feature those internally)
- Amplification: standard operation amplifiers as needed (i.e., NOT a 70V system), or self-powered speakers
- Outputs: main L/R system, possibly with sub(s)
- Converters: as needed depending on which parts of the system (if any) are digital; built-in, not external
- Advanced processing: ideally a DSP, though at this scale, this step is sometimes skipped
- Control: if the console is digital, iPad control via wireless access point
- Clocking: if any system components are digital, usually self-clocking with either console or DSP as master, rarely (at this scale of production) external master clock

UNIQUE FUNCTIONS AND SYSTEM NEEDS

The basic paradigm of concert reinforcement, be it small or large, is a system consisting of left and right main speakers (or arrays). These speakers flank the sides of the performance area, placed downstage enough to avoid much risk of direct feedback from onstage mics. While L/R mains are the default system configuration, many concerts are run in mono or very limited stereo, because as performance spaces widen, the audience to the far house left might hear a keyboard part that is panned too far for the audience on the far house right to hear. As such, much concert presentation is in mono. Why, then, do we continue to use L/R systems? While large concerts often add front fills to help fill the front rows of the audience, and to an extent to center the sonic image on the performers, the simple answer is that L/R systems are the earliest established version of "immersive" sound, and we as listeners like to be enveloped or surrounded by sound. If I'm seated in the middle of the audience, and there is a speaker to my left and one to my right, even if they are playing the same content, the space feels larger than if it all just comes from one spot in the middle. Additionally, we can hang (or stack) speakers lower down, closer to the performers' vertical plane, by placing them left and right. If we wanted center speakers, they would have to either be directly below the performers (which works for front fills on an elevated stage, though signal from those is quickly absorbed by audience members in the front row(s)), or above the stage entirely, which creates a strange lifted image that isn't terribly pleasant. In contemporary large concert systems, we sometimes have a center cluster augmenting L/R mains and front fills, but we still want that sense of size that the split main pair provides.

Small concert systems are unique in that, unlike large concert systems, they are often not tasked with reinforcing every instrument onstage. There are scores of small venues whose sound systems are designed principally to reinforce vocalists in bands, counting on stage volume to carry drums and amplified string and key instruments through the space. In such systems, for example, there is probably not need of subwoofers. On the flip side of these venues, there are many small nightclubs whose sole performers are DJs and electronic acts, and these small venues *do* benefit from robust low-end coverage. Suffice it to say that the first consideration with small concert systems is what actually needs to be reinforced and to what SPL. Small systems sometimes provide only limited frequency response, because that is all that is needed for the anticipated act(s).

Additionally, small concert systems frequently forego stage monitors, and if they do provide for stage monitors, it is likely that they are being fed via aux sends from the FOH console, rather than featuring a full monitor console.

Many small sound systems, by virtue of budgetary lack, also forego DSP between console and amplifiers. This, however, is always a mistake—if output processing can be had, it should be had.

In Figure 37.1, we see the plot of a typical small concert system in end-stage configuration.

It is quite simple, featuring Main L/R speakers, a pair of stage monitors, and no other speakers.

In Figure 37.2, we see the plot of a typical small DJ-oriented nightclub, where speakers ring the dance floor to create an immersive sound.

On empty nights, such a system might only use the pair of speakers along the wall that features the DJ booth (speakers 1/7, 2/8, + monitors), as using 360° of speaker coverage without bodies to absorb some of the signal can easily lead to phase problems from opposing speakers.

VENUE CHALLENGES

Small music performance systems are often installed in less-than-ideal circumstances. Power challenges are common—either lack of enough power for the amps in small (but loud!) nightclubs or power that is not isolated from the lighting system and thus hosts a world of hum and buzz from dimmer backflow. If there is a small amount of power (circuits and thus amps of current) available, then speaker sensitivity becomes quite important to the design. If your power lacks isolation, there's no easy fix without an expensive (and frequently physically large) isolating transformer. Ensure that all your devices are run through robust power conditioners and design a system with the lowest self-noise components possible.

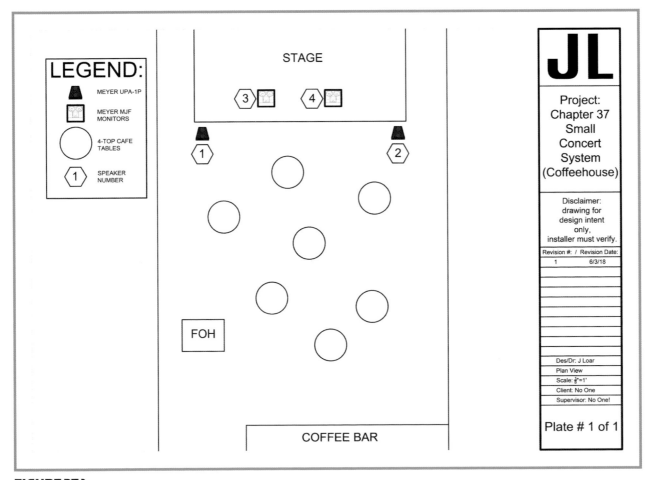

FIGURE 37.1
A Small End-Stage Concert System Plot

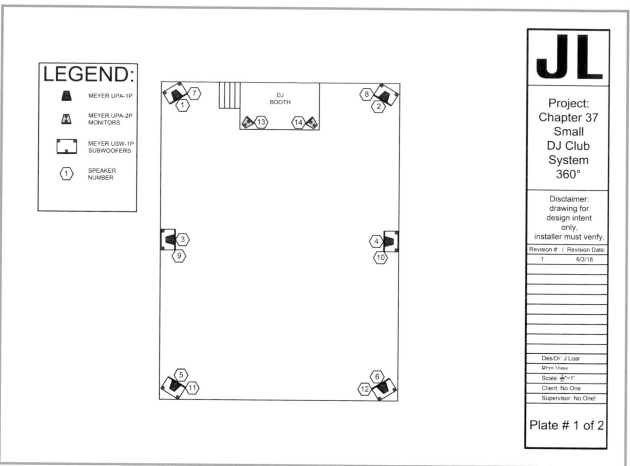

FIGURE 37.2
A Small 360° Nightclub System

Many small music venues offer a mix position that is. . . *less than ideal* (ahem!). In such cases, it is vital to try to specify a console that offers wireless tablet control, such that the mix engineer can occasionally get out of their terrible mix position, and move around the space, hearing what the audience will hear.

Small coffeehouses also face the challenge that we may need to amplify, say, the vocals of a folk duo, without becoming so loud that patrons can no longer have conversations. In this case, we need to pay close attention to vocal intelligibility frequencies, and very little else. Contradictory though it may seem, I have heard successful coffeehouse systems that aimed vocal speakers at the walls of the venue, rather than at the audience—just enough lift to allow you to hear the music if you wanted to listen, but without shooting vocal-range direct sound at the audience (which would then obscure their own vocal ranges and thus their conversations).

Other small concert systems may be used outdoors for summer concerts in the park, and the like. These systems often need a bit more power than indoor systems for the feeling of equivalent loudness, because it is unlikely that such small systems will be using highly directional array speakers, and the point-source boxes don't have the benefit of reflections from walls to help increase total SPL.

UNIQUE TERMS

Not many. Terms used in small sound system work are basically identical to those used for large concerts and many other live productions.

Soundcheck: the process of testing the lines with performers onstage, setting gains, and getting a rough mix in the house.

Tales From the Field

I once served as head engineer for a small concert venue in Brooklyn, that shall remain nameless. The FOH mix position (there was no monitor console) was located in the back-left corner of the room, far from the proscenium stage, against a brick wall, underneath a staircase that ran up to the production offices. It was a nightmarish bass trap. This is long before the era of iPad-controlled consoles, so the only way to figure out what the mix sounded like for the audience was to leave the console and move around the space, making mental notes (it was too dark to make written notes), and return to the console for adjustments. This was fine when the venue was mostly empty or even about half-full. However, when the venue managed to book popular acts, the club would be packed, wall to wall, with audience members (standing room only). On those nights, not only could I not get out into the crowd because it was too thick, but even had I successfully made my way there, I'd return to the console to find some drunk asshole had either posted up in my mixing booth area, set their cocktail on the doghouse (the back of the console case that protects the cable patch points), or both—which happened a few times. As such, I had to use moderately full nights to get a decent sense of how my bass trap mix position altered the sound and adjust based on memory during packed shows. Of course, the acoustic problems would change when it was packed, so I had to also *anticipate* how they would change. If I'd had an iPad to control the board, I would simply have made sure to fight my way out into the crowd, adjusted the mix properly, and fought my way back—and made sure to post a security staffer at the booth to prevent drunken catastrophe. Remote iPad mixing can't fix drunk idiots, but it can help fix mix problems that you can't hear in poor mix positions. *5 stars, would recommend!*—J.L.

Linecheck: similar to soundcheck, but usually (or at least often) without artists onstage, and typically without setting real levels, just checking signal paths. Sometimes referred to as a "scratch test," since onstage crew will often just scratch the microphone grilles in order to send a brief signal up the line (however, this isn't a great practice, as seeing/hearing a scratch test doesn't tell you if there's unwanted noise on the line, since the scratch *sounds like noise itself* . . . it is far better to actually speak or make other instrumental sounds into the mic).

Drink tickets: what you will often be paid in when designing or operating small concert systems.[4]

UNIQUE GEAR

- Small-format digital consoles with tablet control
- "PA Systems"—mixer/amplifier combos with little or no DSP that are sometimes the heart of *very* small systems[5]
- Notable *lack* of com systems[6]

TYPICAL TIMELINES

Not applicable. There are no "tech weeks" like in theater; these projects usually happen fast and loose.

DESIGN PRESENTATION

Technical drawings are always a good idea, but they may be more for the sound team than the client at this level. Often, design ideas need to be summarized in a PowerPoint, as short as possible, with major concepts in bullet points. Usually, you are presenting to a venue owner or series promoter. This is the type of system where you will often need to generate the *least* supporting documentation. You will still likely need to make a block diagram and plot, and often an IO sheet and gear list, but that's about it.[7]

Figure 37.3 shows a sample block diagram for a 360° nightclub system.

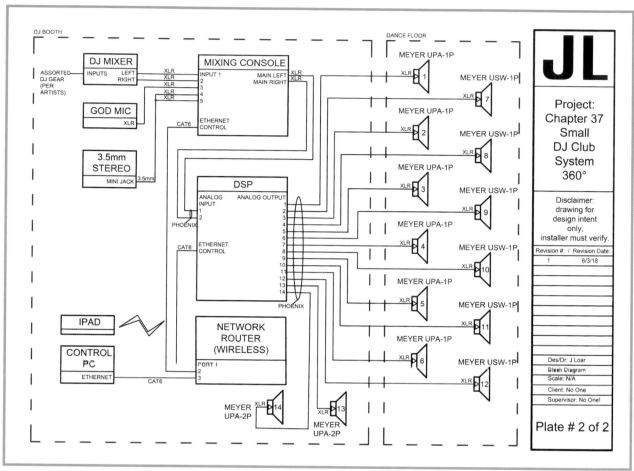

FIGURE 37.3

A Block Diagram for a 360° Nightclub System

DESIGNER QUESTIONS

In listing designer questions for each type of system identified in this and the following chapters, I am assuming that the major questions—those outlined in Chapters 30–36—have already been asked and answered, and thus will not be recapitulated here. As such, the questions listed here will be those unique to each type of system. As always, this does not constitute a comprehensive list, but it is meant as a guidepost to the thought process of working on a given system type.

- Is this system (and the performances that will be amplified by it) central to the overall mission of the venue, or is it a sideline? Many small-format music systems are installed in places like coffeehouses, where music may be an important part of the offering of the business but is by no means the central focus (*selling coffee* is the central focus). Such a system will typically operate at a lower SPL, focus on clear direct sound only close to the stage (often intentionally leaving the opposing side of the venue quieter for those patrons who came to the establishment to have a conversation), and so on.
- Does this system need to get packed away/stored at times? Where a large-scale concert venue is unlikely to take down its house system and put it away (unless, like Madison Square Garden, *all* the concert systems are brought in by the individual acts), many small restaurant/coffeehouse type venues will clear out the sound equipment when they are closed for a rental event that doesn't need sound, or if they are holding a special meal event and need more table space (taking over the stage) or the like. If your system needs to be stored, where can it be stored on-site, and in what fashion?
- Does the system need to do double-duty? e.g., are weddings also held at this venue, and the system needs to be a good enough spoken-reinforcement PA to support speeches and toasts as well as the tiny folk combo playing in the back most weeknights?

* * *

Now, let's examine *large* concert systems—whose design can be substantially more complex . . . and usually more fun!

Notes

1 Note that by definition, these examples will be only that—single examples—and are not meant in any way to represent the totality of possibilities for systems of these types. Particularly when we get into theatrical and themed systems, the complexity of which vary considerably from project to project, it is important to remember that our examples do not attempt to represent anything beyond their solitary selves.

2 I include DJ and electronic performers in the "live music" category, much to the protestation of some classically trained instrumentalists.

3 Certainly, there is no fixed size of venue or audience at which a "small" concert system turns to a "large" scale concert system, and indeed, many such could be easily described as "medium-sized." Such medium-scale systems, however, will typically adhere more or less to the paradigms we associate with either the small or large definitions here, and as such will not be addressed individually.

4 If only I were kidding. . .

5 To be avoided if at all possible.

6 Unfortunately, most small venues are running without any com, even when there is need of it.

7 See Part IV of this book for more info and details on standard technical documents.

CHAPTER 38

Concerts, Large Scale, and Festivals

Large-scale concert sound systems can take a dizzyingly wide variety of shapes. A concert system supporting a performance of the orchestral score of the *Star Wars* films will be very different than a system supporting the main stage at Coachella Festival. The systems will likely have different routing needs, different monitoring needs, and certainly different microphone needs. The types of speakers that will be acceptable for a rock/pop festival stage will encompass a wider range of available units than the range that will be acceptable for high-fidelity classical music. However, while such systems will differ in the specifics, the general parameters will be similar: they will each require high-SPL delivery, full frequency range, and (depending on the size of the venue) possible delay speakers to supplement the mains. They both will need FOH consoles with a large input channel count, and they both will (most likely) need monitor consoles with mic splits, and a built-in (or at least designed-in) option for HD multitrack recording. In this chapter, we'll examine large concert systems, addressing some common variations, and identifying elements shared by most (if not all) of such systems.

DEFINITION

A sound system intended to carry musical performance of one variety or another to a large audience. Such systems will feature dedicated separate FOH and monitor mixing consoles, often will feature delay speakers to supplement the mains, and will frequently feature options for multitrack live recording.[1]

STANDARD ELEMENTS

- Inputs: microphones on performers and their instruments, DI boxes, two-channel playback for interstitial music, talkback mic at FOH, intercom systems
- Processing: FOH mixing console, monitor mixing console, outboard effects to taste
- Amplification: standard operation (not 70V) amplifiers as needed

- Outputs: main L/R/C speakers, dedicated subs, onstage monitor speakers, in-ear monitors, possible house delay speakers, possible recording rig
- Converters: as needed, typically at inputs to console, and outputs from DSP; sometimes, on high-profile artists, dedicated outboard input converters (for mic/DI signals) will be used
- Advanced processing: DSP(s) controlling FOH and monitor system tunings and alignment
- Control: typically (at very least) iPad control of console(s) via wireless access points, computer control of DSP via either direct or VNC connection, scene recall within digital console for settings for opening vs. headlining acts, wireless mic/instrument monitoring via network, speaker monitoring via network, etc.
- Clocking: depending on the scale and complexity of the system, clocking may be handled internally by assigning a console or DSP as the master clock, though external master clocks are very common in large-scale concert systems

UNIQUE FUNCTIONS AND SYSTEM NEEDS

Large-scale concert systems need to deliver full-range, high-SPL signal, distributed evenly across typically large audience areas. In general, such systems call for high input channel count, high monitor channel count, and a wide range of processing/effects. A microphone split is required, whether analog (via transformers) or digital (via something like Dante)—at least two-way (FOH and monitors) if not three- or four-way (for recording rigs, and/or broadcast). Typically, com systems are required, to connect sound operators, lighting teams, and other production staff. Such systems usually feature playback devices, but only of a basic nature (typically two-channel for music between performers). Consoles with programmable scenes are desirable, as the ability to save settings for different musical acts is valuable.

We talked in the previous chapter about the reasons for the L/R main configuration so common to concert systems. At the large scale, this can vary somewhat, particularly for shows in ¾ thrust or the round, but the standard remains, most concerts do the heavy-lifting of reinforcement with L/R mains, whether those are single speakers or towering flown arrays. In large systems, we will frequently augment our systems with front fills, ground-stacked subs, delay speakers, and even sometimes high-center clusters/arrays, but the general system of enveloping the audience remains based on the L/R setup, even when the concert is mixed (as so often is the case at large events) in mono or limited stereo.

Festival systems also often feature two-way video com, with cameras frequently aimed at whiteboards at FOH and mains, such that engineers can write notes to one another without having to put ears in a com set while mixing, and without having to rely on (often overtaxed) data networks (for text messaging or the like).

Any system being installed outdoors needs to consider the weather-resistance of the equipment, as well as specify tents and other shelters for mixing positions. Touring systems must additionally pay particular attention to the size of gear when packed, as maximizing space in the tour vehicles is critical.

In Figure 38.1, we see the plot for a large outdoor rock concert system.

It features main left/right arrays, front fills on the lip of the raised stage, a range of ground-stacked and flown subwoofers, and two delay tower arrays down the field. Such a system is very common at outdoor shows whose audience area is all standing room. This is a common setup at many festivals (including Coachella's main stage), but also can be seen in venues with a seated audience close to the stage and large "lawn section" areas farther from the stage.

Assessing the speaker coverage needs of large-format systems is a process that depends heavily on loudspeaker modeling. While small concert systems typically do not require modeling, any time we're getting into large systems—and particularly any systems featuring vertical speaker arrays, which are difficult to properly configure—modeling is a critical part of our design development process.

VENUE CHALLENGES

Large concert systems can be found in a wide range of venues, from theaters and traditional concert halls to warehouses, stadiums, and wide-open fields. Each of these will present their own sets of challenges (and this is not a comprehensive list of potential venues). One challenge that is typical of most large system venues is providing

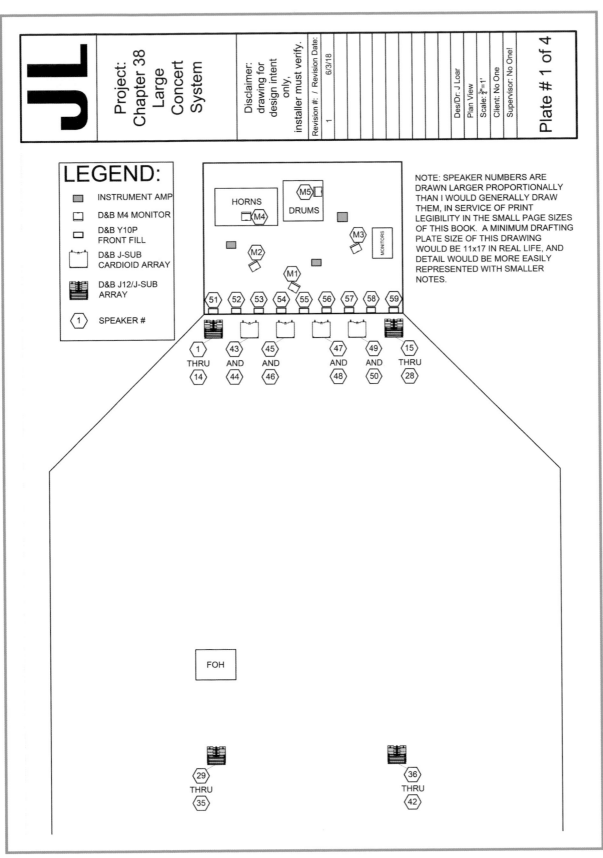

FIGURE 38.1
A Large Outdoor Rock Concert System—Plan View Speaker Plot

high enough SPL to the audience toward the back of the house (or to the coverage area before delays), while not deafening the audience closest to the speakers. This involves, again, modeling the system to determine where delay speakers are called for, such that even coverage can be achieved, rather than simply turning the mains up. In stadiums or other large round venues (like Madison Square Garden, which despite having "square" in its name is basically an oval inside), there is the additional challenge of creating some evenness of coverage for three or four sides of the audience around the concert stage. Large-format line arrays are challenging to configure in end-firing setup, but when we must overlap them in the horizontal plane (as the coverage from a front-facing array will certainly overlap the coverage of a side-facing array in the low-frequency ranges), the need of detailed modeling is even more profound.

Large concert systems (or any professional sound system beyond the small concert scale) must be fed dedicated sound power, and as such may need an entire power distribution system at the front end.[2] Large enclosed concert venues—whether designed as concert halls or as basketball arenas—will also feature long reverberation times, which must be overcome (for most concert production anyway: acoustic orchestral performances—that might benefit from long reverberance—aside) by high levels of direct sound. Otherwise, sound will wash out and details will be lost.

Large outdoor venues often run afoul of local sound level limits and curfews, so care must be taken to investigate potential legal sound limits at an outdoor venue before specifying a system. Outdoor systems frequently make use of line-array style speakers, to help reduce inverse square law loss away from the audience areas.[3] Outdoor venues also face the challenge of the changing speed of sound! When you've soundchecked and set your delay times at 2:00 pm in full sun and heat, and then the show is after dark, the cooling air will reduce the speed of sound, thereby changing what delay times you actually need set in your system. The equation for determining the speed of sound vs. temperature is typically expressed in Celsius and meters, like so:

$$V = 331.4 + .6T,$$

where V = velocity (speed) of sound in meters/second and T = the temperature in Celsius. As we can see, the warmer it gets, the faster sound moves, which is why a show running after the sun has gone down and the air cooled will have to retime delays from their afternoon soundcheck settings. If we soundchecked in weather that was 100° F, that would be 37.77. . . Celsius. Therefore, speed of sound during soundcheck was approximately 354 m/sec, which equals approximately 1,161 ft/second. If the evening showtime has cooled off to 72° F, that means our speed of sound has changed (72° F = 22.222° C). Our new speed of sound is approximately 344.7 m/s, or 1,130.9 ft/sec, which is a lot closer to our standard calculated speed of sound. If sound now travels slower from main speakers to our delays, we'll need to set our delay times longer accordingly, based on distance and sonic travel time.

Situating the FOH mix position is flexible in many large-scale installations. Where should it be placed? This depends, in part, on how your speaker system will be configured. If you are running the concert in mono (as most large concerts are, in order to ensure a similar experience for all audience members), the console should be aligned with one of the main stacks (left or right, it doesn't matter which), and should be within the coverage of the main system (not within the delay speaker range). If you are running your concert in stereo (less common, but not unheard of, for large concerts),[4] your mix position will need to be centered between the left and right mains, otherwise, the panning will be based on a false image. In general, whether or not you have delay speakers, you don't want your console more than about 100′ from the stage.[5]

Large venues benefit from careful site survey when possible, such that acoustic models can be built with an accurate sense of textures and responses. While I am a believer that *every* project could benefit from at least rudimentary advance modeling, it is in large venues that this practice becomes essential. This is both because a large venue will present more opportunities for ruining your coverage if you plan improperly, and because the cost of assembling a system in a large venue (both in terms of equipment and personnel) is high enough that we want to correct any potential errors in our designs well in advance.

UNIQUE TERMS

Concert production on any scale will talk of soundchecks and linechecks. Large-scale concert production will typically see the addition of a stage manager to the production crew, which means there will be more coordination between teams. When a stage manager is calling cues for the production, there is a standard protocol (typically based on theatrical procedure), for setting up and executing actions. The stage manager may well be calling for sound (and other) operators to "standby" for certain actions. The operators then respond by saying "sound standing by" (or the like). The SM will then direct crew to perform cues by such language as "intro playback GO," or similar.[6]

Downbeat: sometimes used in small concert production, but frequently heard in large concert productions, this is a term that refers to the anticipated start time of an act. As in, "The opening act has a 9:00 pm downbeat." This means that our opener is anticipated to be onstage and playing their first note at 9:00 pm. Concert production is typically *far* more flexible with time than is theatrical production—it is rare that a play will start more than 15 min late, but commonplace for a musical artist to come on an hour or two late (so long as they are not playing a festival or a venue with a strict curfew).

Rigger: whenever you are working on productions with substantial amounts of equipment flown overhead (particularly when such equipment is directly above artists and/or audience), it is advisable to leave the actual rigging to a certified professional rigger—someone who has trained and specialized in the overhead mounting of equipment. While it is unlikely that you'll be coordinating with professional riggers on a small-scale concert system (even overhead gear is frequently just hung from lighting-style pipes via yokes and c-clamps), on a large-scale production, rigging plans should always be finalized before installation by real riggers. As a designer, you may need to specify some rigging needs (particularly with line arrays, you will need to detail angles of aim, rigging frame(s), and rigging points—all of which can be determined by using the proprietary line-array calculators provided by most all line-array manufacturers these days) but at the same time, you should never agree to "sign off" on a rigging plan (even one you created) unless you are a certified rigger. I've been flying speakers for decades, but whenever possible, I stamp my rigging drawings with "Design Concept Only, all rigging must be verified and installed by a certified rigger."[7]

Backline tech: on large-scale concerts, there are often technicians whose job is the setup and maintenance of the backline gear (musical instruments, particularly heavy instrument components that may be shared by multiple acts, like drum sets, amplifiers, Hammond organs, etc.). In small concert production, it is likely that the audio crew is performing this role. This category sometimes overlaps with "roadies", the stage techs traveling with the band(s).

Delays: this is a colloquial term used to identify secondary speakers that are meant to extend the throw of the main system and preserve intelligibility over long distances. They are called "delays" because they are time-delayed back to the arrival of the sound from the main PA, such that the sound is generally aligned in the summation zone where both sets of speakers may be heard.

RF coordinator: on very large concert productions (and other large live events, like the Super Bowl, or Broadway shows), there may be so many RF devices (mics, IEMs, coms, etc.) that the production requires a crew chief dedicated just to the management of RF frequencies and equipment. This would never be part of a small concert production team but is frequently found on large events. After all, a concert festival needs to make sure each stage avoids the others' frequencies.

UNIQUE GEAR

- Large-format consoles, often with third-party effects loaded into the control frame, typically with most IO handled via stage boxes
- Dedicated FOH and monitor consoles, definitely with recallable scenes/presets
- Recording systems (not always, but increasingly common)
- Com system
- Playback system—often just a laptop connected directly to the FOH console and operated by the FOH mix engineer
- Measurement microphone and system—it is increasingly common to see a measurement mic placed at FOH, running live analysis of the venue as it fills with the audience . . . by using the actual room sound as the test signal, compared to stored traces from before the audience comes in, the engineer can subtly retune the system (in frequency response, primarily) as the audience thickens, thus adjusting for the absorption coefficient of the crowd
- High-power speaker systems—particularly in the United States, large-format concert systems are designed to get *really fucking loud*. A typical concert spec for a large-format system is to deliver 120 dB SPL to the mix position (and thus to just about everywhere else, as you want the mix position to be representative of the audience, such that the mix engineer has an accurate sense of what they are delivering to the crowd). This is different in different parts of the world—Norway, for example, requires that no 30-min period of a concert exceed 99 dB SPL (A-weighted), with permission for *momentary* peaks to hit as high as 130, but not for longer than a snare hit or other such transient. The United States has no such noise laws for concert production, and as such, systems are designed to deliver enormous power.[8]
- Full-range response—where a small-scale concert system might get away with speakers only designed to reinforce vocals (if the other instruments will not be run through the system), it is to be assumed that a large-format concert system should always be designed to handle as close to the full human hearing range as possible.

TYPICAL TIMELINES

Installation time will vary by the scale of the project. A festival might take a week or two to load in at the concert site, and if it is a well-funded festival, may bring artists in a day or two early to do soundchecks, such that there are no soundchecks (only linechecks of the restored setups) on performance days. A concert tour will have less time, if a system is to be installed at a venue for a one-off show (whether in a field or at MSG), chances are that the team has at most a day in advance of the performance, and in many situations, will be granted only the day of performance—in which case, the crew will begin installation quite early in the morning, such that the system is tested, tuned, and soundchecked before the house opens and the audience is admitted.

Soundcheck length will vary depending on a few factors:

- Availability of the venue and system—if the system install leaves only an hour between the finish of install and the house opening, this won't afford a 2-hr soundcheck (obviously).

- Stature of the artist—the more commercially important and critically revered the artist is, the more likelihood they will be afforded a long soundcheck.
- Complexity of the show—a power trio of guitar/bass/drums with one vocal mic will likely be afforded (and need) less time to soundcheck than a 12-piece band with horns and samplers and other unconventional items. This is closely related to stature, however. A power trio like The Police probably gets however long they'd like for soundcheck, where an up-and-coming 12-piece that isn't well known or selling very well will still have a shorter soundcheck than the headliner.

System tuning is usually done as quickly as possible—and with even the largest concert systems, there isn't usually need of a terribly long tuning call. Seasoned system crews can tune a concert system in an hour or two, generally regardless of scale (assuming everything is working properly). This is in contrast to, say, a Broadway system tuning, which can take most or all of a few nights in the venue, to fine-tune coverage and timing.[9]

DESIGN PRESENTATION

Large-scale concert systems typically require a fairly detailed document package. This package will include the following:

- Speaker plot
- Block diagram(s) (sometimes network and control risers are drawn separately from audio signal risers)[10]
- Rigging details—particularly for flown arrays (Figure 38.2)
- Equipment schedules—lists of all gear, for shop assembly; sometimes these are subdivided by category, such that there will be a "computer schedule," a "hardware schedule," and even a "cable schedule"
- IO sheets—aka "hookups," spreadsheets showing the signal path of each input path to the system, and each output path from the system, including all devices, ports, or channels of patch for those devices, cable types from one device to the next, and often, cable lengths
- Stage plot(s)—drawings of the artist(s) layout onstage, usually detailing specific backline equipment
- Mic plot(s)—stage plots with mics drawn in and detailed
- Power requirements—a sort of simplified Facility Impact Report (FIR) focused only on power needs (sometimes a full FIR is required, but that particular document is found more often in installation work)

These documents should be presented as a unified document set, with plate numbers and a table of contents and the like. Typically, this will be presented by the designer to the promoters/producers/whoever is paying for the work. These design intent drawings, once approved by the producers, will be sent to the production house/integrator, who will then attempt to fill the design requirements as close to specification as possible and will propose substitutions as necessary.

As we can see in Figure 38.3, large concert systems are often far more complex than those for small concerts (compare to Figure 37.3).

This block diagram has included audio signal paths, as well as control and network connections, all in one diagram (it has omitted com systems for the sake of sanity—the diagram already consists of three pages). Combining everything into one document makes for a crowded drawing. If we broke network control signals and the network (as opposed to analog) audio signals out into separate plates, they would be considerably easier to read, but some producers will express a preference for one style of drawing presentation over another, and in concerts and theater

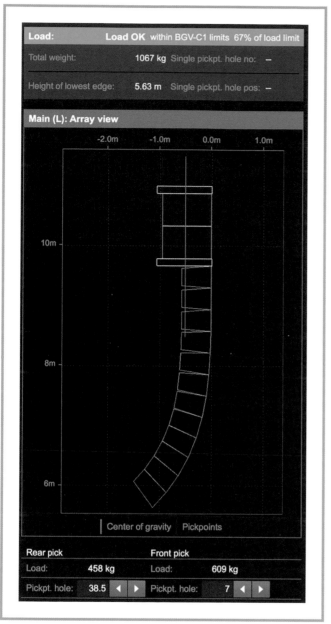

FIGURE 38.2
Typical Line-Array Rigging Detail

we will often encounter producers who prefer to crowd info into a smaller number of plates at a higher rate than in installation work, where breaking out subsystems onto separate plates is the standard.

At the end of the day, a concert system is still performing a relatively simple task. Communicate the music from the stage to the audience. Allow musicians to hear themselves in real time. Ensure that you provide high-resolution, full frequency response signals at a high enough (and a nicely even) SPL across the whole audience. Provide enough direct sound that the details don't get washed out. Develop a system with low enough self-noise that the full dynamic range of the content can be represented.[11]

Design of festivals is typically just design of several parallel large-scale concert systems. Coordination of power needs, RF frequencies, and com infrastructure is usually important, but otherwise, each stage is effectively its own venue. There are some festivals at which the SPL of one stage must be reduced such that it does not impinge upon another stage, but otherwise, this is the same process as designing any large-scale concert system, rinsed and repeated.

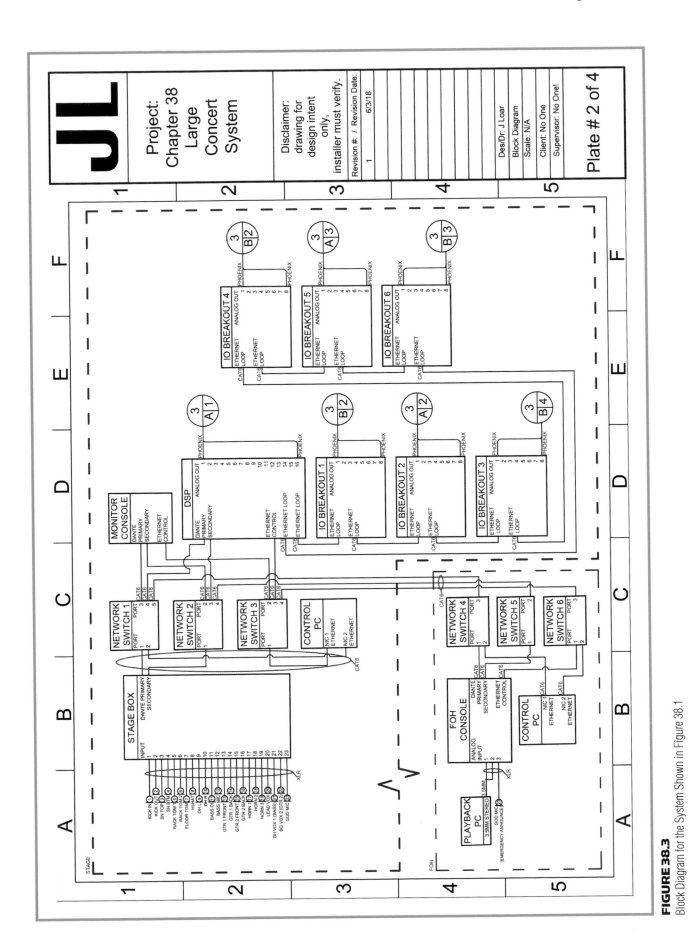

FIGURE 38.3
Block Diagram for the System Shown in Figure 38.1

390

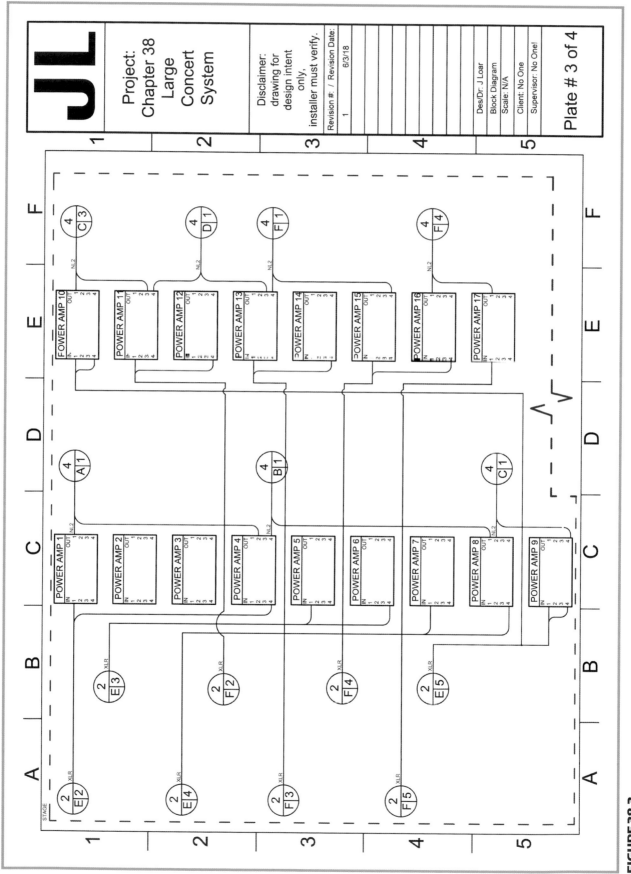

FIGURE 38.3

Continued

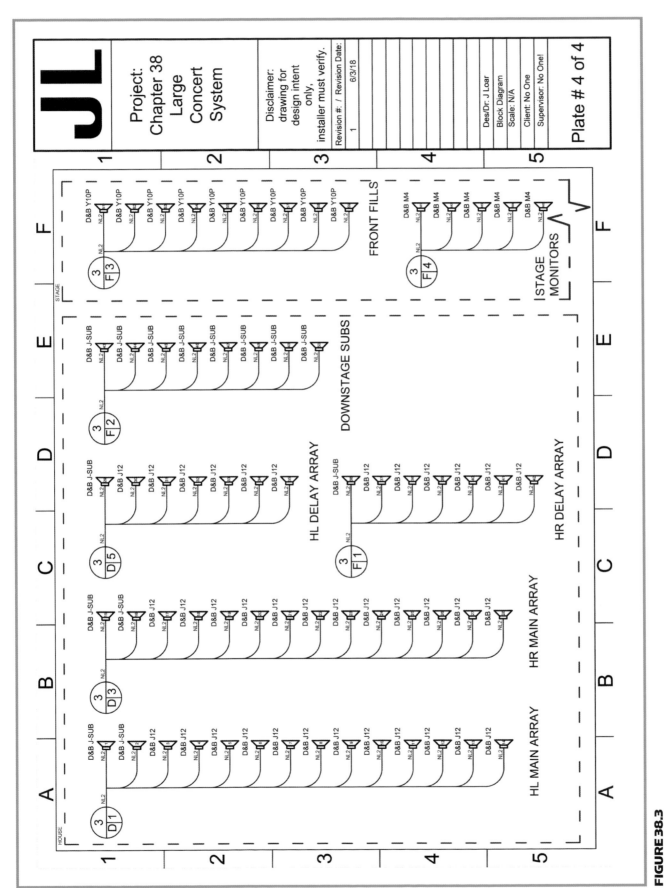

FIGURE 38.3
Continued

391

A note on designing arrays: with the rise of the line array/curved array, early career system designers often wonder how best to design arrays for their project—since array design is complex, this is a natural question. My advice to those just beginning to wrestle with line arrays is this—use modeling software and begin each array with a single box. Aim that single box at the rear of the desired coverage area (since the top box of any array is the one focused the farthest away). Model that. Add a second box, and work with splay until you've got the most even possible coverage between the two boxes (creating a seamless HF patch). Repeat this with box after box until you've determined how many boxes you need, and what the ideal splay will be between them. If you can assemble a single array box by box and really grasp what each new box is doing to your coverage, it will be easier to make good educated guesses as you work on more and more array-based systems, such that for *very* large systems (like those featuring a dozen 24-box arrays, for example), you will be more easily able to start with a full-length array and make fewer adjustments between modeling passes. However, first-time array designers who try to model a 15-box array right out of the gate often find themselves overwhelmed, making too many changes from modeling pass to modeling pass, and not understanding which of their changes actually produced the result they see.

This leads me to another note about modeling speaker systems. Unless you are certain of the changes you make from one modeling pass to the next, I recommend only making one change between modeling passes (such as a splay angle, but not a horizontal aim, or vice versa) so that you can get a really firm grasp on what each change does to your coverage. As you get more confident as a speaker system designer, it will be easier to make multiple changes and know what each has done to your coverage, but especially while you are learning it can be more educational to go one step at a time, until you really understand what each change is accomplishing.[12]

DESIGNER QUESTIONS

- Does this system need to cover the audience evenly, or is it desirable that sound levels fall off toward the rear/sides of the venue? While large-scale concert systems are typically only installed in venues where the concert is the main event (as opposed to small coffeehouse systems, discussed in the previous chapter), concerts still count on selling merch, beer, and so on. Some designs attempt to produce strong clear direct coverage at sales points, to make an audience feel that they are always part of the show. However, many others (possibly a majority) allow for sound levels to decline at merch areas, both to make conversation easier, and to give audiences a space they can be in that is not as intense as the main concert floor. It is important to discuss this issue with producers/promoters to determine what kind of event space you are working to create.
- Do you need transformer-isolated mic splits or can FOH/monitor/recording engineers agree that one of them sets the gain levels and the rest live with those settings (this assumes a digital system where signals can easily be copied and where consoles feature digital trim controls for shared preamps)? A system without isolated splits gives each engineer less control but makes for a cheaper system and faster installation.
- Are the performers likely to damage gear? Do they throw guitars, destroy microphones, climb speaker stacks? If so, you will need to either specify mostly (if not entirely) dynamic mics, and other damage-resistant equipment, OR specify a large number of backup items for those fragile pieces (like condenser mics) that you select as part of your system.
- Are the performers particular enough about monitor feeds that they may demand analog signal paths for in-ears (which avoids latency)? Or will they accept the (generally low-latency) professional digital systems at their disposal?

* * *

Now, we move onto the theater, where systems are often (though not always) smaller than concert systems but are often (though not always) more complicated in terms of signal flow.

Notes

1 As such, even large-room nightclub systems designed around DJ and electronic performance will generally *not* fall into this category (for the purposes of this book)—while they are delivering high SPL, their monitoring systems are usually fed from FOH, and any live recordings are typically of the two-track mix heard in the venue. To every generalization, there is an exception. There is a trend, just beginning as of this writing, of nightclubs featuring detailed surround systems based on Dolby Atmos processing, wherein electronic music performers

are creating elaborate multitrack content, and often recording such in HD multitrack formats. Such venues often require more complex monitoring solutions for the performers (whether requiring a dedicated monitor mixer or no). While these are interesting and exciting trends in electronic music, they are still few and far between, and will not be further addressed in this chapter.

2 Production power distribution systems (often referred to simply as "distro") will typically take high-voltage three-phase power and step it down to usable house-style voltage and current.

3 Indeed, in recent years, there have been considerable advances in LF control for live concerts, with such speakers as Meyer Sound's 1100LFC providing extremely directional sub-bass, with little to no spill back onto the stage. Such devices are increasingly in demand for concert work in general, particularly in outdoor locations surrounded by residential areas.

4 Even those large concerts that are presented in stereo are typically in a limited stereo, such that there is *some* spatialization of sounds, but without panning anything so wide that the opposite side of the audience would lose it entirely.

5 This is, of course, just a general rule, not a hard and fast figure. I've heard shows that were wonderfully mixed from an FOH position that was 120′ from the stage and poorly mixed shows with a console 75′ from the stage.

6 In theatrical productions, these are usually numbered cues with a pre-established meaning. However, in concert work, it is more likely that individual actions will be described in the SM cue patter, rather than having pre-established, say, that "Cue 5" means play the artist's entry music.

7 Riggers are certified by ETCP, The Entertainment Technician Certification Program, via a series of *rigorous* tests. Pun definitely intended.

8 As of this writing, a judgment has just been handed down in the UK awarding a musician workers' compensation damages from hearing loss resulting from performance in an orchestra and being subjected, up close, to excessive SPL. Noise and hearing law is evolving, and one of the jobs of any system designer is to not only research regulations in the location for which they are designing, but also to stay up-to-date as laws change.

9 Tuning a reinforcement (rather than replacement) system that localizes to moving actors is considerably more subtle and can be more challenging than tuning a loud concert system, where we are comfortable localizing to the main left and right speakers, without any illusion that we're localizing to the actual artist on stage.

393

10 Block diagrams are sometimes referred to as "risers," "one-lines," or—incorrectly— "schematics." I prefer the term Block Diagram, as this is truly a description of what is contained in the document—a detail of each device in the system, drawn as blocks with signals passing between devices. Schematic, in technical terms, should be a more detailed drawing—a mixing console might be shown in a system block diagram, but the schematic of the console should be a drawing of the electronic components and wiring inside that console (every diode and resistor and capacitor and so on). However, as we'll see when discussing themed and architectural design processes, there is a standard phase of the design process in these worlds called "schematic design" (SD), which is a middle-stage phase, between "concept design" (CD) and "design development aka detail design" (DD).

11 To this last point, systems for the amplification of classical music (which are not super-common but certainly exist) must have lower self-noise than systems delivering highly compressed hip-hop or electronic music (for example).

12 The exception to this, even for young designers, would be the event where you place speakers just flat-out wrong. If you know your speaker is just totally in the wrong place, approach the next modeling pass as a starting over point, such that you adjust all speaker parameters until you think they are now correct. Model that, and *then* make incremental changes as you refine.

Theater—Straight Plays

Theatrical sound system design is often challenging. While the goal of a large-format concert sound system is very clear (amplify full-range music being performed onstage), the goal of a theatrical sound system can vary from project to project quite substantially. One straight play might call only for a few diegetic sound effects, while another might call for drastically changing immersive acoustic environments, created by virtue of a complex simulated-acoustics system such as Meyer Sound's Constellation. In between, there will be the potential for every combination of sources and purposes.

DEFINITION

Straight plays are plays that do not feature interrupting musical numbers. The vast majority of the theatrical literature is comprised of straight plays of one variety or another.

In assessing the needs of a straight play, we can break the source material into three component parts:

- Speech
- Sound effects
- Music

In a straight play, speech is generally the first priority. If you are in a large enough venue that speech needs to be amplified (or if speech needs to be amplified in order to apply special vocal effects to the performance) we must design at least part of our system for vocal reinforcement (or replacement). Sound effects are present in many (if not most) straight plays, but each play will have different requirements. Close reading of the source material and discussion with the creative team will allow the system designer to properly assess the needs of the effects systems. I've heard effective theatrical sound designs that devoted only two speakers to effects playback, and others that devoted more than 100 (sometimes in very small spaces) to effects playback—it really depends on whether the effects need to be localized, how many locations they need to represent, what other spatial and informational cues the sound effects need to give the audience, how loud they need to be, and so on. Music in straight plays is generally used in three different ways (or a combination between them): interstitial (between scenes), diegetic (music characters hear and respond to), or underscoring (music played under a scene, that characters are not aware of). Once you determine which use (or combination of uses) music will be turned to in your production, you can determine how best to support that content.

Design of vocal/speech systems for the theater is its own technological art. The desire of theatrical productions to have vocal sound localized at the actors and not at the speakers makes placement quite challenging—we can't put speakers right in front of the actors, such that they are producing sound always aligned with their onstage positions. In a proscenium stage, it is very common to see a design incorporating high-center vocal cluster, front fills (to cover those first few rows that won't be hit by the high center, and to help pull the image down to the stage plane for audience farther back in the house), and (in houses that feature these elements) under- and over-balcony fills (Figure 39.1).

In a less-conventional theater, such as a black-box or warehouse, the need for vocal reinforcement can be more challenging. If actors are too quiet, or if the effects and music are by necessity very loud, vocal reinforcement can be needed even in small spaces. However, speaker placement becomes very difficult when there is no frame through which the audience is viewing the performance. If we're in a black box with a ¾ thrust setup and the audience very close to the actors, where can we place vocal reinforcement speakers that will serve to localize to the actors, and not up to the grid?

Looking at Figure 39.2, we can see that the audience is very close to the actors.

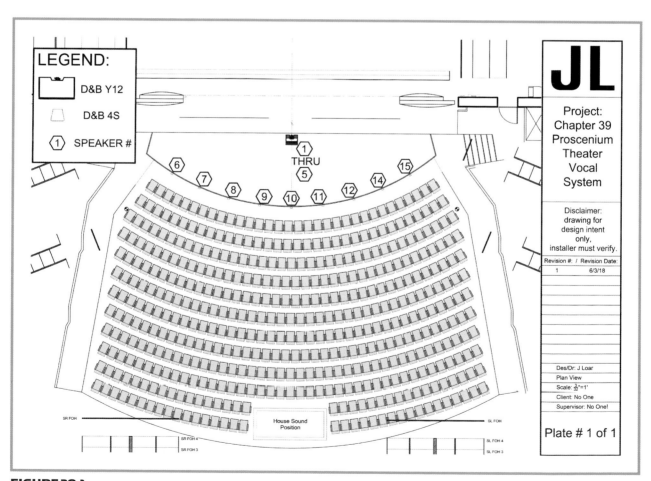

FIGURE 39.1

A Proscenium Theater With Vocal System

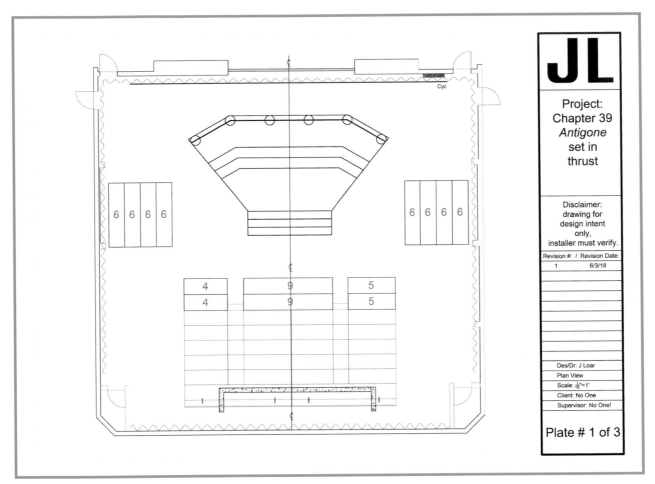

JL

Project:
Chapter 39
Antigone
set in
thrust

Disclaimer:
drawing for
design intent
only,
installer must verify.

Revision #: /	Revision Date:
1	6/3/18

Des/Dr: J Loar
Plan View
Scale: $\frac{1}{16}$"=1'
Client: No One
Supervisor: No One!

Plate # 1 of 3

FIGURE 39.2
A Black-Box Set in 3/4 Thrust
Source: Set for *Antigone* Designed/Drawn by M. C. Friedrich, Courtesy M. C. Friedrich

In such situations, it is common to *not* amplify the voices at all. However, there are still scenarios in which you would be required to amplify the voices. Perhaps you are working with semi- or non-professional actors, who do not possess the skill of vocal projection to the same extent as professional actors. Perhaps the director wants the actors to literally whisper because the scenes take place in hiding, and even a "stage whisper" is too loud for this director's taste. Perhaps you want to alter vocal tone through pitch shifting or other effects.[1]

If we need to amplify voices in such close proximity, we have the challenge of localization. If we use flown speakers, we can easily avoid feedback, but if we amplify very loud, we risk localizing the voices up to the grid, not at the actor's height. However, chances are we can't mount actor-height left/right speakers (and even if we could, proximity means we'd localize to the sides of the performance area, not directly in front of us). If the set is raised off the floor, with steps or platforms, we can put speakers inside the scenery, with grills poking out of the fascia of the steps or platforms. This risks more feedback, as the actors may play downstage of the vocal speakers, but it does improve localization. Often, we may find ourselves placing speakers both in the set and above and using each little enough that we can find a balance between lack of feedback and desired localization (Figure 39.3).

In advanced systems, actors are fitted with radio-tracking devices that allow their individual systems to be re-timed in relation to the reinforcement system, such that actor sound can be constantly in proper phase relationship with the amplified sound. It is well beyond the scope of this book to detail how such systems are utilized, but it is worth knowing that they exist.[2]

Sound effects systems may take a wide variety of shapes and sizes. From a simple practical inside a set piece for a localized telephone ring to an immersive surround system with speakers above, below, and surrounding the

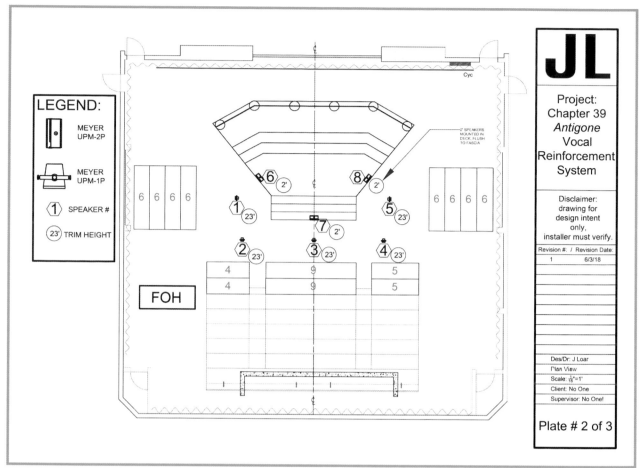

FIGURE 39.3
Example Vocal System in a Black-Box Thrust

audience with dozens (if not hundreds) of speakers, effects systems are designed to suit the production needs, and (unless running productions in repertory), won't be interchangeable. While some theaters will have stock vocal systems that they use (or at least recommend) for any production in a given configuration, sound effects systems by definition should be designed to suit the needs of the production. There is a type of effects system that is sometimes used both for playback sounds and also for vocal amplification. This is the so-called reverb system. In such a system, speakers will often be mounted around the walls of a venue (typically well above audience head heights), aimed at the walls/ceilings of the venue. Signal to these speakers will thus excite the venue walls themselves as part of the amplification system, which can give effects more size and less specific location, and can make vocal reinforcement sound more naturally tied to the venue. Sometimes this is desirable, sometimes it is not, but it is worth addressing—as this is the only situation in which I've seen speaker system designers deliberately aim speakers away from the audience. The signal played through these speakers has a diffuse sound, and generally is more difficult to localize than sound played directly at the audience. Voices are sometimes played through these systems in addition to direct vocal speakers in order to create a more diffuse timbre than is possible in close quarters from direct speakers only. In general, evaluating the needs of an effects system can be done via a basic process: what content needs to be delivered? Where does it need to come from onstage? What is the frequency range and SPL required? Does the sound need to move while being played? What is the arc of motion? And so on.

Music systems, like effects systems, will vary wildly depending on their purpose in a straight play. Is music essentially a sound effect, played diegetically through a radio onstage? Is it loud underscoring for action and movement sequences? Is it some combination of diegetic and non-diegetic? There are a number of uses of music in straight plays, and just like evaluating sound effects systems, music systems must be evaluated against the content delivered—what SPL, frequency range, localization, and style needs to be delivered?

> **Message in a Bottle**
>
> The development of larger-scale system design becomes really essential, especially because the tools have become so sophisticated—there's a number of choices we have, whether or not it's essential for a producer to think that he or she has to track the performer and the reinforced sound of that performer . . . that changes the needs of the sound system very, very dramatically. Whether or not they want to use area mic-ing to create something more naturalistic, whether they want to do something that's completely exaggerated, over-the-top, Cirque du Soleil, or Disneyfied big spectacle production—they all have different needs, different requirements, and different budgets, and require both a sensitivity on the part of the designer to understand what is necessary, and then to have the depth of knowledge of the tools to be able to pick the right tool for the job.—David Budries

STANDARD ELEMENTS

- Inputs: wireless microphones on actors, in-set area mics, playback systems—typically multi-channel, of varying levels of complexity (from two channels to hundreds of channels), intercoms
- Processing: if microphones are used, FOH mixing console; reverb (for effect, but also on occasion to give spoken vocals a sense of being in a different acoustic space than the theater itself)
- Amplification: standard operation amplifiers as needed
- Outputs: a wide variety of speakers and positions—possible vocal reinforcement speakers, effects speakers placed as desired, music speakers if needed
- Converters: if the system is digital, typically found at in/out points to digital system components
- Advanced processing: DSP if the vocal system is needed; for playback-only systems, advanced processing is often carried out in the playback software (e.g., QLab) rather than using a dedicated DSPs, though the larger the system gets, the more likely a DSP will be used even in a playback-only production
- Control: wireless mic monitoring via network, iPad control of console if possible, designer VNC control of both console and playback computer (set up to be run from the designer's tech table—workstation—during tech week, then struck during production)
- Clocking: generally internal, with DSP or console serving as master

UNIQUE FUNCTIONS AND SYSTEM NEEDS

Sound systems for straight plays are often centered around the playback system (as opposed to a concert system which is normally console-centric). Theatrical playback is typically run on the "go button" premise, which is that all pre-recorded content is sequenced and leveled in the software during the technical rehearsals so that during the run of the show nothing is adjusted, but content is merely played back at the specified time according to the script. Such systems are usually software-based, and often (as is the case with QLab) feature a giant GO button, that you can literally press in order to fire the next cue in sequence. Such systems often are configured with a unique signal path to each speaker in use, such that any given sound effect can be routed to just one speaker, to all available effects speakers, or to any combination thereof. This requires dedicated outs from the playback system (whether via hardware interface sending line-level analog signals or via a Dante patch into a network), dedicated amplifier channels for each individual speaker, and enough speakers to effectively communicate the sound as intended.

When vocal reinforcement is needed, wireless mics and a mixing console come into play (and in some cases area mics such as shotguns, boundary mics, and others hidden in and around scenery), but for straight plays without live microphones, there is generally no need for a console, as sound can be run directly from playback systems through DSPs into the amps and speakers (music can be handled this way as well, so long as it is not performed live).

Theatrical systems are often divided into subsystems, such that vocals will be handled by a certain range of speakers, effects by another range, and sometimes music by a third range. Such division allows us to create sonic environments that are rich in detail and spatial depth, without worrying as much about content piling up in a

399

Tales From the Field

I want to talk about a show that I did called "Guards at the Taj" that won an Ovation Award in LA for Outstanding Production. It was something I was really proud of. And from a system point of view, we had a fairly complex system. The space that we were in is fairly small. I want to say it's maybe 25' wide maybe 50' deep, and the ceiling is maybe around 18–19'. So it's not a big space, and the play needed to shift locales very quickly from right outside the Taj Mahal, to a dungeon, to back to the Taj Mahal, and then, ultimately, we go to this beautiful, lush forest at the end of the play. And I needed to build a system that was going to help tell that story because we didn't have a lot of time, we didn't have a lot of space, scenery-wise, so we were doing a lot to suggest locales. So what I ended up doing a couple of things: first—I designed a front presentation system that just gave me good stereo coverage, and those speakers are downstage so that the music and the sound were presentational so that during the transitions [we] would have a more presentational moment where the sound comes from the proscenium—or, from where the proscenium would be. But then the other thing that I did was installed a VRAS (Variable Room Acoustics System) system from Meyer Sound. So we had 20 or 30 small loudspeakers spaced throughout the room on a grid. And then eight different microphones hanging around the room. So I built a VRAS system that was flexible so that we could actually change the acoustics on a *cue by cue* basis. In some situations, we actually changed the acoustic response of the room every line or two. But then also, because I had that system built and I had VRAS in there, I had access to Space Map and to Wild Tracks. So I ended up doing a lot of a lot of work in Space Map sending sounds all around. And then the final thing that I did was I built a jungle—a series of sounds that were to help connect us to the, sort of, nighttime birds, nighttime activity, and when the guards were guarding the Taj—because I had a system of so many loudspeakers I was able to source individual bird chirps to individual loudspeakers so that the room really took on a three-dimensional feel. So it was a small house, but it was a lot of gear that went into it. It was very, very successful, I thought.—Vincent Olivieri

single pair of speakers. Where some of the joy of a live concert is that we hear the mix of music coming from a reduced number of acoustic paths (the mains), theater often takes the opposite approach—though musicals are a different story, as we'll examine in the next chapter.

Theatrical sound systems often need to have control software that can operate both sound and lighting devices or can send cues from one to the other. They also often feature mixing consoles with savable scenes, especially those with crossfades between scenes as a programmable option. Advanced consoles designed specifically for theatrical work also allow the assigning of actor profiles, such that if a performer goes out sick, rather than having to reprogram EQs and other settings in each scene in which that actor appears (for the understudy, who will have a different voice than the lead), you can just update the actor profile with the new actor's settings, and this will auto-propagate through the whole production in one step. *This is very desirable in musicals, but also in any straight play with a long run where it is likely that understudies will be used at least once.*

VENUE CHALLENGES

The signature challenge of designing sound systems for theaters is that (particularly on Broadway or in Europe, but elsewhere as well) so many of the theaters themselves were designed before sound reinforcement was commonly used in theatrical production. As such, they were designed to carry acoustic sound from the stage to the audience, which can be good, but they were also designed without any idea that we'd need to mount a ton of speakers in them, and this can present major challenges. In addition, theatrical directors and producers are often averse to seeing speakers at all, which means that not only may we not have good rigging positions for our speakers, but where positions exist, we might not be allowed to use speakers there because they are too visible!

As such, we find ourselves wrestling with situations we would never face on, say, a national outdoor concert tour. Where to hang our main speakers? Can we tune them to sound clear even though we have to conceal them behind scrim curtain?[3] Much of theatrical sound system design consists of solving the unique challenges posed by the

Tales From the Field

I worked for some time at a prominent regional theater. This theater had decided (long before my arrival) to purchase a pair of line arrays for their left/right mains setup. These arrays were, as most arrays are, large columns of black speakers hung from chain hoists. Persistently, the arrays would be designed to hang in a specific location, everyone would approve this location in production meetings, we would hang the arrays, and then the director and/or set designer would arrive for tech rehearsals and throw a fit because these speakers were too large and too visible. We started building the arrays into every 3D scenic model (including the paper models) built by the set designers, so they could see exactly where the speakers would be, how large they would be, and so on. This (mostly) didn't help! With rare exception, designers that had approved these positions based on the set model would get in the space and say some variation of "I didn't know they would be that big!." Even though they were literally to scale in the model! Eventually, we settled upon a default house position that the arrays could be mounted, a position that would cover the sonic needs decently, and stay (mostly) away from the scenery. Because this compromise position became known as "the stock position," designers accepted the visibility of the speakers there much better than when they had been literally 3" farther onstage at the request of a sound designer. Because "the venue" had decided that was where the speakers needed to be, set designers accepted it. This is a lesson in two things:

1. Speakers in theaters are always an eyesore (unless you're doing a rock musical where speakers are part of the set). We are always working to use the smallest possible speakers and to hide them as much as possible.
2. Institutional politics matter quite a bit. When the speaker position was *just* the decision of a "non-visual" designer, it was constantly contested. When it was the decision of the institution, it was accepted. By getting the venue's production supervisor on board, we were able to place the speakers in essentially the same position we had already been requesting, but with the understanding that scenery was not *allowed* to contest their placement. Making allies among your production staff can save a lot of work calls for the crew!—J.L.

venues (and the ever-changing scenery housed inside them). As such, acoustic/speaker performance modeling is a powerful and increasingly essential tool in the theatrical design process.

UNIQUE TERMS

10 out of 12: a standard long technical rehearsal day, held a few times during a tech "week" process, in which the crew and design teams are called for 10 hrs out of the course of 12 hrs (usually subtracting an hour for lunch and an hour for dinner).

Apron: the part of a stage downstage of a proscenium arch.

Borders: curtains hung above a stage to help conceal pipes full of gear (particularly lights).

Catwalks/galleries: walkways for workers and equipment mounted above a playing space. Catwalks are suspended above the venue, galleries tend to run around the walls, but both are narrow walkways with railings preventing falls.

Centerline: an imaginary line running through the center of the stage and through the house, used as a reference for gear positions.

Crossover: not the audio device, but a space upstage of a set used by actors to cross from one side of the stage to the other without being seen.

Electrics: pipes flown above a stage (particularly a proscenium stage), outfitted with electrical circuits, generally used for mounting stage lights (though also playing host to speakers and mics at times).

Fly system/fly rail/loading gallery: a system by which pipes and equipment are suspended above the stage for the mounting of lights, speakers, scenery, etc. Contemporary fly systems are usually based around a variation of the counterweight fly principle, where arbors with stage weights are loaded (via the loading gallery) that are equivalent to the weight of equipment on the pipe, such that the counterweight allows the pipe to move in and out with minimal effort on the part of the fly operator (who is working at the fly rail, where all such lines terminate at the side of the stage and feature lock-offs to keep lines from moving at unwanted times). Very old "fly systems" will be referred to as "hemp houses" because the ropes are not contemporary synthetic weaves but old hemp lines, and usually are rigged without counterweights, such that whoever moves the lines must pull a much larger amount of weight (though a system of block and tackle outfitting still provides some mechanical advantage over simply lifting the equipment outright).

Fire curtain: a curtain typically mounted just upstage of the proscenium arch, designed to fall in case of fire, to isolate the stage from the audience (thereby protecting the audience if a fire should start onstage).

Grid: the mounting area above a stage. In black boxes, the grid is often literally a gridwork of pipes, installed for the purpose of mounting gear. Above a proscenium stage, the grid is sometimes just an area where the fly lines for the counterweight system travel (above the electrics).

Mic check: at the start of each performance call, actors will come onstage after being mic-ed up and will speak while the FOH engineer verifies that their signal is properly patched, and the mic is working well.

Plaster line: an imaginary line drawn from the upstage edge of the proscenium arch across the stage. Often correlates with the position of the *main drag*, aka the *grand drape*, or the main curtain that is often but not always available to close the proscenium opening. At or near the fire curtain line.

Plenum: a space below an audience area, where cabling is often run.

Pit: orchestra pit, typically a sunken area just downstage of the apron, housing the orchestra for a musical (or being used in other ways in straight plays).

Speaker check: each night, the playback operator will run a special cue list in which a sequence of spoken cues will fire, with a voice saying, "Speaker 1, Speaker 2, Speaker 3," and so on, with each number spoken corresponding to the speaker's number on the plot (and playing only through that particular speaker). This is to verify that the system has been properly turned on each night and that all components are working.

Stage directions: while this is also used in concert production, it comes from the theater, so I've listed it here. Stage Left is the left side of an end-facing performance space, from the perspective of an actor onstage facing the audience. Thus, Stage Left equals House Right (which is based on audience perspective), and vice versa. Upstage is the part of an end-facing stage farthest from the audience, and downstage is the part of a stage closest to the audience. These are often abbreviated, such that DSL means "downstage left," USR means "upstage right," and so on.

Tech week: the process of technical rehearsals, during which the performance is spaced on the finished set, and all details of sound, lighting, video, prop, costume and other cues and changes are worked out in detail.

Tormentors: curtains hung on the sides of the stage to conceal the wings. Tormentors can usually be closed to help frame a set, or to shorten the stage (as a backdrop). Curtains hung to conceal wings that *don't* or *can't* be closed are often called *legs*.

Trap room: a large open space beneath the stage of many theaters—used for scenery, automation gear, storage, and for literal actor appearances/disappearances via "trap doors" in the stage floor.

Vom: aka vomitorium, a passage for audience entry/exit, usually below a set of seats in the form of a sort of tunnel.

Wings: side of stage spaces for staging of performers and gear out of sight of the audience.

UNIQUE GEAR

Area mics may be used to reinforce voices, rather than lav mics. Area mics are usually boundary mics placed at the downstage edge of the stage, and/or mounted to set pieces, and shotgun mics suspended above the stage. These have the advantage of being fixed, such that blending a choir's sound is easy, and you don't have to worry about mic

Tales From the Field

We did *Metamorphosis* which has the big pool, and we did it in thrust, and I wanted sound to come from the center of the pool. So we had some speakers upstage on the back wall. And then we did little vents and did speakers around the bottom edge of the pool. And getting the delays just right to get this ball of sound to just appear in the middle of the pool, I enjoyed that. And nobody saw any speakers—there was just the sound. It's kind of weird that my goal as the sound designer is to not exist, but to make the show just *that much* more powerful—and for nobody to notice that it's me. We're not there, but it's 200 cues and 50 speakers.— Christopher Plummer

changes on actors, but the disadvantage that we'll be hearing a lot of the stage itself—reflections from the deck and scenic items that often sound less than ideal, and that don't give good direct control of vocal tone.

Computerized, live-operated, cue-based playback systems (a la QLab) are also found frequently in theater and not as commonly elsewhere.

Mixing consoles with "theater" software (such as the Digico SD9T—T is for "Theater"!) are increasingly common due to the feature sets they offer designed specifically for stage production.

Raw drivers—often built into sets and props in order to send sound effects that localize to a specific device.

Stagecaller—a TeleQ (see below) but for smartphones, triggered via MIDI or OSC, over a Wi-Fi network.[4]

Sub-miniature lavalier mic elements, coupled with beltpack transmitters, are extremely common in theater due to their ability to be concealed . . . whereas in concert work, wireless beltpack mic systems are usually more worried about sound quality and durability than concealment.

TeleQ—a device that allows the remote triggering of an old-school analog telephone via the press of a button.

TYPICAL TIMELINES

Loadin for a straight play can take anywhere from a couple of weeks to a couple of days, but generally not less. It is extremely rare (though not unheard of) to load in a play the same day it performs.

Tech week is so-called because it is usually a week long (give or take a couple of days). It generally encompasses such events as a "spacing rehearsal" which is to get the cast used to being on a real set instead of tape marks on a rehearsal room floor, "cue to cues" in which the production is worked with only technical cues run in real time, and all other scene work skipped, full tech runs (doing the whole show end to end with all technical elements save for costumes), to dress rehearsals (where costumes, makeup, and hair/wigs are added to the other technical and performance elements). Mix engineers cannot typically get their final EQ settings worked out for actors until dress rehearsals, as mic positions often change until they are hidden in the real costume pieces. For regional professional theater, about a week is common time in the theater before opening. For Broadway shows, tech "week" can be 2 weeks, 3 weeks, or (on major spectacles and especially musicals) even months.

For touring productions, loadin and tech times will be negotiated based on the scale of the production, the needs of the staff, the availability of the venues, and the lengths of the runs at each stop.

Some stage shows (mostly musicals and Broadway straight plays) also hold "preview" performances. These are fully produced performances with an audience, but they are considered "pre-opening." The production team may make adjustments between previews based on audience responses, critics may be invited so that reviews can be written in time for formal opening, and so on. Unlike actual performances, the design team is still often in the theater for many (if not most) previews. Preview audiences are sometimes invited by the company and sometimes pay for their seats, though preview tickets typically cost less than tickets to a fully open and running production.

DESIGN PRESENTATION

Production documentation will include, but is not limited to the following:

- Plan drawings
- Elevation drawings (side and front are common)
- Mounting details if/where needed
- Block diagrams (signal, control, network, com)
- Mic run sheets (which actors have which mics at which times, where do mic or costume changes that the RF crew will need to be involved with take place, both in the script and physically in the theater)
- Hookups/IO sheets
- IP schedules
- Magic sheets—these are unique to theater design. A magic sheet (Figure 39.4) is a simplified speaker plot, where each speaker is shown with a large number next to it. That number corresponds (typically) with the numeric signal path out of the playback system. This is for the designer's use during tech, such that if the designer decides they want a sound sent to the "third surround back from the proscenium on the house left side," they can just send to "17" or whichever number is listed on the magic sheet.

Design paperwork will be presented to the creative team as part of the design presentation process. Most theater companies have a formal round of design presentations prior to the beginning of the "production period" wherein sets are actively being built and so on. If the sound designer is the system designer, they will be presenting their work to the director, producer(s), and scenic design team (while the scenic designer *shouldn't* have any authority over the sound system, we've already established that they often do have some investment in where we place

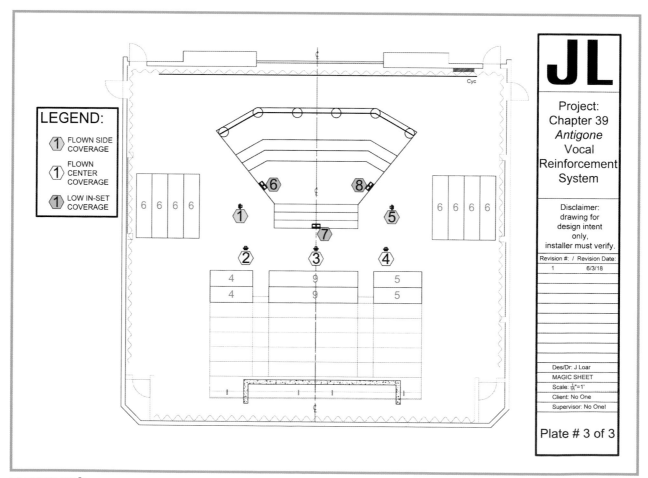

FIGURE 39.4
Sample Magic Sheet

our devices). If the system designer is not also the sound designer, the system will be presented first to the sound designer for approval, and *then* to the team as a whole.

The process of theatrical system design can be difficult and challenging. Unlike concert design (where sound quality is of paramount importance and thus often takes precedence over other departments' desires), theater design is a very collaborative art form. As such, when working with a difficult or disagreeable team, it can be very stressful, but when working with positive and engaged collaborators, can be enormously rewarding. There will be more complex challenges in theatrical system design than in concert design, but the results can be transporting and magical.[5]

DESIGNER QUESTIONS

- Are you going far enough, or too far? I have spent a lot of my career in theatrical system design, and I have often encountered designers who tend toward one or the other extreme in terms of planning their speaker layouts. On the one hand, many young designers feel that adding anything beyond the basic L/R they've experienced in concert systems is too elaborate. This is far from true—most theatrical systems of any complexity at all feature multi-channel playback, localized to varying locations. On the flip side, I have worked with very technologically oriented designers who feel that they need a speaker in every conceivable position in the theater, even if they don't have a plan for using those speakers, just so that "the option exists" once they get into tech. While designing a simple or a complex system can be appropriate for a production, care should be taken to avoid oversimplifying or overcomplicating a design just because you can. Always remember that system follows content. What do you need the content to do, in what way? Let that be your guide in terms of how the system is specified.
- Does the sound designer's impression of the sound level needs match the director's? If you are designing a system but are not the content designer, it is important to get ideas from both your sound designer and your director about how loud a show needs to be. If your designer wants huge levels, but the director wants everything to be whispered, who do you design toward? If budget allows, provide a system capable of high SPL but also capable of low SPL with a very low-noise floor. If a budget does not allow this, this may be a point at which to save money during value engineering, as a quieter system is often less expensive than a loud one.
- If you are mic-ing voices, are you reinforcing or replacing the voices? Do we need to localize to the actors onstage?
- Where do sound effects need to come from (spatially), and what kind of speakers do you need to tell their story effectively?
- Is there music? Is it diegetic or not?

＊ ＊ ＊

Next, we will examine musical theater, which (perhaps obviously) is less focused on sound effects, and more focused on blending lyric intelligibility with musical fullness.

Notes

1. All of these are based directly on my personal experience.
2. TiMax is one of the standard systems, though several exist.
3. A note on scrim—it is often said by theater people that scrim is available in an "acoustically transparent" version that will visually conceal the speakers but allow sound to pass unchanged. Functionally, this is false. There is no obstacle that you place in front of speaker drivers that is truly acoustically transparent, it is physically impossible. That said, there are scrims designed with acoustics in mind that will present the *least* possible obstacle to sound, and these are desirable if/when we have to conceal our speakers. Of course, then the scenery department wants to paint the scrim, for every hole in the scrim filled with paint, our sound transmission gets worse. We can still work around it if it is painted properly, but let's not pretend it won't alter the sound quality. It will, unequivocally, measurably, alter the sound, and always for the worse.
4. Developed by Tony-award-winning sound designer (and all-around mensch) Rob Kaplowitz!
5. Don't get me wrong, concerts can be transporting and magical too, but this is usually down to the music itself, and we are trying to get out of its way—whereas in theater, sometimes the system itself plays a huge role in the magic-making. Think of the 6,000+ speakers installed for Cirque du Soleil's *Love*. They put that many speakers in place so that each and every audience member has a rich and detailed personal experience. I've designed and worked some large concerts, but none of them had, or needed, 6,000 speakers.

405

CHAPTER 40

Theater—Musicals

Musical sound design differs from straight play design mostly as a matter of focus. Where straight plays emphasize dialog and sound effects, with music (typically) playing a supporting role, musicals—perhaps unsurprisingly—prioritize the clarity of the songs. While musicals themselves come in a range of styles—from the so-called light operas of Gilbert and Sullivan, to the contemporary rap musical *Hamilton*, and encompassing many genres and arrangements in between—the general scheme is that, unlike, say, a Radiohead concert, we really need to understand all the lyrics. The music that supports those lyrics is the next highest priority, and sound effects take a distinct back seat.[1]

DEFINITION

Musical theater is a Western genre of stage performance comprising a narrative stitched together with vocal musical numbers, which often advance the plot or provide important details of characterization. It is often separated into the songs and score versus the "book"—the narrative itself. Early Broadway musicals were often less narratively focused—so-called follies shows consisting of popular songs strung together with little to no story element. Today, we still find "jukebox musicals" that are designed to play a bunch of popular songs (think *Mamma Mia!* and the songs of Abba, or *Rock of Ages* playing a host of '80s power-rock songs), but even these jukebox musicals today have narrative stories, as audiences are hesitant to accept a pure "follies" type show today.

Sound systems for musical theater are often based around a subsystem paradigm known as the "A/B System." In an A/B system, the designer has provided two parallel sets of speakers to cover the audience, each carrying distinct content. In the earliest A/B systems, the idea was that if two actors were onstage, both wearing mics, we could avoid some of the in-system comb filtering that happens in the summing electronics by carrying one actor's voice along one signal path to one speaker (or set of speakers), and the other actor's voice along a *discrete* signal path to another speaker (or set of speakers), thereby ensuring that the actors' amplified voices only combine acoustically in the space and preventing one possible layer of summing artifacts. As time has gone on, the term "A/B system" has also come to be used for sound systems featuring discrete subsystems wherein one set of speakers is carrying the vocals/singing, and another is carrying the orchestra or band. When I mention "A/B systems" from here forward, it is the latter usage to which I am referring.

This type of system provides a few distinct advantages in musical production:

1. Vocals can more easily be localized to the stage, while the orchestra can be located in a different spatial location, which preserves a feeling of natural voice from the actors.
2. By tasking each set of speakers with only vocals or only music, each set is driven more efficiently, since each subsystem is required to do less. This can result in more overall clarity.
3. By summing the vocals and music acoustically, in the space, rather than electronically, in the console, a more natural-sounding effect *can* be achieved.[2]

This kind of system design assumes that sound effects are minimal and/or typically handled by a third subsystem of speakers—yet we still refer to this as an A/B system, not as an A/B/C (or A/B/C/D) system. Figure 40.1 shows an example of an A/B system.

The process of designing for a musical is a combination of the process of designing for a genre-specific concert (where vocals are the preeminent concern) and the process of designing for a straight play, with some particular wrinkles in equipment and process thrown in.

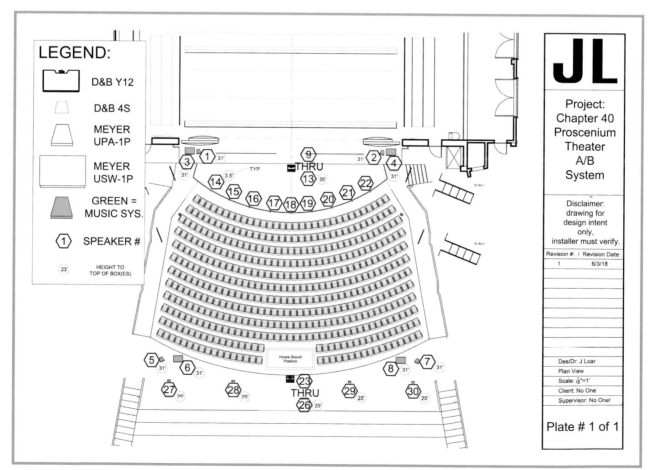

FIGURE 40.1
A Broadway-Style Theater With A/B System

STANDARD ELEMENTS

- Inputs: wireless microphones on actors, area mics, instrument mics/DIs on orchestra/band instruments, playback for sound effects, intercom systems
- Processing: FOH console, occasional monitor console, personal mixing systems for orchestra musicians, occasional outboard effects, plug-ins
- Amplification: standard operation amplifiers as needed
- Outputs: often A/B system, with one set of speakers for vocal reinforcement, and one for music (in effects-heavy shows, it is possible that there is a third subsystem for effects, or in shows with minimal sound effects, they may be played through one or both parts of the A/B system)
- Converters: at I/O of digital equipment, typically
- Advanced processing: DSP, certainly; sometimes flexible acoustic system like Meyer Constellation
- Control: iPad control of console(s), VNC control of playback systems for designer, wireless mic monitoring via network, speaker monitoring via network, VNC control of DSP, scene recall in console(s), etc.
- Clocking: often internal, though on large productions (think Cirque du Soleil) external master clocks are common

UNIQUE FUNCTIONS AND SYSTEM NEEDS

Musical theater once again typically centers the production system around the mixing console. Mixing a musical is an art form unto itself, and many musical mix engineers make lifelong careers of it (though many graduate into designing sound for musicals, as the role of "sound designer" on musicals often largely consists of designing the system). Mix engineers prize consoles with flexible programming, "theater"-based actor profiles, robust DCA/VCA control (for groups of channels, such as a host of drum mics, that don't need to change their level relative to one another but as a group need to change their level in relation to other instruments).

Orchestras (or other live bands) for musicals typically monitor their performance via in-ear monitors, fed via personal mixing systems (as monitor consoles are a luxury for which most Broadway houses have no space)—which will be explained in more detail next in "UNIQUE GEAR." The exception to this is sometimes the conductor, who will often have a personal monitor speaker whose feed the FOH engineer controls (unless it is the rare show with a dedicated monitor console).

When assessing the needs of a musical, the first thing that must be done is to assess the music. If this musical has been performed before, there is probably a cast recording available, which can be very helpful in understanding orchestrations and genre/style. If this is a new musical, hopefully, you can attend some orchestra rehearsals and make audio recordings to study so that you can get a strong sense of the score. Regardless of whether the musical is extant or brand new, it behooves any musical sound designer (even just a system designer) to read music. If you can examine the printed score and understand what the music is likely to sound like, you will be in a strong place to begin the design process.

Once you've heard and/or evaluated the score, you will let the genre of songs and the overall feel help you determine your system goals. If you are designing *The Pirates of Penzance*, you are unlikely to need a system that bowls the audience over with its power, though you will, of course, need to deliver crisp clear response with a low-noise floor and full frequency range. However, if you are designing Green Day's *American Idiot*, you will need to create a system that has the perceived power of a rock concert system, while prioritizing vocal and lyrical clarity above what you might produce if actually designing a rock show.

From there, the position of speakers becomes very important. It is difficult to deliver clear vocal and orchestral coverage without acoustically favorable speaker locations, so speaker positions need to be planned and predicted as soon as a rough set design is available. Registering early your desired speaker positions will help ensure that you are actually allowed to place speakers in those positions (though there is no guarantee, as we've previously discussed on other projects with scenic design elements). Negotiating speaker placement in theater is often tricky, but at least in musicals, the production team is typically aware that the music itself has to be a priority. Thus, with proper planning, we can often (though not always) position speakers in acoustically desirable positions.

VENUE CHALLENGES

Broadway theaters are often smaller than young designers (especially those who have never been to NY) imagine. They were built in eras before sound systems were common, so mounting positions may be challenging to secure (unless the theater has been well retrofitted). Any venue with an orchestra (floor) audience section, and a balcony section, presents specific challenges as well—the seats under the balcony will be shadowed from direct coverage by flown main speakers, so under-balcony speakers are common. Main speakers must often be flown rather high, and the audience is often seated very close to the lip of the raised stage, such that the main speakers can't adequately cover the front rows, particularly for vocal systems, as localization would pull right up above the actors. Thus front fill speakers are common. Some older theaters are also historical landmarks, meaning there are strict rules about what can and can't be altered (or drilled into, or mounted on, etc.)

UNIQUE TERMS

All of the unique terms from the previous chapter apply here as well. Many terms from concert production apply as well (soundcheck, linecheck). Additionally, there are a few terms specific to musical production that we need to be aware of.

Seating the orchestra: like a spacing rehearsal, but for the orchestra. Theatrical pits are notoriously small (again, having been designed for previous eras before we added mics and amps and other gear to the setup). It often takes a few hours to properly situate the orchestra and verify that all members can see who else they need to see and that they have room for all of what they need for the performance.

Sitzprobe: often abbreviated as "sitz." This is the first rehearsal in the theater with the orchestra (early tech rehearsals are often done with piano accompaniment—or another minimal accompaniment—only). There is no actor blocking at such a rehearsal, just singing. The sound production team is often running mics during sitz so that they may use it as an extended orchestra soundcheck.

Message in a Bottle

Musicality is the most important developed skill so that there's a sensitivity and a perspective. One of the things that seems most lacking in large-scale system designers right now is that of perspective. They don't know when too loud is too loud. They've lost track of having a naturalized initiation of voice, or instrument, and complementing with reinforcement versus replacement of sound. To listen and to know for example "maybe the particular voices of this ensemble are not being expressed properly." That ability to observe, and then having a depth of experience to make choices for the right gear, the right microphones, changing from a crown to an ear-rig mount, to a head-worn mount [for RF microphones]. What are the things you need to do, what are the compromises, how do you make it through that? That—about the ear—is clearly the most important, and of course judgment associated with that. If you can pull that off, if you can—in your mind's ear—know where you want to take it, then the rest of the system design is a matter of mechanics. Yes, there are some choices regarding the gear, and the aesthetics, of the choices you might make about what gear is appropriate for a particular situation—when do you need balls without accuracy, and when do you need accuracy and sensitivity? When do you need everything—so you're talking about this incredible ability to reinforce a bunch of beautiful simple small sounds, and yet at the same time create these thunderous gestures through the ensemble or sound effects—all of those things are big, big challenges. They all require the ear to have some sensitivity, some perspective about what is achievable in the context of a particular venue. If you do have that in your head, then I think it is absolutely possible to design and deliver a sound system—even at a number of different scales, that fit different budget categories—there's a lot of things that are possible. If you're honest with somebody and say, "look, we don't have enough money to do all of this, but here's the things we can do really, really well with what you've got"—that ability to both use your ear, to communicate faithfully, with intention, and to understand the politics well enough to be able to negotiate for your particular department, all of those things are really important primary skills.—David Budries

Message in a Bottle

I would say the other major skill which I see honestly sadly lacking in a lot of younger designers is how to read a play and how to talk about it. You know some people say, "well, you know, it's reinforcement, I don't need to read the play, I need to know how many actors there are, and the pit, and then what the room is shaped like." And I feel like that's very much only half the conversation. And you can't know if you can't connect to the text. Also, reading sheet music. I'm surprised at how many reinforcement-based designers don't read music. I always respect when we can request a complete score. And they're always like "why do you need that?" Well, because I need to be able to listen to a piece in the show and know that there's something wrong with the way it sounds. I can know that emotionally, but I need to be able to open up the score and see why it's wrong. Like, "Oh my God, there's these dynamic markings here that I'm inverting. So this moment is supposed to be all about this really strong acoustic bass bowing and I've foregrounded the bongo player, what's wrong with me?" You don't even have to be fluent at it. I think there's often an intimidation factor with sheet music, with notation, where people feel like "well I can't play it, so I can't read it." You know I'm never going quiz my engineer about whether that's an A or an A flat. But I feel like I want you to be able to follow along and understand the *movement* of the music at the very least.—Robert Kaplowitz

Wandelprobe: (pronounced "vahn-del-probe") like a sitzprobe, but with actors performing minimal blocking (also known as "marking" the scenes—not performing full out, not doing dance numbers, but moving from place to place on stage to give a sense of position at various moments in the script).

UNIQUE GEAR

Multi-mic-ing is a common technique in musical theater, particular on lead characters. This is the tactic of placing two microphone elements and two distinct transmitters on a performer, such that if one should go out mid-performance, there is another redundant mic that can be immediately used, without having to rush the actor offstage for a complex changeover.

Personal mixing systems are very common in Broadway orchestras.

As seen in Figure 40.2, personal mixing systems are small mixers, situated with every performer in a pit orchestra (or at least the leader of each section).

Each mixer features prepared submixes (stems) of musical content—sent via auxes from the FOH console. A typical personal mixer features 16 submixes, so on a standard musical, each performer might be receiving submixes broken out as follows:

1. Vocals
2. Drums
3. Percussion
4. Bass
5. Cellos
6. Violas
7. Violins
8. Low Brass
9. High Brass
10. Low winds
11. High winds
12. Acoustic guitars
13. Electric guitars

FIGURE 40.2
An Aviom Personal Mixing Station
Source: Courtesy Aviom, Inc.

14. Sequenced elements (synths)
15. Click track
16. Conductor talkback

The aforementioned, of course, is just an example, and how the tracks are broken out will depend on the orchestration of the show, and what the music director, conductor, and ensemble decide they need to be able to control individually. At any rate, these submixes are pre-blended by the FOH engineer and saved as part of the show programming. Then, during the run, each musician in the pit has wired in-ear monitors (or in rare cases, over-ear headphones, if they need isolation from musicians around them), and can raise or lower the levels of each channel as they need in order to derive pitch and timing for their own performances.

Onstage monitors are increasingly common in musical theater. While typically not as loud as, say, rock concert stage monitors, they are necessary for productions where the actors may need to hear some part of a tricky or dense musical arrangement more clearly, and particularly on productions where—as is happening more and more these days—the orchestra is not in the pit but is instead somewhere backstage in acoustically controlled performance rooms. In fact, for some productions, not only is the orchestra not in the main performance space, the orchestra is not altogether in one space at all—horns are in one room, strings in another, percussion in still another, etc. This is done for enhanced acoustic control, but also ensures the need of accurate monitoring onstage. In addition, occasionally vocals will need to be fed into the stage monitors, to help actors match pitch with one another, though with the prevalence of omnidirectional mic elements, this can be quite challenging to mix without feedback.

Mixing consoles with dedicated theater software are in high demand for musicals, as are consoles with robust cueing features, since the contemporary practice of mixing musicals often relies on scene recall for every major exit, entrance, or scene change. It is much easier to hit "next scene" than to have to remember which 25 mics to unmute each time.

Dedicated "reverb" speakers, aimed at walls and ceilings of the venue, and used to excite the space itself, rather than to aim directly at the audience, are sometimes found in musical design (and straight plays) as a way to create a diffuse sound field that an audience perceives indirectly.

Message in a Bottle

If you're mixing a musical, and you add reverb to the vocal and it's coming out of a reinforcement speaker, it doesn't sound like reverb. It often just messes up your mixing and masks things and takes away clarity. But if you actually have reverb speakers that are specifically designed to excite the reverberant field [aimed at walls/ceilings/etc. in the venue] as opposed to the way a PA speaker is [aimed at the audience], suddenly you can add reverb, and you can do things that you just can't if you don't have those speakers.—Christopher Plummer

Orchestra mic-ing often relies on the smallest form factors possible, clip-on mics for horns and percussion instruments, at the sacrifice of sound quality (this is the case anyway when the orchestra is in the pit . . . one of the advantages of situating the orchestra in a space other than the pit is the ability to mic with whatever size and form factor is desired for sonic quality).

Whenever tap dancing is a part of a musical (as it often is, even in the 21st century), it is common practice to mount contact mics or boundary mics to the tap surface (often they are mounted to the underside, so they are visually hidden). This allows the tap dancing to be amplified as its own signal.

Occasionally when a musician needs to move around stage with his or her instrument (as a "wandering minstrel with violin" in a commedia dell'arte production might), it may be desirable to not see a wireless pickup on the instrument. In such a situation, a wireless mic might be run down the sleeve of the performer, such that it can act as instrument pickup, but be more easily concealed.

"GO buttons" are often given exaggerated form factor in musical production. Rather than require a console operator to find a small button mounted on a console surface in order to recall the next scene, often, that scene advance is fired by a device routed into the GPIO ports on the console—frequently a large "mushroom button" type control (Figure 40.3) that is easy to hit without having to visually search for it each time.

FIGURE 40.3
A Mushroom-Shaped "GO" Button
Source: Courtesy Honeywell International, Inc.

TYPICAL TIMELINES

Tech week is still a standard for musicals, though as identified in the previous chapter, large-scale Broadway musicals and/or Cirque du Soleil-style spectacles will often tech for longer periods (any show that anticipates running at least a year doesn't generally feel compunction about paying for a little more tech rehearsal time).

Sitzprobe and wandelprobe, respectively, each take one rehearsal period. Music rehearsals with orchestra members are typically the strictest part of any schedule. While union productions of even straight plays will need to deal with Actor's Equity[3] and IATSE[4] rules about call lengths and break times, the musicians' union (American Federation of Musicians) is the strictest about work periods, call lengths, break times, etc. Musical rehearsals are often scheduled around the musicians themselves *as if they are made of plutonium*—they are rare and powerful materials that need to be handled with care lest we all suffer the wrath. Their call lengths, break times, and so on are tightly proscribed.[5] On non-union musicals, this lightens up, but most musicals that can afford to pay a proper sound system designer are probably employing professional (and thus typically union) musicians. While it is not the job of a sound system designer to schedule musicians, it is the job of the sound team to be on point during times when the musicians are available, such that we maximize the value of those calls.

DESIGN PRESENTATION

All of the paperwork standards and processes as listed in Chapter 39 repeat here. Speaker predictions and modeling are *very* common for musical systems.

In Figure 40.4, I show some alternative drafting methods to what we've seen so far.

Instead of using small arcs to "jump" lines over one another wherever they meet, I've used color-coding of signal lines so that different types of signals can pass over one another but it is still clear they are not interconnected. This figure shows only the first page of what would be a larger drawing set block diagram. I have not shown any amps or speakers (since we saw plenty of those in previous chapters) and have instead focused on certain aspects of a musical theater system that we have yet to see. For example, the Dante-fed Aviom personal mixing system for the orchestra—I only show one of what would be 13 personal mixers in this setup (and in a full drawing set I would show them all) but

414

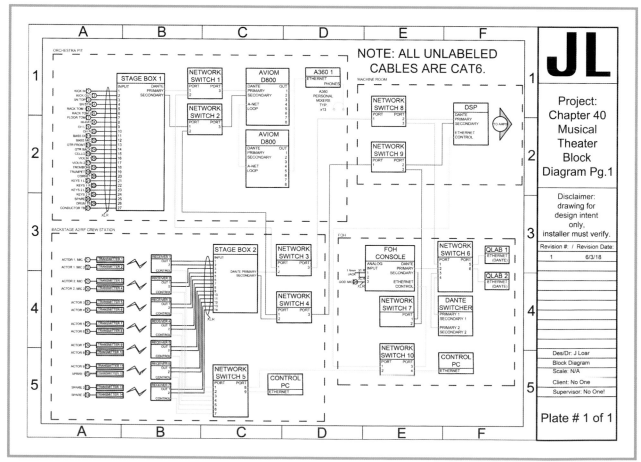

FIGURE 40.4
Block Diagram (First Page Only) for a Musical Theater System

the principle of how that system works is important. I show a redundant Dante-based QLab playback system, feeding Autograph Sound's Dante Switch.[6] I show control computers at both FOH and at the A2/RF station. And, as would be expected in most contemporary sound systems, I show a ton of network switches.

As we move through this book, we are approaching an ever-more-realistic vision of our technical documentation; however, keep in mind, this section's drawings are still intended primarily to illustrate system configuration *concepts*. If the system represented in Figure 40.4 were to be drawn for production, we would need to include many other details. For example, none of the equipment models are listed. No computer is shown to have a connected monitor, mouse, or keyboard. I have even (for the sake of clarity, since the print is bound to be small when reproduced in this book) omitted our aforementioned GPIO-connected mushroom button.[7] In the final section of the book, we are more concerned with such details.

The process of designing for a musical is often simpler in some ways than for a complex straight play, as most musicals don't demand extensive playback effects,[8] and thus are reduced to vocal and music systems. The tuning, calibration, and mixing of content in those systems is typically vastly more challenging than doing the same for a straight play, and thus high priority is placed on mix and system engineers with very sensitive ears, narrative sensibility, and ability to read and understand scores.

Since most musical productions are more expensive than most straight play productions (in part because the popularity of musicals often dangles the promise of greater returns at the box office), there may be more producers and other higher-ups that must approve your design work than on a straight play. Sometimes, these people will have no sense of what different sound equipment does, so your presentations must explain clearly why each system choice is needed, and particularly, what each will gain your audience and the overall production.

DESIGNER QUESTIONS

- Where does the line between musical and non-musical get drawn? Sure, *Death of a Salesman* is a straight play, and *West Side Story* is a musical, but there are shows that blur the lines. In these cases, it can be advantageous to work with some design principles from both straight and musical theater paradigms. In that case, what are the narrative priorities?
- Is this show intended to run for 6 weeks at a regional theater? Or 6 months? Or 6 years? The level of investment in system equipment should scale with the intended run length. A show designed for Disney on Broadway will always be intended to run for years. A show by an upstart producer in a regional market is unlikely to do the same unless everyone is *very* lucky.
- Is there an existing (or multiple existing) cast recording(s) for this show? If so, are the producers and director of your production intending to live up to the sound captured in that recording? If so, how best can you recreate those tonal qualities and level balances? If not, what are they intending to change?
- What kind of music is being delivered? How should it make the audience feel? Are we supposed to feel like we're in a West African nightclub, sweating and drinking while we listen to *Fela*? Or are we supposed to feel like we're in a very refined and elite space, surrounded by *The Song of Norway*?
- Are we reinforcing or replacing the voices? I know I mentioned this in the last chapter, but it bears repeating— for a musical, the vocals are always paramount, and it is important to understand early what we're trying to accomplish with them.

* * *

Live events are not limited to concerts and theater. In the next chapter, we'll examine conventions and other multi-room live events. While there is typically less artistry involved in designing such systems, the engineering challenges can be formidable.

Notes

1 Some musicals really have no room for sound effects of any kind. Today, there are some musicals that are also dense with sound effects, but these are the exception rather than the rule.
2 Note the word *can*. This process can also result in ugly sonic blurring, in the hands of inexpert mixing staff.

3   The actors' and stage managers' union.
4   International Alliance of Theatrical Stage Employees, the technicians' union.
5   Check with AFM for the latest rules.
6   Autograph is one of the most famous theatrical sound design firms in the world.
7   I will admit an aesthetic fondness for mushroom-buttons, giant metal-handled company switches ("Franken-stein switches"), and other such ephemera left from the last mechanical age.
8   Though sometimes they require a ton of mixing effects, either via the console or an external computer-based processing software, like Apple's Main Stage.

CHAPTER 41

Conventions and Other Multi-room Live Events

Multi-room installations are often referred to under the general heading of "distributed" sound systems. This generally means any system that has a central "brain" that is controlling sound and events for multiple sub-venues. While many distributed systems can be found in installation work, from hotels and casinos to theme parks and cruise ships (all of which will be addressed, however briefly, in the following chapters), we do find distributed systems for some types of live events—particularly massive conventions, whether they are gatherings of industry or gatherings of comic-book fans. Sometimes, the subsystems are all tied together, sometimes, these are merely collections of small satellite systems with no interconnection. Either way, when tasked with a large multi-room event, we must assess it as both a coherent event and assess the sub-event/space needs.

DEFINITION

A multi-room live event is one in which there are multiple parallel spaces being visited/engaged by audience/guests simultaneously. A convention is the most obvious example, featuring (typically) a large open show floor with exhibitor booths, and many side rooms for presentations, seminars, meetings, and the like. A less obvious example might be a warehouse party, with a main space featuring a DJ set, and auxiliary spaces either featuring other electronic/DJ performers, or art gallery installations, or the like.

STANDARD ELEMENTS
- Inputs: mics for presentations/performances, playback for BGM (BackGround Music), emergency announcements (either canned or live), intercom systems
- Processing: DSP, with consoles as part of subsystems around the venue
- Amplifiers: often a combination of standard amplifiers for subsystem events, and 70V amplifiers for overall venue systems

- Outputs: small subsystem mains (typically L/R config, run in mono), 70V installed systems for overall venue coverage
- Converters: at I/O of any digital equipment, not outboard
- Advanced processing: DSP may be programmed to combine rooms or separate them into zones, may feature subsystems that are or are not integrated into the main venue system, paging stations, and control
- Control: zone by zone paging/announce, as well as "all-venue" paging/announce
- Clocking: internal to any digital equipment, DSP as master for a venue system

UNIQUE FUNCTIONS AND SYSTEM NEEDS

When planning a system for a large, distributed installation, the first thing we generally need to do is evaluate the venue and see how it will be subdivided for the event in question. This usually involves talking both with venue staff and your own production team to assess how the needs of the event will be constrained by the venue. It is typical of such multi-room events that the venue is preexisting (if it is not, chances are you are part of an architectural design team, and this is a permanent installation, not a temporary one).

In Figure 41.1, we see the main exhibit floors and a group of meeting rooms of the Los Angeles Convention Center.

When we are designing a project in a venue so large, the first question we may have is "what, if any, sound system(s) is already in place?" Looking through a convention center's booking and promotional materials, we may easily be able to establish first that such venues typically own speakers and small consoles and mics that can be set up in meeting rooms. Will these suit our needs? It largely depends on what type of event we're holding. If the meeting rooms will be simple discussion panels, as at many professional conferences, these stock items may be just fine—particularly if we're not recording events. However, if we're designing ComiCon, we should be aware that such meeting rooms often show footage and teasers from upcoming spectacle films. These events are often packed to the gills by superfans and are often discussed widely on the Internet as soon as they are finished. Media companies

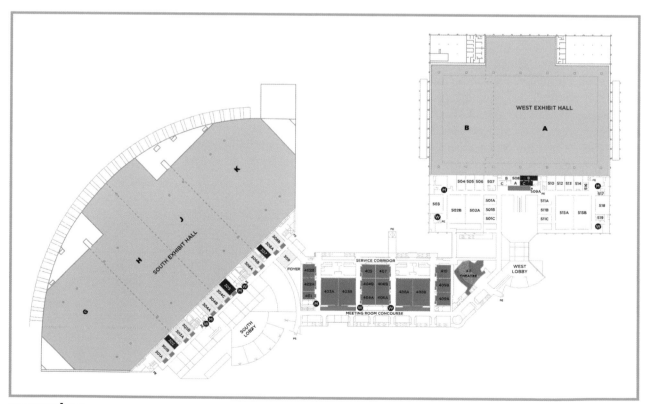

FIGURE 41.1

Los Angeles Convention Center—View Showing Main Exhibit Halls and Meeting Rooms
Source: Courtesy LACC

spend a lot of money on these events, and as such, meeting rooms will be temporarily outfitted with cinema-quality surround sound systems. Depending on the meeting room configurations—most of these rooms are separated by temporary partitions and will usually be sized differently according to the anticipated popularity of the events programmed into them—these systems can vary wildly in size and scope. If events are being recorded, room systems must be coordinated with the recording team.

One of the early questions that needs to be asked is how fire alarms or other emergencies are to be handled. Does your particular large event space feature an alarm system that consists merely of blaring klaxons? Or are there verbal announcements programmed, telling the audience where to go and what to do? If the latter, do the individual rooms need to be tied to that system, such that the sound in each meeting space is overridden automatically by the fire announce? Or will we count on the fact that the fire system will be sufficiently loud to alert local room staff that they need to mute signals in their individual spaces? The latter scenario is more common, but newer multi-room venues are being designed with automatic ducking of all subsystems in the space if/when the emergency systems go live.

Another question to be asked is "what events are taking place in the main exhibit hall?" Conventions with hundreds of exhibitor booths often sound like a calamity—competing small sound systems at many booths, salespeople working their game, visitors chattering and laughing and eating chicken strips.[1] Some conventions will have no "main events" on the main exhibit floor, preferring to hold keynote addresses and the like in ballrooms (typically on another floor of the building). However, some others will have at least one if not multiple stages on the exhibit floor, with events taking place throughout the day. In some of *those* instances, events on one of those stages will be treated as "main events," such that when they are taking place, the sound from them must be audible throughout the venue. How we reinforce these events will depend on their nature. If these are speeches, we merely need enough direct sound sources around the venue to communicate the speech frequencies, though we may need to work out a complex delay system, such that we don't have too many out of time multiple arrivals at various listening areas. If these events are musical in nature (sometimes conventions hire bands to play, even if the convention has nothing whatever to do with music), we must provide as close to full frequency range as possible. Key to main event sound in such environments is that even if the audience is all supposed to be keyed into the central event, the overall SPL should never approach real concert level. Exhibitors will still be doing business, trying to make money, and sound that is too loud will prevent these conversations. In such cases, it is critical to have a ton of direct sound sources, distributed widely throughout the venue, running at moderate SPL, such that guests can hear clearly in all locations, without ever being overwhelmed by loudness.

We need to determine how interconnected, or not, our overall system scheme will be. Conventions, multi-room parties, and the like are increasingly trending toward the theme park model of large integrated systems, with central processing and control for at least the major functions, but—in part because many convention centers are old, and in part because many of them aren't as worried about presentation as themed installations would be—there are still lots of multi-room events taking place with small analog subsystems that don't interconnect in any real way. Figuring out the scope of the system(s) will be very useful in determining how to go about designing them.

VENUE CHALLENGES

Any large multi-room space that needs a central sound system begins by presenting the mechanical challenge of the need of very long cable runs. Even with balanced signal cables, we sometimes find ourselves exceeding recommended maximum run lengths, and as such, we need to convert our signals to another format with a longer maximum distance. At the low and mid-budget, this often means converting audio signals to network signals, but once we exceed network run lengths, we either need signal repeaters/boosters or to convert our signals to fiber-optic lines, whose effective run lengths outpace any network cable by a long shot.

Convention centers often have strict rules about how installation must work (including, often, union rules—as their house installation crew may be union). Contact your center in advance of your event, as knowing their in-house rules will help determine your operation plan.

Convention centers are often large and cavernous, and thus good direct sound coverage can be challenging as well—necessitating more sources, placed fairly close together (but not so close together as to create horrible unwanted combing interference). This is a pattern we will see repeating in many large distributed systems—favoring multiple smaller sources at lower SPL, in contrast to the "giant L/R mains" paradigm found in concert production.

UNIQUE TERMS

Air walls: temporary partitions between meeting rooms

Dead-case land/boneyard: designated place for empty road cases that will be re-filled during a strike

Exclusive contractor: someone hired by the venue or overall producer to provide services across the entire event as needed

Exhibitor approved contractor: someone hired by an individual exhibitor(s) to provide service for his or her booth(s)

Floor manager: like a stage manager, but running the exhibit floor

UNIQUE GEAR

Hanging pendant speakers are extremely common for exhibit floors (especially when music is not needed). Stand-mounted, point-source speakers are typical in meeting rooms—often self-powered to avoid the need for an amplifier rack.

For large, interconnected, recorded events, there may be a broadcast-style control facility set up temporarily, with camera and sound control for rooms throughout the venue. Broadcast-style facilities typically take multiple digital signals (often in redundant pairs in case one fails) into a vast routing matrix, often software-based, that then sends a signal to recording systems and other such endpoints.

Com systems are often more like retail-style com, functioning without one central base control station, but simply connecting multiple channels of staff. Though, if there is a broadcast-style control room, this is where com base station control will be found. Sometimes com needs are handled, not by theatrical *or* retail-style intercom systems, but just by long-range walkie-talkies (which themselves often allow multiple channels of communication)

Small (often analog) submixers for the various subsystems are found frequently.

TYPICAL TIMELINES

Loadin usually takes place over several days. It is also common for the exhibition floor to not be open on the first day of a multi-day convention, such that exhibitors with less to install can arrive on the first day of a convention and still open along with all the other exhibitors on day two.

Unless there are elaborate musical acts, or this is an audio convention (like the AES National Convention), soundchecks are unheard of. Linechecks will suffice, and for the smaller systems, it is common that individual sources (a presenter's laptop, for example) won't be checked until 5–10 mins before the presentation itself.

If your multi-room event is *not* a convention but instead a warehouse party, elaborate wedding, or other event, then timelines go out the window. Each project of this sort must be evaluated individually, and the timelines will be determined by the scale of event and systems, venue availability, budget, etc. There are no good general rules to be made about these sorts of events—I've seen weddings that took weeks to set up, and others that took less than 30 mins.

DESIGN PRESENTATION

If you are designing a multi-room event with a central control system, then a host of standard production documents will be required:

- Plots
- Block diagrams
- IP/IO/equipment schedules
- Et cetera

If you are designing a multi-room event with no central control, you may not produce drawings for many—if any—of the systems. If the meeting rooms are small, separate, analog systems, there is generally no paperwork generated. If there is a main event space—ballroom or exhibit floor or what have you—this may demand some standard paperwork to guide the crew in installation, but it would model after small concert system paperwork.

DESIGNER QUESTIONS

- Do the subsystems in the separate presentation areas need to integrate in any way? Or are they truly just small individual systems, operating in parallel, with no connection to one another?
- Are there offsite events that are part of this convention or event? Are you expected to handle sound systems for those offsite locations and events as well?
- Is there any call for high-fidelity sound? Most convention-style events (with the possible exception of those focused on the audio/entertainment industry) don't care about high-fidelity, just high SPL and speech intelligibility, but you should be sure of this before making gear selections.
- Is there any possibility of getting power isolation (sound power separate from lighting, hoist, or other show power), or will you be forced to clean up power as best you can with conditioners on the way into the audio system? We generally want power conditioners for our systems anyway, but when dealing with convention-style venues, it is often challenging to get true isolation from the lines powering lighting rigs and the like. Know this in advance and plan accordingly.

* * *

Multi-room events often consist of many small sound systems running in parallel. It is not till we get to themed entertainment and architectural installation projects that we start to see coherent, large-format, interconnected distributed systems.

In our next chapter, we will begin our tour through permanent installation systems by examining performing arts facilities and houses of worship.

Note

1 Why does almost every convention center around the United States sell the same pre-packaged chicken strips? I went to a conference in Germany where there was homemade kaesespaetzle, roast pork, and other delights available every day!

Performing Arts Venues and Houses of Worship for Permanent Installation

Now we begin our examination of permanent installation systems. We will attempt to address both systems that are installed in preexisting venues (as renovations) and systems that are installed in new buildings or facilities as part of an architectural project, but our focus will be on the latter as the process is more involved, requiring more unique terminology and the like.

DEFINITION

Permanently installed systems are intended to be part of a venue's facility offerings long-term. Performing arts venues here are defined as any venue whose primary business is offering live entertainment (from theater to concerts to stand-up comedy and so on). Houses of Worship are defined as religious spaces that have multimedia system needs (and that—at the largest end—are very similar to performing arts venues, hence their grouping in this chapter).

Designing permanently installed systems is a bit different (in most cases, though there are notable exceptions) from designing systems for events or tours. When designing systems that will be installed in a space and remain there for some time, there are a few things we need to consider that we haven't had on our radar for temporary projects:

- Life span: how long will this system be expected to operate with only minimal routine maintenance? For some venues, the answer is "in perpetuity"—which is unreasonable, but at least tells you that you need the most robust and future-proof system possible. For some, the answer might be 5 years, 10 years, 20 years. The client should provide this information, and if they don't know it, prompt them to estimate. If they can't, or if they provide an objectively ridiculous estimate, provide them with yours and explain why that is your estimate.
- Range of performances: most venues will host more than one type of event or act. Even a concert hall might host a seminar sometime, and a nightclub might host a gala fundraiser. Assessing the full range of possible events is key to producing a system design that will cover most of the venue's needs without extra expense.[1]

- Expendable schedules: operating venues will need to regularly buy new expandable items (tape, batteries, etc.). A good system designer anticipates this, and not only stocks the venue for its initial opening, but estimates—based on standard operation—how often the venue will need to restock expendables, and what impact that will have on their operating budgets. This would be on the radar for a Broadway designer, of course, or the designer of a concert tour system, as well.

STANDARD ELEMENTS

- Inputs: microphones of all types, wireless and wired, depending on the production; DI boxes; playback systems (often multi-channel); talkback mic at FOH; intercoms
- Processing: FOH console, monitor console, outboard effects
- Amplification: standard operation amplifiers as needed
- Outputs: FOH system speakers, monitor speakers, lobby speakers, backstage/dressing room speakers, possible recording rigs
- Converters: generally onboard at I/O stages of digital equipment
- Advanced processing: DSPs, sometimes handling zone routing for lobby/backstage/etc. as well as FOH and monitor systems
- Control: iPad control of consoles, zone paging via intercom system
- Clocking: generally internal, occasionally via external master clock in performing arts venues or megachurches

UNIQUE FUNCTIONS AND SYSTEM NEEDS

First, we must assess the production needs of our given venue. The unique functions of each will correspond to the work being presented within. We've addressed many common performing arts types in the previous chapters, but to recapitulate:

- Theater—straight plays or musicals are the typical Western fare. A theatrical venue that has equipment in stock to handle both of these forms will likely be able to adapt to handle non-Western performance forms—dance drama and the like. As we established in previous chapters, so much of theatrical sound system design is complex and drastically changing show to show that venues will often specify a baseline "house system"—vocal reinforcement for the whole standard audience configuration, and some small number of additional music and effects speakers—that productions can choose to use or supplement with their own rentals or purchases.
- Concerts—a general "concert" system may be slightly more standard than a general "theatrical" system, but even still, there may be vast differences between a system made to reinforce the Kronos Quartet and one made to reinforce a Jay-Z tour. However, most venues tend to have certain styles or scales of acts that they are focused on booking, and the venue house system can be designed to suit. The biggest change in a venue system versus a one-off system is merely the range of options—monitor speakers and mixes, mics available, etc.[2]
- Nightclubs—while previously we've joined this with concerts, when considering permanent installations, nightclubs are often distinct from concert venues. Here "nightclubs" are places where there will never be acoustic or electric instruments performed—where the only music performed will be delivered by DJs or other electronic acts, such that there will be very few live microphones, and most everything is intended to be loud, immersive, and not localized remotely to a stage. Such environments sometimes use speaker systems found in concert venues, but sometimes use speakers that would never be used by traditional concert presentations.[3] Such systems are often designed for 360° coverage of the dance floor.
- Stand-up comedy/seminars/lectures—the simplest form of live entertainment, people talking into mics, with some interstitial music.

Now, there are also multimedia spectacles and interactive productions that are harder to classify—each of these will need to be examined individually, and typically will rent in any unusual equipment that they need.

As to Houses of Worship (HoW), these fall somewhere between a lecture, a concert, and a giant theatrical production (if you're in a huge megachurch). Small HoW generally have simple needs, a mic or two, maybe a little reinforcement for a choir or band, and/or some simple playback. Mega-HoW often produce enormous multimedia spectacles, with wireless lav mics, huge bands, video systems, and so on. Each venue will need to be examined based on the scale of the location and the goals of the producing entity. The signature important feature of HoW systems is that the main speechifying location (a "pulpit" in many churches, though this is called different things in different religious disciplines) is almost always in need of vocal mic amplification. Care must be taken to ensure that wherever the default pulpit location(s) will be, vocal amplification can be driven very loud without any substantial risk of feedback. While theaters presenting seminars or lectures may have the same concern, it is common for such to position lecterns at different spots for different events (or, TED-talk-style, to feature speakers in wireless headsets who wander the stage at their will).

For any venue installation, we will need to assume that our systems need to have longevity. Of the aforementioned, nightclubs will be expected to have the shortest shelf-life, simply because nightclubs are notorious for pummeling their gear night after night, so replacement and upgrade rates are (generally) higher.[4] For other performing arts venues and HoW, it should be assumed that we are aiming at a *minimum* life span of 10 years. It would be ideal to aim for a 20-year life span, but realistically, digital audio protocols have been evolving quickly enough that there is no guarantee that, say, Dante (which we rely on constantly at present) will even be supported in 10 years.[5] Being realistic with clients about these challenges is important.

If we are hired to design a new system for an existing venue, we must first perform detailed site surveys. We need info about space for equipment racks, available power and cooling, floor space for consoles, noise floor measurements (at different times of the operational cycle, with HVAC and any other standard equipment running, etc.), acoustic measurements of the space(s) for reverb times, and so on. If we cannot perform these site surveys ourselves (perhaps we are in a different state or nation and cannot get to the venue in a timely fashion due to other projects), we at least need to employ a reliable surrogate to collect the data we need.

If we are hired to design a new system for a new venue, then we get the fascinating experience of working with architects and design firms. The world of architecture is rife with a host of confusing technical terms and acronyms, which are not found in the standard event production world (the essentials of which we will address next in "Unique Terms"). The earlier we are engaged in the design process, the better, for architects (at least those who employ no acousticians or AV designers of their own) can sometimes get too far along before we join the team—committing to choices that present significant obstacles to our work without being aware of the implications of their choices. If you have the good fortune of being employed early in an architectural process, make sure to claim a "seat at the table."

Whatever version of the process you are involved in, any design for a performance venue must consider (at minimum) the following:

- At some point, *someone* will need to give a speech. Even if this is the DJ talking from the DJ booth, your system should always provide for *at least* one person (and for anything other than nightclubs, typically several more) to deliver a speech or presentation with sufficient loudness, vocal clarity, and lack of feedback.
- Any of these venues will need provision for stage monitors. Some may require dedicated monitor consoles, while others will not require such, but all will require at least one, and often many more, monitor mixes. For a mid-sized concert venue, it is reasonable for a touring act to expect to be able to demand 12 mixes onstage without a problem. Most of these will be stage wedges, though drummers often require both a wedge and a "drum sub"—a 15" or larger single driver, fed only with kick, bass, low synths, etc. The larger the performance stage, the more likely additional monitor solutions will need to be presented, such as side-fill speakers aimed across the stage to deliver a general blended mix to the band to help them feel more natural while performing.
- FOH systems, with quite rare exception, are designed by setting SPL and frequency response goals, and then ensuring that these targets are met for every seat in the house (or every audience position, in standing room venues with no seats).
- Small and mid-sized concert venues and most HoW typically provide a well-balanced stock of microphones suited to the material they are presenting. Theaters vary wildly as to whether they provide microphone stock. Many university theaters and mid-sized regional theaters will own a stock of wireless and wired mics, whereas high-end Broadway houses usually assume everything will be rented in.
- While backline is not technically part of a sound system, it is often a sound system designer's job to provide a specification for backline stock at concert venues and nightclubs. Nightclubs in particular will often own stock DJ equipment—direct-drive (as opposed to belt-drive, which cannot be used for scratching and cutting)

turntables, a couple different styles of DJ mixer, CDJs, etc. Small and mid-sized concert venues often provide the same kind of stock, in addition to a range of instrumental equipment: at least 1 drum set, a guitar amp or two, a bass amp, often a stock 88-key weighted keyboard, etc. At the large scale, or for high-level touring acts, everything will be brought in.

- DSPs and consoles will typically be chosen that *exceed* the generally expected needs in the venue. If a concert hall regularly expects to book acts with 40 channels of mics and DIs onstage, a 48-channel console is cutting it rather close and specifying a 64-channel is a safer bet. There is nothing promoters and producers hate more than not being able to book an act because the technical package of the venue is *just shy* of being able to support said act (to rent a console to buy you four extra channels is maddening). If you are designing for a venue where producers and promoters have already been contracted, talk with them in detail about the typical acts they plan to program.

- Wherever possible, specify a provision for recording the shows. It is a rare contemporary system that does not provide for recording in some fashion. This is least common in theater, as there are complicated rights issues surrounding the recording of live theatrical performances, but for any concert or HoW venue, it should be a default choice.[6]

VENUE CHALLENGES

It depends on the venue. These tend toward the same challenges found for any concert or theatrical productions (see previous chapters in this section).

UNIQUE TERMS

Standard performing arts terms apply, but there are additional terms unique to HoW and architectural projects, some of which we will introduce here.

Ambulatory—space for walking (aisles)

AVC (Audio Video Control)—these three system design disciplines are often grouped together under one designer or team in the architectural world.

BIM (Building Information Modeling)—this is a 3D platform for developing architectural plans, that can model not only 3D surfaces and rooms, but also electrical systems, plumbing, HVAC, and more. As of this writing, this is becoming an increasingly common method of developing architectural projects and is becoming common in themed entertainment as well. While 2D documents may be prepared in AutoCAD (or Vectorworks, if you're only a theater designer), Autodesk (maker of AutoCAD) also makes Revit, which is the most-used BIM platform.[7]

CD phase (Concept Design)—an initial design phase where the teams are throwing ideas at the wall, speculating based on the general "program," and getting rough shapes in place. Often very general. Followed by SD and then DD (see below).

Charrette—an intense, brief, design brainstorming meeting.

DD Phase (Design Development, or depending on who you are working for, Detail Design)—the third and final main design phase, where general ideas become very specific design plans, often down to the exact models of cable connecting each device. DD documents will be used by integrators to prepare bids for the system.[8]

Message in a Bottle

Sometimes the usage of the venue mutates over time. This happens way more in traditional concert hall work because the concert hall is dying, and to design a system specifically for classical music—it's not happening anymore. So I think having a system that can mutate through different programmatic materials and iterations of a venue is going to make you a lot more successful.—David Kotch

Groundplan/RCP (Reflected Ceiling Plan)—while a show production crew typically refers to an overhead view of equipment positions as a plot, or plan view, architectural teams break these into at least two types of documents. On a theatrical plot, it is common to see grayed out scenery shown below speakers that are flown. However, on architectural projects, the plan view (GP) should only show items actually mounted on the ground. Anything suspended or flown from the ceiling would be shown from overhead in an RCP—a drawing that is essentially the same view as a plan, but only showing items that are NOT on the ground. Of course, we often need to overlay GPs and RCPs to make sure our overall plan is coherent, but these will still typically be separate documents.

Nave—the general area where the congregation sits in HoW.

Pews—the literal seats where congregants sit in HoW.

Program—while sound designers think of "program material" which is any specified content playing through a system, architects think of the "program" as the client's overall wish list for the resulting edifice.

RFI (Request For Information)—a standard document used during architectural projects. This is a formal request from one contractor to another for an official report of whatever information is in the request. If, for example, the architecture firm is not the same firm as the HVAC engineers, which are not the same firm as the AV designers, it is common for RFIs to pass between these companies to acquire information that is needed. This process seems a bit formal to anyone used to show production and not architectural work, but the key is that information passed along in this way is official and is—importantly—transmitted in writing, such that a company or team member that fails to live up to info transmitted in a response to an RFI can be held accountable for such failure.

Sanctuary—where the altar sits (the altar is where ceremonies are performed, and sometimes doubles as the pulpit) in HoW.

SD Phase (Schematic Design)—the second main design phase, where designs get more specific than concept designs. If a concept design specified that a sound system needs "main speakers, mixing console, mics, etc.," a schematic design might begin to specify what type of main speakers, and possibly a rough idea of how many: "left and right main line arrays, approx. ten boxes each," for example. Individual models of gear will still not be specified at this stage, but a basic system flow will be developed that begins to be specific enough to communicate clearly how a system is intended to function, even if it is not yet explaining what that system actually is.

There are many more terms used in architectural projects—I've done a lot of this work, and yet on almost every project, I find a new term or acronym to learn.[9] There are some other particular terms that come up more frequently in themed or other project types, and those will be addressed in subsequent chapters.

UNIQUE GEAR

See Chapters 37–40.

TYPICAL TIMELINES

Architectural projects happen in a wide variety of timelines, most of them much longer than production timelines for events. While a design for an existing venue may have a narrow installation window, the planning phase still may take months if not longer. Design for new buildings (or new venues in existing buildings) typically takes at least a year, and often can take between 2 and 5 years, depending on the scale of the project, the complexity of the facility, and the amount of money involved—the more money involved, the more likely the process will take a long time, as people are understandably hesitant to rush into spending millions of dollars.

What is important for a designer on projects such as these is that they learn what the schedule and anticipated deadlines are at the start of the project. The more clearly a timeline can be laid out in advance, the better. That said, I have seldom worked on an architectural project that didn't run over deadlines at some point or another. Always endeavor to make sure your work is delivered on time because that is a huge motivation for you to be hired for the next project. Designers who run past deadlines better have damned good reasons why they couldn't meet the target, and even still, this is a big reason that designers don't get asked back for round two.[10]

All this is fine and well in theory, but in practice sometimes things get a bit crazy. I have been on architectural projects where *theoretically* there was a space of 3 months for turning CD drawings to SD drawings, but we were waiting on the other teams to get us updated floorplans before we could expand our CD set, and at the end of the day, by the time we had the background documents we needed, we had *two weeks* to do this same work. If there's ever a time when you might acceptably run late (while communicating clearly about this with your bosses/clients) this is it, but if you can still pull off the work in the shortened time period, clients tend to be impressed. Work begets work, as they say.

DESIGN PRESENTATION

Design documentation for architectural projects is frequently more detailed and intense than for any other type of project. Sound system documents will become part of the overall project document set, which often runs into the hundreds of pages. Each drawing you produce for an architectural project typically will be broken out by subsystem, location, or other such divisions. A plan view that might be one page for a theatrical design might be shown as one overall page, but then broken out into 10 detail plates, each at a larger scale, to show more detail. Block diagrams, which in theater and concert work often attempt to communicate all (or at least most) of the system information in a single plate, will be broken up over several pages for ease of reading and clarity.

Figure 42.1A shows the concert system design from Chapter 38, with callouts to various detail plates. Figure 42.1B, C, and D shows the detail plates deriving from plate A.

This is a common way of illustrating a design, though details are often called out in a number of different fashions. Sometimes they are just listed as separate plates in the list of plates that start the drawing set (a common table of contents found in large document sets), sometimes they are blown up over the main drawing as a floating window—there are many styles. Always check with your project team as to how they would like detail plates broken out and shown.

It is also common (as we will cover in more depth in Chapter 50) to break out block diagrams into multiple plates based not only on keeping a large system clear, but also on the principle of separating subsystems by function—network control on a different plate from com, from audio signal flow, etc.

In addition, mounting details, facilities impact reports, cable schedules, and more are quite common in architectural projects (all of which will be discussed in more detail in Part IV).

Design documentation is typically presented to a range of people on such projects. First—whoever hired you directly. Sometimes you will be working as a subcontractor or employee of an AV design firm, who themselves will have been hired either by the architects or directly by the end clients. Sometimes, you will be working for the architecture or acoustics firm. Sometimes (rarely) you will be working directly for the end client. Whoever is your direct supervisor first must sign off on your work. From there, it is presented to either the main contracting firm (if, for example, you are working for an AV firm who is working for an architecture firm), and once the main firm has approved, it will be presented to the client. This can get frustrating, and at times confusing, and can result in designs being misunderstood or watered down without your control. Fight for what you know is right but pick your battles—it's always better to work again than to die on the hill of a single project.

An important note on large-scale projects with multiple firms—typically, a drawing standard will be set at the start of your work on the project. What size plates are being drawn? What fonts are approved for use and in what sizes? How should drawing layers/classes work? We'll talk more about the details of drawing layout in Part IV, but for now, suffice it to say that you want to know what is expected of your paperwork from Day 1, so you don't draw in such a way that the client won't accept it.[11]

Architectural presentations are also generally more formal and "professional" than event design, meaning there's a lot more PowerPoint and khaki slacks. Be warned.

DESIGNER QUESTIONS

- Does this venue plan on showing videos/films as part of its productions? As such, do you need to incorporate the potential for 5.1 or 7.1 (or Dolby Atmos) surround?

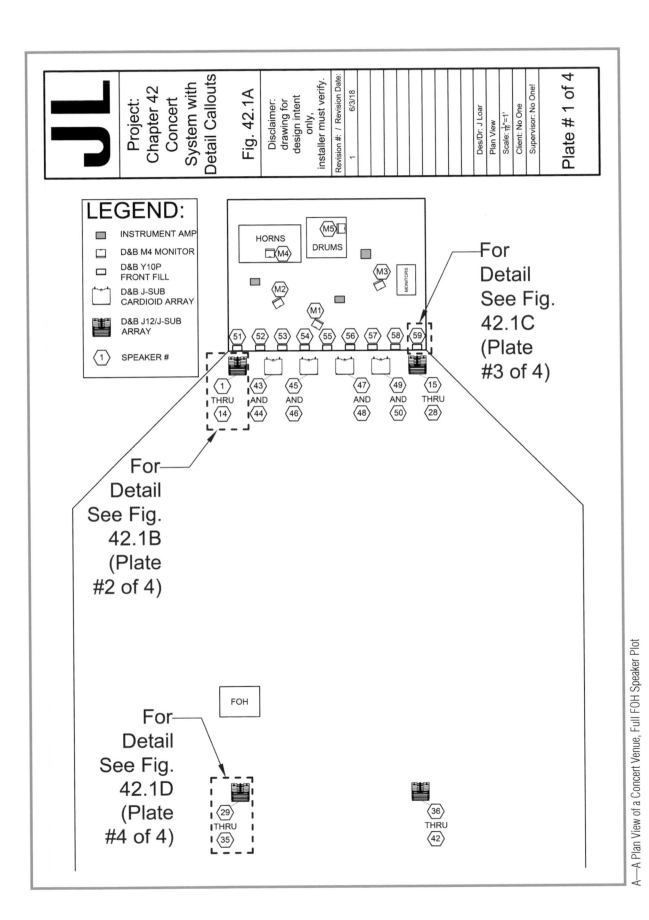

A—A Plan View of a Concert Venue, Full FOH Speaker Plot

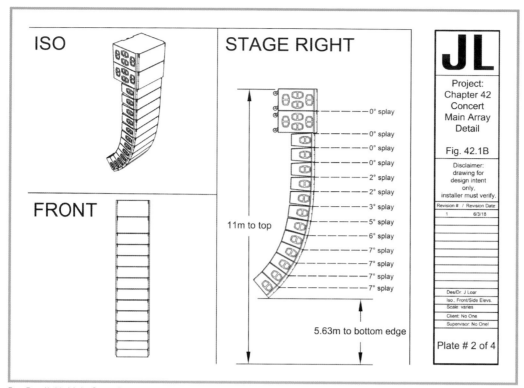

ISO

STAGE RIGHT

FRONT

0° splay
0° splay
0° splay
0° splay
2° splay
2° splay
3° splay
5° splay
6° splay
7° splay
7° splay
7° splay
7° splay

11m to top

5.63m to bottom edge

JL

Project:
Chapter 42
Concert
Main Array
Detail

Fig. 42.1B

Disclaimer:
drawing for
design intent
only,
installer must verify.

Revision # / Revision Date
1   6/3/18

Des/Dr: J Loar
Iso., Front/Side Elevs.
Scale: varies
Client: No One
Supervisor: No One!

Plate # 2 of 4

B—Detail #1: Main Stage Right Array

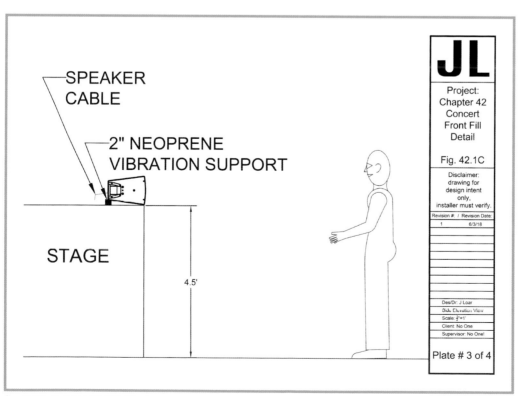

SPEAKER
CABLE

2" NEOPRENE
VIBRATION SUPPORT

STAGE

4.5'

JL

Project:
Chapter 42
Concert
Front Fill
Detail

Fig. 42.1C

Disclaimer:
drawing for
design intent
only,
installer must verify.

Revision # / Revision Date:
1   6/3/18

Des/Dr: J Loar
Side Elevation View
Scale: ¾"=1'
Client: No One
Supervisor: No One!

Plate # 3 of 4

C—Detail #2: Front Fill DSL (Typ.)

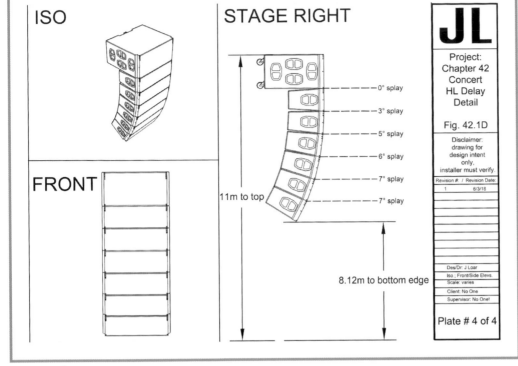

ISO

STAGE RIGHT

FRONT

11m to top

0° splay
3° splay
5° splay
6° splay
7° splay
7° splay

8.12m to bottom edge

JL

Project:
Chapter 42
Concert
HL Delay
Detail

Fig. 42.1D

Disclaimer:
drawing for
design intent
only,
installer must verify.

Revision # : / Revision Date:
1   6/3/18

Des/Dr: J Loar
Iso ; Front/Side Elevs.
Scale: varies
Client: No One
Supervisor: No One!

Plate # 4 of 4

FIGURE 42.1
D—Detail #3: Delay Tower, House Left

- Does this venue plan on using live cameras on jumbotron screens to convey the action? If so, how far away are the viewers, and how does video need to be delayed so that it matches up with audio (given that light travels much faster than sound)? Is video sync necessary, and does it need to talk to the audio clocking system? Ideally, you will be working out these questions with a video system designer—though sometimes, if video is new to the venue, you as the audio system designer may be tasked with solving this yourself.
- How many events in a given day does a venue intend to produce? How much turnaround time is there between those events, and how much system configuration change is required for those changeovers? Your system plan needs to be able to accommodate these venue goals.
- If recordings will be regularly made in-house, what is the data storage and backup plan for the venue's operations?[12]
- And, of course, all the questions implied by this whole chapter (and the previous section of this book): what kinds of acts will be hosted in this venue? What does the audience expect from this venue? And so on.

* * *

Now, let's examine another kind of installation: retail and commercial environments.

Notes

1   Within reason. There will always be some event that demands a particular setup or piece of gear that you don't own. If it's in the event's signed contract rider, the venue will need to either rent that gear for the event or negotiate a substitution. No system can truly prepare for *all* eventualities; we just want to cover about 85% of them at least.

2   There certainly are venues that break these rules—I have worked at several. In the case of one such venue in Manhattan, we had worked with the design firm to create two parallel sound systems, one of which was typically in use for rock and pop shows, and another of which would be implemented for classical and avant-garde shows. This was essentially the venue equivalent of an A/B system, except that both A and B were for music.

If your venue has the money, and the ambition to produce multiple types of events, there is no reason *not* to provide parallel systems with distinct styles of coverage and operation. Meyer Sound's Constellation systems are found in many smaller classical music venues, and while they provide wonderful natural-sounding lift for such shows, if those venues host a rock band or a DJ, they still typically set up conventional directional main speakers.

3   Consider the speakers made by Void Acoustics, which are found in many high-profile electronic clubs around the world (including the legendary Tresor in Berlin). These are visibly strange speakers, and they are not designed to deliver flat, reference-quality performance so much as they are designed to give a listening experience suitable for dancing: enormous, pounding lows, crisp attacking highs, and HIGH SPL.

4   In addition to which, nightclubs are often the riskiest business endeavor, coming and going quickly with the whims of the musical public's taste.

5   Though, given how many Dante systems are installed around the world, I sure hope it continues to be viable for a long while!

6   It also makes a decent item to cut during value engineering if push comes to shove.

7   AutoCAD and Vectorworks are capable of 3D drawing, but they are not made for it—it's easier to use dedicated software like Autodesk Inventor for simple object 3D, and Revit or other BIM platforms for advanced architectural projects. We'll discuss these platforms more in Part IV of this book.

8   Note that while these three design phases are very common, they are not universal. Check with your firm(s) at the beginning of any project to establish how they plan to break up the stages of work.

9   Seriously, architects are like the military in that anything that *can* become an acronym *will* become an acronym. If you are intimidated and don't understand what someone on your team has said, quickly Google it. If you can't clarify it readily, just ask. Most people will be happy to clue you in, so long as you only ask once per acronym (or other specialized term) and remember thereafter.

10   This is true for any project—show designers usually have an opening date that they cannot change, so there is some limit built into the process, but architectural projects often only have a "target opening date," nothing hard and fast, so the temptation is to be more lax about time. This is always a mistake. Meet your deadlines whenever possible.

11   This is generally not an issue for theatrical or concert productions, with some exceptions (*cough* Disney *cough*).

12   See Chapter 45 for more on recording system planning.

CHAPTER 43

Supermarkets, Airports, and Other Retail Spaces

What do a supermarket and an international airport have in common? In both places, there is near-constant sound being played over an installed (and thus designed) sound system, and in both places, you're generally not supposed to ever notice that sound—right up until an announcement is made, at which time you are *definitely* supposed to notice it. As such, the sound systems in these, and other similar retail facilities (shopping malls, big-box stores, auto parts stores, and so on) share similar characteristics. They are capable of playing relatively full-range content (background music, aka BGM),[1] at moderate levels (allowing conversation), but also capable of quite loud announcements, with enough sources spaced around the venue that any shopper will hear clear direct sound when announcements are made.

DEFINITION

For the purposes of this chapter, we are examining a general category of "retail spaces." Retail spaces will be defined as anywhere that consumers spend time primarily engaged in retail transactions, and where there is no central entertainment or "show" element present. As such, retail spaces will encompass:

- Stores of any kind
- Malls—both indoor and outdoor
- Airports[2]
- any retail space enclosed within another overarching business—shops within hotels, casinos, etc.

Retail spaces are in many ways the common areas and town squares of many regions in the 21st century.[3] When we enter a retail space, we are meant in most cases to respond with positive emotions, to feel simultaneously comfortable and yet provoked to spending. Sound increasingly plays a hefty role in that design—there are whole consulting firms whose sole job is specifying music and/or soundscapes for your particular retail environment, targeting the demographic you believe is attending your shop (or that you hope to attract), aligning with different times of day/densities of crowds, and so on. As this industry has increased in prominence, so have sound systems for retail environments gotten more elaborate and more customized—particularly in major urban or wealthy suburban areas, where retail competition is stiff. As sound system designers, it is our job to assess the particular needs of the retail space and market—the content to be delivered, the moods the content intends to create—and create systems that will support those goals.

STANDARD ELEMENTS

- Inputs: paging station microphones, BGM playback, automated playback announcements, intercoms
- Processing: DSP, no consoles
- Amplification: 70V (almost exclusively)
- Outputs: 70V installation speakers, distributed
- Converters: I/O at DSP(s) (airports and other large spaces often require multiple interconnected DSPs as the processing power gets overtaxed with all the routing and zone control)
- Advanced processing: elaborate zone control and paging via DSP
- Control: access-controlled paging stations and touch panels, programming interface to DSP in EER/machine rooms
- Clocking: internal with DSP as master

UNIQUE FUNCTIONS AND SYSTEM NEEDS

First, we need to understand what type of retail establishment we are designing for. While the actual variety of retail shops is as variegated as the items, humans can think to sell to one another, in broad terms, we can break these down into a few subcategories:

434

- Basic stores:
  - Clothing
    * Men's vs. women's
    * Jewelry
    * Shoes
    * Formalwear vs. casual
  - Furniture
  - Electronics and media (music, movies, games)
  - Sporting goods
  - Hardware:
    * Mom 'n' pop vs. big-box
  - Automotive
    * Mom 'n' pop vs. big-box
  - Other/specialty
- Big-box:
  - True big-box—e.g., Target, Walmart
  - Department stores—the original "big-box," multi-item stores, but typically laid out in a different configuration and for a different effect on the shopper
  - Discount stores
- Grocery:
  - Small/local/co-ops
  - Supermarkets
  - Discount/dollar stores
- Malls:
  - Indoor
  - Outdoor
  - Inside other venues (e.g., mall located inside a casino)
- Airports—shopping malls inside a high-security environment

This is not a comprehensive list; looking around your town I am certain you can find examples of retail establishments that don't fit one of the earlier categories, but this is enough detail to allow us to examine the thought process of designing for retail spaces. We will now examine each of the aforementioned in terms of sound needs and client/consumer expectations.

The first category I listed earlier is what I am calling "basic stores." These are stores that are not selling more than one broad category of item, and as such are focused on selling a particular type of item in a particular type of environment (unlike big-box and department stores, that are focused on creating either a more generic experience or several segmented sub-experiences, respectively). Basic stores tend to have an audience in mind, as opposed to big-box stores that tend to aim at "everyone" as their audience. For example, a women's clothing store generally is worried about women, or about people who are buying for—or dressing as—women.

CLOTHING STORES

The purchase of clothing items is a process and topic about which much has been written. Clothing is tied intrinsically to the customer's sense of identity, of self-worth. As such, clothiers of any kind are very particular about the mood they create, as they find that it impacts sales quite a bit. People are placing themselves in a vulnerable state of mind when shopping for clothing—depending on the items, they may be disrobing in order to try things on, and so they are in a highly suggestible state.

Sound systems for clothing stores generally have two goals: music playback, and (depending on the size of the store) emergency announcement. In addition, very large clothing stores (think Old Navy) will outfit their operations staff with headset systems for communication, such that cashiers can call to stockroom staff in the back of the house, and vice versa. Such systems are typically decentralized, in that there is no "stage manager," there are merely teams who are interconnected on party lines (though there may be more than one channel of communication available).

As to music, what type of music, at what intensity, varies across the types of shops. Looking at our subcategories, a shop that serves both men and women will typically be more defined by the style of clothing, how formal or informal it is, and the prices of the clothing, than by gender. A Banana Republic, for example, is aimed at shoppers of all genders, who are typically buying "business" attire that lives in the realm between formalwear and true casualwear. The pricing is not low (it's no discount shop), but neither is it luxe (it's not a high-end boutique). As such, it is a good case study for a baseline "clothing shop" sound system. Music played will generally be inoffensive pop hits of the era (and recent eras), amplified at moderate volume such that the content is audible all the time (not subliminal—explicit), but is still not totally overwhelming to average conversation—somewhere in the 65–75 dB SPL range, depending on the size of the shop and how crowded it is at any given time. Such a system will be unlikely to deliver true full-range performance, generally leaving out sub-bass and low bass but covering something like 75 Hz and up decently. Little attention will be paid (if any) to "flattening" or tuning the system response (in this or most retail settings, truthfully).

Now, if our shop in question has the same demographic targets and price points, but is only aimed at male or female customers, you will hear some difference in what gets played in the shop. Stores aimed at female customers will feature playlists with a lot of female vocal artists, and stores aimed at male customers will include more beat-heavy material that may require lower bass extension.[4]

If the clothing trends toward the casual, for either gender, the SPL will be higher as a general rule. As clothing trends toward the casual (which includes any shoe stores focused on sports shoes and casual shoes, as opposed to dress shoes), low bass also becomes more important in the system. It is common for casual clothiers to pump popular music quite loud—often above average conversational levels—to make the space energetic and "fun," and also to intentionally disorient the consumer a bit, to make them feel caught up in the rush of high-energy fast-paced music, and thus more likely to make an impulse purchase.[5]

Most jewelry stores and formalwear shops will still feature sound systems, but often they are playing "high-status" content: classical music, jazz, anything that makes the consumer feel fancy (and like spending a lot of money). Such music is usually played quietly, subtly, at such a level that allows discrete conversations about expensive items.[6]

Most retail systems worry more about total SPL than direct SPL. So long as there are enough sources to produce direct sound for emergency announcements without blurring the signal beyond recognition, there isn't as much

435

concern for evenness of direct coverage for these music systems. In fact, some spaces make use of reverberant spaces and blurred sound for increasing the sense of disorientation which (in some versions of retail theory) may help encourage shoppers to purchase things they might not need.

FURNITURE STORES

In today's market, there are small-ish furniture stores, there are big-box stores that also sell furnishings, and there's IKEA. The era of big furniture stores that are *not* IKEA is largely past, but I set furniture stores aside as their own category because they warrant a short discussion about how they differ from other retail spaces. Furniture stores are typically configured in one of two ways:

- They cram as much furniture into the space as possible, without regard to visual appeal or any kind of rhyme or reason; or
- They set up small scenes or vignettes meant to illustrate how their items might be configured or used in the home.

This is true both of large and small furniture stores. IKEA, for the most part, falls into the latter category, with most of the store staged as a series of scenes—though that goes out the window when you enter their "marketplace" and warehouse sections.

For the former style of shop, the chances are that if there is music playing, it is music playing throughout the whole store, regardless of the store's size. There will be one sound source, it blankets the space, and that's it. SPL is generally lower than in casual clothing stores, without being as low as might be found in a jeweler's shop. Music styles may range wildly depending on the style(s) of furniture offered by the store,[7] the owner(s), their preferences, etc. Any business (furniture or otherwise) that is part of a chain is more likely to have corporate-mandated playlists, as opposed to small privately owned shops, where individual staff may have some leeway to play what they want.

In shops that are staged like living areas, the sound systems can get interesting. Typically, a very small shop that stages items like this will still feature one central sound system, but when you get to the scale of IKEA, sometimes the system gets more distributed. Different "rooms" and styles of furniture may be accompanied by different musical styles—though this will also depend on how the spaces are divided from one another. If two different styles of "room" are actually adjacent to one another with only a corkboard partition between them, it is still likely that they will be hearing the same content, whereas if there is a long passage between them, or if they are on separate floors, chances are the sounds will be different in each space. At any rate, the larger and busier a furniture store gets, the more it can represent a small theme park environment. IKEA is essentially a theme park—the theme is "what your house could look like." As such, large distributed systems with multiple playback zones, and stable playback servers designed to run all day every day with no hiccups are common here.

High-end furniture boutiques are very like high-end clothiers, quiet, unified, designed to give the impression of luxury. Low SPL.

ELECTRONICS AND MEDIA

In all but the largest cities (or those particularly focused on the entertainment industry), electronics and media stores have mostly been replaced by the Internet. However, I have included them here for the reason that there are some unique system considerations for these shops that are worth briefly mentioning.

Any store that sells recorded music will require a full-range sound system (if not several), and often one that—rather than being tuned flat like production systems—may actually be designed of components designed to sound warm and pleasing: high-end home equipment is often in play. Stores that sell electronics for media playback often sell speakers, and in that case, they often use their own floor models as the main music sound system(s), in order to show off their most expensive models. If they don't sell speakers (or only sell a very limited range of speakers), they will sometimes install a higher-end system to play full-range audio aligned with the wall of televisions each playing the same content.[8] Game-focused shops will vary between playing loud (often electronic) music, showing promo reels that have accompanying sound, or showing gameplay with accompanying sound. Whether gameplay is played through headphones for individual test stations, or through the whole room system, depends on the individual business, but both are common.

SPORTING GOODS

Sporting goods stores tend to fall into two basic categories: loud and soft. There is one school of sporting goods store—often focused on younger adult customers, and sometimes with particular outfitting for so-called extreme sports—that is usually loud, either playing upbeat music or showing sports with sound. The other school of sporting goods store is aimed a bit more at families and tends to be very quiet, sometimes with no BGM whatsoever. The louder version of these systems is designed quite like clothing stores.

HARDWARE

Almost always lacking BGM, but at the big-box scale often featuring distributed com systems, a ton of paging stations, and a wide-area announcement system (vocal only). Paging stations are sometimes DSP-based custom stations, and sometimes are telephone-system based. Small hardware stores sometimes have a basic two-way intercom in order to call to stockrooms in the back, but not much else.

AUTOMOTIVE

Large auto supply stores often play hard rock at moderate volumes.[9] They often share the paging infrastructure of large hardware stores. Small automotive shops sometimes have music playing, but seldom any formal sound system.

SPECIALTY/OTHER

As always, there are examples we haven't covered. I have been to some strange upscale boutiques that had interesting custom sound systems because they were trying to create a unique shopping experience. Also, anytime you get into themed shopping—like a Disney store—the rules for design start to mingle with theme park design in terms of effect speakers built into scenery and more show control elements. Examine the requirements and budget of each job individually, and deliver a sound system that will create the desired SPL and feel, based largely on the BGM and announce requirements.

BIG-BOX

True big-box stores are giant warehouses, usually with corrugated metal roofs, exposed conduit, and HVAC ducting, and little or no care for acoustics. The sound systems are often taxed with playing moderately loud BGM and being able to transmit announcements to both the whole store and to individual zones in an articulate manner. As such, these systems often feature a lot of speakers each playing not terribly loudly, in order to create distributed coverage without gaps that is capable of rising above large crowds for announcements without washing out consonants. Such systems often must be capable of 90–95 SPL, in order that in an emergency announcement can be made even over large crowd conversations. In any very large retail establishment, the announcement system should also be designed to operate in case of emergencies, making use of powerful uninterruptible power supplies (UPSs—large, purpose-built battery backups) to keep the signal processor and amps on if the city power goes out.

Department stores often divide their goods into physically separate areas—isolating those areas with walls and partitions. As such, many sub-areas within department stores feature their own sound subsystems. The formalwear department of a department store might be playing "highbrow" music like classical or jazz, where the young men's section may be playing pop-punk, and the makeup area playing techno. This requires more robust playback systems, more channels of DSP processing, and often more complex DSP programming for multi-way paging/com systems. Each sub-area will need to be evaluated individually based on its particular materials and layout in order to determine proper speaker choice and placement. Each sub-area may have its own SPL targets, and certainly its own moods.

Discount stores are like community theater: they want to do everything that the big players do, at 1/100th the cost. This is not generally possible, and so shortcuts are taken. Don't let a low budget get in the way of necessary safety systems for large shops, but compromise is often the name of the game when designing for discount stores. Hot spots and gaps in coverage, overly reverberant spaces with fewer direct sources—these are common features of the discount store sound system.

GROCERY

Almost all grocery stores play music. It is seen as key to making the environment friendly and comfortable—the longer the shopper stays in the store, the more money they are likely to spend. Music is not played terribly loud, but coverage is typically fairly consistent. Every grocery store also has some form of paging system—a way to call for price checks or backup on the register. The main difference between small grocery stores, supermarkets, and dollar stores is one of scale and audience. Small stores (co-ops, "natural" food stores, and the like) are often catering to specialty or high-end clientele and as such are essentially the boutiques of grocery—providing specialty product to a wealthier clientele, so the intended feel is one of luxury. Systems may be higher-resolution in such shops (and in upscale supermarkets, like Whole Foods), and coverage may be particularly even. Standard supermarkets will be designed essentially like a large hardware store, but with a (slight) emphasis on BGM (though most will still present nowhere near a full-range system). Dollar stores—see discount stores noted earlier.

MALLS

Shopping malls vary considerably in size and scope of customization. There are very large malls that feature themed and interactive decorative elements, and there are small and mid-sized standard shopping malls that operate more like giant warehouses. Indoor shopping malls are generally characterized by their clattery echoing acoustics. Since malls generally play BGM (in addition to the stores) and sometimes need to make emergency announcements, many of the design suggestions for a warehouse store apply—lots of low-level direct sound sources.

Outdoor malls have similar goals (BGM, announcements), but are unique in that they are not contained, and thus don't have the reverberance to blur signals, but also don't have the reverberance to help increase total SPL. As such, there are often even more speakers placed throughout outdoor malls. Unlike indoor malls, which (like much retail design) rely on pendant and in-ceiling/in-wall speakers, outdoor malls often make use of production-style point-source or even column/line-source speakers, mounted atop towers/posts. We will discuss this type of coverage more in Chapter 44 (on theme parks).

AIRPORTS

Airports are sort of like shopping malls, in that there are lots of retail spaces, and they often need BGM and announcement speakers. Unlike malls, however, individual store sound systems will typically be a part of the overall sound system—such that in emergencies all BGM can be overridden by central announcements. Additionally, airports need a ton of zone-independent control, meaning a page call can be routed to just one gate, to a whole terminal, to the whole airport, or (often) to other combinations of these areas. Also, because of the heightened security environment in airports, systems installed in these venues must have robust access control features, such that a random stranger cannot easily dial into the system and make announcements. Many DSPs are outfitted with these options, but it's important to verify before selecting any system equipment for an airport.

Let's look at a typical sound system for a big-box store. Figure 43.1 is a generic discount big-box store floorplan (simplified).[10]

As we can see in Figure 43.1A, we have specified mostly pendant speakers (Figure 43.2). In big-box spaces, pendant speakers are extremely common. We have been limited by hanging speakers where metal infrastructure was already going to be in place, in between lighting fixtures (not shown). Coverage is asymmetrical, as the store itself is asymmetrical, and because (being a discount store), we have placed as few speakers as will create decent coverage without spending a lot of money. Back of house is not shown. This store is smaller than a typical big-box store, but the main point of the illustration is the principle. If this were 1/5th of a larger store, so long as the layout doesn't change drastically, we can basically rinse and repeat. Notice the bottom left corner of the drawing, the "café" area with round tables for eating. If this were not a discount store, we might ask for added pipe or mounting positions so that the speaker could be centered in between those four tables. The placement we've chosen is as close to the center as possible, but it is not perfect. We will pick our battles, and this isn't one of them.

As we see in Figure 43.1B, the system is rather simple. Since this is a discount store, there is no bother with redundant systems, and the 70V tap speakers are fed via a single amplifier channel, with their 30W taps, which allows each speaker to hit a rough maximum of 100 dB SPL

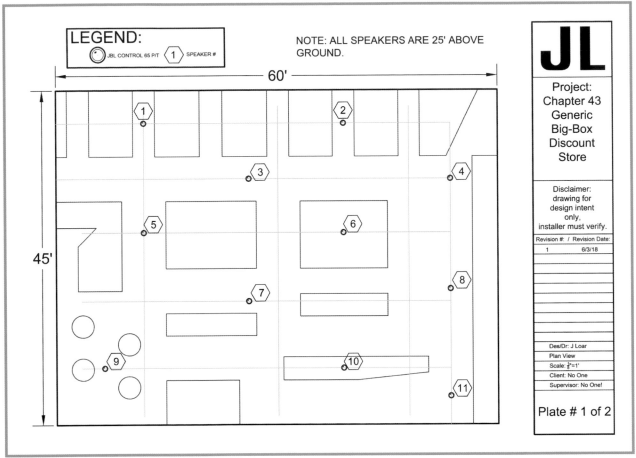

A—Plan View of Big-Box Store With Speakers

We will often see other ceiling-style speakers in such spaces but mounted on stiff metal poles instead of flush to a drop ceiling (because most big-box stores do not install a drop ceiling). In a department store, that more typically *would* feature drop ceilings to conceal conduit, HVAC, etc., flush-mounted ceiling speakers are the most common.

Examining Figure 43.1B further, we can see that the heart of a big-box system (or any large retail system) is the DSP There is a playback server feeding BGM, and otherwise, there are paging stations distributed throughout the venue. All of these are routed through (and controlled by) the DSP. This is true for a big-box store, a department store, an airport—most large distributed systems of this type. The DSP then feeds amps, which feed speakers. The DSP and those amps deemed "critical" (the bare minimum for safety announcement purposes, which in the case of our example is a single amp, as there are no zones in this store) are all connected to quite powerful UPSs, while the BGM system may be connected to a slightly less robust UPS (or in a discount store, not at all).[11]

VENUE CHALLENGES

Retail stores are easiest to work on before they are open to the public—projects for new shops are far easier than those for existing shops. Design for existing shops often requires quite a bit of compromise with elements already in place (and the installation crew will generally have to work overnights, while the shop is closed). The signature challenge of retail venues is acoustics. It is rare that retailers even considered acoustics in the design process, and if they did, they still may not have wanted to spend any money on acoustic treatments. Most stores (especially those that are *not* aimed exclusively at very rich people) have poor to middling acoustics, and it is also common that retailers are loath to pay the fees to have acoustic models built. Design for such spaces often must rely on the designer's knowledge of speakers and their coverage, and good estimates based on specs. The other signature

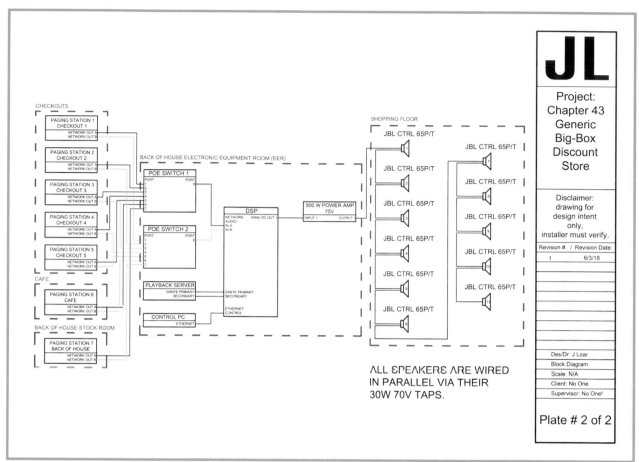

FIGURE 43.1
B—Block Diagram of Big-Box Store System

challenge of many retailers is that unlike entertainment projects, where sound quality is a factor that can help raise budgets and win arguments, sound quality almost never trumps other design concerns in a retail store. Thus, when defending design choices, your arguments must be focused on function and utility, and little else (for the majority of projects).

UNIQUE TERMS

Beacon: sensors that typically use Bluetooth to sense people's mobile devices as they move around the store. Sometimes, beacons will send a message to local mobile devices about a sale on the item they are walking past, but sometimes, these sensors will communicate with sound devices to play content from the station the shopper is near. Motion and proximity sensors are also used for this kind of playback triggering. RFID tags are sometimes embedded in items, such that when, say, a customer is purchasing a razor, it can send a message or sound to alert the shopper to the presence of nearby shaving cream, hopefully increasing the likelihood that the cream will be purchased alongside the razor.

Planogram: a term used in retail for "groundplan" or "floorplan." These usually include placement of items for sale, and in some stores, this plan will change frequently, while in others (large department stores, for example) the overall layout remains the same for quite some time.

POS: point of sale—stations in various retail environments where purchases can be made (also known as "cash wrap," or "cashiers"). These are usually points where paging stations are installed.

UNIQUE GEAR

Retail systems are known for their pendant speakers, and their in-ceiling and in-wall speakers (Figure 43.3).

They often feature paging stations and touch-panel controls, which are typically selected to go with the DSP that is the central engine of the system. They often feature playback servers, some of which are network-connected such that a music playlist vendor can easily push new content to the store.

TYPICAL TIMELINES

More than most architectural installation projects, retail jobs can definitely vary in terms of timeline. Sometimes a new retail establishment will secure a location and before 3 months have passed, the building is complete, and the store is open (this is especially true of chains working from stock designs, and discount/big-box style stores that don't pay much attention to décor). Other times, it may be a year, 2 years, even more (a large-scale mall may be in process for several years while it continues to attract investment, structure payments to contractors, etc.). I have yet to see a typical timeline for retail, but I can say that many still go through the Concept>Schematic>Development phases as outlined in previous chapters.

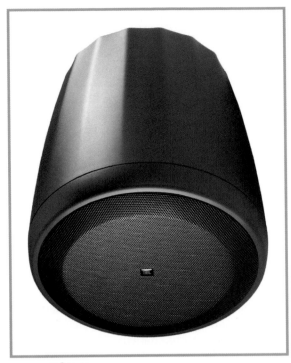

FIGURE 43.2
JBL Control 65 Pendant Speaker
Source: Courtesy Harman Professional Solutions

441

DESIGN PRESENTATION

For any architectural process, as has been mentioned in earlier chapters, drawings will adhere to the standards set by the architecture firm. They may include the following:

- Plan views
- RCPs
- Elevations
- Mounting details for any and all speaker models
- FIRs
- Budget estimates
- Et cetera

FIGURE 43.3
(Left) JBL Control 16 C (Without Grill), (Right) JBL Control 126 W
Source: Courtesy Harman Professional Solutions

The presentation will be made first for the firm that has contracted your services (unless you are freelancing and working directly for a business owner). As with any architectural design project, there may be a chain of presentations, starting with the AV firm, then the architectural firm, then the end clients. Presentations should be fairly formal, with a minimum of technical "sound" language.

DESIGNER QUESTIONS

- Who needs what levels of access to the system, and what operator station(s) are they likely to be accessing from? Some stations may restrict all access except to the zone in which the station is located (local paging only), some may allow local and "regional" access (paging, say, a gate at an airport versus a whole terminal), and some may allow some measure of access to the entire venue. Determining a control hierarchy and access plan with the client is a very important step before selecting gear with which to implement this plan.

- Is the store broken into zones? If so, what are they, how many are they, and who needs to be able to contact each zone? Do zones operate entirely separately, or are they sometimes combined into larger areas? Who is allowed to page which zones, from which stations?
- Is the client working with a music consulting service for their playlists? If so, does that service/firm provide a dedicated hardware player as part of their service? Or is the playback network/streaming based? These answers will determine how to interface BGM with the DSP.
- Is the venue large enough (like an airport) that they will be employing a system engineer to maintain the equipment? Or, like smaller retail stores, does the integration plan need to include periodic service calls?

* * *

Now, if we take the standard processes we use for general architectural projects, combine them with the artistic design processes we use for theatrical and concert design, and add a heaping helping of advanced show control, we get theme park design! Also known as the focus of our next chapter.

Notes

1. I say *relatively* because the supermarket doesn't need subwoofers, whereas a trendy upscale clothing boutique might have powerful subwoofers. The overall SPL for the BGM in both cases will diverge as well, which we'll get into later in this chapter.
2. Yes, I acknowledge that ostensibly the actual purpose of an airport is to depart or arrive on an airplane, but while you are spending time in that airport, it is essentially a giant shopping mall with very intense security rules.
3. This could be the topic of a very different book, on the erosion of common, unmonetized, public space in contemporary society, but I am getting off-topic.
4. This doesn't happen in every shop, but it does happen a lot. The art of manipulating (or pandering to) retail shoppers is . . . well . . . kind of gross at times, but it is still the business that we're in when we're designing for these establishments.
5. At least in theory—personally I am always put off by shops with pounding music, even though I love concerts delivering the same, but given the preponderance of this as a design technique, I think I am in the minority here.
6. An exception to this rule would be a store like Men's Warehouse, which does focus on formalwear, but focuses particularly on discount formal wear in an explicitly warehoused setting, and thus follows system rules more commonly associated with warehouse/big-box retailers than with other smaller formalwear outlets.
7. I once visited a furniture shop that sold only tropical and tiki-related furnishings, but they were playing *black metal* when I walked in. This contrast was so jarring I had to leave immediately.
8. Though in big-box and many other stores selling TVs, they will often just tune the TVs to the same channel, or feed them all from the same DVD player, which creates a multi-time arrival nightmare. If you stand near one TV but hear the delayed arrival of 30 other units all in a row, it is dizzying and odd. This happens at gas stations where they've installed TVs in the pumps—the multi-time arrivals slap over the concrete of the station and create all kinds of phasing/flanging weirdness.
9. It's weird, but . . . prove me wrong. I've been in auto parts stores all over the United States, and I've heard a lot of Journey at about 60 dB SPL.
10. In this chapter (and others like it), details of the building architecture have been simplified, both to enhance understanding of the sound systems themselves, and because designs of projects like this are typically proprietary and/or subject to non-disclosure agreements. As such, I have created floor plans extrapolated from experience. I am not an architect, as will be immediately evident from my floor plan drawings.
11. Power connections are not shown in Figure 43.1B. They are typically shown in a separate plate, if a block diagram is needed at all.

Theme Parks, Cruise Ships, Casinos, and Other Themed Environments

Themed entertainment design can be enormously complex. It could be the subject of its own book this size if we really went into vast detail about how it operates. In this chapter, we'll examine overarching processes, but keep in mind—even more than theatrical sound system design, themed entertainment design is no one thing. Each project is different, each company is different, and on every new project, you must take the time to clearly establish design goals because they will be *different* every single time (and they may change as the project evolves). Themed entertainment design can be a lot of fun—it's high-budget work that demands creative technical solutions. It can also be quite high-stress—when budgets get into the millions it tends to create a bit of a pressure cooker.

DEFINITION

Themed entertainment can be defined as any entertainment environment or project that creates a heightened environment by use of elaborate scenic, costume, prop, lighting, video, and sound elements. Such environments are said to be "themed" as they are often focused around individual themes (such as "Fantasyland," or "Tomorrowland"), though today those broad themes may be replaced by specific intellectual properties (IPs), such as the Harry Potter films and the Universal parks dedicated to those worlds. At the largest scale, themed entertainment consists of elaborate parks, whole worlds of imagination brought to life via intense technology and design. At the smallest end, we might speak of a "themed" restaurant decorated to look like a pirate ship, or the "theming" of a particular courtyard in a shopping mall by addition of some animatronic figures and sound playback. From the smallest themed corner of an establishment to theme parks larger than many towns and cities, from casinos to cruise ships, themed entertainment is one of the biggest, booming-est areas of the entertainment industry today.

What kinds of businesses and environments may feature themed elements?

- Theme parks (of course)
- Cruise ships

- Casinos
- Hotels
- Restaurants
- Shopping malls
- Any tourist destination (may include retail shops, clothing stores, anything, really)

STANDARD ELEMENTS

- Inputs: playback servers for multi-channel and multi-zone BGM, effects, etc.; microphones (for guides on certain types of attractions[1] or performers on attraction stages or parade floats); DIs and mics for occasional live music, intercoms
- Processing: DSPs everywhere, consoles where there are live performance venues within a themed environment
- Amplification: standard operation amplifiers, 70V systems, any other conceivable variety
- Outputs: speakers of all types, both hidden and visible; speakers made out of vibration transducers coupled to props or other scenic elements; in-ear monitors for performers, etc.
- Converters: wherever analog and digital are converted, which mostly applies to mic signals in or speaker signals out. Typically built into the gear (not external)
- Advanced processing: every kind imaginable—elaborate DSP programming, touch-panel interfaces, paging stations, plug-in effects, custom code, trade-secret prototype devices, etc.
- Control: show controllers, triggering events in DSPs, consoles, playback servers, etc.
- Clocking: often internal, with DSP or show controller providing a master clock

UNIQUE FUNCTIONS AND SYSTEM NEEDS

Ha ha ha ha ha ha ha ha ha ha ha ha ha ha ha ha ha!

(Catches breath.)

OK, in all seriousness, I laugh because the signature feature of themed entertainment sound systems is that there are a *ton* of unique system functions and needs, and for all but the simplest stock installations (say, a Ferris wheel), they vary project by project. Themed entertainment is predicated on creating larger-than-life events and environments that (for the most part) work to conceal their operations from the guests who experience these environments. True, in some fairly basic or low-end theme parks, you will see loudspeakers visibly mounted on posts or truss, but in a majority of themed environments, speakers are hidden, control equipment is stowed away in back-of-the-house rooms that no guest has access to, and data (both control and signal) is flying around at alarming rates in order to make it all happen. There are, however some generalizations we can make about themed systems and the types of things we'll find there.

- Robust control architecture: most themed environments will require show control. It often (though not always) falls to the audio system designer to also design the control system for running said audio (and video, and for interacting with the ride systems if such are part of the project). As such, any sound system design for themed entertainment must, from first concept through final development, keep in mind how the system will need to be controlled—particularly if that control is to include any measure of interactivity, as is increasingly the case. Show control may be something as conventional as a Medialon device running dedicated show control programming, or as unique as completely custom-coded AMX devices running scripts that are original to the project itself. In Figure 44.1, we see an example of a small themed attraction control system—as usual, keep in mind that this is a concept-level drawing. For a fully realized design, drawings would get down into the details of each installed junction box and patch panel (which would need to be specified around the attraction). For this diagram, we are simply looking at the actual control devices involved, and the basic flow of control signal between them.
- Hidden speakers: while we've discussed previously theater's desire to make us hide our speakers, this amplifies considerably in a lot of themed projects. In a theme park *ride*, it is often the case that we can determine exactly where a guest will be facing and what their sightlines will be, but in themed *environments*—parkwide

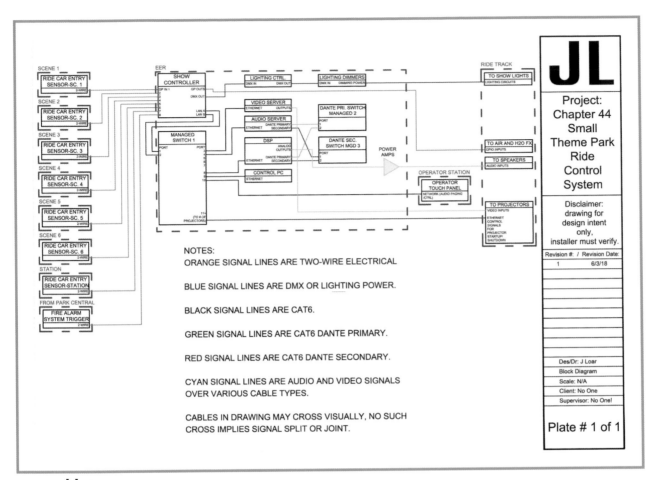

FIGURE 44.1
Block Diagram of a Small Themed Attraction's Control System

spaces, walk-thru exhibits, themed restaurants or malls, or what have you—the ability to determine sightlines diminishes substantially, as guests may wander in a variety of places and positions. Thus, the hiding of speakers gets more intense. I've hidden speakers in every manner of prop imaginable, and those speakers range from raw drivers to very substantial enclosures. While the design of the prop is not the sound system designer's purview, it often falls to the sound system designer to work closely with scenic and prop designers to determine where a speaker can be placed, how large it can be, what physical openings for sound propagation will be available, and how those speakers will be mounted. I've had to come up with unique mounting solutions for speakers involving all manner of hardware that was not at all designed for speaker mounting, in order to get my speakers to sit comfortably within (and not rattle) bizarre props and scenic items.

- Life-safety redundancy: any time guests come into an environment bereft of the normal environmental cues for danger (or, as is the case in many themed spaces, where they actively suspend their panic response to such cues, assuming them to be part of the show), life-safety systems become critical. We must design sound systems with emergency features such that
  - The show sound will be interrupted by fire alarms and other such problems;
  - Operators can easily override show sound to make critical announcements; and
  - Such life-safety features are redundant (in terms of power source, and in terms of DSP signal paths) and are supported by robust dedicated UPSs.
  - If you are a guest on a ride, and part of that ride's show features an explosion that makes it look like the track has ruptured and there is a fire ahead, it's hard to tell when an *actual* explosion happens and there is an *actual* fire ahead. We must design systems that impart this information clearly with no room for misinterpretation.
- Lack of live inputs: often (though again, not always), themed systems will not have any live input sources in the system. I've never worked on a theme park ride that used live mics or musicians (though, again, since themed

work is variable, I'm guessing such a thing either exists or will at some time). That leads to systems without mixing consoles. A lot of young designers, who come up in live events, have a hard time wrapping their heads around systems without a mixing console, but without live inputs, the DSP and playback systems are the heart of the enterprise.

- Multiple zones: themed attractions are seldom contained in one activity zone, and as such feature distributed systems delivering (typically) different content to different areas at different (often overlapping) times.
- GPS: many theme park parades these days use actual GPS transmitters and receivers to track vehicles, and to tell sound subsystems when a particular vehicle is in a particular zone, and thus when its theme music should be playing.
- Playback servers that are impervious to fast motion and shock: for "onboard" (inside ride car) playback on fast rides.

When working on themed projects, we also must be aware of the preexisting technological environment in which we're creating. A design for a brand new park has the opportunity to set paradigms for that park, whereas the design of an attraction for an existing park may be constrained to select gear and methods of operation that correspond to the existing rules and structure of the park, even if there is a better way to execute the design. In some parks, there will be dedicated staff who oversee system designs, and who won't allow units to be specified within the parks unless they preapprove the decision—so system designers working for these entities still have a design supervisor, in essence.

VENUE CHALLENGES

This is the whole thing. Themed design is about creating or customizing a venue. As such, venue challenges can vary wildly from project to project. That said, each subtype of themed environment will pose distinct challenges that we can address in a general way here.

- Theme parks: theme parks range from low-rent collections of pre-built rides assembled in one spot, with little attention paid to décor, to elaborate architectural marvels that make us feel transported to fantastic lands beyond anything we would encounter in real day-to-day life. At the cheapest end, such parks often still specify speakers for each attraction, though this may simply be a stereo pair for BGM and emergency announcements, aimed at the overall ride space or structure. At the highest end, each attraction and area will feature its own dedicated system, which themselves may both be divided into subsystems and also connected to the overarching "parkwide" system.
  - Attractions: an "attraction" is any featured element, ride, show, or event within a theme park. Each will have different considerations.
    - Rides: there are different stock subtypes of rides, each of which features their own signature venue:
      - Dark ride: a ride typically taking place entirely indoors, with a slow-moving ride car traveling a winding path through different scenic environments. *Haunted Mansion* is a dark ride. *Pirates of the Caribbean* is a dark ride/water ride. Dark rides are generally housed in very nondescript buildings (that may be themed on the façade and in the queue, but not overall). As such, the acoustics are often warehouse-like. While installing scenic elements can help dampen the acoustics, these are still often clattery spaces (unless drop ceilings/false walls are part of the interior design). On the flip side, dark rides provide a ton of places to hide black speakers. It's very easy to hide even quite large speakers in a dark ride environment where the whole space is painted black and only certain scenic elements are lit (and even then, only at certain moments).
      - Roller coaster: a ride that foregrounds speed of travel and novelty of rider orientation over scenery. There are some rides that are on the border between roller coaster and dark ride style scenic ride (think *Matterhorn* or *Big Thunder Mountain Railroad* at Disney), but hardcore coaster fans (and many designers) don't consider these to be coasters precisely because they still feature a lot of scenery and vignettes that the riders take in along the way. "True" coasters are thought of as the giant steel or wood behemoths with cars whipping around at highway speeds. As such, they don't typically feature as much in the way of sound effects, though there is increasingly a trend toward so-called onboard content (content played in the car, which has its own dedicated sound system)—a design feature that

calls for a host of considerations (onboard players, vibration isolating mounts, speaker SPL vs. track/ car SPL at high speeds, etc.). Roller coasters often feature elaborate theming and sound systems in the queue, and minimal sound systems along the ride track itself. However, these systems are getting more elaborate every day, and there is still a need for safety sound systems at any of the so-called block brake sections: where track brakes are installed that can completely stop a full-speed train or car (and fully restart said car)—sometimes these speakers are used as part of the ride experience, sometimes they are just there for emergency measures, but either way, these block brake sections normally accompany emergency exit platforms for the rides, and emergency announcement systems should be found at any emergency exit.

- Flat rides: aka "sit 'n' spin," the type of rides that typically sit on a fixed platform and feature basic motion effects and generally minimal theming. Think the *Scrambler* from seemingly every traveling fair ever, the *Mad Tea Party* or *Dumbo the Flying Elephant* at Disney, any giant swinging "pirate boat" pendulum-style ride, carousels, Ferris wheels, and the like. The venue for these is generally just a small plot of real estate—often outdoors, though this kind of ride certainly exists indoors, like at the Mall of America—and as such generally features a simple sound system for BGM and announcements. The only real challenge of these rides is that they are often quite loud, between the machinery itself (which is sometimes exposed), and the screaming of the patrons. High SPL is generally required. In Figure 44.2, we see a Ferris wheel at a theme park (not just a traveling carnival—hence the fire system interrupt, and individual channels for each speaker). Note that in this drawing, I've shown yet another method of labeling speakers.

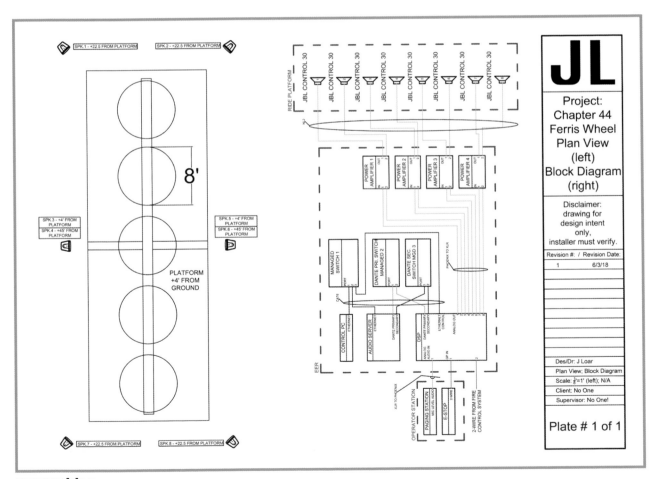

FIGURE 44.2
(Left) Plan View of a Simple Ferris Wheel Sound System; (Right) Block Diagram of a Simple Ferris Wheel Sound System

- Motion simulator rides: these rides place guests in a "vehicle" (in quotes because often, this vehicle doesn't actually move from its home location) and shows them immersive video content with accompanying audio. The vehicle itself jerks and bucks along with the video, such that riders feel they are in the actual projected environments. Some rides (such as *Transformers* at Universal) combine traditional dark ride features—moving through a dark space on a track—with motion simulator features like minimal scenery (framing the projection), and lots of motion-synced video and audio. These rides are sometimes referred to as "roving motion base" rides. In addition, they overlap with a category of ride called "4D" rides, that often incorporate 3D video rather than standard 2D, as well as so-called 4D effects like misters, smells, and other sensory excitements. In addition, there is a category of ride loosely grouped in the motion simulator category, which is the "flying theater"—essentially a motion simulator ride with a large audience section, designed around the idea of making the audience feel, unsurprisingly, like they are flying. As such, motion simulator rides often feature elaborate surround systems, combining cinematic system design principles with effects like seat rumbling vibration transducers (like the Clark Synthesis TST series). Playback is generally part of a fixed timeline along with video, both of which require very close communication with the ride control systems. If any video/audio seems to lose sync with the motions of the vehicle, not only will the ride fail to deliver the desired experience, it can also be very disorienting to passengers (particularly when 3D video is included).
- Drop tower rides: from the parachute jump attractions found at many outdoor parks to the *Guardians of the Galaxy—Mission Breakout!* tower ride at Disney California Adventure, the premise of these rides is simple: go up a long way in the air, and drop down, usually quickly. While old-school parachute jumps don't usually feature much theming, the newer variations may feature elaborate theming, right down to terrifying and intense sound effects during the drop. How much or little theming there is will control how elaborate the system needs to be.
- Water rides: from rapids/raft style rides to log boat flume rides, there are a variety of different types of water rides, but the general premise is always the same—you are in a boat of some kind, floating along a path, and things happen. Sometimes (like Disney's *Pirates of the Caribbean*) there are mild thrill elements, like some drops down waterfall chutes, but many of these rides still center on a tour through a story and scenic environment, much like a dark ride. Sometimes (Disney's *Splash Mountain*) there are themed and story elements, but much of the "point" of the ride is about a couple of extreme drops down steep slopes. Other times—most river raft rides—there is little or no story, and the whole point of the ride is an exciting trip down an intense rapid, more like a roller coaster than a dark ride. At any rate, the usual design challenges for the associated types apply, coupled with the fact that, even if the ride is entirely indoors, everything exposed to the ride path must be highly waterproof.
- * Stunt shows/effects shows: these are shows with an audience in a fixed location, watching either a narrative show featuring stunt performers or a display of special effects meant to mimic film production effects. While effects shows have gotten a lot less common these days,[2] due in part to the high expense of operation relative to how many guests can experience them at one time, stunt shows are still fairly common (and they often incorporate special effects—explosions, water effects, etc.). Systems for these shows are often more like regular theatrical systems, however: it is generally assumed that we are replacing voices rather than just reinforcing; we often are allowed to see the main system speakers; SPL is often much higher than in traditional theater (in part because the actions shown onstage may involve crashes, explosions, buildings being broken apart, and the like).
- * Character meet 'n' greets: these are generally small corners of a park's open areas dedicated to periodic appearances from actors dressed in full head to toe costumes. These range from someone appearing as the relatively normal Cinderella, to someone wearing a giant foam carved suit to appear as Mickey or Minnie Mouse. There are often small speakers in meet 'n' greet locations to play BGM, and in the case of some character performers, there may be a live mic inside the headpiece that also is sent (at relatively low level) through the local speakers—though oftentimes the full headpiece characters just don't speak.
- * Walk-thrus: any environment where guests walk through a series of exhibits, tableaux, or vignettes. Some of these will be highly interactive, encouraging guests to touch things, play with things, etc. Some will be strictly one way, delivering moments to guests as they move at their own pace through the environment. The systems, as such, will vary widely depending on how interactive the attraction is meant to be, and what kind of experience is being delivered.

* Parades: almost every theme park has a parade or two that they run. Parade floats usually have onboard audio, playing canned BGM, and sometimes vocals (often canned as well, which the performers mime along with). Rarely, there will be live mics on parade performers. There will also be speakers lining the parade route, as sound effects and supporting tracks will often change as floats enter or leave different zones along the route. As mentioned earlier, one of the industry-standard methods of controlling such events now is GPS, which allows real-time tracking of each vehicle, and crossfading of content as vehicles with different themes and characters travel along the route. Outdoor parade zone speakers are sometimes allowed to be visible, though they are sometimes hidden inside scenery (or both). Parade vehicle speakers usually need to be hidden.
* Parkwide areas: not attractions per se, but just themed areas of the park in between attractions—often playing BGM and needing to be accessible for area or parkwide emergency announcements. Also called "Area Development" locations
* Restaurants: almost always themed and generally providing BGM, sometimes live or animatronic shows and emergency announcement systems.
* **A special note on animal parks: while they have fallen out of favor in the wake of the documentary *Blackfish*, parks focused on animal shows still exist (and are still being built in China and elsewhere). Such parks often feature animal shows—which are designed basically like stunt shows, including the typically visible speakers. Marine mammal parks will often specify truly waterproof wireless mics for the performers.[3] Such parks will sometimes feature "pass-through" attractions, like an aquarium through which guests pass via a Plexiglas tunnel, in which they are surrounded by water and animal life. Such attractions often feature BGM or narration, but not very loud, and generally mounted either at the ends of the tunnel or built into water-sealed enclosures in the tunnel walls. Otherwise, such parks also feature rides and other standard park attractions.
* **A special note on waterparks: parks filled with waterslides often feature parkwide BGM and announcement systems, and some attractions (particularly slow-moving attractions like lazy rivers) will feature dedicated BGM/effects. Rarely, some "thrill slides" will have sound systems at the launchpad of the attraction, playing exciting sound effects just before a rider gets into the slide and takes off. It is very rare to have any sound system whatsoever on the slides themselves—not unheard of, just not common. The same principles apply when designing boat-based water rides, speaker choice depends on content, speed of traveler, SPL, and waterproof status.
* Cruise ships: ship work is essentially *defined* by the venue challenge—your system will be at sea for long periods of time and will need to be robust enough to run for years at a time with very low-touch maintenance. Additionally, any system or production that calls for constant replenishment of expendables must budget for the length of the longest cruise, and then ensure there is ample storage for these expendables (plus contingency) aboard. Cruise ships today often feature performance venues (which would be designed like any other installed performance venue, with the aforementioned caveat about low maintenance), interactive 4D experiences, themed dances/balls, and generally themed décor. Each hallway may feature lighting and BGM being run by a show controller, every restaurant may be themed individually, and so on. On cruise ships, systems place even higher priority on emergency systems and redundancy than standard theme parks, due to the inherent isolation and danger of sea travel. Much like airports, and theme parks, wide-area data/DSP networks are likely to be in place, and access control is very limited, so should a guest encounter a control panel, they wouldn't be able to do anything dangerous or damaging.
* Casinos: some casinos are very basic—slot machines, card tables, BGM. Some are essentially theme parks with gambling, including the performance venues. The signature venue challenge with casinos (at the higher-budget end particularly) is that the main goal of casino design is to keep guests playing at all hours of the day and night, without feeling the passage of time. As such, BGM needs to be present but not obtrusive—the more guests notice which song is playing, the more they notice the passage of time, but conversely, when BGM is too quiet they may start thinking about the money they are spending. Different casinos have different target markets and demographics, so this will define SPL levels and need for full-range BGM versus narrowband. A casino aimed at high-rollers in their 20s and 30s is more likely to feature louder BGM with more sub-bass, where a "classic" casino focused on families or older patrons may be more mellow. Casinos often feature particular spots with more themed activity—an indoor rainstorm in one part of the building, or a ride itself. Each of those needs to be designed individually while still interacting with the main venue systems.

449

Tales From the Field

One of the least enjoyable parts, but it's a necessary part, is to be the "buck stops here" guy. Be the one who says, "No, no we cannot settle for that. No, a giant piece of Plexiglas blocking the speakers: No! I'm going to get on a plane unless you remove that Plexiglas, and I will disavow that I ever came here. No." They said, "You're the famous system tuner. I know you can find a solution." I said, "I can, but the solution starts with a hacksaw, not with a filter." It was a theatrical production in [a foreign country], and if you know anything about the politics [there], it's a go-along, get-along culture. The set designer was a friend of the president of the casino, who was a friend of the president of the country, and they told me [about the Plexiglas], they said, "Somebody put them there, but nobody's allowed to tell the set the designer that they're wrong." So I said, "OK, well, then I'm leaving." This was a two-day standoff and then they said, "What about if we pick up the PA and move it over here?" I said, "That's not great, but that at least can work." And, 4 hours later, I was tuning that PA. A really important part is to know when to stand your ground because inevitably someone's going to say *you were there*, and if that thing stays like that and your name is attached to it, it will damage you for a long time.—Bob McCarthy

- Hotels, restaurants, shopping malls, etc.: any site that features themed elements, whether they are animatronic creatures in a particular spot, or just elaborate A/V decorations and enhancements, can be called "themed." The more we have to hide our gear and the more the décor is based around a particular narrative element or location, the more this is truly "themed" design. The venue challenges for each of these will vary, but are generally the same as in any hotel, restaurant, or mall, plus the added challenge of show control and (sometimes) delivering content meant to be noticed (like the dialog of animatronic figures) in such a way that punctures the ambient sound levels without overwhelming guests who may not be invested in participating. Speakers with narrow coverage areas are often employed to allow guests in proximity to a themed display to enjoy the experience without swamping, say, every table at the restaurant with the story of whatever creature has come to life in the corner. On the flip side, some theme restaurants (think *Medieval Times*) are basically dinner theater, so the show system must cover all the audience, just as any show system would intend to.

UNIQUE TERMS

Themed entertainment possesses a rich lexicon. The following is a list of some major terms, but as this is a cutting-edge field, both techniques and terms evolve quickly. Many of the following terms also apply in other architectural projects.

Animatronics: robotic animated three-dimensional figures

Area development: a term used to refer to inter-attraction open areas in the "lands" of a theme park. This would be the streets between attractions, the open courtyard areas, and the like. Sometimes, area development is done according to a strongly themed design aesthetic, and sometimes you're just mounting speakers on poles out in the open, but either way, area development design usually involves BGM and emergency systems, and often integrates with parade systems.

BGM (background music): BGM can be found in almost every themed environment. Background music may be playing constantly on a loop, it may be triggered and stopped based on fixed wall clock or timecode triggers, it may be triggered by a ride car entering a scene, or really by anything. BGM isn't even always just background music, as the term is sometimes used to refer to *any* music playback in a themed attraction.

Bollard: a short, thick post—often holding speakers and lights.

Bump door: a door that helps separate one dark ride scene from another, that is opened by a bump from the ride car (usually the bump just triggers the show control system to fire actuators that open the doors, it's not just pushed open by the car).

Creative narrative: a document that describes, in simple elemental terms, the sequence of story events for a given themed installation. For a ride, this would describe scene by scene the action of the attraction, with general notes on what is seen, heard, and experienced by the audience. Concept design stage drawings are based on initial narratives, and as the narrative is refined, so the technical systems to support it will evolve.

Cycle time: the time it takes for a ride vehicle to go from one dispatch to the next dispatch (with load, ride, unload, and reload in between).

Dispatch: the departure of a ride vehicle from the station. Designers often need to know what the dispatch interval is, in order to time certain effects.

EER (electronic equipment room): the space that houses equipment racks. Climate-controlled, located far away from guest access, and locked, these rooms hold playback servers, computers, DSPs, amplifiers, lighting dimmers, and more.

E-Stop: emergency stop controls, built into a particular operator control station (or even as a feature via a paging station), and also automated based on certain dangerous or failure conditions.

Gamify: to make a themed attraction highly interactive, and usually competitive. Gamified attraction guests may participate in a contest for who "shoots the most dragons" in an interactive dark ride,[4] they may interact with controls that help "steer" the ride vehicle, and the ride is "more or less successful" (i.e., the narrative changes) depending on how the ride vehicle is steered. More and more attractions today are "gamified" as it provides a new layer of interaction and thus of novelty for riders, encouraging more re-rides for guests who have already been through the attraction, since they can influence the experience and outcomes. Such attractions require very elaborate show control.

J-Box (junction box): patch panels for various signal types around an attraction. These may be for speaker-level signals, network cables, line-level audio signals, fiber-optic cables, etc.

Loading platform: the area used for boarding and disembarking from ride vehicles. Also known as "station."

Onboard: audio (and/or video) played from a subsystem within a ride car or on a parade float.

Ops: ride operations staff

Point source: used both to refer to speakers in point-source format, but also to refer to effects speakers providing a localized source, often tied to a specific scenic element or prop.

Pony wall: a short wall, over which people can easily see, but which separates, say, parts of a queue from other parts of a queue, or themed elements from guests, or the like.

PP (programming port): a hardware network juncture for computer access. PPs are placed throughout themed attractions or spaces, such that programming can be done locally in any major area or zone of the attraction (or within reasonable cable-length distance). PPs will be locked to guest access, and only used during commissioning and maintenance of the attraction.

PPM (people per minute): how many guests load into an attraction per minute. Also expressed as THRC (Theoretical Hourly Ride Capacity).

Pre-show: a themed "performance" for a group of riders prior to boarding a ride. *Haunted Mansion* at Disney has a famous pre-show in which guests are assembled in a group in a large room, which quickly "stretches" as story elements are presented (the stretch is a visualization of the fact that the room is an elevator to another level of the attraction, where the station is located).

Queue: the line that people wait in for their turn on a ride or in an attraction. Queues are increasingly being seen as part of the themed experience, and some feature quite extensive sound systems.

Redlines: drawings that have been marked up with red to indicate needed changes or cuts.

Ride envelope: the physical clearance needed around ride vehicles on their tracks, such that the tallest theoretical rider extending their arms all the way out still cannot hit or touch anything (including our speakers!).

Ride timing: a schedule of when a ride car will pass certain points on a ride track. Used to place sensors for various effects, as well as set delay times for said effects being fired.

Rock speaker (Figure 44.3): literally a speaker themed to look like a rock. Available commercially from several manufacturers.

Scope: the range of work expected as part of a design contract. What documents and services are to be provided, in what format, to whom, on what timeline. Must be agreed to between the designer and employer prior to the start of the design work. Any substantial change to the pre-agreed scope is referred to as a "change order" and may accordingly call for additional fees to be paid to the designer. This term in particular applies to all architectural projects.

UNIQUE GEAR

I could repeat my long, forced, cackling laugh here, but I think just saying that has made my point. Gear is very unique in themed work and changes a ton depending on what you are trying to do. That said, there are some categories of gear that are found frequently.

FIGURE 44.3
Klipsch Pro-650-RK "Rock Speaker"
Source: Courtesy Klipsch Group, Inc.

Audio to fiber converters: themed attractions often require very long signal line runs, longer than even balanced or network cables can comfortably handle without signal repeaters, so today many systems convert audio signals to fiber-optic signals, which feature very long maximum signal lengths. Fiber-optic repeaters also don't suffer the same signal degradation that audio repeaters induce, nor do they have as much potential for signal corruption as network cable repeaters. As such, the major AV traffic at the world's largest parks[5] is all run over fiber.

Contractor-grade speakers: these are speakers that are designed for permanent installation and are often manufactured in a range of unique shapes and sizes.[6] These include ceiling speakers (often referred to as "backcan speakers" for the housing cans that cover the crossover electronics of the speaker), wall speakers, columnar speakers, and many other form factors.

Operator consoles: controls for operators, often in the form factor of customized touch panels or the like.

Paging stations: extremely common.

Message in a Bottle

You want to get out there and fail as quickly as possible. You shouldn't worry about failure. You learn as much from your failures as you learn from your successes; in fact, your successes tend to . . . you tend to just focus on trying to repeat your successes. The reason for your success is a number of things, and you think, "OK if I set my arrays like this, this is my solution that's my panacea." That's the thing: you really should not fear failure. As long as you know you're always trying to do your best, and you're always trying to do right by the people that you're working for. I mean . . . don't *kill* anybody . . .—Jamie Anderson

Show controllers, media servers, network infrastructure: just about every themed attraction requires each of these elements—with the network infrastructure typically being very robust.

Weatherproof equipment: lots of themed attractions are outside. In Florida. *Y'know, with the hurricanes and alligators and such*. 'Nuff said.

TYPICAL TIMELINES

These vary quite a bit but are normally on the order of at least several months, and often somewhere between 1 and 5 years, depending on complexity and other factors (usually financial).

I have been hired to provide concept designs 2 weeks after my first meeting, and I have also been involved with projects that took many years to develop. It is very important to set the schedule (as much as possible) in advance of doing the design, as part of the scope, as it impacts the level of intensity of the work quite a bit.

DESIGN PRESENTATION

Documentation for themed attractions is enormously detailed and complex. A CD package will be considerably simpler than an SD package, and an SD simpler than a DD package, but the overall set of documents is often massive.

- Plan view drawings (floor and ceiling) *^%
- Elevation view drawings ^%
- Block diagram (aka "one-line," aka "riser") ^%
- Budget worksheet ^%
- Technical system narrative *^%
- Source destination list (aka "hookup," aka "I/O sheet") ^%
- Facilities impact report ^%
- Equipment schedule %
- Cable schedule %
- J-Box schedule %
- Trigger schedule %
- Tech book % (a collection of PDFs of spec sheets for all specified equipment)
- Coverage study/speaker prediction/acoustic model ^% (sometimes all three as separate documents—the first being a CAD drawing showing basic cones of coverage from each speaker, the second being a prediction of the speaker coverage, the third being an acoustic model of the venue(s) without speakers—sometimes all as part of one 3D model).
- Mounting details ^% (drawings of speaker mounts and rigging, showing exact hardware and connections)
- Operation narrative ^% (describing what operators are needed for the sound system, and what their roles are)
- Operation manual % (describing how to operate the system as designed in the DD phase, sometimes provided by integrator based on an as-built system)
  - *=Usually needed for CD level submissions
  - ^=Usually needed for SD level submissions
  - %=Usually needed for DD-level submissions

The design will be presented in phases, and to teams according to hierarchy. If you are working for a design firm, your individual work will be shown to them before the firm shows to the client(s). If you are freelance, you may be showing directly to clients on the park (or hotel, or mall, etc.) creative/technical teams. This is generally done like any other major architectural installation project.

Themed work is often covered by NDAs (non-disclosure agreements) such that it is difficult to show much of it in detail without violating contracts and revealing trade secrets. Suffice it to say for now that these systems are often large, distributed, creative, highly interactive, networked, and (usually) redundant. These are some of the largest and most elaborate sound systems in the world, and as such require a ton of time and effort to design well.

In Figure 44.4, we see an example of a hypothetical dark ride track.

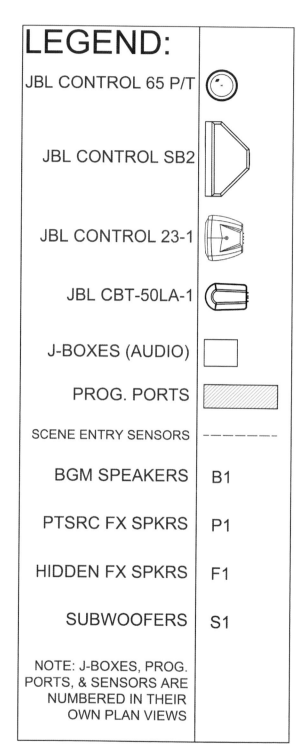
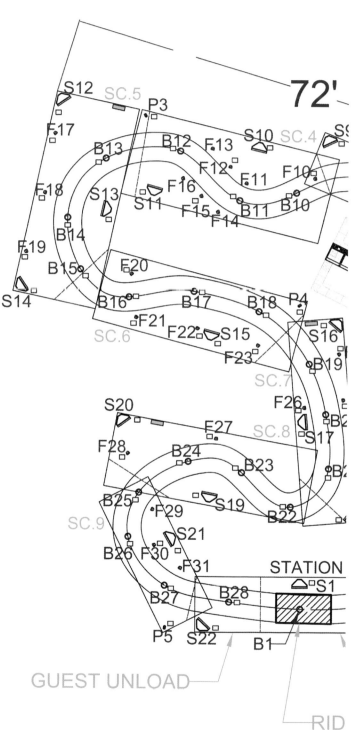

FIGURE 44.4
Plan View of a Hypothetical Dark Ride

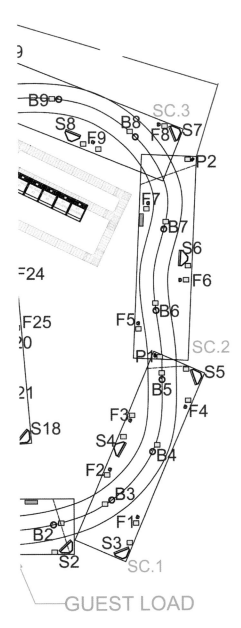

B9

SC.3

S8 B8

F9 F8 S7

P2

F7

B7

S6

F6

=24

B6

F25 F5

SC.2

P1

S5

B5

F3 F4

S18

S4 B4

F2

B3

F1

B2

S3

S2

SC.1

GUEST LOAD

E VEHICLE SHOWN FOR SIZE

JL

Project:
Chapter 44
Dark Ride
Plan View

Disclaimer:
drawing for
design
intent
only,
installer
must verify.

Revision #: /	Revision Date:
1	6/3/18

Des/Dr: J Loar

Plan View

Scale: $\frac{1}{12}$"=1'

Client: No One

Supervisor: No One!

Plate # 1 of 3

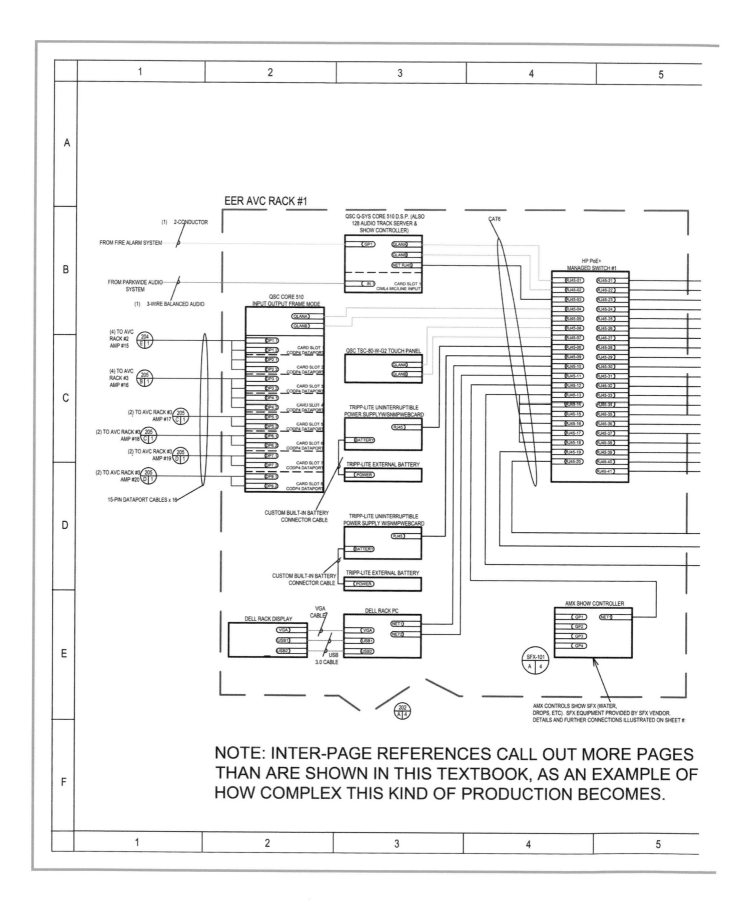

NOTE: INTER-PAGE REFERENCES CALL OUT MORE PAGES
THAN ARE SHOWN IN THIS TEXTBOOK, AS AN EXAMPLE OF
HOW COMPLEX THIS KIND OF PRODUCTION BECOMES.

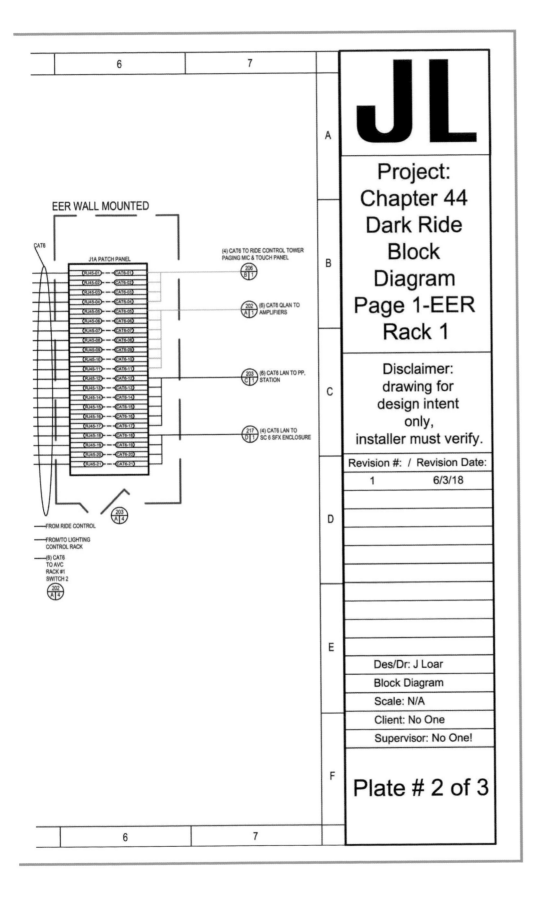

6	7

EER WALL MOUNTED

CAT6

J1A PATCH PANEL

RJ45-01	CAT6-01
RJ45-02	CAT6-02
RJ45-03	CAT6-03
RJ45-04	CAT6-04
RJ45-05	CAT6-05
RJ45-06	CAT6-06
RJ45-07	CAT6-07
RJ45-08	CAT6-08
RJ45-09	CAT6-09
RJ45-10	CAT6-10
RJ45-11	CAT6-11
RJ45-12	CAT6-12
RJ45-13	CAT6-13
RJ45-14	CAT6-14
RJ45-15	CAT6-15
RJ45-16	CAT6-16
RJ45-17	CAT6-17
RJ45-18	CAT6-18
RJ45-19	CAT6-19
RJ45-20	CAT6-20
RJ45-21	CAT6-21

(4) CAT6 TO RIDE CONTROL TOWER PAGING MIC & TOUCH PANEL
206 B 1

(6) CAT6 QLAN TO AMPLIFIERS
202 A 1

(6) CAT6 LAN TO PP, STATION
203 C 1

(4) CAT6 LAN TO SC 6 SFX ENCLOSURE
217 D 1

203 A 4

—FROM RIDE CONTROL

—FROM/TO LIGHTING CONTROL RACK

—(6) CAT6 TO AVC RACK #1 SWITCH 2
202 A 4

JL

Project:
Chapter 44
Dark Ride
Block
Diagram
Page 1-EER
Rack 1

Disclaimer:
drawing for
design intent
only,
installer must verify.

Revision #: / Revision Date:	
1	6/3/18

Des/Dr: J Loar

Block Diagram

Scale: N/A

Client: No One

Supervisor: No One!

Plate # 2 of 3

6	7

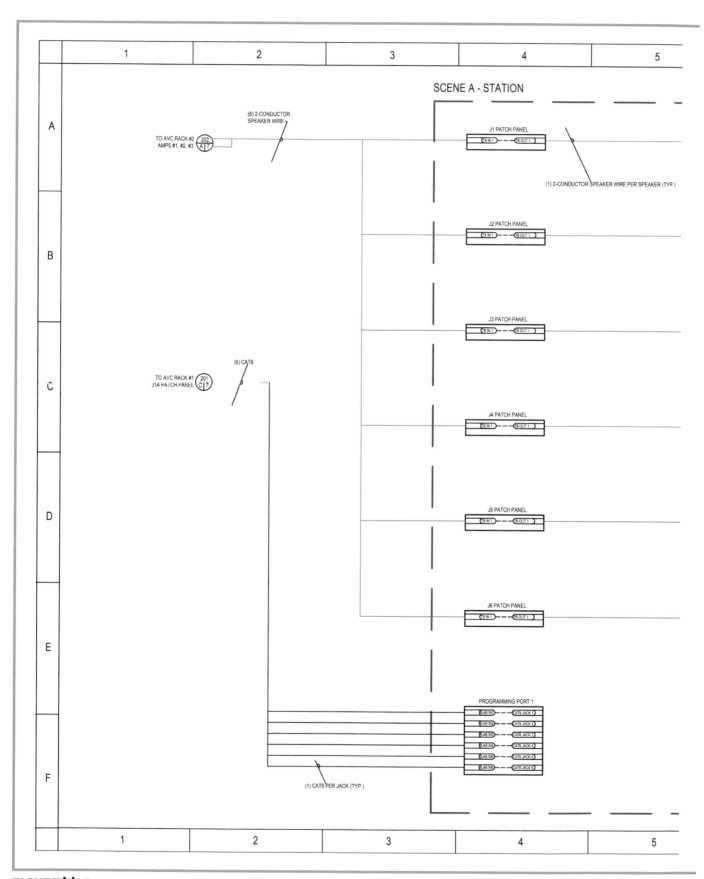

FIGURE 44.5

(A) First Page of Block Diagram for Hypothetical Dark Ride; (B) Detail Page for Dark Ride Block Diagram (Scene: Station)

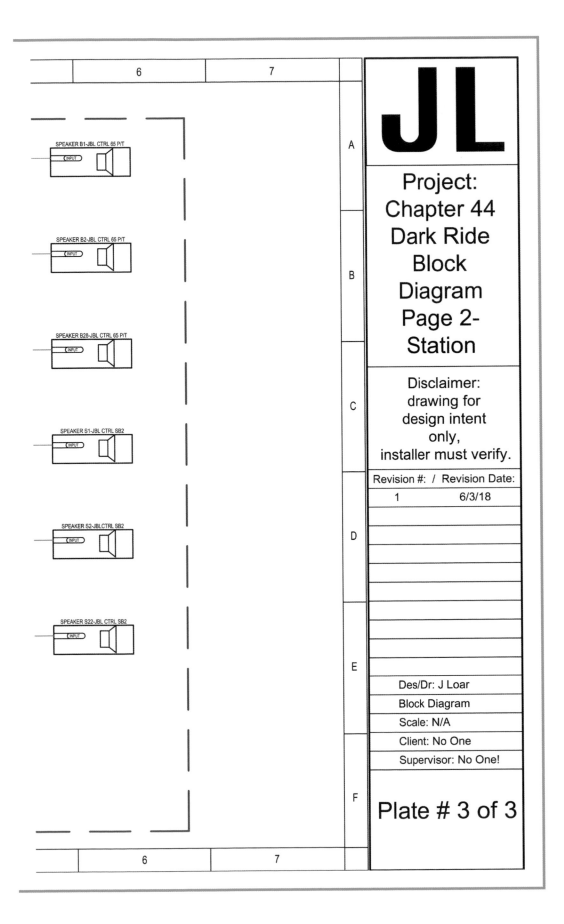

	6		7	

SPEAKER B1-JBL CTRL 65 P/T
INPUT

SPEAKER B2-JBL CTRL 65 P/T
INPUT

SPEAKER B28-JBL CTRL 65 P/T
INPUT

SPEAKER S1-JBL CTRL SB2
INPUT

SPEAKER S2-JBLCTRL SB2
INPUT

SPEAKER S22-JBL CTRL SB2
INPUT

A

B

C

D

E

F

JL

Project: Chapter 44 Dark Ride Block Diagram Page 2- Station

Disclaimer: drawing for design intent only, installer must verify.

Revision #: / Revision Date:	
1	6/3/18

Des/Dr: J Loar

Block Diagram

Scale: N/A

Client: No One

Supervisor: No One!

Plate # 3 of 3

	6		7	

In this plan view, we have overlaid both ground and ceiling plan information about speaker positions (heights might be listed on this plan view if this were a larger plate size, but would definitely be shown in detail plates and elevations), junction boxes, programming ports, EER, sensors, etc. The ride is divided into scenes. Each scene has its own small subsystem, delivering content for when the ride vehicle is in that spot and that spot only. Some speakers are hidden in the dark, some are mounted in scenic props (noted here by type as per the legend, props not shown).[7] Each speaker is labeled by speaker type and number (which would correspond with the block diagram)—speaker prefixes are B = BGM, P = Point Source for general FX (ambience, etc.), F = FX synced with animatronics or other automation (these are hidden in scenic props), and S = Subwoofer (this acronym scheme is just an example, on some projects "P" might designate projectors . . . just make sure your team is all using the same shorthand).

In Figure 44.5, we see in (A) the first page of the dark ride's system, focused on the EER.

You may notice several different drawing styles here than in previous drawings. In an attempt to represent some of the variety of drawing types you will see in the industry, I have drawn this more in accordance with standards seen in themed engineering drawings than theatrical standards.[8] This includes the use of "sockets" for each IO connection to a device, lists of signals traveling from one page to another, and so on. You will also notice that, at least in the EER, the pretense of a signal traveling left to right in the drawing is largely abandoned in service of clarity of reading (rather than having too many signal lines cross one another). This drawing is *more* like a real block diagram, and as such, it reflects a great deal of complexity. In (B) we see a typical detail page that shows signal paths from EER to a scene. Many of the drawing conventions shown here will be detailed in Part IV of this book but note the inter-page references and zone borders to help trace signals between plates in the set. While theatrical systems often attempt to show all of the system on one page of block diagram when possible, themed attractions break each physical location out into its own page (at least).

Theme park design is not usually a beginner's task. If you want to work in themed design, you must typically work for one of the few firms that handle this work globally, and often need to work your way up from assisting the engineers, to engineer, to designer. It is a complex master craft, relying on the cutting edge of production technologies.

DESIGNER QUESTIONS

- Ask every question you can think of! These are complex systems, head off problems in advance! All the usual questions about the type of show, SPL, feeling we want to create in the audience, etc.
- How will fire or emergency systems need to interact with your sound system? What kind of message or trigger will they provide to tell the sound system to enter emergency mode?
- Which operators at which stations need what kind of access? Who has e-stop privileges?
- How should day and night mode (startup and shutdown for an operational day) work? Or is this a themed installation in, say, an airport terminal or hotel casino, that runs 24/7 and thus has no night mode at all (or maybe their definition of "night" mode is a separate set of programs running at night than during the day, but still in operation)?
- What am I forgetting? Honestly, this is a great question when working on themed installations. They are projects designed by huge teams over long time periods, and after a good summary meeting, it is often a wise idea to ask if you've missed anything, if there are any functions of the design or project you failed to discuss. I've seen projects get off the rails because this question was never asked, until, come bidding time, there is some section of the project that no one designed because everyone assumed it had been handled by someone else and thus hadn't ever actually been discussed.

* * *

Theme parks are getting bigger and more elaborate every day. Some media production facilities (the focus of our next chapter) are also getting bigger and more elaborate, while others are getting smaller and smaller every day (until some "studios" are just a bro in boardshorts with a MacBook). We won't be examining project studios, but we turn our attention now to professional production facilities.

Notes

1 Think *Jungle Cruise* at Disney.
2 Many of these have closed in the last decade. Universal's *Backdraft* was a good example of this type of show, where an audience crowds into a warehouse set and fire effects surround them to mimic the experience of the film *Backdraft*. However, the show couldn't hold very many people at once, it took several minutes, and thus the PPM served (people per minute, a standard measure of guest service in theme parks) was quite low versus the cost of operation. The cost is one reason motion simulator rides have gotten so common, as they are cheaper to make than most other ride types of similar complexity (many flat rides are cheaper, but not as exciting to patrons).
3 As of this writing, Lectrosonics is the only company that manufactures genuinely waterproof mic systems for this purpose.
4 For example, "Dragons Wild Shooting" at Lotte World in South Korea, for which I designed and engineered AVC systems as part of a team at Technical Multimedia Design, Inc.
5 Like Disney Shanghai Resort, the world's largest park—I was one of the engineers planning the parkwide fiber infrastructure (among other design and engineering tasks prior to opening).
6 For example, among many other form factors, JBL manufactures ceiling speakers (speakers to be mounted in the surface of a drop ceiling, the grate flush with the drop ceiling itself) in sizes from hi/mid speakers measuring just a couple of inches across to sizable subwoofer drivers.
7 Hey, I designed a phony ride to show you this stuff, I wasn't about to start drafting detailed props to go in it.
8 These standards derive from the work I've done for a certain major company who shall remain nameless.

CHAPTER 45

Recording/Mixing/Mastering Studios (Project and Commercial) for Music, Film, Etc.

Including a Brief Note on Sound Stages for Visual Media

There are many excellent books already dedicated to production studio designs, and this brief chapter won't attempt to dive into too much depth on the topic. In production studios (particularly music production studios), acoustics are at the forefront of design priorities, as a poor-sounding live or mix room will impact the quality of work done there as much as (if not more than) inferior gear. There are some basic system functions and rules we can examine here, but if you are interested in this topic, see the "Bibliography and Further Reading" section at the end of this book.

DEFINITION

A production studio is any space whose primary function either is or includes the recording of performances, the mixing and/or mastering of recorded content, or some combination of these activities.[1] Music studios typically consist of live room(s), often themselves partitioned such that there may be one main tracking space in which an ensemble can perform together, surrounded by isolation booths for vocals, drums, amplifiers, and more. These live rooms feature "tie-lines," signal paths (typically balanced) from the live rooms to patch points in the recording control room. The control room in an expensive studio may feature HD computer recording interfaces connected to a computer (or computers), a recording/mixing console, racks of outboard gear, multiple sets of monitor speakers, and so on (Figure 45.1). Certainly, there are studios that are as simple as a single computer audio interface, a laptop, and one set of monitor speakers—but these are generally referred to as "project" studios and are not typically operating with budgets that allow the hiring of a sound system designer.

Film studios are divided into two categories: soundstages (for filming and recording of performances, either against green screens or on sets built into the stage space), and post-production studios (for editing, mixing, and mastering cinematic projects, be they film, TV, web-based, whatever). Soundstages often feature control rooms adjacent to, but isolated from, the stage, such that recording equipment can be contained there, and thus not make noise on the

FIGURE 45.1
Vienna Symphonic Library's Synchron Studio (Control Room A for Synchron Stage), Studio Designed by Walters-Storyk Design Group
Source: Photo by Heinz Zeggl, Courtesy VSL/WSDG

stage itself . . . but there are also plenty of soundstages with no control room, in which case all recording equipment is brought in per production needs.[2]

Post-production houses often feature several "suites," different workstations, each with their own computers, monitors, etc. They often feature vocal booths, for recording ADR,[3] or voice-overs (Figure 45.2).

STANDARD ELEMENTS

- Inputs: mics and DIs from performers, talkback to live rooms, intercoms (for film/video production), playback (certainly musicians need to hear what they are performing along with, but also video production systems frequently rely on music/fx playback to help cue actors/dancers/etc.)
- Processing: recording/mixing console, DAWs,[4] tape machines, outboard effects
- Amplification: standard operation amplifiers as needed (often Class A)
- Outputs: studio monitors (often multiple sets), playback speakers in live rooms/on sound stages, headphone sends for musicians (headphones, headphone amps/mixers)
- Converters: either very nice I/O at interfaces to DAW rig, or outboard converters, or both . . . studios often feature boutique converters for both input signals to record and output signals to monitor
- Advanced processing: DSP for tuning monitors in a control room
- Control: remote controls (via tablet or other means) for starting and stopping playback/recording
- Clocking: often features outboard master clock (such as Apogee Big Ben or others), though sometimes still relying on an internal clock, in which case recording computer is often the master

UNIQUE FUNCTIONS AND SYSTEM NEEDS

Obviously, the signature need of a production studio that isn't represented in all live systems is the ability to record! While designing live concert systems increasingly means specifying a Pro Tools HD system and signal path as well (it has become so easy to record concerts that lots of bands do it by default, just in case they want to release a live record at any point), many live systems (particularly for theater) don't *ever* incorporate recording. Recording systems can be as simple as a MacBook running Logic Pro, connected to a hardware interface with microphone inputs. In practice, however, professional studios are often considerably more complex.

FIGURE 45.2
A Cinema Post-Production Audio Suite at Sound Lounge (NYC)
Source: Courtesy Sound Lounge

Let's examine studios by type:

- Music:
  - Recording/mixing: professional recording facilities often attract clients (in part) by offering a wide range of expensive boutique gear to shape the sound on the way to the recording medium. This begins with microphones—professional studios need to possess a stock of mics that can cover any instrument and produce a range of tones. From dynamic to condenser to ribbon, from small to large diaphragm, from cutting-edge contemporary low-noise-floor units to vintage items with a particular sound coloration (and sometimes relatively high noise floor), a good professional studio will offer a wide selection of mics. Mics then connect to preamps, and likewise, a high-quality professional music studio will offer a range of preamp selections. Often, even in studios where everything is recorded digitally, a medium-to-large-format recording console will be installed in the control room. If this is a high-quality console (e.g., SSL, Neve, API, Audient), it will provide preamps on every input channel as a base for the recording process. In addition, it is common to find several outboard preamps of different tones and colorations—again, to offer the widest possible range of sound production options. We might find a very transparent sounding set of Avalon pres right next to some very warm and tonally colored Drawmer channel strips. Any professional console will offer a range of processing options, from the stock EQs (and often dynamic effects) of analog consoles to a host of plug-ins in digital consoles. It is often very common for studios to offer an additional range of outboard effects—rack-mounted processors that can vary from standard compressors and gates and such, to esoteric units as the Eventide Harmonizer, the Empirical Labs Distressor, and many, many others. What types of mics, pres, and outboard gear a studio offers will tell you a lot about the type of work they produce—a studio focused on esoteric indie rock will generally feature more oddities than a studio focused on pop-country production.[5]

Most recording studios are, at this stage in history, recording digitally[6] (typically to Pro Tools HD, though sometimes to Logic or other DAWs).[7] Some studios are set up to mix through their analog consoles, and some are outfitted with mix control surfaces—virtual consoles that control settings in a DAW. Some studios are only set up for "mixing in the box," which is to say entirely with the mouse and keyboard, inside the DAW software. Most professional studios will feature at least two, and often three, sets of monitor speakers: one set of large mid-field monitors—set into wall soffits—for loud playback, a large focal area (for the whole band and production team to be able to hear), and for impressing clients; a smaller set of nearfield monitors, often mounted on stands just behind the console or on the meter bridge of the console itself, for general mix work; and very frequently a smaller still set of "ugly" monitors that will sound closer to the less-than-ideal listening setups of your average consumer.[8] Such systems require a monitor controller, not only for main level control but also for easily toggling any source (live or recorded) through any set of speakers at the press of a button. In addition to the setup in the control room, there will usually be a talkback system so that the engineer can speak to the musicians in the tracking room(s). This is usually a single mic in the control room, run either through speakers in the live room, or through headphone mixes, or both. The musicians will need dedicated headphone mixes of what they are tracking (and playing along with, in the case of overdubbing), so a range of headphones and headphone amplifiers is also needed.[9]

- Mixing only: some studios are not set up to track but are just mixing suites. These studios are often software-only, though some do feature mixing consoles. They still feature multiple sets of monitors, often a range of outboard processors, and a host of nice mixing plug-ins.
- Mastering: these studios are reduced in function. They don't need a ton of outboard or a ton of different plug-ins. They need a system set up to perform very specific finishing tasks and as such need a clear and clean monitoring setup, and either the software or hardware processors (or sometimes both) they need in order to level mixes to commercial media standards. Monitoring speakers are often multi-amped (I have heard a lot of five-way monitoring speakers) in order that the engineer hears the fewest artifacts possible.

- Film/video production:
  - Post-production: these facilities are most similar to mixing studios in the music realm. It is common for them to feature limited (if any) tracking facilities, but to feature robust editing/mixing control rooms, with many plug-ins (and sometimes outboard). Most contemporary post-production studios are set up to work in surround sound, because almost all film, TV, and video games are released in at least 5.1 surround,[10] if not 7.1, or even Dolby Atmos (a very elaborate object-oriented spatialization platform). That said, stereo mixes are still produced (or at least, the automated pull-down to stereo that many systems are encoded to do from surround sources is monitored to make sure there are no egregious errors before release) as web browsers still largely do not support surround playback, and many people now get their video content via web. Surround rooms are unlikely to feature multiple monitor sets for A/B monitoring in the way that music studios will. Post houses also need to have robust video software, that can interact with audio production software for creating sound-to-picture mixes and the proper output formats, and a host of software that is related to what is used by mastering engineers, for cleaning up tracks, noise reduction, removing clicks or pops, and achieving final signal levels.[11]
  - Soundstages: some soundstages have audio recording facilities built in, but many do not, and rely on each individual production team to bring in its own gear. Film set recording involves typically wireless recorders (Nagra has long been a standard, along with Marantz, though there are many excellent recorders available). Boom mics, lavalier mics, and sometimes area mics will be used. The key feature of any production studio (sound isolation from the outside) becomes critically important in soundstages, hence their name.

VENUE CHALLENGES

The production studio environment is, first and foremost, concerned with creating a quiet space in which to work. As such the challenge is primarily one for the acousticians and architects, rather than for the sound system designer. Most professional studio facilities create what is called a "floated room," which is to say that the production spaces are rooms within a greater room, set apart from the outer shell by neoprene shock-reducing supports, air cavities filled with absorptive material, and so on, in order to create an internal space that not only prevents unwanted sound from entering or exiting the space, but is vibrationally isolated enough that (ideally) even sub-sonic rumbles (like a truck or train passing by) are not transmitted into the space. In some geographic regions, this is relatively easy. If your facility is in Los Angeles, and not directly under a freeway, it's challenging but not at all impossible to reduce or entirely remove highway noise from the inside of a facility. However, in NYC, isolation can be more difficult. Between the common stricture of having to update older existing buildings (rather than create wholly new construction, as would usually be the case in LA), and the existential fact of the NYC subway system, it's not a simple matter to create a truly isolated venue. I have worked at studios in NYC that were otherwise perfectly isolated until the train came by, and then (sometimes even just dimly in the distance) that telltale rumble penetrated the room.

Besides the acoustical challenges, studio venues get more challenging the more they need to combine filmed production with audio support. A multi-camera sitcom stage will need to have a boom-mic deck above the studio audience, with long, robust, articulated boom arms available so that mics can be sent over the performers without obscuring either the audience's or the cameras' views of the action (and without creating shadows). Not to mention broadcast studios, which are much more specialized and complex. Music studios often soffit-mount[12] large monitor speakers that have a coverage area of most of the control room, for use playing back to clients and producers, while also supporting at least one and many times two pairs of additional, smaller, monitor speakers on which the critical work is done (but whose coverage area is typically limited to the engineer's seat alone).

The studios that present the greatest venue challenges are those spaces that are repurposing other older spaces for use in studio work. It has become fairly common, for example, to use old churches or barns as recording studios. In some cases,[13] there is enough money for the renovation, and acoustics are handled as part of the redesign. In lower-budget cases, the studio may be forced to either create their own kludged solutions to acoustic problems, and those problems may involve placement and tuning of the speaker system. However, any production facility strives for flat speaker response, so if you find yourself faced with any sort of extreme changes to speaker performance in the name of making a room "sound good," you probably need to take a step back and address the acoustic problems before tilting your speakers toward what may measure flat on a chart but may sound unrealistic and false, and lead to mistakes in the recording and mixing process.

UNIQUE TERMS

Each sub-discipline of production that will be undergone in studio facilities has its own lexicon of dedicated terms. To say that cameras and sound recorders have reached "speed" on a soundstage set used to mean that the physical magazines rolling film and tape through the respective devices had started recording, and that enough seconds had elapsed that the machinery was now running media through the devices at the appropriately selected speed, such that the time domain was now stable and action could begin without the action being unusable because the machine was still ramping up. This term, for example, is still used on some productions, even though almost everything is digital now,[14] as a matter of course—a way to ensure that we know that all the machines are rolling and that the crew operating them is ready to engage in a take of the scene at hand.

My point here isn't really about the malleability of the term "speed" itself, but that just that one simple word has a practical background, a history, and a complex relationship to current production standards. Most production terminology applies to the act of producing the work itself, not to the design of the systems capturing that work. As such, I will identify next a few terms relevant to the design process, but this is of necessity not a glossary of actual studio production terminology, merely a list of a few terms that come up when specifying the equipment for such spaces.

Control room: aka "booth," the room in which the engineers work, listening through monitor speakers. This is as contrasted to the "live room" (or the "stage" in film/video production), where performances take place.

Control surface: music studios for generations were centered on the mixing console. Nowadays, some music studios don't have any mixing console at all, preferring the simplicity (in both cost and footprint) of recording via interfaces directly "to the box," into a computer system with no console signal paths. There is, however, a halfway point, which is the device known as the control surface, a mixer-style device that does not actually pass audio, but instead controls parameters inside a DAW. This sort of physical interface is preferred by some engineers and can be very desirable, particularly for projects that, while recording digitally, may be attempting to capture something a little looser and more human in performance terms than the typical top 40 pop song. It can be advantageous to be able to "ride the faders" on productions of this type, without needing those faders to actually pass signal.

ISDN line: Integrated Services Digital Network—a type of data line connection that provides a dedicated trunk for communications transfer. These lines have been extensively used in the broadcast industry but are now being supplanted by faster solutions. Formerly very common for voice-over actors performing roles out of town from the recording studio. These are still found in studios around the world, though as of the last few years, Verizon and other providers are no longer taking orders for new ISDN lines, shunting customers instead toward Internet-based transmission protocols.

Machine room: the room in which computers, hard drives, amplifiers, and other such equipment is housed. Such equipment is generally sequestered from the control room because it is associated with the loud sounds of their internal fans. This fan sound is disruptive to the production process and can cause engineers to miss small details in the work that they need to catch. Such rooms are generally nearby to the control room, such that in the event of a hardware problem, they are easily accessed, but such rooms are also isolated acoustically, and generally feature robust HVAC so that gear is kept at optimal operating temperature, regardless of the season.

Mix position: the spot where the recording or mixing/mastering engineer sits. It forms the apex of an isosceles triangle, the other two points of which are the nearfield monitors. There is an old adage in studio design: "There is only one mix position," and as such much care must be taken to ensure that the mix position sounds as clear and neutral as possible, so the engineer has the best opportunity to craft excellent work.

Redundancy: most studios are designed such that at least two capture devices (hard drives of some sort) are simultaneously storing any recorded data as it is being performed/created. It is very expensive to hire union film crews, multi-millionaire performers, and the like, and as such, redundancy ensures that should one capture device fail, the other has generally succeeded when everyone happens to get "that perfect take" of a scene, a song, or whatever.

Soffit: in architecture, this term refers to "the underside of any construction element," whereas in studio design, soffit always refers to a cavity in the wall of a control booth or production space, set into which are large-format monitor speakers. Such mounting (when done correctly, with literally no gaps around the speaker enclosure, and the face of the speaker exactly flush with the wall into which it is mounted) reduces issues of edge diffraction on the corners of loudspeakers and sets the speakers into a large boundary condition. These are desirable for large-format speakers (so-called mains), as they enable a wider coverage area with robust bass power for impressing upon clients the vast and impressive nature of a sound production.[15] While actual mixing and recording is often carried out with nearfield monitors of smaller size (as they will reveal problems in the audio more readily), many studios feature large soffit-mounted speakers both so that a band playing at 100 dB SPL in the live room can have playback at 100 dB SPL in the booth (so that they don't feel the recording is "lacking energy") and so that impressive sound can be spread around the room, especially to the "producer's couch," which is usually a comfortable sofa set up some feet behind the actual mix position of the space.

TDM: time-division multiplexing—a standard of data processing that splits tasks out over a series of dedicated processors, such that tasks that would take a single-bus linear system a lot more RAM to accomplish (and thus

Tales From the Field

The thing that trips me up the most currently is that if you work in broadcasting, they have completely different words for the same things that we do in theater—just like, "you want me to get the what?" IEC power cords. This is the three prong that goes in the back of your computer monitor or whatever; the broadcast companies I've worked through have called them OSHA cords. I'm like, "What is this?" There are these times where you're in with a new company, and they just use, you know, *looms* versus *snakes*. Especially now, because I'm like, "I know what I'm doing," and then there's something simple that I'm having trouble communicating. I'm having trouble finding a power cord, and I'm getting these weird words thrown at me. Why are we having trouble communicating?—Christopher Plummer

would generate more system latency) are instead carried out with lower overhead and in much closer to real time. In audio production, TDM systems are often referred to as "HD" systems, after the popular recording platform Pro Tools HD. HD systems have dedicated hardware for audio processing that goes beyond what is possible with simple USB or Thunderbolt interfaces in non-HD systems. HD systems are the standard for professional recording, because the latency is so low that performers can hear clear real-time foldback in their headphones without the challenge of added system latency. This is critical for overdubbing, particularly when it comes to vocal performance, where even a small time offset can be read by the performer as pitch issues and can make singing on key very challenging.

UNIQUE GEAR

Studio gear is typified by two opposing classes of equipment:

1. Reference-quality gear: loudspeakers and other equipment designed to deliver as neutral or "flat" a response as possible, to produce the actual sound being captured or mixed, rather than adding or subtracting from that sound by virtue of its own characteristics. Speakers are the number-one type of reference-quality gear in studio systems. We also seek amplifiers that fall into this class, for any passive speakers. In some facilities—for example, commercial voice-over studios—front-end equipment may be desirable in this category as well. A studio aiming to capture truly neutral voice-overs may seek microphones and preamps that "color" the sound as little as possible.

2. "Colored" gear: aka "gear with character," or any of a dozen other ways of describing equipment that imparts its own specific audio signature to signals that pass through it (*tasty!*). Studios often specify a long list of varying outboard equipment—everything from preamps to dynamic effects to strange hybrid effects—selecting a variety based largely on the sonic characteristics that the particular gear will add to the sound of the tracks being processed. Especially in music production, it is very common to see studios with a wide variety of gear selected just for the *difference* that the gear will make to the sound—striving in this case for the exact opposite of "reference-quality" sound, working to change the sound distinctly either on its way to "tape"[16] or from tape to the mixdown. Such coloration may be imparted by mics, preamps, outboard processors of all kinds, plug-in effects, etc.

Note that even in studios where it is desirable to feature a ton of sonically colorful outboard gear, we still want reference-quality monitoring systems on which to listen, for how else would we know what exact coloration we're adding to the sound?

Headphone mixers: studio music production often relies on a ton of headphone mixes sent to performers in the live room so that they can properly monitor the playback along with which they are performing, as well as their own performances. Headphone mixes are sent via aux/bus sends (in pre-fader mode) from the DAW or recording console, with each mix being customized to the given performer's needs. Each performer, however, may need independent level control of the overall mix, and thus headphone amplifiers/mixers are often used to easily give overall level changes to performers' mixes (sometimes these are given to performers in the live room, so they can adjust overall level—if not mix—themselves).

RAID systems: studios often use RAIDs (Redundant Arrays of Independent Disks), as the backbone of their recording storage. A RAID, in the simplest terms, is a pair of drives each acting as the same drive, such that they are both recording designated material simultaneously. This is a simple way to ensure that if one side of a data system is corrupted by error, the other generally will not be. Simple RAIDs are housed in single enclosures, with single power supplies, there are also many professional RAIDs sold with separate dedicated power supplies for each side of the array, such that if one dies, the other may continue to live (assuming the problem is with the circuit, or the power supply itself, on one side of the array). Any recording media will, of course, want to be powered via a UPS, such that in the event of facility power loss, you at least have enough time on your battery backup to save the work in progress and shut down properly before a power loss interrupts data being written and corrupts your drive(s). RAIDs come in a variety of formats and styles, and it is beyond the scope of this work to go into great detail about how they are configured. Suffice it to say, RAIDs are very common in any recording production work.

Studio consoles: in-line vs. split. Live production consoles typically are not concerned about monitoring return signals from a recording medium, however, in studio work, this is very important. You must be able to monitor the signals running "to tape" and "from tape" to ensure that not only are you sending clean signal to the recording medium but also that what the recording medium is returning is a clean signal having been captured "to tape." As such, analog (and now digital) studio consoles have had to be designed in one of two basic formats to allow for this: in-line and split. An in-line console (Figure 45.3) is one that features two rows of faders, one set directly up the channel strip from the main faders. One set of faders is monitoring/controlling signals sent to the recording medium, while the other is making a mix of the signals coming back from the recorded medium as a monitor feed. Generally, such consoles have simple toggles so that the engineer can monitor either set of signals at any time, and many such consoles allow for flipping which set of faders is used for which set of signals, such that during recording, the main faders may be controlling signals to tape, and during mixing, the main faders may be controlling playback signals.

A split console, on the other hand, is one that simply returns the recorded signals to a separate set of main faders, down the console from the *input to recording medium* set of signals. In an analog console, this means that you need a console with at least twice as many faders/channel strips as the number being recorded. In a digital console, this is often accomplished via banked channels, meaning that in one setting, the faders control the inputs to tape, and in another, the same faders are reassigned to control the reproduced signals from tape. Any console with sufficient channel count can be configured as a split console, though those designed for this purpose will allow for easy toggling between the two sets of signals, via mute groups or other such automation.

Timecode generator: any studio working with analog tape at any stage in the process will need to feature a timecode generator that can "stripe SMPTE," or generate a standard SMPTE timecode signal (a square wave timing signal, that when routed to speakers sounds like an obnoxious buzz) that can be laid to a track of the tape. This code lays time synchronization information onto the tape, which then allows for multiple tape machines to be used simultaneously for transfers or large projects, for tape to easily be laid into Pro Tools for editing with original timestamps still in place, and so on. Timecode is particularly important in synchronizing video and audio in multimedia production. All timecode is based on the SMPTE standard which breaks time information into Hours:Minutes:Seconds:Frames, where Frames represents the frame rate of the video in question (anywhere from 23 fps to 60 fps—frames per second—unless shooting in slow motion, in which case very high frame rates are used). In audio equipment, such timecode data is also transmitted by MIDI devices, which adds a further division of quarter-frames to the timecode—such timecode is called MTC (MIDI timecode). There are many varieties of timecode generator, from so-called blackburst generators for locking video sync, to simple SMPTE generators for audio production, but the general principle remains the same—transmitting a standard time value to all devices, such that they can lock in synchronization, and then be used together interchangeably.

FIGURE 45.3
Audient ASP8024-HE In-Line Console
Source: Courtesy Audient

470

TYPICAL TIMELINES

Studio projects, like most architectural projects, will vary depending on size and scope. Most studio design projects will take at least several months, and some can take several years. There is no one timeframe typical of studio design—a complex studio will take longer to design than a simple one. These projects often follow the general CD>SD>DD plan outlined in previous chapters for architectural projects.

DESIGN PRESENTATION

The presentation of design work for a production studio often begins with the architects and acousticians. While a sound system designer may be on board during this process, to help guide those parts of the architecture and acoustic treatment that directly impact sound system choices, studio designers often focus on the space itself first. That said, most studio designers have (or *should* have) at least a passing familiarity with studio needs for speaker placements, equipment footprints, and so on. Many studio design firms employ AV designers as well as acousticians and architects, and the finest studios are often designed in tandem, with system needs defining footprint needs and thus defining physical spaces and solutions.

The design documentation package for studio work is similar to any large architectural project—it will be presented first internally to the firm or firms responsible for overall design, and then to the client. The package will include, but may not be limited to, the following:

- Plan view layouts of speakers, consoles, racks, etc.
- Elevation views of speakers, particularly those soffit-mounted
- Custom mounting details—particularly, again, for soffit-mounted speakers
- Block diagrams, for both data and audio (and video, in an increasing number of cases)
- IP schedules for computer equipment
- Data storage schematics. Signal flow and plan for storage of recorded data, as well as for operation and maintenance of same, including anticipated data storage needs and turnover rates (drives being considered as part of operations cost of a facility)
- Tech book for any selected equipment: all specs, and sometimes operation manuals, for selected gear—collected in one document such that client can evaluate gear choices of their design team compared with their intended operating plan

Studio design is a lucrative field for a very small and select group of companies. Most new studio design is either being undertaken in expanding economies overseas (China, Brazil, India, etc.), or domestically and in Europe for either new educational facilities or for high-profile artists creating their own dedicated production facilities. The era of rampant mid- and large-scale studio construction in the United States has largely drawn to a close, as the market is fairly saturated, and the demand has been somewhat hollowed out by the proliferation of the "laptop studio"—small production facilities, designed by their users, that can accomplish surprisingly high-quality work for much lower costs than could have been executed 20 years ago. For much pop/rap/electronic music production, large live ensembles are not being recorded, and as such, all the studio needs is a booth that can accommodate nothing larger than a drum set, a few mics, and some inexpensive hardware interfaces to their laptop and hard drive(s). While not exactly cheap, the cost of entry for this kind of production is still orders of magnitude lower than what it would cost to design and build a real mid-sized production facility, and as such most American mid-sized professional studios have closed, leaving the big rooms for cinematic orchestra tracking, and the small rooms for mixing/post work.

DESIGNER QUESTIONS

- What resolution (bit depth/sample rate) is necessary in this studio's work? Does my intended equipment support this maximum rate?
- How many musicians are going to be performing at once, and how many mics and input paths do we need for them?
- How many separate headphone mix paths do we need to support?

471

- How many pairs of monitor speakers in each control room, and what kind of master monitor controller will we need to properly toggle between them?
- Are any special production-oriented tools (like a ReAmp box, or custom pedals for guitarists, or other such) needed?
- What is the anticipated range of material to be produced in this venue, and what genre-specific tools might we need in order to support it (e.g., DPA bridge mics for string instruments in a room doing a lot of scoring, versus an MPC for a room doing a lot of hip-hop)?

* * *

Next, we will examine corporate environments, from offices to hotels to restaurants—spaces that may not generally prioritize audio as part of the business model, but that still often require audio systems, all the same.

Notes

1. There is a notable exception here—we are not going to discuss broadcast facilities in this chapter. The reason for this is twofold: firstly, broadcast facilities are essentially recording studios but with the addition of extraordinarily robust data transfer lines for disbursing signals, and as such, the same rules can apply to the recording parts of their infrastructure as are written about in this chapter (regarding other media production facilities). Secondly, it has quite honestly been more than a decade since I have worked on a broadcast facility, and since the data transmission technology is leaping ahead in terms of speed and practice, I would not want to mislead you, gentle reader, into believing that my impression of data line needs from 10 years ago was anything but already out of date. Suffice it to say that there are wonderful books on broadcast engineering and data transfer, and if you are working on such a project, you would be well served to seek them out. See "Further Reading" at the end of this book for more information.

2. I was *going* to show a photo of a professional soundstage here, but after 2 months of contacting various soundstages, I was no closer to obtaining permission to use their images. If you'd like to see what a soundstage looks like, Google "Elstree Studios," which are the soundstages on which much of the original *Star Wars* saga was filmed.

3. Automated Dialog Replacement—the performance of actor dialog in sync with picture, which is necessary either when the set sound was interrupted—say by an overflying airplane—and is unusable . . . or in the case of action movies, where the director had to continually announce cues out loud for explosions and other such effects, in which case it is assumed all dialog will be done via ADR.

4. Digital Audio Workstations

5. This, of course, is a broad generalization.

6. There are still studios tracking (and mastering) to tape, but even those studios still typically feature a Pro Tools rig, as no one wants to edit on tape—because it is one of the levels of hell.

7. Studios that operate Pro Tools HD still often feature other DAWs and production software, both so that they can operate any session files that a client brings to them, and because some other software (like Ableton Live, for example) can do a range of things that Pro Tools is not capable of.

8. For a very long time, the preferred "ugly" speakers of the industry were the Yamaha NS-10s, now discontinued (though you will find them operating in many studios to this day). Today, new studios are sometimes using alternate newer Yamaha models, little boxy Avantone speakers, or the like.

9. Some musicians will bring their own headphones, but it's usually a good idea to stock some sets of closed-back over-ear phones, some open-back over-ears, and some in-ears.

10. The 5.1 refers to the count of output channels in cinema, with five "top speakers"—left, right, center, left surround, right surround (the "surrounds" are side speakers aimed at the audience, often more than halfway back from the front of the audience), and the ".1" refers to the LFE (Low-Frequency Effect) channel, used to enhance explosions, add sub-bass to music cues, and the like. True cinema surround systems feature top speakers that all must reproduce 20 Hz to 20 kHz, with reasonably flat performance, at the same SPL for the mix/listener position. As such, regular content bass is not routed to the LFE channel (only "effects," special low rumbles, explosions, etc.). However, home surround systems often send some of the top speakers' bass content to the LFE channel, because the tops are small bookshelf speakers and as such don't properly reproduce 20 Hz–20 kHz. The process of making a mix for home listening that doesn't become unbearably muddy in the LF

range is known as "bass management," and there are some very good books that deal with this topic. The 7.1 surround is basically the same as 5.1, except the surrounds are supplemented by L/R rears. Atmos is a system based around at least 9.2 channels but can feature a flexible number of speakers . . . 30, 40, whatever, scaled to the size of the cinema. For more on cinema surround, see Tomlinson Holman's excellent book on the topic (Tom was the chief engineer for Skywalker Sound who developed the THX standard—which was named both for George Lucas' first film, THX-1138, and for Tom himself, it was the "Tomlinson Holman eXperiment").

11 I conducted an informal survey of my Facebook groups of audio design and engineering friends, to ask who among them listened to films in surround at their homes. *Almost none of them did* (I do, but to be fair, I only bought my home surround system when I was contracted to mix my last feature film in surround, as I wanted to be able to check mixes at home). The fact that even among audio people, home surround was so rare, leads me to believe that engineers need to pay more attention to their stereo releases, even today, as there are *way* more people listening to home media in stereo than the industry likes to think.

12 Flush-mount into walls themselves.

13 See Paul Epworth's Church Studios, designed by WSDG.

14 OK, so high-end movies are still often shot on physical film for the look of it, but the vast majority of production is now done with digital audio and video recorders. Even on productions using film, it is still common to record the audio digitally.

15 A friend of mine used to call large-format soffit-mounted speakers "client bait"—they were used to get clients feeling confident, not for any actual detailed production work. They sound too nice for production work and cause engineers to think their products are finished when they still need refining.

16 A term that many still use for recording media, even when the medium is a hard drive.

CHAPTER 46

Corporate Offices, Hotels, Restaurants, and Other Architectural Environments

In this chapter, we will briefly examine some non-themed architectural environments that nevertheless still sometimes need professional sound systems designed for them. It is rare that the job of sound system designer is hired independently for such non-themed installation work—typically this will be handled by in-house (or at least subcontracted) designers at whatever architecture firm is handling the rest of the design.

In addition, establishments like hotels and restaurants that are not themed but that have the resources to have sound systems designed for them typically fall into one of two categories:

- Boutique: high-end establishments with custom design elements throughout, catering to a wealthy clientele
- Chains: establishments that repeat the same (or a collection of similar) designs over and over again, at different physical locations, such that they achieve a consistency of customer experience, and also such that they minimize design fees by essentially commissioning one design and copy-pasting it all over the country/world.

The corporate and architectural spaces we are examining here are a grab-bag of spaces that are unified simply by virtue of the fact of featuring no narrative or performative elements, no theming, and no special focus on audio.

DEFINITION

- Corporate offices: any office building or complex that needs a sound system (generally for paging, emergency announcements, or the like), but does not require any attention to artistry or sonic "quality." Such spaces are very common, and while many large corporate office buildings are engaged in a flashiness war for who has the shiniest and most exciting lobby, many others are just content to serve as a traditional office space.
- Hotels: any temporary residence for travelers (including so-called motels or motor-hotels) that again does not feature any theming or other "artistic" design elements.
- Restaurants: any eatery within which diners sit and eat. This includes fast-food restaurants with seating areas, "fast casual" sit-down restaurants with table service, and any other eat-in establishments.

- Other architectural environments: any space not covered here or by previous chapters, and not a dedicated artistic space or installation. Factories, warehouses, bus stations, etc.

STANDARD ELEMENTS

- Inputs: playback for BGM, paging stations for announcements, intercoms
- Processing: DSP, no consoles
- Amplification: typically, 70V systems
- Outputs: typically, 70V distributed speakers
- Converters: at I/O for DSP
- Advanced processing: zone control via DSP
- Control: zone control via DSP
- Clocking: internal, with DSP as master

UNIQUE FUNCTIONS AND SYSTEM NEEDS

Most of these environments need very little in terms of a sound system—they need speakers to cover most public areas (e.g., hotel lobbies but not inside hotel rooms), with speech reinforcement frequencies and sometimes minimal music playback. Music playback, when present, is handled by minimalist playback servers, sometimes directly connected to the web—since music playback is not mission-critical, it is often decided that local storage is unnecessary, so playback can be handled by services like Pandora or the like.[1]

Most such systems require at least one paging interface for announcements, either via dedicated paging station(s) or via the telephones installed in the space.

476

Restaurants, in particular, pay more attention to sound playback than other establishments—sound level, mood, and coverage are important to the patrons' enjoyment of the space. If patrons enjoy their time in a restaurant, they are more likely to spend money. Some restaurants will aim for an upbeat, "party" atmosphere—typically their systems make sure they can produce thumping bass, and higher overall SPL levels—as opposed to other restaurants that cater to a more "refined" atmosphere—counting on low-SPL delivery, speaker hot spots aimed off of tables, and usually very little bass response (though not so little as to make music sound tinny, which is a sound that makes patrons feel a space is "cheap"). Still, others aim for a "family friendly" environment, that often targets the middle ground between the two previous extremes.

VENUE CHALLENGES

Corporate offices are often very low-noise environments. Gone are the days of noisy adding machines and fax printers—in the computer age, most office work is done quietly. This has gone so far to reduce the noise floor of some office environments that the quiet becomes a problem for workers—in open-plan or cubicle-based layouts, everyone can hear everything that everyone else says or does. As such, sometimes sound system designers are called in to *increase* the noise floor of a space to help mask the everyday sounds that become distracting when transmitted so clearly from one workstation to another. In some offices, this may involve a music playback system, distributed throughout the office and playing at very low volume (just high enough to mask transmission of voices from one desk to another but not high enough to impede clarity of phone calls and the like). Even more common is the placing of simple vocal-band noise generator systems. This noise is not technically white noise or pink noise but is basically a version of this same randomized scattered static, focused only on the vocal band, intended to conceal conversational sound.[2] Such noise generators can either consist of dedicated systems, with a noise generator base feeding a collection of 70V speakers, or integrated systems, with noise generators coming from within a DSP program, routed to speakers that may otherwise also carry announcements or music at different times. The calibration of such systems can be tricky, involving a ramping up of level over the course of many consecutive days in tiny increments, such that each increase is not noticeable perceptually, and by the end of the adjustment period, the noise is loud enough to mask the intended sound but still quiet enough not to impede work—the sound has been raised so gradually that workers have a chance to acclimatize and don't feel it is intrusive.

Hotels often feature lobbies that need some kind of sound support, and depending on the configuration of the hotel, may also need BGM/announcement speakers through hallways, next to pools, or other such common areas. Hotels that feature bars, swimming pools, and other such areas often feature dedicated subsystems in those spaces that are typically capable of fuller-frequency reproduction, at higher SPL, than the generic lobby or hallway systems. Like any festive environment, these spaces are designed for people who are celebrating, and as such, the upbeat "party" atmosphere may be in full effect. Again, this will vary by type of establishment—an upscale hotel bar where the cheapest drink is $18 might be capable of loud dance music reproduction, even if it commonly runs at low SPL except for special events, where a low-end bar focused on cheap Bud Light may prioritize loud reinforcement of the sports playing on the accompanying television screens. Such venues vary so widely in architectural terms that there is no generic single way these spaces are designed, and as such, no typical challenges beyond the use of the venue and intended clientele.

Restaurants have the challenge of creating a lively atmosphere without alienating customers. Such spaces again will vary depending on the intended effect. Some restaurants that intend to "turn tables" quickly—to make money by handling the highest volume of customers possible—will intentionally make the environment loud enough to entice diners in, and then just a little louder at tables, such that guests feel the need to dine and depart quickly. Other establishments that want to keep guests for as long as possible (seeing profits from added courses of drinks, appetizers, desserts, etc.), may take the other tack, creating a waiting area that is energetic and loud, but reducing SPL a bit at the tables, to enhance guest comfort and encourage long meals. Architecturally, restaurants vary quite a bit, so there isn't one "challenge"—though care must be taken to allow spots for servers and back-of-the-house staff to exchange information that are not themselves so swamped by sound that conversation is impossible.

There are many other spaces that sound systems will accompany, and each presents their own dedicated set of challenges, but none should ignore these basic questions:

- Who needs to communicate in the space, and where does the communication take place? How does communication happen (conversation, in writing, com system, something else)?
- Does music need to be played? What kind, at what SPL, and when?
- Do emergency announcements need to be played or made live? Who needs to hear them? What is the ambient noise floor of the environment, and thus how loud do announcements need to be in order to be heard clearly (keeping in mind that an emergency announcement must generally be between 10 dB and 15 dB louder than the environment to be noticed, unless the noise floor can be reduced along with the announcement—which *is* the case in a restaurant playing music, but *is not* the case in a factory producing auto parts)?
- Does the system need to be full range or just vocal band?

UNIQUE TERMS

See previous chapters on architectural projects for terms associated with architectural installations. When working on such projects, a general familiarity with architectural terms is advised, but it is beyond the scope of this book to define every common architectural feature.[3]

Specifically, restaurants (like a lot of retail, and some hotels) will use terms like "front of house" and "back of house," referring to the areas where the hosts welcome guests and the areas where only staff members are welcome (respectively).

The hospitality industry features plenty of dedicated terminology, but like the world of studio production, much of it is keyed to the *operation* of the space, rather than the design.

UNIQUE GEAR

Nothing that you wouldn't find in other architectural or retail projects.

TYPICAL TIMELINES

Ranges wildly, from a few months to a few years, depending on the type and complexity of the project. As with many architectural projects, the process for spaces of this type often (though not always) will follow the CD>SD>DD path as outlined in previous chapters. Sometimes, for simple systems, you may skip the middle step, proceeding directly from CD to DD (or even leaping directly to SD, with no CD phase at all). On occasion, CD may be less of a formal design stage and more of a technical narrative stage, describing the intent and function of the system, and then proceeding directly to DD.

DESIGN PRESENTATION

These projects will follow typical architectural project demands, and as such will be presented first to the design firm(s) under which you are working, then to the clients.

Design documentation will include but will not be limited to the following:

- Ground and ceiling plans
- System block diagrams
- Equipment and cable schedules
- FIRs

. . . and so on (see previous chapters for more info on these documents).

DESIGNER QUESTIONS

- Do I need even coverage, or is spotty coverage (avoiding full dropouts) acceptable? Is fidelity *at all* a desire of the client, or just speech intelligibility?
- Is the venue hard-surfaced and reflective? Or soft and porous? If hard and reflective, we may need more speakers at lower levels to accomplish speech intelligibility, if soft and porous, we may be able to get by with fewer units (unless absorption is too high).
- Does there need to be access control for the paging system? What tiers of access will there be, and where will those stations be located?
- Is there a need for dividing the venue into zones? Or is it OK to operate as one big zone?

* * *

Finally, in our breakdown of systems by type, we look at an oddball category: art installations. Design for these spaces can overlap with performance system design, themed entertainment design, retail design, and other sub-disciplines, which is why I have placed it last in this section.

Notes

1   Pandora is a popular internet radio service. In order to legally play back music from such a service in a public location, a host of rules apply: a "Pandora business" license must be paid, there must be no cover charge for entry, etc. These kinds of rules apply to any playing of music in public establishments, though check with your individual streaming service/provider for their particular rules for usage. Many small restaurants and cafes forego this charge, assuming they will never be caught, and stream music via an employee or owner's personal account. Others still play CDs or other physical media. Regardless, such establishments are technically required to license the material they play—via a blanket license from ASCAP or other such—and occasionally, establishments that do play music without permission are fined quite heavily for failing to secure proper licensing.

2   For those who have ever worked in mastering, this solution is like an acoustic version of dither—low-level randomized noise not intended to be noticed, but just loud enough to conceal errors, which in this case are *the sounds of your coworkers.*

3   What's an atrium? What's a balustrade? There are many good books on architectural terminology. See "Bibliography and Further Reading" for more info.

Creative Systems

Multimedia Art Galleries, New Media, New Uses

Finally, in our tour of system types, we come to a category that often defies simple categorization: art and new media installations. There are a wide variety of projects that get classed as "multimedia art," that can include everything from a simple audio creation playing on a loop over headphones, up to a multi-room interactive themed environment (Figure 47.1). As such, we will be sticking to generic statements about such systems here, as a means of general entry to the topic, but must acknowledge up front that this kind of project might include elements of sculpture, theme park design, theatrical performance, electrical engineering, computer programming, and more.

DEFINITION

For the purposes of this chapter, we will limit our focus to those installations using more of a system than a simple looped player through headphones, as no one needs a chapter written about such presentations from a system design standpoint. We will also limit ourselves to projects that reproduce sounds via some manner of amplified loudspeakers, so no "Playing the Building."[1]

A multimedia installation piece in these terms is any piece that an audience member experiences in a particular space, with content either triggered by sensors and interactive controls, or that is on some manner of fixed schedule (looping, wall-clock triggered, etc.). Headphones may be involved, if there is some kind of interactivity/triggering, but if playback is looped, for the purposes of this chapter, the system must also feature speakers.

STANDARD ELEMENTS

- Inputs: microphones are sometimes used in installations, but most commonly, installation work will feature either playback systems, generative music[2] systems, or both in some combination. Such playback is typically computer-based (rather than hardware server-based), as the flexibility of computer systems is usually necessary

FIGURE 47.1
Reactive Digital Installation at Terrell Place in Washington, DC, for Beacon Capital Partners, Experience Design by ESI
Design and Sound Design by Bruce Odland
Source: Photo by Caleb Tkach; Courtesy ESI Design

to produce such content. Max/MSP is a very commonly used software for installation pieces, and generative music particularly.

- Processing: sometimes a DSP is needed, sometimes processing is all handled via software in a laptop/PC, sometimes playback or generative music is played directly through speakers *without* processing (though in the case of generative systems, or anything with aleatory potential, it is wise to at least set good limiters for your speaker lines, such that some variation in the code doesn't blow your system).
- Amplification: ranges based on speaker types
- Outputs: speakers of every conceivable type (and sometimes strange custom speakers, as well); headphones
- Converters: at I/O of digital devices
- Advanced processing: DSP or other software-based (Max/MSP, PureData, and other such software are very common in the art installation world)
- Control: this is one of the most variable areas of art installation work, a lot of installations feature interactive elements—sometimes interactivity is central to the piece itself. That interactivity is generally engaged by guests via some form of control inputs. Whether you are using simple relays built into the space, elaborate sensor arrays, tablet- or computer-based GUIs, or what have you, how the content is controlled is a major part of the creativity involved in multimedia installation work.
- Clocking: usually internal, either computer or DSP as master

UNIQUE FUNCTIONS AND SYSTEM NEEDS

What is the artist trying to achieve? What kind of experience do they want for their audience? How long will it last? Does the audience have control over it?

As you identify these answers, you can begin to assess what kinds of unique equipment are needed in order to execute the plan. There has been a revolution in recent years in the low price and availability of small, efficient microcontrollers (Arduino, Raspberry Pi, Beagle Bone, etc.). These have become fairly common in art installations because they are an easy means for transforming the output of some trigger devices—be they entryway sensors,

Cirque du Soleil is the one who enabled me to be able to express sound in a different format. It wasn't intentional, just working on the productions, being part of something being grown from the blade of grass up to a huge orchard per production, and not expecting where it should go. Allowing all that to go and going along with it became something that was such a creative environment, and as it grew—as the company grew, I did the second production for them, I was their first sound designer—as it grew, so did the budgets. Then it got into the Vegas stuff, and the budgets became pretty big, the time for tech and research—both beforehand and while in production, to try things—you could stay through the night . . . to develop the art of Cirque du Soleil, was to develop the art of sound. To develop this immersive—immersive is sort of the tag word these days—but not doing it just to do it. It's really boring doing something just because you can—it might be fun, but it becomes really boring and tedious to everyone—but as long as there's intention and reason to everything you're doing--it's one of those things where you might sit back and think, "Oh, I didn't mean for this to do this, but this is what this has become." That can start in a meeting when you're sitting in a meeting listening to people talk, and you're just drawing or sketching ideas—and in my case, it's not that I'm intensely drawing and placing loudspeakers on plots . . . not at all . . . just drawing things that happen, and thinking, "How would that sound; how would that be?" It might be a speaker, or a seat, or a forest. As meetings go on, it becomes a very collective and collaborative sport. By the time you've worked out "oh, this is really cool; this is what I should do," then you think, "I wonder what the building's like," because now you have an idea of what it might be and think about how it might fit into a building. Then it's "I wonder what the budget is?" And usually I would—certainly in those days—design without a budget in mind, and then have that all worked out what it would be, and have everyone agree, hopefully, to move forward with the ideas. And then we look at "how much is this gonna cost." Then it's up the producer to turn around and say, "Well, I totally want this, but I can't do it like this . . . can we do it another way, and if so, what would I lose from that?" Or, "This is great, let's go farther, what would we gain if I gave you another million?" So it went both ways, it wasn't just about a spreadsheet. It was about the desire and the intention, so—imagine doing that across every department. So you get something that is totally unique and special, like *O*, like *Love*, like *Ka*, like those big iconic productions that sit in Las Vegas—that were not cheap to do, but that made their money back within 2 years. The investment on doing something as unique and special as that was huge, but the payback was vaster than most Broadway shows, who work on a spreadsheet, and who can't make ends meet, and who are terrified the show's going to close before it's even opened. It's like those red Solo cups that have been jiggled around in this order or that order, but it's like "they're still just red Solo cups, people!" Where's the art? Where's the passion? Where's the conception of this that leads you somewhere else?—Jonathan Deans

483

proximity sensors, physical switches, whatever—to a message format usable by whatever audio engines your main computer(s) are running. Take numeric proximity sensor data, turn it to an OSC command, use that to tell Max/MSP to change parameters in a composition or the tone of an instrument, or use it to tell QLab to trigger or stop a sequence, or whatever you desire. This is much the same as using real robust show controllers as the front end for theme park design, but it is much smaller scale and much less expensive (both of which make this a popular solution for artists, who are not—with some notable exceptions—known for being flush with cash).

There is no generic "effect" I can identify for art installations—some aim to tell a story, some aim to discomfit an audience, some aim to merely be interesting and interactive with little agenda beyond pure sound manipulation. The range of possibilities is as wide as your imagination.[3]

VENUE CHALLENGES

One of the signature challenges of this kind of multimedia work is that it very frequently is designed not to be installed in art galleries or performance venues, but in commercial spaces, public spaces, abandoned buildings, and other spots that were initially built with no thought that there would ever be a sound system installed in them. As such, these spaces are often quite challenging: is there enough power? Is it clean? How can we safely mount speakers? Et cetera.

> **Tales From the Field**
>
> One of my favorite systems was just this little parade float that we put together as part of our local town thing. We borrowed some dual 21″ subwoofers from Fulcrum, powered them each up with 3,000 W a driver from PowerSoft. We set it up in cardioid so that basically it was like a figure eight. So we were hitting people with some amazing sub on either side of the road but up and down the parade route we were basically silent. It was really fun, I don't think people knew what hit 'em. Most of the other floats had, like, blown-out [JBL] Eons being run off a little generator, so it was orders of magnitude beyond, but that's kind of fun. My favorite stuff is when you provide really high-quality stuff for people who don't know that it can be like that.—Jamie Anderson

Again, because of the complexity and variability of such work, there is no magic bullet answer, but I would recommend a series of "best practices":

- Always ensure your electrical power system is operationally safe and poses no risk to the audience or equipment.
- Remember the cardinal rule of rigging, that you do not want any items overhead mounted on a single point of failure—every item must have at least two points of structural connection, such that if one fails, the other acts as a fail-safe preventing equipment damage and audience injury. This might mean a speaker c-clamped to a pipe, with an aircraft-cable safety line around the yoke to a different structural member, or the like.
- In situations where the art will be installed alongside other works of art, communicate early and often with the curator(s) to let them know what kind of SPL your work will be delivering, over what range of space, such that there is no surprise when they hear your work bleeding into the viewing area for an adjacent sculptor's work (or whatever).
- If there are interactivity points in the work whose operational process is not evident immediately to guests, be prepared for the audience to misinterpret, miss entirely, or damage, the controls. Be prepared for the idea that even very clear interactivity points may be abused and damaged by an audience. Plan for spares if possible, particularly if this is an installation that will be visited and interacted with by children.

UNIQUE TERMS

The art world, like the world of architecture, has a large lexicon of terms used to describe works. However, unlike the world of architecture, most of these terms are creative and analytical terms used for the content of the work, rather than for anything technical. There is one dichotomy I've heard addressed many times regarding art installation pieces, which is "product-" versus "process-" based work. The idea of this duality is that some works of art are more focused on creating a fixed product or experience that most visitors will experience in either identical or mostly identical fashion. On the other hand, there are works of art that are more focused on the interactivity, or the evolution—the process—of creation, not with what actually results from said process.[4] Determining which (if any) the given work of art prioritizes can help in the design of the system. If your piece is product focused, then strict attention must be paid to sound delivery methods, frequency response, SPL, and so on. If it is more process oriented, more attention is often paid to the control system and methodology. Sometimes the priority is both, but asking the question is still a good entry point to understanding what a system needs to accomplish.

UNIQUE GEAR

Anything under the sun.

TYPICAL TIMELINES

No such thing. Some pieces of this nature take years to develop, some are put together in a matter of days or weeks. The more elaborate the piece, the longer it will take.

DESIGN PRESENTATION

This really varies. If you are employing an integrator to install the system, you will need to present typical system documentation to the integrator. However, a lot of art installation work is installed by the artist(s) themselves, or a small team of assistants. As such, whether there is *ever* proper technical documentation created is a bit up in the air.

Figure 47.2 shows the block diagram of a multimedia gallery piece I created and designed. This document was only ever used by myself and my team of collaborators.

DESIGNER QUESTIONS

- What is this piece trying to accomplish, emotionally, procedurally, etc.?
- Is this a more process- or product-centered piece?
- What are the unique issues presented by the installation venue, and what non-standard solutions may be necessary in order to accommodate them?
- How long does the piece have to run? Does it need a day and night mode, à la theme park rides, or does it run continuously?
- Are guests interacting with the piece? If so, via what means? Switches, sensors, tablets, etc.

Multimedia installation art is some of the more exciting work being done both in the fine art world and in terms of sound system design, as it is a field in which the rules are malleable, and new structures and methods of manipulating sound are frequently in play.

We thus conclude Part III, on system types. I am certain there are system types I've missed, some intentionally (automotive sound) and others unintentionally (if I knew, it wouldn't be unintentional!). In Part IV, the final Part of this book, we examine technical documentation and the process of presenting your work.

Notes

1. With sincere apologies to David Byrne, whose piece "Playing the Building" involves triggering a host of mechanical sound-making devices mounted to the bones of old buildings via relays attached to a musical organ as the "controller." It is a quite exciting piece, and I have much respect for Byrne's work overall, but such a system is more a lesson in electromechanical instrument design than *sound system* design.
2. Generative music is a category of software-based music that is variable, constantly changing, and whose musical parameters are based on some incoming data/control stream, possibly from a graphical user interface, a set of triggers or sensors, a real-time dataset, or the like. For example, I have heard generative music that shifted and evolved according to the average snowfall in a region, or the solar flare activity of our sun, or the like.
3. This world moves pretty quickly. For a good selection of contemporary sound installation art, see www.everyday-listening.com.
4. Personally, I am often most fascinated by works that prioritize *both* an ingenious process and a product that is at least well-considered, if not at all consistent, but that's just my taste.
5. I have included a block diagram of my own multimedia gallery piece *Noise Floor*. It is difficult to describe, so I will just encourage you to take a look at a short film tour of the piece: www.youtube.com/watch?v=4gvvJL8gGWg

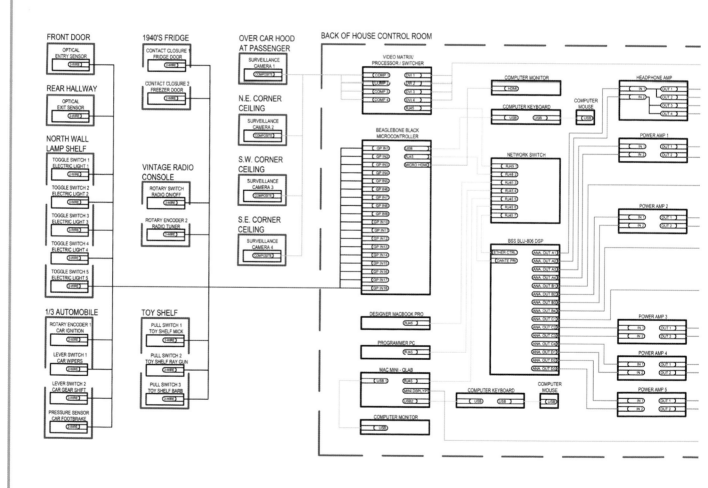

NOTE: BLACK = TWO-WIRE ELECTRICAL

BLUE = COMPUTER (CAT6 OR USB)

GREEN = VIDEO (COMPOSITE, DVI, DISPLAYPORT TO VGA)

ORANGE = AUDIO (PHOENIX TO XLR, NL2, TRS$\frac{1}{4}$")

FIGURE 47.2

Block Diagram of *Noise Floor*, a Multimedia Gallery Piece by the Author (Installed at the Rozsa Center for Performing Arts, at Michigan Technological University—Custom Controller Programming by Mike Schmitz and Paul Bristol)[5]

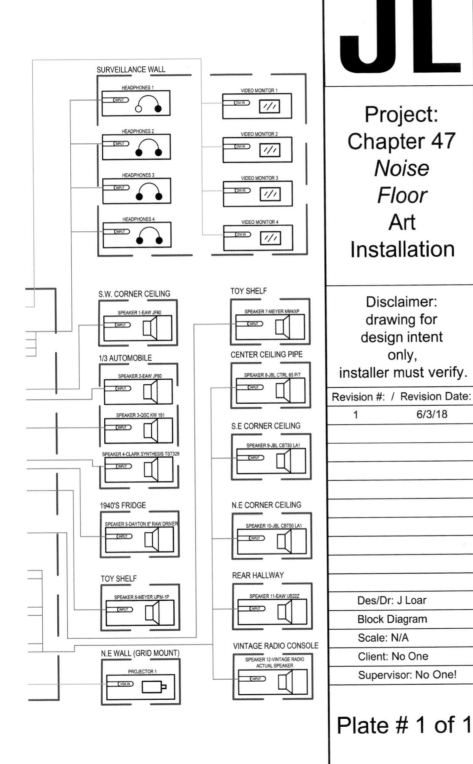

JL

Project: Chapter 47 *Noise Floor* Art Installation

Disclaimer: drawing for design intent only, installer must verify.

Revision #: / Revision Date:	
1	6/3/18

Des/Dr: J Loar	
Block Diagram	
Scale: N/A	
Client: No One	
Supervisor: No One!	

Plate # 1 of 1

SURVEILLANCE WALL

HEADPHONES 1 — INPUT
HEADPHONES 2 — INPUT
HEADPHONES 3 — INPUT
HEADPHONES 4 — INPUT

VIDEO MONITOR 1 — DVI IN
VIDEO MONITOR 2 — DVI IN
VIDEO MONITOR 3 — DVI IN
VIDEO MONITOR 4 — DVI IN

S.W. CORNER CEILING
SPEAKER 1-EAW JF60 — INPUT

1/3 AUTOMOBILE
SPEAKER 2-EAW JF60 — INPUT
SPEAKER 3-QSC KW 181 — INPUT
SPEAKER 4-CLARK SYNTHESIS TST329 — INPUT

1940'S FRIDGE
SPEAKER 5-DAYTON 8" RAW DRIVER — INPUT

TOY SHELF
SPEAKER 6-MEYER UPM-1P — INPUT

N.E WALL (GRID MOUNT)
PROJECTOR 1 — VGA IN

TOY SHELF
SPEAKER 7-MEYER MM4XP — INPUT

CENTER CEILING PIPE
SPEAKER 8-JBL CTRL 65 P/T — INPUT

S.E CORNER CEILING
SPEAKER 9-JBL CBT50 LA1 — INPUT

N.E CORNER CEILING
SPEAKER 10-JBL CBT50 LA1 — INPUT

REAR HALLWAY
SPEAKER 11-EAW UB22Z — INPUT

VINTAGE RADIO CONSOLE
SPEAKER 12-VINTAGE RADIO ACTUAL SPEAKER — INPUT

How Do You Document and Present Your Design?

What are the standard documents in a sound system document set? What do we need to consider as we prepare them? Part IV examines the role of documentation and presentation in the sound system designer's work.

Basic Technical Drawing Standards

Organizing and Labeling, Drawing Setup, Notes on Computer Drafting, Etc.

Technical documentation for sound systems, like systems themselves, changes constantly. When designers are learning how to document their work, they usually learn from one of a few existing sets of "standards,"[1] such as the United States Institute for Theatre Technology (USITT) Sound Graphics Recommendations, the ANSI Standards, or the like. These sets of recommended styles are updated every few years at fastest, while the technology we are documenting changes every week. As such, these drawing standards are never 100% up to date, and never will be.

In this and the following chapters, I will be discussing and recommending a set of best practices that synthesizes my own experience in theatrical design, concert design, and themed/architectural design. Please be aware of the following:

a. These chapters are written (and should be read) with full awareness that these recommendations are not hard and fast rules, and as such will endeavor to identify where practitioners differ in their approaches to the documentation (and why they differ, or why one approach might be better suited to one industry or type of project than another). That said, it is inevitable that, in service of making these chapters a readable length and not a whole book unto themselves, there will be topics where professional practices differ that I present here as done in one basic way. It is not my intent to state that my way is the correct way, just to present a path through for designers learning the practice.

b. These recommendations reflect not only my own professional practice, but a series of extensive conversations I have been conducting with a range of professional designers and assistants from Broadway, the AV/consulting world, and elsewhere.[2]

c. Particular projects (and types of projects) may not only be better suited to one or another documentation style, but there are also regional and company differences that must be observed. For example: in the United States, and in most of the international AV/consulting world, creating some version of a system block diagram (SBD) is standard practice. On the West End, and throughout the United Kingdom, such a document is rarely generated, and designers instead rely on a document set called a "workbook" that is a combination of our "hookup" sheets and some minimal signal flow drawing for isolated subsystems. Both will be discussed here, but we will focus on typical US/international practice. In addition, it is very common, especially when working for large corporations, for them to have "house styles" that you must adhere to when working on their projects. For example: in my work on Disney theme parks, I found that WDI[3] had specific fonts and sizes they demanded through whole packages, specific connector representations that I haven't found in other companies' work, and other such specific demands that engineers and designers working for them must meet.

All of which is to say, take all of this with a grain of salt, and remember that the overall purpose of any technical documentation set is to communicate clearly:

1. What equipment makes up the design
2. How it is to be installed (both physically in the venue and how it is to be connected)
3. How signal (of all types) flows through the system

In order to do that, we create a range of standard documents:

- System block diagrams: a diagram showing all audio (and audio-related control) devices in a system and delineating the wiring connections between all such devices.
- Plan and elevation views: venue drawings showing (at bare minimum) speakers and their positions, and sometimes racks of equipment, control stations, etc.
- Mounting details: drawings of any item mounting that needs specific attention. A speaker mounted with an included c-clamp and yoke may not be shown in individual detail, but any line array will be shown, such that details of speaker splay, rigging points and weight loads, etc. will be clear to the installation team. Particularly relevant when custom mounting must be used.
- Prediction data: for client perusal, and as a reference for tuning.
- Technical narratives: a written document that describes all of the basic functions of a system, in the most non-technical terms possible (for clients).
- Budget: estimates of the cost of your system.
- Facilities Impact Report (FIR), equipment schedules: lists of all equipment in a system. FIRs total heat, weight, and power loads, while simple schedules are lists of quantities.
- Hookups: also known as I/O sheets . . . spreadsheets detailing all of the signal connections and cable types (and often lengths) of the connections (this becomes a central part of the workbook in absence of a signal flow diagram). Some hookups will be divided into two parts, with one part detailing physical signal connections, and another representing signal routing data within digital systems (so, in the former, output Card Slot 1, Output 2 might be connected to Amp 4, Channel 1; in the latter, main right will be patched to Card Slot 1, #2 in the console system).
- Magic sheets: common in theatrical design, a simplified speaker plot, showing numeric values that correspond typically to playback channels, such that a designer can easily see which numbered output from a playback system will correspond to which speaker in the venue.
- Tech books: on AV installation jobs, this is a collection in PDF form of the spec sheets (and sometimes manuals) of every piece of equipment specified in the design.

Even though this is a detailed list, there are sure to be documents needed for particular projects that are not represented here.

Now, before we dive in deep to the specific documents, there are some general principles we need to discuss.

ORGANIZING A DOCUMENT PACKAGE

To begin with, a document package should be created with the intent that it be legible as a full document set, including all drawings, spreadsheets, narratives, etc. As such, styles of representing information should be consistent throughout, whatever those styles may be. It is very important that all pages of a document set (generally referred to as plates) feature a *Title Block*.

Title Block

a graphical table that provides basic orienting information about the document set. Such information includes, but is not limited to, production/project name, designer name, draftsperson name (particularly when different from the designer, or when multiple draftspeople are working on the same set), producing entity/company name, location design will be installed (venue), revision number, date drawing was created/revised, disclaimers,[4] plate title, and plate number (to orient the reader in the overall set). Title blocks are found in many formats, but probably the two most common are a rectangle in the bottom-right corner of a page or a tall rectangle along the entire right side of the page (Figure 48.1).

Documents that take the form of spreadsheets are (in installation/architectural projects) to be plated as images or tables in a CAD frame, such that they form part of the overall document set, at the same size as every other plate. For theatrical and other show work, this is less common, though most such documents will still include most of the relevant title block information in a header at the top of the spreadsheets. Plate sizes will vary considerably from project to project, from the relatively small 11x17 to the larger "blueprint-style" Arch D.[5] 11x17 has become a very common plate size, as it makes large multi-page drawing sets easy to print, and it blows up easily to Arch D or other large plot sizes when integrators want a large working document for installation. Whatever the size, the SBD should not be a different plate size than the plan views, which should not be a different plate size from the elevations, and so on. Consistency is key.

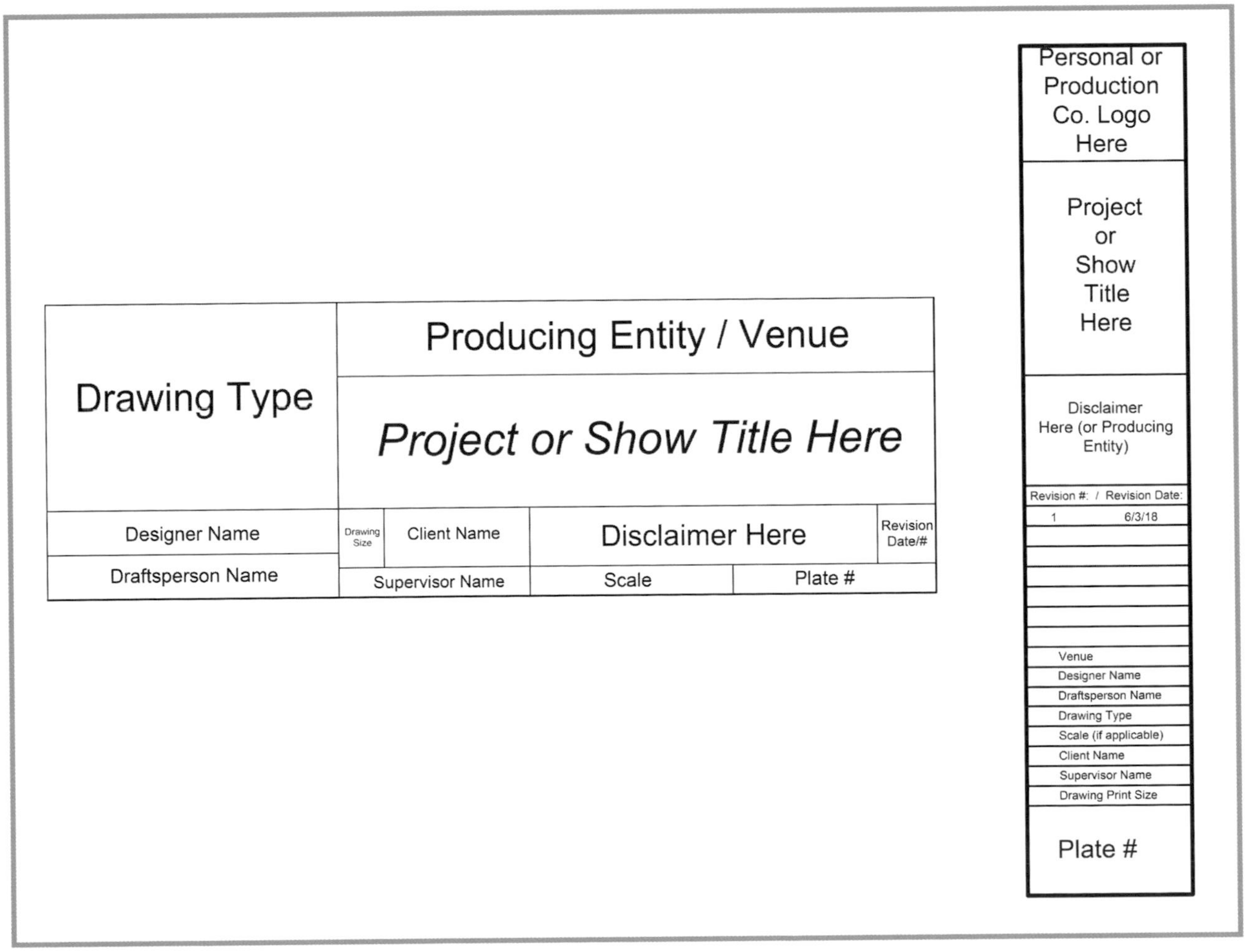

FIGURE 48.1
Example Title Blocks: (Left) Bottom-Right Corner; (Right) Right Side

The front page of any given document set often contains three items that allow the overall document set to be an easier read:

- Plate index
- General notes
- Legend

A plate index is just what it sounds like: a list of all plates in a given drawing set, with plate numbers and titles listed in order, such that someone seeking a particular part of the drawing set can easily see which page to turn to.

General notes are any notes that apply throughout the whole document set—for example, if every speaker in a design is intended to be run in 70V mode, this might be noted here, and then never mentioned again, so as to simplify the representation of items later in the package.

A legend is a visual key to symbols used in the document set. For smaller system designs, the front-page legend may be able to identify all symbols used in SBDs, plan/elevations, mounting details, etc., all on that first plate. For larger systems, it is common to find several legends through a drawing set, one for SBD conventions, one for plan view conventions, one for elevation view conventions, etc. A legend is a critical part of a drawing set, as it not only identifies what drawing conventions will be used but also clearly represents any conventions used that deviate from recommended practices. This includes images of items represented, so we know what we're looking at, it includes drawing choices (like color-coding), and any other info that is essential so that we can understand the plates that follow (or, on a small drawing, the plate that is next to the legend). It is fairly common to deviate from recommended practices, especially on large projects, in order to illustrate particular information that might not be well-suited to the conventions as written by USITT or others. However, such subversion of conventions MUST be clearly identified in your legend(s), otherwise, your drawings are at high risk of being misinterpreted.[6]

Legends should also identify notation styles used in the document. In Figure 48.2, we see in the SBD example not just a sample device, but how it, and its associated cabling, will be labeled, how callout notes will be drawn, etc.

You may notice that all the text is written in ALL CAPS. This is the conventional standard for drafting. Drafting software has fonts that mimic the particular handwriting style that hand drafters taught and used for decades (a very particular type of block lettering), but many projects now either use a custom font or a simple sans-serif font for clarity. I have used Arial throughout this book. For large block diagrams, we often need to connect signals across multiple drawing plates. As such, there needs to be some system of identifying which signals from one plate carry over to another plate, and where exactly they pick up. There are a number of different means for showing this.

In Figure 48.2, we see the first version of this. This is what I refer to as "zone-based" inter-plate references. In these objects, we have a circle divided into three parts. In the top half of the circle is the plate # where the given signal will appear next. In the bottom two chambers, we have zone IDs. These are based on the idea that multiple-plate SBDs will all feature zone borders around them, with letters down each side of the plate, and numbers across the top and bottom. Block diagrams throughout this book have used this style.[7]

Each inter-plate reference will thus seek to be placed in its own dedicated zone (or at least, if it must share a zone with other references, be placed such that each inter-plate reference is jumping to a clearly different location on other plates, so there is no confusion tracing between them—as each reference is labeled reciprocally). As much as possible, the goal is to make it easy to trace from one page to the next.

However, when there are too many signals running between plates to easily accomplish this (if, for example, you have 64 signal lines between plates and can't easily give them each their own zone of arrival, maybe because they are not easily grouped together visually), you might use the example from Figure 48.3. In this example, the circle is bounded by an arrow, which indicates which direction in the drawing set the next connection is to be found. Within the circle, which is divided into two sections, the bottom field now indicates the plate number on which the connection is to be found, and the top number is simply a sequential number, unique to each plate. In this situation, ideally, a reference with "3" on top will be the only reference numbered 3 on the sending plate, and the only reference numbered 3 on the receiving plate. However, when signals from multiple plates are arriving at one later plate (as we see in Figure 48.3), or vice versa, this numbering scheme gets very complex. In this case, we see two references numbered "3" on the receiving plate, and the only way to tell which sending #3 they refer to is

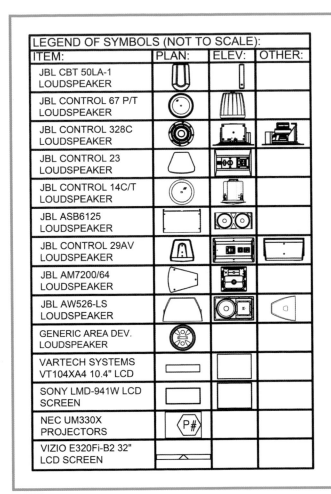

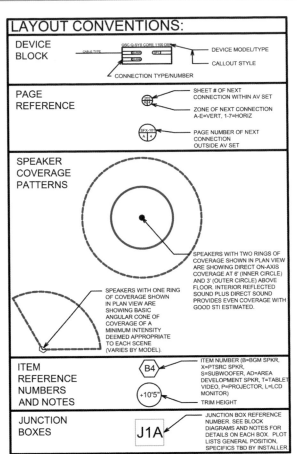

FIGURE 48.2
Example Legends: (Left) Plan/Elevation View Legend; (Right) SBD Legend and Further Info

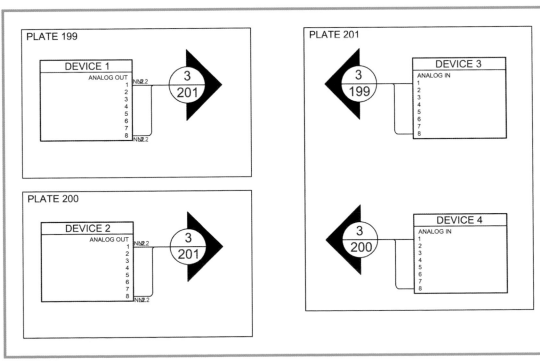

FIGURE 48.3
Inter-plate References, Numbered (No Zones)

that (hopefully) there will be only one #3 on the sending plates. With just two references in our example, it is easy to trace, but it is easy to extrapolate to the idea that in a 25+ page SBD (which is common on large installation projects) this type of reference poses its own challenges.

Personally, I find that whenever possible, the zone system is clearer, and easier to read, as it tells you exactly where on the subsequent plate to look for the signal in question. However, this is not always practical, and some people prefer the other style. Maybe you will invent a new style that is even clearer! If you do, put it in your legend so that we understand it! After all, when connecting signals across plates, it is critical to explain clearly how this will be shown.

A document package should be assembled in a logical order. Mounting details, for example, don't generally get placed before the plan and elevation view drawings—it is helpful to first see where all the speakers are in the venue before thinking about what mounting has been designed for them. Different productions and projects will ask for drawings in different orders, but the following is a relatively logical order:

- Narrative(s) *
- Plan and elevation view drawings
- Speaker predictions *
- Magic sheet(s) *
- Mounting details
- SBD(s)
- Hookups
- FIRs and schedules
- Budget *
- Tech book *

Items marked with an * are items that are generally presented to clients, not to integrators or shops. Additionally, some clients will not want to see overly technical documentation like SBDs, hookups, FIRs, schedules, and the like. As we discussed in the previous section of this book, each type of project and client will have different needs and desires as to your technical documentation set—and different documents may be appropriate at different design phases. I would not generally expect a detailed SBD at the concept design phase, but I would definitely expect a generic plan view drawing demonstrating basic locations of major equipment.

A NOTE ON COMPUTER DRAFTING

Almost all professional paperwork today is generated by computer—with the exception of a few old dogs,[8] no one is drafting by hand anymore. As such, it behooves us to take a brief moment to address the organization of computer drawings, and the platforms common to different industries.

First, all professional computer drafting programs will give the designer the opportunity to draw in "layers," whether the software calls them "layers" or something else.[9] One of the first steps of generating a computer drawing is to set up your layers to reflect meaningful groupings of types of drawing objects. In a plan view drawing, for example, I will often set up one layer for "speakers,"[10] and a different layer for speaker numbers (which identify individual units, such that we can track them across plans, SBDs, hookups, etc.). That way, I can decide, for example, to temporarily render invisible all my speaker numbers, and just show the speakers themselves. This is enormously useful when comparing design drawings with other departments, when trying to get a clear view of one set of items, or whatever. Some designers "go hard" with layers, separating callouts from device labels from speaker numbers, where some designers might just call all of these "notes" and put them in a single layer. However, you choose to organize your drawing, *make use of this functionality*. A well-organized drawing is so much easier to use, and share between departments, than a poorly organized one.[11]

In the following chapters, we'll describe all the major items in a document set, and how they are typically shown, but keep in mind, there are many varieties of these concepts and forms that get used in the industry. One key principle we must always keep in mind is ensuring that our documents are readable by anyone at any time. For example, in the hand-drawn era, much was made of *lineweights* in a drawing set.

> **Lineweight**
>
> the heaviness (in the computer age, the *thickness*) of a given drawn line, which is often used to encode information—such as a heavy lineweight represents X, but a light lineweight represents Y.

However, in the CAD age, some programs (like AutoCAD itself) default to turning lineweight view off, so even if a drawing features lineweight distinctions, they may not appear visible to the reader (without toggling this setting back on).

In addition, it is becoming increasingly common to make use of color-coding to communicate information. Color-coding is sometimes used to communicate signal type (e.g., red lines are analog audio, blue lines are digital/network audio). It is sometimes used to distinguish system functions (e.g., red speakers are part of the vocal system, green speakers are part of the music system). However, care must be taken to ensure that information is available in the document set in another way than *just* color-coding. The reason for this is threefold:

1. A majority of drawing sets will be printed, as printed docs are easier to mark up during installation. A majority of those that are printed will be printed in black and white, because color ink is very expensive. As such, any info that is *only* represented as color data will be unclear when printed.
2. Some people are colorblind. If your info is *only* represented as color-coding, any colorblind reader will be unable to understand the information.[12]
3. Many drawing programs (especially those made by Autodesk) default to a drawing interface of black backgrounds, and colored lines that show clearly against black (like yellow) may wash out entirely when transferred to the white background of a page printed in color.

This is not to say that color should never be used—and in fact some clients will demand that color be used—but that information shown via color should always be available in a drawing set in at least one other fashion, and you are always wise to make sure your drawing still reads properly if printed in black and white.

For many decades, sound systems have been drawn in 2D, but today, there are many projects (especially in the AV/installation and architectural worlds) that are drawing everything exclusively in 3D so that real dimensions are modeled.[13] While sound system drafting is just beginning to join in the 3D revolution, and many projects (like theater and concerts) are still generally drawn in 2D, the day is coming when 3D will likely be the standard for any project of even moderate size. Standards for 3D drawing are not entirely different than those for 2D, and many times, 2D views will be exported as plates from a 3D model, but designers should be ready to think about a speaker in a drawing not as a symbol representing an object, but as an object in real dimensional space.

In addition, even in 2D drawings, there is a trend in some projects (again, mostly complex architectural or themed projects) to using what AutoCAD calls "dynamic blocks" to represent information in SBDs. Such items are visual representations of the devices in question, which also include metadata giving more information about those

> **Message in a Bottle**
>
> I think everybody develops their own process for what they do. I think if you said there was just a 100% standard process—I just don't believe that. I think everybody's got a quiver of tools they can use to get the results they want. That's why there are different designers and different people get hired for different jobs, not that I'm better or worse than anybody or vice versa—that the next guy's better or worse than me. It's just that you're going to get a different result. I think that in terms of the tools, the one thing is—even though sound isn't necessarily a visual discipline, having good graphic communications through drawing or through modeling—to convey to the client, and to have accountability—is really important. I think having good graphical communication skills is very important in our field.—David Kotch

devices than could be easily represented on a page. Such data can be extracted for use in spreadsheets and other documents throughout the set. This kind of metadata is also a feature of 3D/BIM-style drawings. The future of technical drawing is getting even more data-rich than it is today, and that data is being encoded in the drawing documents in such a way that it can be easily ported to other forms and documents that are needed. It behooves today's designers to begin thinking about all the information that would be needed on any document in a set about a particular device, and, where practical, to encode it as such in the main CAD drawings.

A brief note on software used in different industries—different parts of the sound world typically rely on different software. While there are exceptions, the following is a list of software commonly used for basic documentation:[14]

- Theater—Vectorworks (often used because Vectorworks Spotlight offers robust tools for lighting design modeling and paperwork, and sound people are fitting into a paradigm begun by lighting designers);[15] for spreadsheets: Excel or Numbers[16]
- Concerts, AV installation work, themed, architectural (2D)—AutoCAD; Excel for spreadsheets
- Themed and architectural (3D)—Revit for full buildings or projects; Inventor for custom mounting or enclosures
- Speaker Prediction—AFMG EASE, or where using only one brand of speakers, proprietary systems like Meyer's MappXT.

A brief note on converting CAD drawings from one software to another. In *theory*, AutoCAD and Vectorworks drawings should be able to be exported from one to the other platform and used more or less interchangeably. They each have features to export to the other's formats, and being vector graphics, should scale and adjust with no problems. However, in practice, at least some of the time[17] the drawings have noticeable and problematic errors when re-imported into the new platform. A typical example: I exported a 3D drawing of a building from Vectorworks to AutoCAD. When I opened it in AutoCAD, it looked fine, except that *one entire wall* of the building was missing, later to be found floating in space, disconnected from the rest of the building. This was the only error I spotted, but, of course, there may have been tiny errors throughout that I hadn't noticed. As such, caution must be taken when shifting a drawing from one platform to another, and the details must be rechecked and verified in the new platform.

Also, when emailing CAD drawings of one platform or another, it is fairly common for whatever reason for email to corrupt the CAD files. As such, industry-wide best practice dictates that CAD files be compressed as. zip archives before being emailed (even speaker manufacturers package their CAD files for each unit as. zip files for download). I haven't seen this issue with shares via Dropbox or Google Drive or other online file sharing services, it seems to be primarily something particular to email platforms—but it still happens often enough that compressing the files is the standard for transfer, even if it's via online share drive rather than email.

A good document set tells a story about how equipment is configured and used—as you create your document packages, always keep communication of the overall ideas forward in your mind. If you create a document that technically shows your system but is very difficult to read or understand, you need to rethink how you've done things.

When we're drawing our system, the first thing we do is generally to lay out what speakers and other major equipment we're going to use, and where, so let's dive into the world of the plan and elevation view drawings.

Notes

1 Some such standards are not actually standards at all, but just sets of recommended best practices, hence the quotation marks. Standards are properly defined as a set of parameters all drawings should follow for a given discipline, where recommendations are looser, more of a suggestion than a command. In this text, I will use both interchangeably to refer to rubrics providing an overarching structure for drawing packages.

2 As of this writing, I am serving as chair of a Sound Graphics Working Group for USITT (United States Institute for Theatre Technology), and that working group is made of professional designers from a range of fields (who all have extensive theatrical experience). That group will be issuing the first new Sound Documentation Recommendations from USITT in more than a decade—estimated delivery date is sometime later in 2019. I spend a lot of time with that group discussing both overarching style and purpose questions,

as well as every bit of the minutia of contemporary sound documentation. So, while no one plan will suit every project, I will do my best to recommend practices that won't leave my colleagues aghast.

3 Walt Disney Imagineering

4 Such as "drawings are for design intent only, electrical and rigging details must be verified by integrator," which is a fairly common type of disclaimer, particularly for large installation designs.

5 While these are US print sizes, the principle holds true overseas as well, from a "small" print that can still include a fair amount of info (A3 in European sizes) to much larger "blueprint-sized" working documents (A0, for example).

6 On the importance of title blocks and legends—I have, for nearly a decade, taught university students how to design and document systems, and one of my rules is that a drawing submitted without title block and legend is grounds for immediately failing the project. If you can't be bothered to give me information about how your drawing is set up, why should I waste my time trying to figure it out?

7 For an example, see Figures 38 A–C, in Chapter 38.

8 With apologies to the dogs. I mean no disrespect.

9 AutoCAD refers to the function I am describing as layers, Vectorworks calls them "classes" (and their *layers* mean something altogether else), but the general idea remains the same.

10 On a complex project, this may be further subdivided by purpose: one layer for music speakers, one for vocal speakers, one for effects speakers, etc.

11 Our upcoming USITT recommendations will suggest best practices for layer divisions in sound drawings.

12 I know several sound designers who went into sound because they loved the theater but couldn't see color, and so found that sound design was a better place for them than, say, costumes or lighting.

13 The basic paradigm for such drawings is known as BIM, Building Information Modeling (e.g., documents created using Revit). While BIM has yet to incorporate acoustic/speaker modeling, that day can't be far off, and when that happens, more and more sound system designs will be realized in 3D. BIM allows the user to model weight loads on actual structures, electrical system loads, HVAC system performance, plumbing, and more.

14 Different software platforms also save documents in different versions, depending on the year the software was issued. AutoCAD, for example, updates its software every single year but updates its *save format* every 3-ish years. At the start of a project, it is inclement on the designer/draftspeople to identify what file format/version you will use throughout a project—as certain formats will not be backward compatible with others, and thus will be unable to be opened by people running older versions of the software.

15 Theater designers often feel like Vectorworks is way more common than it is. While it is still very common in theater design (though AutoCAD is used as well), it is not terribly common in any other part of the professional sound system design world. Theater designers are occasionally found using the inexpensive and not terribly robust software OmmiGraffle. The less said about Graffle, the better. It is only used because it is inexpensive, and it is not accepted anywhere in the industry outside of some theatrical jobs.

16 Also, theater designers will often use Mac's proprietary Excel-copycat software Numbers. In most other areas of the business, Excel is the standard, and Numbers documents would have to be converted to Excel.

17 Lately, I find it is about 5% of the time, though some would argue this is at least 90% of the time.

Plan and Elevation Views

Including Rack Elevations

Preamble: as we get into the specifics of drawing in the following chapters, it is assumed that you have already taken a moment to read over and familiarize yourself, at bare minimum, with USITT's recommended graphics practices for sound, scenery, and lighting. This is so that you have a basic familiarity with a version of representing equipment in a venue. There are plenty of variations from this style throughout the industry, but of all the standards out there, these are the shortest to read and as such are a good entry point. I will not simply recapitulate these or any other standards in the following chapters, but, instead, I will discuss the use of these standards, variations from them, and other such topics for consideration. If you haven't read these, take a moment and do so—they are online, and they are free.

✳ ✳ ✳

In technical documentation sets, the plan view drawing is the cornerstone. It shows the venue (or *venues* in the case of large distributed systems) from a bird's eye view and demonstrates the position of every item that will be mounted in that space. As mentioned in an earlier chapter, in a lot of theatrical and concert documentation, the plan view is one coherent drawing, with different lineweights or shading to indicate what items are on a stage deck, flown above a stage deck, mounted beneath a stage deck, etc. In larger architectural-style projects, these drawings will be broken out by true ground plans (everything on or below the ground level), and RCPs (reflected ceiling plans), which look like ground plans, but show only items that will be mounted on the ceiling of a space, and their relative positions in comparison with walls and such. Figure 49.1 shows the groundplan and RCP of a theme park EER.

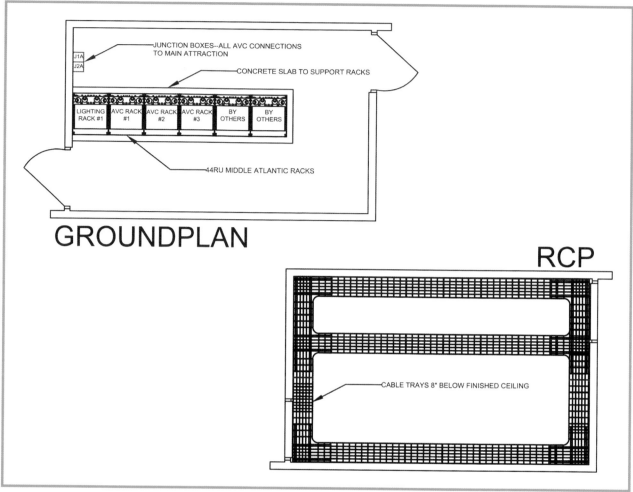

FIGURE 49.1

(Left) Groundplan; (Right) RCP of Same Space (Both of Theme Park EER)

Elevation drawings (Figure 49.2) are simply cross-sections of the same space, facing from what is called a "cutting plane" into the rest of the room. Such drawings give us the chance to see directly how high up an item is mounted (even though it will be noted on the plan view, it is helpful to see it in elevation, as we can then draw sightlines from audience seats and make sure that we are not obscuring audience view, lighting/projector throw angles, or the like). While plan view drawings are *de rigueur* in any document set, elevation drawings are merely common. If I am designing a show in a black-box theater, and all of my speakers are small and hung directly from the overhead grid, I will not likely draw elevations for them, but if any of those speakers hang extra-low, or are extra-large, I will need to show this in elevation view.

Elevations and sections, by the way, are not the same thing. An elevation is any elevated view, from any side. If we're working on a whole new building, an elevation might, for example, show the outside face of a building from ground to roof. A section (short for cross-section) is cut from within a structure, showing a view from a certain plane (the cutting plane, again) across the rest of the space. For any of our work done indoors, we are concerned with the interior of a space, and thus will be drawing section elevations from a given position. A centerline section (such as the one shown in Figure 49.2) takes as its cutting plane the *centerline* of a venue (a position drawn in plan view that shows the imaginary line cutting through the center of a performance space—most relevant in symmetrical venues) and facing one way or another from that centerline (Figure 49.3). You can draw a section from any cutting plane, along any angle, so long as you indicate it on the plan view.

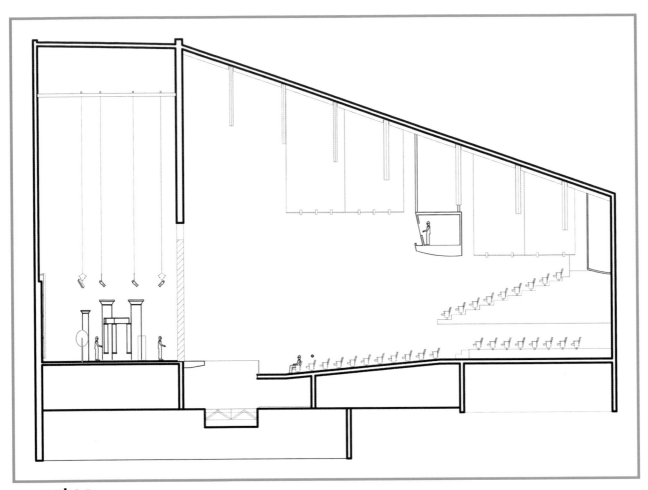

FIGURE 49.2
A Centerline Section (Elevation) for a Proscenium Theater

In Figure 49.3, we see a proscenium theater with a cutting plane set through the venue, with arrows at the end-facing stage right. At the point of the arrows, we see page jump references that indicate where in the drawing set we will find the section elevation that this cutting plane references. This indicates the "cutting plane" that our CL (centerline) section in Figure 49.2 was drawn from.

Plan view and elevation view drawings are quite common in document packages.[1] But how do we go about preparing them? What conventions do we need to observe? How do we ensure that our intentions are communicated clearly in our drawings? First, we need to understand a bit about where our conventions come from.

Scenery drawing is the baseline set of practices that initially laid out how we show production elements on a venue architectural drawing. Before the age of computer drafting, the principal methods used to indicate the position of an item on a shared plan view drawing were variations in lineweight and linetype. In such drawings, the heavier the lineweight, the closer to being placed on the stage deck an item was. For items that were flown above the deck, lineweight decreased. For items mounted below the stage (such as scenery hidden in trap doors that would be revealed partway through a show), a dashed line was used.

As we see from a hand-drafting shown in Figure 49.4, the "roof line" ceiling items in this set are drawn with dashed lines, walls are double-thick heavy lineweight, and a bench resting on the deck is solid but not a super-heavy line.

Before sound system design was a discipline, lighting design came of age. As such, lighting designers were the first to update the drawing standards from scenery, to show their instruments overlaid on sets and stages. The general

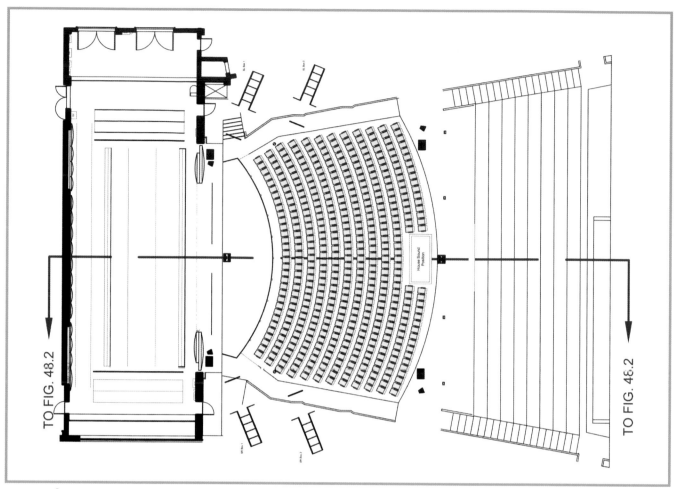

FIGURE 49.3
A Proscenium Theater with Cutting Plane at Centerline

principle of lighting drafting with regard to lineweight is that all scenery and venue architecture would be shown in lighter lineweights than in a true scenic plate, such that the heavier lineweight would be given to all lighting instruments, thus making clear that the lighting plot was a drawing of relevance to the lighting team in particular. Everything else shown was shown merely for reference purposes—as lights need to focus on parts of the stage, the stage must be clearly shown, but the lights must be the element in the drawing that catches our attention (Figure 49.5).

With the advent of complex lighting designs featuring many instruments, systems of notation had to be invented (Figure 49.6) so that designers could number the instruments and show other relevant information about them (color of gel, area of focus onstage, etc.). As instruments get more complex in the era of DMX-addressed moving lights and LEDs, more and more information gets encoded in a plot, which can make for crowded drawings. Note fields have been standardized for many years in lighting.

Along comes sound design, and suddenly there is even more demand for space in the rigging above theaters and concert halls. Lighting designers were at first not terribly pleased by this, because they had grown used to having their pick of spots in which to mount lighting instruments (now, of course, lighting and sound are both having to accommodate projection/video design, which demands space *and* clear throw angles to the stage). When lighting design first became an art, there were only so many types of lighting instruments available. Early icons for lighting

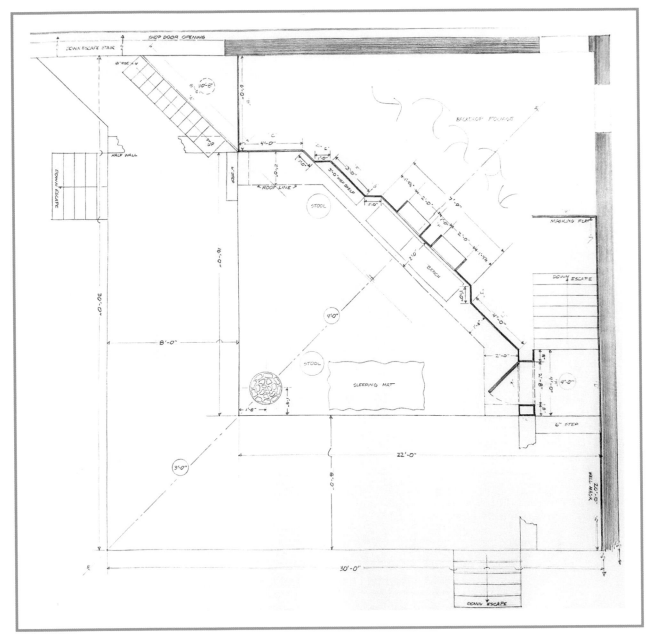

FIGURE 49.4
Example of Scenery Items With Position Identified by Lineweight and Type
Source: Design and Drawing by M. C. Friedrich, Courtesy M. C. Friedrich

instruments tended to be a bit generic—you would draw an ERS (Ellipsoidal Reflector Spot) that was closest in size to the one you actually planned to use and call it good. Now, however, in the age of computer drafting, it is expected that each item shown on a plan view is shown in 100% accurate dimensions—a practice made easier by the fact that most professional manufacturers of speakers, lighting instruments, and projectors make free CAD objects available for download of each of their models.

When drafting a plan view, there are a few essential facts that we must always address, and while these may seem bluntly obvious to anyone who has encountered such drawings before, I will state them here so that even beginners may get up to speed.

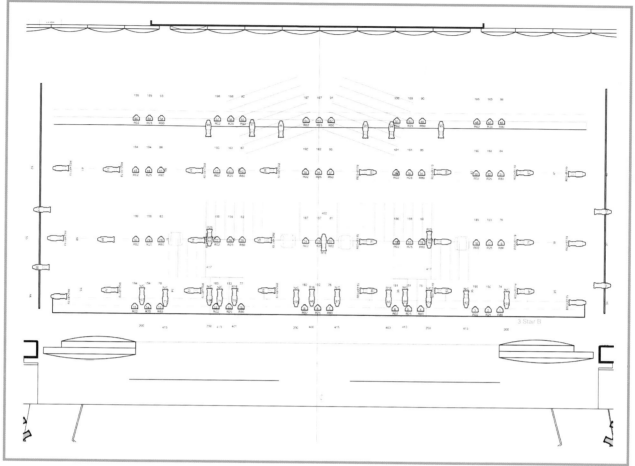

FIGURE 49.5

A Light Plot With Scenery Grayed Out in the Background

Source: Design and Drawing by M. C. Friedrich, Courtesy M. C. Friedrich

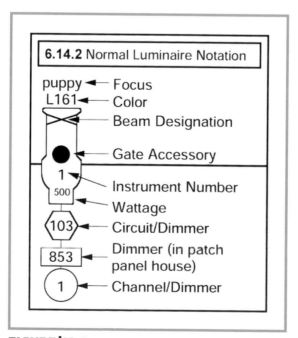

FIGURE 49.6

USITT Standard Notation for Lighting Instruments

Source: Courtesy USITT

ESSENTIAL CONSIDERATIONS FOR DRAWING A PLAN VIEW

▪ The goal of a plan view is to show any items that will occupy space in the given venue. Show everything that will take up space, and don't assume that someone will understand your space needs unless you document them. One of the key mistakes I see young designers make is to show, say, a speaker, but not to demonstrate the fly hardware. On many speakers, this wouldn't be a huge issue (and for some shows, without tons of gear in the air, it wouldn't matter very much), but some speakers (particularly array speakers) feature large fly frames, and that stuff has to be addressed in the venue view drawings; otherwise, a lighting designer may assume he or she can place a lighting instrument closer to your speaker than will actually be possible and will be irritated when this proves to not be the case.[2]

▪ Make sure everything is drawn to scale! Again, for anyone with any experience of architectural drawings of any kind, this is obvious, but for those who haven't experienced this kind of paperwork, let me explain: *scale* is the manner in which we reduce the size of an architectural drawing from 1:1 (a drawing which would be the actual size of the venue in real life) to something more manageable. As such, we typically work with a scale in which some fraction of a measurement unit (inches in the United States, centimeters or millimeters elsewhere) is equivalent to a fixed measurement unit in real life (generally 1 foot in the United States, 1 m elsewhere). As such, a typical architectural scale might be 1/8"=1', which means that 1/8" on the drawing equals 1' in real-life space.[3]

▪ In terms of lineweight, the computer age has wreaked havoc on our standards. Even in scenery drafting, the old concepts of "heaviest lines closer to deck" are often discarded in favor of using layer tools to impart the same information. One of the standard settings in most CAD drawings allows a layer of objects to be "grayed out," such that they are still visible, but are just ghosted in the background. As such, lighting, sound, and projection designers typically set scenery layers to show as grayed out beneath their plots so that it is very clear we are looking at lights, speakers, or projectors—but we still have scenery as a reference.

▪ If there will be operator stations in the venue (such as an FOH mix position) these need to be shown as part of the plan view drawings. Many early designers only show speakers on plan view drawings. In an old proscenium theater with a fixed FOH position, it might be OK to leave your console off the plate (though, conversely, it might also be nice to make sure your selected console will fit where you want to place it), but in black-box modular theaters, where the FOH position often moves, it is critical to stake out the needed space early in the planning process, lest it be allocated elsewhere before we can claim what we need.

▪ All items in a plan view drawing must be labeled. This is an area of contention—not the need to label items, but the method with which to label them. As of this writing, USITT is in the process of writing their first-ever recommendations for sound plan view drafting,[4] and it is anticipated that they will roughly follow lighting's standards for how to label units (shown in Figure 49.6). One challenge to this in the modern era is that the CAD objects for contemporary sound equipment show all the details, every bolt and handle. As such, the information that traditionally might have been noted *inside* the speaker object (as some info is traditionally labeled inside a lighting instrument) now has to reside in floating note fields adjacent to the speaker—which is fine but can get cluttered when there are a ton of units in the air, all very close to one another. Once the new recommendations from USITT are out, I (being the chair of the working group writing them) highly encourage using them, or at least referring to them, as a baseline. In this book, we have often used variations on the lighting notation scheme but have also shown some alternatives. What information needs to be noted? First and foremost, unit number. If you have a design with 60 speakers in the air, each speaker should have its own unique numeric ID. There are a range of fashions in which the notation scheme could work:

- Straight numeric: in which each speaker is simply given a sequential number, so in a 60-speaker design, you would have units 1–60.
- Sub-grouped by type/purpose, then numbered: in a scheme like this, BGM speakers might all be labeled with the prefix B and then a number, so, if you have 15 BGM speakers among your 60 total, then they would be labeled B1, B2, etc. FX speakers might be labeled F1, F2, etc. There are many ways to divide speakers like this, but whatever you choose must be clearly stated in your legend.
- Color-coded: color-coding speakers by purpose has become more common in recent years. While this can be very helpful, it is worth mentioning that this risks being misinterpreted by anyone with color vision problems, and so information must still be encoded numerically/textually in addition to the use of color.

FIGURE 49.7
A Speaker With Unit # and Trim Height Shown

- It should go without saying (but unfortunately does not) that any speaker numbers shown in plan view must remain consistent throughout your document package. Thus, whatever you label as Speaker #1 in the plan view should be labeled as Speaker #1 in the SBD, the hookup, and anywhere else it appears.
- Speaker number is often shown in a floating note field. For the time being, we will use a hexagonal container for this note, as shown in Figure 49.7.

- Also in Figure 49.7, we show trim height, which is the height at which the speaker is to be flown. Technically speaking, "trim" should only be applied to the term when the speaker or other item is mounted on a flown pipe, a hoist/chain motor, or another rigging point of flexible height. That said, "trim height" is sometimes used colloquially referring to any item height. As of this writing, the industry has not settled on a fixed answer as to whether the height listed refers to A) the bottom height of the hung speaker, meaning the clearance height, B) the height of the rigging hardware (clamps, etc.), or C) the height of the center of the box (the acoustic on-axis center). In addition, in spaces with a raised deck of any kind below the flown speaker (such as scenic platforms), we have not yet standardized whether the listed height refers to height above the stage floor itself, or above the decking/platforms. On the former question, I recommend either A or B (C might be relevant in studio design, where you are measuring to the listening axis of speakers, which is typically centered between tweeter and woofer in a two-way speaker but is generally not useful for larger production work). On the latter question, I tend to feel that height above the actual stage floor is the more relevant information. Whichever you choose, make sure the clearly identify which measurement you are listing in your legend. As shown in 49.7, the trim height symbol is typically just a circle with the height listed numerically inside it (sometimes, when speakers are mounted both above and below deck, the "+" symbol is relevant, often it is just included by habit). Trim height is generally not shown when speakers are mounted to already fixed pipes (such as over-stage electrics whose height is predetermined, or fixed grid points), but becomes very relevant when, say, designing an outdoor concert and you are specifying flown height of your main arrays.

Any plan view drawing should follow the aforementioned prescriptions, and then there are some special cases that need special attention. When building speakers into scenery, for example, close collaboration with the technical director[5] is often necessary in order to ensure that not only will your speaker mounting plan work, but that you are drawing it in such a way that makes sense not just to the sound team, but to the scenic team who will be helping with this installation.

Now, let's walk through the process of generating a plan view document.

1. Obtain the plan view drawing(s) of your venue from the appropriate parties. On a theatrical or concert project, this will usually mean getting scenic and staging info from the technical director. On a large theme park project, these drawings might come from the TD, or from the Production Manager. On retail and other straight architectural projects, these drawings will likely come from the Project Manager working for the architecture firm.

2. Set up the drawing to be in a format that is useful for your work. Sometimes this is easy—you are handed a CAD drawing, already set up with layers and associated with proper units (inches/feet in the United States, millimeter/centimeter/meter *everywhere* else), and you will simply be placing your equipment atop this drawing. In other cases, this is not so easy. I've designed projects where the last technical drawing done of an extant venue was a hand-draft from many decades prior. In that case, you will need to either cajole someone in the scenic/architectural wing of your project into converting this to CAD for you, or you will need to do so yourself.[6] Regardless of the process, you need to make sure you have a workable CAD drawing, and you've then set up some basic drawing layers for yourself: speakers, notes, etc.

3. Select speakers, and run your predictions, so you can be sure the positions you show on your plan views are correct. This may require drafting the space in another context (or at least parts of the space). Even EASE®, which operates on a 3D *CAD-style* platform, does not actually read AutoCAD or Vectorworks or Revit drawings. There is an import from AutoCAD process, but it is lengthy and difficult and doesn't always work well (and it is heavily dependent on which version of "3D" drafting the originator of the CAD document was using—are these 3D solids? 3D meshes? Wireframes that are neither? All of these make a difference when importing to EASE®). On some projects—simple theaters and halls for example—it may well be easier to just build a new model of the venue in EASE® from scratch. For others (such as a very asymmetrical and unusual theme park attraction) you may need to try to adapt a CAD drawing to EASE® format. It is beyond the scope of this book to go into this process in detail, but know that it is difficult, and can take quite a bit of time to get a usable result.

4. Once you've planned your speakers, download all the relevant and up-to-date CAD objects for those speakers.[7] Store them in one central folder (either per project or in general) on your computer so that when you are looking for a speaker to put into a plan, you have easy access to them.

5. Begin to add your speakers to the plan view drawing. With each new model inserted, verify (using the measurement tools within your CAD program) that each model is shown at proper scale.

6. Number/label your speakers. Some designers like to list speaker model names next to each speaker, but this only works on small projects and becomes cumbersome on projects with dozens (not to mention hundreds) of speakers. When numbering your speakers, it is typical to follow a few basic conventions:

   a. Begin with the speakers that are most important to your design. Speaker #1 on an outdoor concert stage would not be a delayed speaker focused on the lawn seats, but generally would be one of your main PA speakers/arrays.

   b. When speakers are paired, especially, when those pairs are L/R pairs, number the left speaker first and the right second (or the left array 1–8 and the right 9–16, or something like that). So, for a theatrical system with two main point source speakers on either side of the proscenium arch, the house left speaker would be #1, and house right #2.

   c. Number from FOH mixer's vantage point (where applicable), so house left is #1, not *stage left*.

   d. Where possible, number speakers that will share amplifiers with adjacent numbers. It is frustrating to see a block diagram with Amp #5, whose Channel #1 powers Speaker #15, and Channel #2 powers Speaker #37. This, of course, necessitates that you develop your SBD in tandem with your plan views, which is frequently the case. That said, it is better that the numbering system follow some kind of readability on the plan views than that it remain sequential on the SBD, and you can't always accomplish both . . . it is just the ideal we strive for.

7. Add title block (if not provided by your TD/PM), and legend. Add any notes/callouts (callouts are notes with arrows pointing to the area being notated) that are needed in order to clarify anything that might not be immediately evident.

One further note on plan views—when drawing vertical/J-arrays, make sure to draw them as taking up the full dimensions they will take up when configured at their anticipated splay. Don't just show the top box, if the footprint of the bottom box extends past it.

Now, once we've drawn plan views, we will draw any relevant elevation views. The basic principles from plan view drawing remain the same, with a few additional considerations:

- Make sure that your axial positions are shown exactly as they are in plan view in your elevations. If your plan view shows a speaker mounted 1' offstage from the proscenium opening, but your elevation view shows that same speaker mounted 1.5' offstage from the opening, you have just rendered both of them useless as technical documentation, because the lack of coherence means neither can be trusted as authoritative by the installation crew.

- Wherever possible, show speakers tilted as you anticipate they will actually be tilted in installation. If a speaker is aiming 30° down, rotate the CAD object 30° so that it is shown correctly. Again, this is important for other departments, but also for your own work, when doing sightline studies.

> **Sightline Study**
>
> A drawing in which lines are stretched from the extreme audience positions (seats at the rear of the house, at the left and right edges of the house, etc.), in comparison with objects mounted in the space. The goal of sightline studies is to examine what the audience can see from any vantage point and whether any of our objects will block their views. Center cluster speakers in proscenium houses with raked audiences are a key item that demand sightline studies from sound system designers. Sightline studies are also used by other departments, to place borders and tormentors, for example, in order to properly conceal the wings of a space.

We've already seen one sightline study in this book. If you need a refresher, visit Figure 34.8!

Plan and elevation view drawings should always

- Show as much info as is necessary to make the position and size of items clear.
- Notate as much as is necessary to cross-reference these drawings with the rest of the document set.
- Remain as uncluttered and visually easy to navigate as possible.

These might seem like desires in opposition to one another, but that's the key to good drafting—it communicates a ton of information, without being difficult to read.

A NOTE ON THEMED AND ARCHITECTURAL PROJECTS

Some projects outside of the performing arts will demand one further piece of information on the drawings: a so-called coverage cone for each loudspeaker (shown in Figure 48.2). These cones take different forms (a point source aiming laterally into a space will actually look like a cone, where a ceiling speaker shooting straight down looks like a circle). These cones are meant to impart to the client (a non-sound person, inevitably) the coverage areas handled by each speaker in a design. However, if we know anything about speakers, we know how inadequate a measure this can be. First, any coverage cone will be showing the angular coverage of the speaker's sound within the directional frequency range only. There will be sound sent omnidirectionally from most speakers below the frequency cutoff for directional throw. As such, the drawn cone is already a lie. Second, what is the SPL in this coverage area? When I am tasked with drawing such cones, I will specify a given distance from the speaker at which I curtail the drawn cone and state in my notes that that border represents "X SPL," where X equals whatever I have determined is the lowest acceptable SPL in this design. At times, clients will even ask for multiple cones, one showing a high SPL, and another showing a reduced SPL (which can be shown by making one of them a solid line, and the other a dashed line, for example). This gets into some tricky estimating, particularly in enclosed spaces. If you have modeled your design, you can mimic the model in these representations, but if you haven't had time to (or aren't being paid to) model the design, the natural boundary effect of speakers mounted near walls and surfaces will change this coverage from the nominal listed coverage, but we typically draw nominal coverage anyway.

Why is this a thing? Why do clients want this and not speaker prediction? The truth is that on a lot of projects, the clients won't pay for the predictions. EASE® is not cheap software, and the running of advanced models for architectural projects is sometimes skipped (especially when working retail spaces or the like, where sound quality is not of paramount importance). On these projects, it is up to you whether you use your own software licenses and time to model the system anyway. When high sound quality is desirable, it is always a good idea to model and predict. However, there are plenty of projects, even those where sound quality is a priority but not the *highest* priority, where the demands of deadlines and budgets make speaker prediction and modeling impossible. In those cases, designers use their best knowledge of speakers, specs, and architectural acoustics to make their gear selections—and these are the main projects that tend to request "coverage cones" be drawn. I personally feel that these are only useful as a pre-modeling visualization tool (as we saw in Figure 34.4). However, they are sometimes required, and for young designers working on first projects (sometimes without access to EASE® or the like), they can still be useful guides for how your coverage may function—but always remember, they are just guides and will fail or be limited in the ways discussed earlier.

Finally:

RACK ELEVATIONS

Rack elevations are drawn for any equipment racks that will be ordered, built, or modified from a venue's default configuration, for a project.

In Figure 49.8, we see a 45-RU rack enclosure with lots of gear.

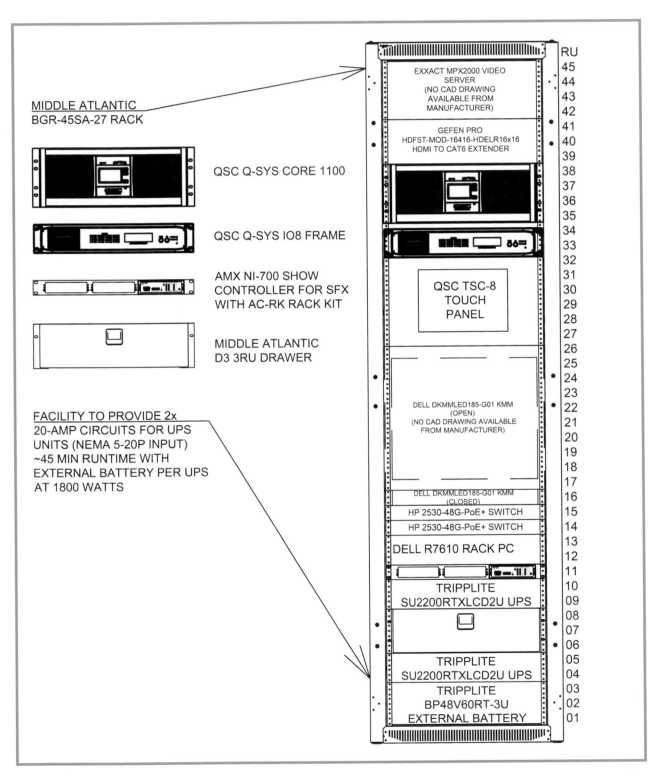

MIDDLE ATLANTIC
BGR-45SA-27 RACK

QSC Q-SYS CORE 1100

QSC Q-SYS IO8 FRAME

AMX NI-700 SHOW
CONTROLLER FOR SFX
WITH AC-RK RACK KIT

MIDDLE ATLANTIC
D3 3RU DRAWER

FACILITY TO PROVIDE 2x
20-AMP CIRCUITS FOR UPS
UNITS (NEMA 5-20P INPUT)
~45 MIN RUNTIME WITH
EXTERNAL BATTERY PER UPS
AT 1800 WATTS

EXXACT MPX2000 VIDEO
SERVER
(NO CAD DRAWING
AVAILABLE FROM
MANUFACTURER)

GEFEN PRO
HDFST-MOD-16416-HDELR16x16
HDMI TO CAT6 EXTENDER

QSC TSC-8
TOUCH
PANEL

DELL DKMMLED185-G01 KMM
(OPEN)
(NO CAD DRAWING AVAILABLE
FROM MANUFACTURER)

DELL DKMMLED185-G01 KMM
(CLOSED)

HP 2530-48G-PoE+ SWITCH

HP 2530-48G-PoE+ SWITCH

DELL R7610 RACK PC

TRIPPLITE
SU2200RTXLCD2U UPS

TRIPPLITE
SU2200RTXLCD2U UPS

TRIPPLITE
BP48V60RT-3U
EXTERNAL BATTERY

RU
45
44
43
42
41
40
39
38
37
36
35
34
33
32
31
30
29
28
27
26
25
24
23
22
21
20
19
18
17
16
15
14
13
12
11
10
09
08
07
06
05
04
03
02
01

FIGURE 49.8

A Rack Elevation

511

It is standard to number rack spaces, though whether you count from 1 at the bottom to 45 at the top, or start with 1 at the top and count to 45 at the bottom, is not standardized across the industry. I prefer labeling from the bottom, as then every rack in your project, whether a 2-space or a 45-space, has "space #1" in the same spot on the floor. Each piece of rack gear must be shown, in real size.

Care must be taken to plan the orientation of gear within a rack according to the needs of that gear. Some pieces of equipment will need 1RU of open space around each unit to ensure proper air flow. Some can be packed in tight with other gear. Remember always that heat rises. Examine the physical depths of your equipment, if an amplifier with a giant heat sink is shorter in depth than a rack-mounted PC, then you may not want the PC right above that amp, such that all the heat flowing off the sink at the back of the amp runs directly into the PC, causing overheating problems. On the flip side, if the amp is deeper than the PC, but not by very much, and the rear of the rack-mounted PC has an intake fan on it, you *still* may not want the PC directly above that amp, as heat may be drawn into it from the fan.

Racks themselves sometimes need fans in order to circulate air among mounted devices. Fans are often placed at the top of a rack, to pull air out that has risen and prevent it being trapped by the top lid of the rack (if your rack *has* a lid, some are open at the top). Some racks are fully enclosed, and have mini-HVAC temperature control units, to prevent wild fluctuations either above or below the desired operating temperature of the gear. Additionally, we sometimes need accessory items mounted in racks: shelves that can be pulled out, drawers for storage of supplies, and whatnot. Rack configuration and design must be carefully considered, and your drawings must reflect that consideration.

There is some debate as to whether the best practice is to show the graphically realistic rack gear in each slot, or whether it is better to just draw a rectangle of proper size and label the unit inside the rectangle by name. I personally prefer the graphical method, as I feel it makes an easier-to-read document for installation work, because you can see exactly what to put in what spot. However, some rack gear isn't available as CAD objects, in which case a rectangle of proper size with the named item inside it is appropriate. Always make sure any labels are numbered in such a way as to correspond with your SBD—so Amp #5 in the SBD is labeled as Amp #5 in your rack drawings.

There is also some debate about whether or not rack drawings should attempt to indicate how the equipment in that rack interacts or is interconnected. I have yet to see an example of this that was elegant and easy to read—for my money, this info belongs in the SBD and hookups.

A NOTE ON ORDER OF EVENTS

While I have listed rack elevations here, because they are technically elevation drawings, rack elevations are not typically drawn until after the SBD (and/or workbook) is complete, as it is hard to know what equipment will go into that rack before laying out your signal flow.

* * *

Now, you've selected your speakers, and maybe your console. You've drawn your plot, and maybe your elevations. It is time to figure out, and draw, how our whole system comes together. It's time to make a System Block Diagram.

Notes

1 Including rack elevations, which we'll touch on toward the end of this chapter.
2 Again, I am always throwing lighting designers under the bus, but since I've also worked as a lighting designer, I feel it is my right to annoy my own. You will notice that I have not shown speakers with hardware in most of the plates in this book, as I've been showing speakers in isolation, and wanted to focus the readers' attention on the speakers themselves. Now I'm saying show the hardware, and yes, that's correct. Whenever possible, always show the hardware.
3 Scale works very differently in different CAD platforms. In Vectorworks, a drawing or view is set to a given scale, and everything drawn within it must be drawn to that scale, which is very like traditional hand drafting. In AutoCAD, while there are many ways to configure the units of a document, the general practice is to draw in

"real units" in the model space—so a speaker that is 6" deep in real life is drawn 6" deep in AutoCAD, and then to adjust the scale of the *view* in the layout pages. This has the advantage that you can draw an item once and show it in many different scales on different plates quite easily. Be aware, some manufacturers have created CAD objects that automatically come into a document at the right size, and some have not. I'm not sure why this is, actually, but if you open certain manufacturer's drawings, they will even show dimension lines that describe the proper size, but if you measure the drawing with a tool in the drafting program, the measurement is different. There are tools and techniques with which to fix this, and they vary by manufacturer, but my point is that you should always be aware of the real-life dimensions of a speaker, and always check the first copy you place in a drawing to ensure it has been copied/inserted/placed in the proper scale.

4 Prior to this effort, the only recommendations they made for sound drawing were for block diagrams.

5 In theater, a technical director is mainly in charge of planning and building scenery from the scenic designer's drawings. In themed and multimedia projects, a technical director may be in charge of oversight of all technical areas and their execution. Check with your own production company to ensure you are working with the proper person in the proper situation.

6 As you might imagine, this can take a very long time and considerably more CAD ability than just plopping speakers atop existing drawings.

7 Up-to-date is important, unless you know you will be using old models. Vectorworks, for example, provides some stock libraries of speakers (as it does quite well for lighting instruments), but their speaker libraries are often out-of-date even in a new Vectorworks release.

CHAPTER 50

System Block Diagrams

Audio, Control, Network Risers

A System Block Diagram (also known, in various segments of the industry, as a One-Line, a Riser, and other terms) is a diagram that shows every physical wire interconnection between devices in a sound system. At its most basic, for an analog system, it is meant to represent the path of physical system interconnections, in signal flow order—audio starts here, travels there, etc. In the modern age, there are many interconnections that—while technically carrying *signals*—are not actually audio signal lines. We have network connections for monitoring, control signals from triggers to a GPIO port, we have analog (one way) audio signal and digital (often two-way) signal. Representing all of these can be confusing, and on large projects, can require a large number of plates.

Before we get into any detail, it is important to identify what an SBD *is* and what it *isn't*. First, it is a misnomer that the S in SBD stands for schematic. It does not. A schematic drawing is one that would represent the so-called electrical truth[1] of the system. Thus, if a signal is shown flowing from an amplifier over NL4 to a speaker in bi-amped mode, inside the speaker device block, we would show the NL4 breaking into its component two-wire leads and powering the tweeter with one and the woofer with the other. Some designers prefer to show this level of detail in SBDs, but emerging consensus is that this is not useful, and tends to take up too much plate space—systems are getting larger every day and keeping unnecessary information off of our SBD while including all the necessary information is important. If a drawing shows an NL4 from an amp to a single speaker, and the cable is labeled as NL4 (not NL2), it is probably assumed that the speaker will be in bi-amped mode—why else would we be sending it 4 wires?

A true schematic, say, for a mixing console, would show every electrical component inside of it—every diode, capacitor, resistor, etc. A block diagram for the same console would show available signal paths through the console, and the major pieces that will alter that signal (gain pots, faders, etc.). In a *system* block diagram, the console (and its stage boxes), act as single entities.

An SBD is used as an installation (and shop-building) document—therefore, any connection to be installed needs to be illustrated, such that the crew using this document can easily see what is plugged into what, with what cable. As such, when drawing the patch from, say, a digital console's stage box to an amplifier, we need to know that the XLR cable is plugged into "Stage Box 1, Card 3, Output 1," not that it is deriving its signal from "main right" within the board patch. Part of the hookup can tell us about internal console patching, but an SBD focuses on *physical* details.

Now, with that out of the way—how do we draw an SBD?

BEGIN AT THE BEGINNING

First, we must understand (again, I know!) a bit of history. In the analog era, system block diagrams were always drawn like Westerners read, from left to right. Inputs on the left, processing in the middle, amplification just to the right of that, and outputs on the right. When you have an all-analog sound system, this is easy.

We've seen several block diagrams throughout this book. Let's take a look back at Figure 10.2. In this system (an *almost* entirely analog system, save for the DSP), we see that microphones are on the left, running into an analog console, into power amps, to the DSP, and then to speakers. Very straightforward. Before we make things more complex, it is worth examining a few things in this example more closely. First, devices are shown as rectangles, with labels inside and above them. These are called "device blocks" in the parlance of our industry. A typical device block shows only the connections that will be used by the given design. If your console features 48 mic inputs, but is only using 4, then only 4 are shown. Inputs are always drawn on the left of the device block, and outputs on the right, which again leads easily to our whole "reading left to right" scheme. Now, the microphones in this plate are drawn using the USITT-recommended symbol for microphones, however, the amplifiers and powered speakers are drawn *without* the USITT-recommended triangles inside them.[2] At present, there is a lot of motion in the industry to abandon amp triangles in our SBDs—we always label the devices as amps anyway, so why waste time with an extra symbol? However, even as designers have largely abandoned the triangles, they are holding on to mic (and speaker) symbols that are different from the standard device block. So what is correct?

First, as I've stated in previous chapters, different segments of the industry have different expectations for graphical representation. Theater, for example, is much more likely to show mics and speakers as semi-graphical symbols than architectural design, which is often likely to demand that every item, whether a mic, a speaker, or anything else, be shown simply as a device block. Beyond differences by industry, each project or production manager may have standards they'd prefer the team to adhere to, and it is worth establishing whether that's the case before spending a lot of time drawing SBDs.

Symbols aside, the main point is that each item is labeled properly. Mics should always be numbered. Any items of which there are more than one per category (mics, amps, speakers, network switches, whatever) should always be numbered. This is for ease of cross-referencing between documents, and also for problem-solving in the field (it's much easier to say, "Speaker #7 is out" than "that speaker over there, up in the catwalk, no not that one the other one."). In addition, wherever possible and appropriate, items should always be labeled with model information.[3] These unit numbers will correspond to the numbers on the mic plot (if your show requires one—any show with hanging mics will likely need a mic plot, or any large concert with many mics will need a mic plot, but small shows often forego the step of drafting mic positions), in your hookup, etc. The signal lines running from these mics into the mixer are all labeled by type. The inputs to the mixer are labeled as "Inputs" and are also numbered. It is assumed that those numbers or names correspond to the real labeling on the console, as would the outputs, and any two-way connections.

We've also already seen considerably more complex designs in this text (Figure 40.4, for example). In all of these examples, we have attempted to represent signal flow in this system from left to right, but that hasn't always been (and isn't always) possible. Dante is a two-way protocol, for example, and so, while our DSPs and consoles have often featured two Dante connections (Primary and Secondary, a redundant pair), the signal is traveling both in and out of the devices over those lines. So how do we represent them? As cleanly as possible.

Again, each device is labeled, with numbers, ideally with the model, and sometimes also with system function.[4] It is all fine to show a network switch labeled #1, but it is more helpful to *also* label it as the Dante Primary switch, because (as we recall from eons ago in earlier chapters), Dante Primary and Secondary signals cannot be sent over the same switch, or they crash. Thus, if we know that a switch is the Dante Primary switch, then every

Dante-enabled device in the design can connect to it from its Primary port, without having to continually trace a signal line back to a device to remember which signal set it takes. Like any technical drawing, we endeavor to encode as much information as possible, in as easy to read a format as possible.

In many of these drawings, we also see that signal lines cross one another. When drawing an SBD, it is always a good idea to try to avoid signal lines drawn as crossing one another whenever possible, as those crossing points make for possible confusion. In old-school drafting (and still today in much theater drafting) the convention of the "jump line" is used when signal lines are shown crossing, but the signals do not interact. This is the little arc in the signal path shown in Figure 10.2. However, in much of modern drafting (especially architectural and themed drafting) it is considered cumbersome to have to draw these little arcs on every intersection point. Instead, it is to be assumed that signal lines crossing one another do *not* mean signals interacting, unless they are shown with a thickened dot at the intersection point, as we see in the insert cable split point in 10.2

Note that when a cable is run alongside other cables of its type, the group may be noted with a single callout, but when a cable is unique to its position on the page, the cable will be individually labeled over the center of the run (or, depending on its twists and turns, wherever is easiest to read)—except where a cable is terminated in two different connectors at each end, in which case the cable is labeled at each end. In large-scale architectural and installation projects, it is common to label the ends with their connector types, and the center with the actual model of cable used (like Belden 3612 for a particular type of Cat 6 cable)—however, this level of detail would generally only be found in DD-level documentation.

Every device that passes, alters, or allows manipulation of a signal should be represented in a block diagram. Every mic, DI, stage box, console, playback device, outboard effect, amplifier, speaker, recording device, network switch, computer, processor, etc. This can start to get difficult to read, especially when we're trying to keep everything visible in one plate. Imagine Cirque du Soleil's *Love*, a show in which there are more than 6,000 speakers. This show, perhaps obviously, does not represent every device on one page (it would be insane to try).

When faced with drawing a very large system, there are two basic standard tactics we take for ensuring readability:

1. Separating the system into plates based on function. This would mean, for example, drawing the analog audio connections on one plate (or set of plates), the digital audio connections on another plate (or set of plates), and the network/control information on a third plate (or set of plates).
2. Separating plates by physical location. So, equipment at FOH for a large concert would be on one plate, but everything in ampland or the machine room would be another, the onstage monitor speakers would be a plate separate from the house speakers, and so on. In any SBD, it is good to group together items that will be in the same physical location, as they are bounded by dashed boxes and labeled with where the items are situated in the venue. When we can't show everything that is in one location *physically* in one spot on the *drawing*, we close the bottom of the location box with a "break line" (of which there are several standard types), and ideally label the break spot with the inter-plate reference guiding us to the next spot in the drawing set. We've seen this several times in the drawings in this book.

There are advantages and disadvantages to each of the above plans, and sometimes we use both in tandem. Let's take #1, separating by the purpose of the gear.

If we separate a document into network plates, analog audio plates, digital audio plates, and so on, these plates are considerably less crowded for reading. However, they also pose a challenge for installation, in that completing one plate does not complete the installation of a device on that plate (necessarily). When a *device* is shown in multiple locations in a drawing set (to represent different connections, for example), a break line is drawn across the bottom of the device, just like when we must break location boxes in our plates, as described earlier.

For very large systems, it can help the designer to separate functions onto separate plates, as it allows the designer to see clearly that the network is connected properly, without being confused by the audio signals, and vice versa. It also requires more tracking between plates and especially more attention to detailed notation. The only branch of our work that should always have its own dedicated plates by function is intercom design. Of course, we could then argue that a stage manager's God mic into a DSP should be shown on the com diagram, not the audio diagram. This is incorrect, but you can see how complex systems become challenging to represent.

If we followed plan #2 and separated FOH speakers from monitor speakers, placing them on distinct plates, it is easier for a crew to install a given location all at once (as tends to happen in work calls), but it is still more cluttered, as multiple signal and cable types must be shown on one drawing (even if that drawing is many plates long). This makes it considerably more difficult to avoid crossing signal lines in the drawing, as the dawn of two-way communications mean that many signals travel to different devices in parallel, rather than from device to device in a straight line as in analog days. Crossed signal lines always mean a more difficult read, so there are drawbacks to this plan as well.

Of course, if your system is large enough, you may have to employ both tactics noted earlier just to have drawings that make any sense at all.

In general, when a system is small enough to be able to be represented coherently on one SBD page, it is a good idea to do so. However, there are many large-scale projects for which that will simply not be true.

One more note on separating plates—when designing for larger projects, you will sometimes need to draw power system block diagrams, showing which devices connect to which UPSs, and power distros, and the like. These should ALWAYS be separate plates from any signal line drawings—no matter how you have or have not separated by signal type in other plates. We never show power configurations on the same plates as other signal and control lines because A) it is damn near impossible to keep a drawing uncluttered at that point, and B) the electrical engineers building and installing the power system are probably not the same people who are performing the AV integration, so each set of drawings needs to focus on what is relevant for the given installation team.

It is my nerdy, nerdy dream that much like we have software to model speaker performance with, someday soon we'll have "SBD modeling software" with a huge database of actual AV devices. Such a software would allow you to insert objects with all the performance features of the real device already assembled, connect devices in a fashion that won't let you connect incompatible signals or connector types without converters, and that will track your signals between plates for you, automatically generating all your inter-plate references, and updating them if a device moves from one plate to the next. There are a couple of programs out there that pretend to do some fraction of this work, but they are prohibitively expensive, and without the full needed feature set, so there are very few people who have adopted them in the industry.

At any rate—the SBD should be the comprehensive interconnection roadmap. Where signals can travel, and how they get there. A good block diagram is peaceful to read (even when it is very complex) because the organizing intelligence behind it is evident. A poor block diagram is a headache. *Don't give your crew a headache!* It's not nice.

Now that we've done our main drawings: plan view, elevation, SBD, rack elevations, we can take a moment on specialty drawings: mounting details.

Notes

1 I am grateful to Sam Kusnetz for this phrase.
2 These sideways triangles are the universal electrical *schematic* symbol for the amplifier.
3 It is not appropriate to list models for most designs at the CD stage. Sometimes, depending on the project, it is not appropriate even in SD—not until DD. Establish the expectations of your company at the start of a project.
4 In this book, I have erred on the side of showing unbranded devices, except for devices where the model of device is relevant to the understanding of the principles at work in the example. However, in professional final drafts, models are almost always listed.

CHAPTER 51

Mounting Details

Can I be honest with you? We've come a long way through this book, and I feel we've become friends, so I'm going to level with you: I don't like drawing mounting details.

Before you accuse me of being unprofessional,[1] let me explain. There are two primary reasons we as sound system designers submit mounting detail drawings for a project. The first is that we've specified line arrays of some kind, and arrays always need mounting details. How many boxes, what is the splay between the boxes, what points on the rigging frame do we pick for hanging points, what is the weight load, etc. This kind of mounting detail I am fine with, because for the most part, line-array manufacturers provide calculator software that allows us to configure our arrays as we desire and will then draw the mounting setup for us with all attendant details.

However, the second reason we need to draw mounting details is because we need to mount something (generally a speaker, but sometimes a playback server on a parade float, a microphone array, or something else) in an unconventional location, and in an unconventional manner. In such circumstances, it is our job (sometimes in collaboration with our project TD) to come up with a way in which our devices will be mounted and to draw it clearly so that the team can build and install it.

Now, why would I find that latter case less than enjoyable? Because I am not a scenery builder, a mechanical engineer, or a hardware designer. Don't get me wrong, decades ago, in a distant past life, I worked as a theatrical TD. I have built all manner of sets, I have cut and welded steel, I have glued wood and sewed costumes and performed all manner of props artisanry. That said, once I started my deeper focus on sound, lighting, and video, and then went even further in paying attention to sound above other departments, I let those other production skills . . . not *disappear* exactly but atrophy a bit. As such, whenever someone asks me to design mounting for one of my devices, my first response—as most sound system designers' responses probably should be—is to go and talk to the head of the scenic or props[2] department and explain what I need and ask very nicely if they will create a custom solution for me. Sometimes this works, and I rejoice! Other times, either there is no such person to ask or their solutions won't be viable for one or more electroacoustic reasons, and I am once again forced to dig deep and create something.

Let me be clear—I am exaggerating a bit. My custom mounting solutions always work, and I am always *very* clear on any such drawings that they are for *design intent only* and should be vetted by whatever team is going to actually fabricate and install these bits. It's not even that onerous a task—the drawing is simple, generally, and in all honesty, if you have training and experience with scenery, props, costumes, fabrication, and so on, it can actually be exciting to design a mounting solution that works just exactly how you need it to work, rather than depending on stock hardware that doesn't quite meet the requirements of the design. However, the vast majority of sound designers I know and work with (and I know a lot of them) aren't scenery people and generally prefer to have someone else step in and design any hardware solutions that need to come into play.

I begin this chapter in this way, not to discourage you from designing mounting solutions, if your skill set allows and the project demands, but rather to console those for whom this request would seem unusual or out of your general wheelhouse. It is unusual, and you must always exercise great caution when designing anything that must take weight, particularly if that weight is suspended over the heads of guests/audience members.

There is a third case in which you may be asked to draw mounting details—on large architectural projects, where, even though every item may be mounted with standard hardware, a team may still want to see your drawing of that mounting setup. In these cases, most often there will be some kind of CAD document available from the manufacturers showing the standard hardware, and what remains is for you to position it at the proper angle according to your design, label it with aiming info, and call it a day.

Mounting details, as such, are plates that show info on the mounting of given devices—almost always speakers. For line arrays and other large installations, a single array may take up a full drafting plate. For smaller speakers (like 4" flush-mounted ceiling speakers), you may be able to fit several units on a single page, in order to save space in the overall package.

Let's begin with line-array drafting.

In Figure 51.1, we see an array of d&b audiotechnik J-Sub and J12 speakers.

I have created this array within d&b's proprietary (and very useful) ArrayCalc software. Within that software, I selected the type of boxes, the count of those boxes, their splay angles, and any other relevant details (I can gain shade, and more, but that's not relevant to the mounting detail). Below the array itself, I can select which holes in the array's flight frame I want to use for rigging cables. The drawing shows the center of gravity of the array, and shows the weight load on each cable, which shifts if I change the holes I am using (it will also render an *invalid* warning if I select holes that won't allow my array configuration to be hung and aimed as I intend).

This kind of software makes array configuration very easy. I can simply export this drawing in a range of document formats (including CAD) and add it to my drawing package. I can export the configured array as an EASE® file and model the setup in EASE® without having to rebuild and shape the array. It's very useful. If you are using Meyer speakers, their MAPP XT software allows the same kind of configuration. If you are using L-Acoustics, their Soundvision software does the same. If you are using JBL, they have different calculator software for different product lines, but the principle remains—if you're working with a vertically arrayed speaker set, chances are the manufacturer has a configuration tool that will allow you to properly distribute weight and provide rigging details. Easy as pie![3]

Now, if you're tasked with drawing mounting details for a speaker that is in a custom configuration—first you must determine how to mount that speaker. As I said in my long-winded intro to this chapter, your best bet (as long as you yourself are not an expert in fabrication) is to go to a professional on your team. Talk with them about what you need. Most speakers come with some manner of rigging points, whether bolt holes for attaching a yoke and

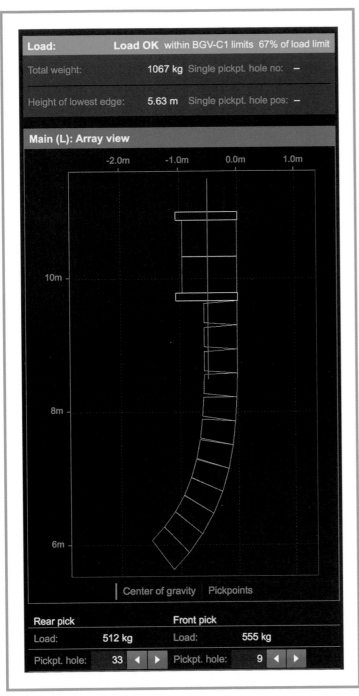

Load:	Load OK within BGV-C1 limits 67% of load limit	
Total weight:	1067 kg	Single pickpt. hole no: –
Height of lowest edge:	5.63 m	Single pickpt. hole pos: –

Main (L): Array view

Center of gravity   Pickpoints

Rear pick		Front pick	
Load:	512 kg	Load:	555 kg
Pickpt. hole:	33 ◀ ▶	Pickpt. hole:	9 ◀ ▶

FIGURE 51.1
A d&b J-Sub and J12 Line Array
Source: Courtesy d&b audiotechnik

clamp, threaded holes for attaching stock wall-mount brackets, eyebolts, or other such. One very important note is that whatever you attach to on the speaker must be *rated to take weight*. Many large speakers, for example, have built-in handles. It is tempting to lash some rope or chain around these handles, attach same to a grid, and call it good. However, if you look closely at the documentation for these speakers (and sometimes at the speakers themselves) you will often find that the handles are in no way supposed to be used for rigging the speaker. They are there for your temporary use, with your actual hands, when you are lifting or moving that speaker. As such, they are designed to take the weight of the speaker only briefly, and they are made for the distributed load of your hands, which are

FIGURE 51.2
JBL Control 29AV
Source: Courtesy Harman Professional Solutions

wider and comparably softer than a length of aircraft cable (for example). So, once you've identified points on the speaker that can be used for rigging, you can find appropriate hardware.[4]

For example, let's imagine the JBL Control 29AV (Figure 51.2), a very popular speaker for themed installations, due to its loud delivery, reasonably robust construction, and inexpensive price point.

Let's imagine that you need to mount this speaker so that it is hanging from a pipe and angled down roughly 35°. It is not really designed for pipe mounting. Its default hardware is a wall or ceiling plate, designed for architectural installation, attached to a ball and socket armature (Figure 51.3).

While the ball mount would allow the down tilt you desire, there is no chance that the wall plate will attach to a 1.5" schedule 40 pipe.[5] Now, we look into the technical documentation of the speaker and see that there are weight-rated mounting points—threaded holes—on the top and bottom. Thus, what we need to do is create a custom vertical yoke, that connects to a c-clamp either by standard bolt connection or via welding. The downward tilt can be basically set in the weld, if need be, but since the unit is pipe-mounted, it can be further tilted by virtue of the yoke's angle to the rigging pipe.

Figure 51.4 shows my rudimentary solution:

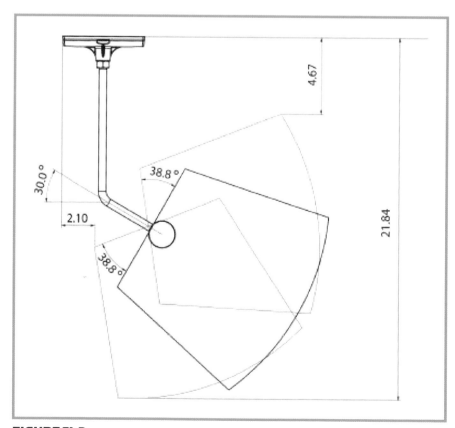

FIGURE 51.3
JBL Control 29AV Ball and Socket Wall Plate Mount
Source: Courtesy Harman Professional Solutions

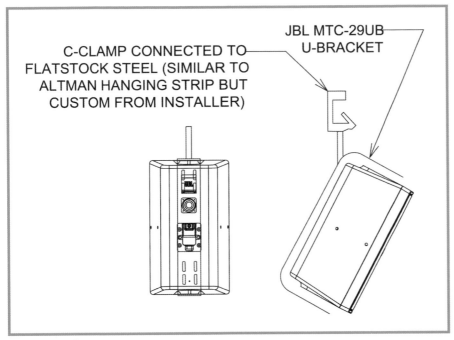

C-CLAMP CONNECTED TO
FLATSTOCK STEEL (SIMILAR TO
ALTMAN HANGING STRIP BUT
CUSTOM FROM INSTALLER)

JBL MTC-29UB
U-BRACKET

FIGURE 51.4
JBL Control 29AV Custom Hardware by Author

In Figure 51.4, we can see that I've specified a solution that could even make use of a stock JBL bracket in combination with a custom weld. It could also be fabricated out of flatstock steel instead. The drawing is simple, clean, and illustrates the point of the rigging plan.

Now, I could have gone into greater detail—I could have specified how thick the flatstock was. I could have illustrated the exact bolt I wanted from c-clamp to yoke.[6] This drawing provided just enough detail to indicate my design goals, without reaching the level of what is called a "build drawing" or a "shop drawing." I am not a fabricator. I have no intent to become a fabricator, and I am guessing that many of you reading this are in the same boat. Don't pretend to know things you don't. Indicate what is needed and pass it to someone who will know best how to build it. Had my project team required more detail, I would have endeavored to provide it, while still protesting that there were better people to tackle this particular task.

Let's take a quick look at a drawing of another stock hardware solution (besides line arrays)—which might need to be provided to an architectural firm for a project that isn't entertainment-related. Figure 51.5 is a drawing of a JBL Control 14C/T, a small ceiling speaker.

It is drawn to scale, with the intended factory mounting indicated. The drawing of the speaker was taken from JBL's CAD library, and the mounting (which in this case is a thin, ceiling-flush ring) is indicated, not shown in detail, because it is very thin, and the idea is communicated (plus, the spec sheet for the mounting ring would be in the submitted *tech book*).

Now, there are some things to consider when designing custom rigging—any speaker delivering high SPL, especially in low frequencies, like subwoofers built into scenery, may need to be vibrationally isolated from the items around it. I have designed custom subwoofer mounts to go into props, where the sub was elevated on a custom wood or steel frame, and the frame was inset with heavy industrial neoprene shock absorbers (like the kind found under heavy industrial machinery or roof-mounted central air conditioners) so that the frame itself wouldn't rattle. I have mounted high-SPL midrange speakers into props that sometimes needed similar neoprene isolation from the mounting hardware, even without LF content going through them, because the SPL was so high (or the prop was naturally so filled with rattling parts). None of this was perfect. A prop that has a rattling sound to it may still rattle when a loudspeaker is vibrating inside it, but we do the best we can, and we inform our teammates of the potential risks of placing a speaker in that given place and for that given purpose. This is, again, why it can be super-helpful to work with fabrication professionals who are not only more knowledgeable about build options, but who (ideally) also have an investment in not having their scenery, props, or what have you, rattle unnecessarily.

523

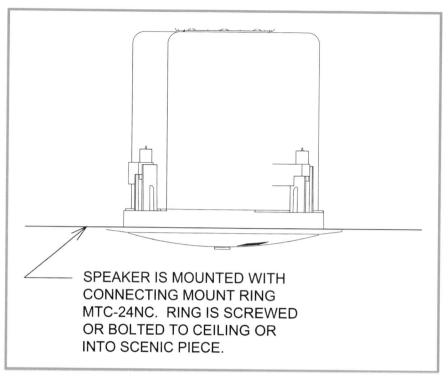

SPEAKER IS MOUNTED WITH
CONNECTING MOUNT RING
MTC-24NC. RING IS SCREWED
OR BOLTED TO CEILING OR
INTO SCENIC PIECE.

FIGURE 51.5
JBL Control 14C/T Mounting by Author

524

This may be the most cynical chapter in this book. For that, I apologize, though I hope the reason for my tone is clear. It is truly founded out of a desire to mount our devices well, and to not injure anyone. When parts of those tasks are outside of our wheelhouses, we need to seek collaborators who have more experience in those areas.

Finally, to wrap up this section, let's address a drawing that only gets used (generally) for theatrical design (and occasionally for theme park design, but less often)—a simplified speaker plot used during programming, called a magic sheet.

Notes

1 Maybe I am!
2 Or sometimes costume department—we do occasionally have to mount things inside costumes.
3 I don't know why this is a saying—making a good pie is a lot of work and can be quite complex!
4 The sourcing of rated materials is another area of difficulty, unfortunately. There have been major scandals, for example in the rebuilding of the Oakland-San Francisco Bay Bridge (www.sacbee.com/news/investigations/bay-bridge/article2589402.html accessed June 23, 2018), wherein bolts or other hardware manufactured, typically, overseas and sold as "rated" hardware was in fact not manufactured to proper standards and failed much sooner than anticipated. American manufacturing is not immune to these problems but is generally a safer bet (when purchasing in the United States, at least)—the best practice is to purchase materials from well-known and long-established hardware manufacturers, and to make sure the materials are rated by organizations such as ASTM or ANSI. It is beyond the scope of this book to go into this topic in great detail, which is another reason I always recommend checking with a TD or other professional whose area of expertise covers this material.
5 Lighting grid pipe is the basis for most entertainment-related pipe size and is generally 1.5" in diameter. It is often referred to simply as "schedule 40," as many believe (erroneously) that the schedule relates to the diameter. The schedule, instead, relates to the thickness of the pipe walls. Schedule 40 steel is specified because it can hold up to standard c-clamps, allowing entertainment devices to be mounted on this pipe and tightened without puncturing or bending the pipe, but without the pipe being unnecessarily heavy.
6 Had I done so, it might have behooved me to work in Inventor instead of AutoCAD, but I did not own an Inventor license at the time of the project this example is drawn from.

CHAPTER 52

Magic Sheets

You are sitting in tech rehearsal for a play. Your system design consists of 57 speakers, placed in various locations in and around the theater and set. You need to send a sound to a *special* mounted underneath the couch onstage (it's haunted!), so you turn to the Arch D printout of the speaker plot and proceed to pull out your magnifying glass to figure out which speaker number that couch special is, so you can route an effect from QLab to it.

That doesn't sound terribly practical, does it? Magnifying glass? What are we, Victorian-era sleuths? This will never do. So, instead of looking over our detailed speaker plot, which might include mounting info, details about trim heights, and other such info, we turn instead to our magic sheet.

The term "magic sheet" comes directly from the world of lighting design. In lighting, a magic sheet is one of a variety of "cheat sheets" that allows a designer to quickly identify needed lights by color, zone of coverage, and channel for programming. If a designer needs to bring up "Zone A, Warm, from Stage Left," they can look at the grouping on their magic sheet of lights that hit Zone A, identify quickly, via color-coding, which lights are warm-colored, and then from there find their stage left unit or units. Such groupings can be drawn in a variety of fashions, from a simplified plot with easy, overlarge notes, to a spreadsheet organizing units by zone and color within the cells.

As the age of sound system design came to maturity, we found ourselves using more and more speakers in our designs. Where, in the past, it might have been common for a play to feature 75 lighting instruments, and between 2 and 5 speakers, now it is not surprising when we've got as many speakers (or more) than lighting instruments in some productions. As such, we've had to figure out our own shorthand for easily identifying speakers while programming. We've stolen the term *magic sheet* and created our own paperwork that allows a designer to see output paths quickly without having to read a full plot or hookup.

In Figure 52.1, we have an example of a simple magic sheet.

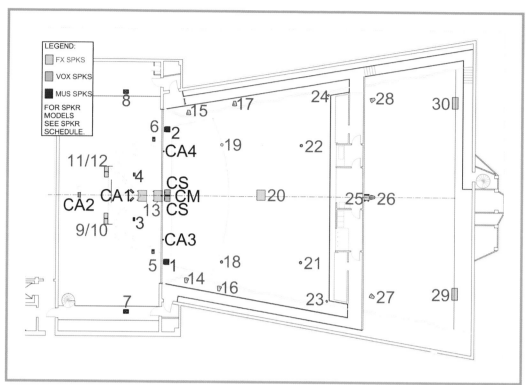

FIGURE 52.1
Simple Magic Sheet

We have encoded information in this sheet in two main ways to make reading easy during programming:

1. The speakers are color-coded. Vocal system speakers are green, music playback speakers are red, and sound effects speakers are cyan. Thus, if our designer is playing a sound effect cue, she can quickly identify which speakers she has set out for effects playback use.

2. There are very large numbers next to each speaker. The vocal system—being routed through the FOH console—has speaker labels "CS" for "Console Stereo," "CA1" for "Console aux 1," and so on. Music and effects playback are both being handled by QLab in this example, so the remaining speakers just have individual numbers with no letters next to them. These numbers correspond directly to the numbered outputs in our QLab patch, such that sending to that numbered value will send to that speaker alone. Note that all the text is color-coded, with black text for console-fed speakers, red text for music system speakers, and purple text for all fx speakers.

Additionally—not shown in our figure, but sometimes done in magic sheets—we will show speakers at larger than to-scale size, to make them even easier to see.

Note that even on this document, which is to be used during production but not typically submitted to any client, there *is still a legend!* The legend is unobtrusive but is there so that in the exhaustion of tech week, if a designer forgets how they encoded the colors of their speakers, they can easily remind themselves.[1]

Now, this works well for most theatrical projects, especially those with relatively low speaker counts. However, once we get into very complex designs—either for theater or themed entertainment—it can be difficult to have one magic sheet for the whole project, as it risks being overcluttered and impossible to read.

In this case, we take a page[2] from our SBD practices and separate out speakers by purpose and location. In a large single-room theatrical situation, we might have separate magic sheets for any subsystems—vocal, music, effects, sure . . . but also further subdivided, depending on how the system is actually configured. On the right production, I can imagine an entire plate of the magic sheet set dedicated solely to subwoofers.

In themed entertainment, it is very common to subdivide rides and the like into scenes, in which case our magic sheets would also be separated by scenes—so the first room of the entry queue might be Scene A, the next Scene B, etc., followed by Scene 1 as the first scene that riders pass through once they've actually boarded the ride, and so on.[3]

Now, we've covered the basic *drawn* elements of a document package. As with the rest of this book, we've hit the key points, and though I am sure there are some projects that require specialty drawings I haven't mentioned here, most system design projects won't require more technical drawing than what we've demonstrated in this section. It is time now to move into Section 14, covering the rest of the standard paperwork for a production process—material that is not drawn but is no less essential to documenting our work.

Notes

1 You may notice that the text is color-coded, but differently than the speakers themselves. This is due to the fact that the green and cyan text was difficult to read, even at large size, and AutoCAD doesn't provide a drop shadow option, so, readable colors were chosen.

2 Or . . . a *plate*? Let the Dad jokes continue!

3 This division is not just performed for the magic sheets, theme park rides with any sort of travel path (i.e., not flat rides) are often divided into scenes when being designed—it helps clarify the whole team's communication about the project. See Chapter 44 for more on this.

Hookups and Workbooks

In their simplest form, hookups (also known as *I/O Sheets*) are simply spreadsheets that track the signal flow of a system, from (you guessed it) input to output. In a basic analog system, this paperwork is very easy to generate, easy to read, and, generally, all of your signals can be shown on a single page.

As shown in Figure 53.1, we are documenting a system with four wired mics, patched into an analog console's mic inputs, which then connects via its main left and main right outputs to a single two-channel amplifier and from there to two speakers.

Input Devices:	Cable Type:	Console Inputs:	Console Outputs:	Cable Type:	Amp Inputs:	Amp Outputs:	Cable Type:	Speakers:
1-SM58 1	XLR	Mic In 1	Main Left	XLR	1	1	NL2	1-JF60
2-SM58 2	XLR	Mic In 2	Main Right	XLR	2	2	NL2	2-JF60
3-SM57	XLR	Mic In 3						
4-D112	XLR	Mic In 4						

FIGURE 53.1
Simple Analog System Hookup

This is easily laid out.

However, as we've spent countless pages detailing in this book, most contemporary sound systems are a good deal more complex than this example—so how do we lay out our connection info in a spreadsheet (or sheets) that is easy to parse, and more importantly, *why do we need to do so?* If we've already drawn a comprehensive SBD, what is the point of the hookup?

A well-organized hookup is helpful to a crew. Instead of working from a blueprint-sized plot, a crew can simply break a system down into component parts, hand sheets to various crew members, and tell them to plug things in as shown in the spreadsheet. As the crew works, they can easily check off any connection they have installed, and thus the hookup makes a very useful working document.[1] Now, of course, if your system features more than one mixing console, multiple stage boxes, outboard effects, wired and wireless mics, playback systems, computers, DSPs, and more, it can become challenging to show all of this info on one spreadsheet. In fact, it can be a terrible idea to try to show this much info all in one place, as it can easily become a soup of letters and numbers that not even the sharpest crew can utilize easily. Since the purpose of our paperwork is to make the work of actually installing the system easier, we must ask ourselves two basic questions before preparing the hookup:

1. What info is necessary for such a document?
2. How can it best be laid out to make it most useful?

Surprisingly for such a mundane document, there is wide disagreement throughout the industry about the answers to the earlier questions. Unlike drafting standards, where some designers will always vary from the recommendations because they believe their methods are better, in the world of hookups, the situation is the reverse. First—no professional organization has presented a coherent standard or recommendation for writing these documents.[2] Second—in talking with myriad professional designers, I've yet to meet one who feels their hookup is the best, and in fact most designers have expressed that they feel the hookup is a difficult document to make clear and utilitarian, as the temptation is always to include more on a page than perhaps is necessary.

As such, for the examples I lay out here, the practices I will recommend come both from my own experience and the best examples I've synthesized from a host of professional designers throughout the industry.[3]

First, let's consider what information is relevant to include on a hookup:

- Each device being connected—this might seem obvious, but it bears repeating: any device you need to connect as part of the system should in some fashion be represented in the hookup.
- Both physical patching info (input as labeled on the hardware itself, for example) and signal path info (console channel that corresponds to physical input)—on a smaller show, these pieces of info might be on the same sheet. On a large show, they may be divided into separate sheets within the overall spreadsheet workbook, such that, for example, all the wireless mics' physical connections from receiver to stage box are on one sheet, and all of the patching of those stage box inputs to console channels is represented on another sheet that only handles in-console configuration.
- Cable type or format—if you are patching wireless mics into a console system, and the patch lists the mic connecting directly to the stage box, it is generally assumed that the connection will be a microphone-level XLR cable. However, as equipment advances, some of these certainties get squishy—mics are now available that output Dante directly, for example, and the connection at the stage box might be to a Dante card. Therefore, it is important to state what type of signal/cable you are patching. If you identify a signal as "mic," we will assume XLR, but "Dante" and we'll assume network cable. Also, if you list Dante, we're *probably* connecting to a switch before going into the Dante card, so the switch connection sheet should make clear what's happening.
- Names of signals—mics should be labeled not just numerically (and the numbers should correspond to SBD, etc.), but should be labeled by actor or character (or both—depending on what the team is most comfortable with/how the A1 wants the channels labeled in the console). Speakers should be labeled by not just number (though those are important to include!) but also by use—main left, main right, front fill SL#1, etc.
- Signal paths—if we're working on a concert system, it's generally assumed that a microphone's signal will make its way to the house system. However, if a concert system is also serving as a recording system, feeding each mic out not just to the house and monitor systems, but via direct outputs to a recording rig, those direct out numbers should be labeled on a sheet with all of those mics.
- Notes—any other relevant info.

Now, in the UK-style workbook, we will include not only the hookups, but also the RF plans/schedules, IP schedules, and other information beyond a standard hookup. We will address the workbook-style document package more toward the end of this chapter, and we cover schedules in the next chapter.

In Figure 53.2, we have four sheets, where the input and output connections for both a console's stage box and a DSP are shown.

A

Input	Actor	Mic Receiver	Cable Type	SD rack input (rack. Card. Jack)	Direct Out	Mic Trans/ Element	Notes
1	Joe Schmoe	3532 RX #1	XLR	2.1.1		5212 TX w. 4061	
2	Jane Schmoe	3532 Rx #2	XLR	2.1.2		5212 TX w. 4061	
3	Fake Namerson	3532 RX #3	XLR	2.1.3		5212 TX w. 4061	
4	False Nomen	3532 RX #4	XLR	2.1.4		5212 TX w. 4061	
5	Not Real	3532 RX #5	XLR	2.1.5		5212 TX w. 4061	
6	Totally Real	3532 RX #6	XLR	2.1.6		5212 TX w. 4061	
7	Denise NoName	3532 RX #7	XLR	2.1.7		5212 TX w. 4061	
8	Joke NotFunny	3532 RX #8	XLR	2.1.8		5212 TX w. 4061	

B

Console o/p path	Detail	SD rack output (rack. Card. Jack)	SD rack output secondary (rack. Card. Jack)	DSP Destination	Notes
Aux 1	Prosc Music Left	3.10.1		BSS BLU-806 1 A1	
Aux 2	Prosc Music Right	3.10.2		BSS BLU-806 1 A2	
Aux 3	Prose FX Left	3.10.3		BSS BLU-806 1 A3	
Aux 4	Prose FX Right	3.10.4		BSS BLU-806 1 A4	
Aux 5	House Subs Left	3.10.5		BSS BLU-806 1 B1	
Aux 6	House Subs Right	3.10.6		BSS BLU-806 1 B2	
Aux 7	Surrounds Left	3.12.1		BSS BLU-806 1 B3	
Aux 8	Surrounds Right	3.12.2		BSS BLU-806 1 B4	

FIGURE 53.2
(A) Console Stage Box 1 Input Sheet, (B) Console Stage Box 1 Output Sheet, (C) DSP I/O Sheet

		Inputs				Outputs		
			BSS BLU-806-806 1					
#	**Card/in**	**From SD**	**Detail**	**#**		**From**		**Detail**
1	A1	3.10.1	Aux 1	1	C1	A1		Prosc Music Left
2	A2	3.10.2	Aux 2	2	C2	A2		Prosc Music Right
3	A3	3.10.3	Aux 3	3	C3	A3		Prosc FX Left
4	A4	3.10.4	Aux 4	4	C4	A4		Prosc FX Right
5	B1	3.10.5	Aux 5	5	D1	B1		House Subs Left
6	B2	3.10.6	Aux 6	6	D2	B2		House Subs Right
7	B3	3.12.1	Aux 7	7	D3	B3		Surrounds Left
8	B4	3.12.2	Aux 8	8	D4	B4		Surrounds Right

C

FIGURE 53.2
Continued

In this example, we've broken the I/O info down both by device and then further by input versus output. Each device is labeled and numbered, their subdivisions (in this case, card slots) are also labeled and numbered, and each connection is detailed. As such, this document becomes an easy key for connecting the hardware.

This provides an answer to our question #2 (from earlier in the chapter). In a system that is anything beyond extremely simple, the clearest paperwork tends to divide the connections up first by location (so, within an overall Excel document, all gear at FOH would be on sheets grouped together, all gear in the machine room likewise, etc.), then by piece of equipment, and then by input and output.[4] I have often seen young designers attempt to represent every single connection in a system on one long sheet, without dividing by location and then piece of equipment, and these documents become unreadable in short order. Even if you will be installing your own work, and don't feel like dividing info onto separate pages (sheets) of an Excel workbook, it is still helpful to subdivide by location and piece of equipment within your long list, as it will be easier to track the work as it is completed.

In addition, a hookup in the digital age needs to feature some info about routing. Particularly when it comes to digital consoles, since any physical input can be routed to any input channel or path, and likewise any output path can be routed to any physical output, there needs to be a map of how this gets configured.

In Figure 53.3, we see each physical input, and what channel path it corresponds to, and each output path in use, and what physical output it corresponds to.

SD rack input (Rack.Card.Jack)		Source	Input channel	Output path	SD rack output (Rack.Card.Jack)	
2.3.1	Another Actor	5212 TX w. 4061	Channel 13	Aux 15	Surrounds Down Left	4.10.1
2.3.2	Actor deActist	5212 TX w. 4061	Channel 14	Aux 16	Surrounds Up Left	4.10.2
2.3.3	Actor McActorson	5212 TX w. 4061	Channel 15	Aux 17	Surrounds Sub Left	4.10.3
2.3.4	Acting Againist	5212 TX w. 4061	Channel 16	Aux 18	Surrounds Down Right	4.10.4
2.3.5	Actist Actology	5212 TX w. 4061	Channel 17	Aux 19	Surrounds Up Right	4.10.5
2.3.6	Acty Actorist	5212 TX w. 4061	Channel 18	Aux 20	Surrounds Sub Right	4.10.6
2.3.7	Sucha Greatactor	5212 TX w. 4061	Channel 19	Aux 21	Plenum Rumbler	4.10.7
2.3.8	Another Actor	5212 TX w. 4061	Channel 20	Aux 22	Underbooth Special	4.10.8

FIGURE 53.3
Console Routing Sheet

A hookup, at the end of the day, is exactly what it is named—a description of all the hookups between system components. A good hookup is easy to read and clearly shows each piece of signal equipment and what physical connections need to be made to it in order for it to function as intended in the design.

A *workbook* (in technical documentation terms, not just an Excel workbook, which is what they call any working spreadsheet document) is a document set that contains the hookups, the equipment schedules,[5] and despite their provocations to the contrary, some simple SBD-like documents, subdivided by function (e.g., a MIDI signal flow document, a network signal flow document). The workbook may also contain such info as the RF mounting details, speaker rigging descriptions (if mounting details are not drawn, a list of them is often produced), intercom assignments, etc. A good workbook is basically a set of documents that represent a comprehensive system plan, minus the plan and elevation view drawings. Anything that is needed for the production paperwork will be included. On some productions, this might also be called the "show bible." As can be gleaned from that latter moniker, the "workbook" style of paperwork is very theatrically oriented. Instead of a workbook, an architectural project will simply feature the entire document package, which will include all relevant details, as well as details well beyond the AV department's purview. Where a theater production's sound team will not generally collect all the scenic, costume, prop, lighting, sound, and video design paperwork in one place for printing, architectural design projects always work toward this goal.

At any rate, however, the information will be packaged, a hookup can be a very useful tool. Construct it carefully, subdivide it into meaningful trunks, and share it!

Next, we'll examine . . . more spreadsheets! The wonders never cease!

Notes

1 If you've done much theater or concert work, you might wonder why a large plot would be onerous to work from. In fact, this is still how most lighting teams work, with the Master Electrician highlighting/checking off units as they are installed, cabled, and focused. Some sound teams do still work from a master plot, but it's still often easier, with the myriad of complex connections in a contemporary sound system, to give crew members their own section of the hookup, from which they will work and then check back in with the A1, who keeps a master list of what has been connected either on their own comprehensive copy of the hookup, or the SBD, or both.

2 This is about to change, as we're including one in the upcoming release of the USITT Sound document recommendations.

3 My gratitude to Joanna Lynne Staub and TSDCA, who have been collecting examples of paperwork for our USITT recommendations project, and as such who have provided me a valuable cross-section of working practices in the paperwork field. My gratitude to Joanna also for her own paperwork for the 2018 remount of *Angels in America* on Broadway, which contains very elegant hookups, upon which I've based much of my writing in this chapter. *Angels in America* credits—Sound Design: Ian Dickinson for Autograph, Associate Sound Design: Joanna Lynne Staub & Maggie Burke.

4 In the case of two-way network connections, the I/O distinction is irrelevant, and so connections are simply broken down by device.

5 See Chapter 54.

Equipment (Hardware, Cable, RF, IP, Etc.) Schedules and Facilities Impact Reports

Equipment schedules are lists of equipment. Chapter finished!

. . . OK, so I brought that feeble joke back. It's been a long book, cut me some slack.

Truly, though, equipment schedules are indeed lists of equipment. In their simplest form, the hardware schedule, they are just lists of all the hardware devices used in a design, with unit counts for each item. In their most complex form, they may represent settings that each device needs to be configured with in order to operate as intended, as in RF schedules, which list frequencies of transmission and other info. Equipment schedules of various types are used in all manner of sound system projects, from performance design to architectural installations.

A Facilities Impact Report (or FIR) is a document that identifies the "loads" that the system presents in each of its sub-locations. These documents are found most often in permanent installation projects, though also occasionally on large-scale performance projects like a massive Cirque du Soleil-style spectacle.

In this chapter, we'll identify the most common types of equipment schedules and what info is generally presented within them, as well as explain the process of assembling an FIR for an installation project. All of these documents are typically presented as spreadsheets.[1] Well-organized schedules and FIRs can make a big difference to how organized a production process will be, and poorly organized versions of these documents can cause major headaches.

SCHEDULES

There are a few basic types of schedule documents that are commonly created for sound system projects:

- Hardware Schedule—aka "Equipment Schedule."[2] This is a list of each piece of hardware to be used in the system, with unit counts. This list is used in a few different ways—it allows the designer to compare their own paperwork (SBD, for example) with a simple list, and make sure they've got everything both represented on the drawing and listed in spreadsheet form. This list makes price estimation easier for the bidding shops. It makes packing and delivery of the ordered gear simpler, as it can act as a simple checklist. As with all of our schedules, it is a working document that eases certain pinch points in the production process. We see an example in Figure 54.1. The C and O in the "Responsibility" column are "Contractor" and "Owner." These acronyms mean "__Furnished, __Installed" and delineate who provides a piece of equipment and who installs it.

Equipment Schedule			
Equipment Type:	Qty:	Responsibility:	Notes:
HP 2530–48G-PoE+ Switch	2	OFCI	
Dell R7610 Rack PC	1	OFOI	
Dell DKMMLED185-G01 KMM	1	OFOI	
VisionTek Connect PCIe card 4x USB 3.0, Dell Part # A6246530	1	CFCI	
QSC Q-Sys Core 510 DSP	2	CFCI	
QSC Q-Sys MTP-128 128-channel audio playback package	1	CFCI	
QSC MD-S "130 hr" audio playback storage drive	1	OFCI	
Q-Sys CIML4 Mic/Line Input Card	1	OFOI	
Middle Atlantic HBL6 machineable blank panel for EER TSC-8 rack mounting	1	OFOI	
Middle Atlantic BGR-45SA-27 rack 45 ru	1	CFCI	
Middle Atlantic D3 3ru rack drawer	1	CFCI	
Gefen Pro HDFST-MOD-16416-HDELR 16x16 HDMI to Cat6 matrix/extender with (16) ELR-POL-R 1-channel Cat6 to HDMI receivers included	1	OFCI	
Gefen EXT-DVI-1CAT6 DVI over Cat6 transmitters (receivers are included in the price and are situated in Scene 3 for Tablets 1 and 2)	2	OFOI	
Tripp-Lite B140–101X DVI over Cat6 transmitters (receivers are included in the price and are situated in Scene 3 for Tablets 1 and 2)	2	OFOI	
Tripp-Lite B126–1A1SR HDMI over Cat6 transmitters (receivers are included in the price and are situated in Scenes 6 and 7 for video devices)	16	CFCI	
QSC CXD4.2Q-NA Power Amplifiers	11	CFCI	
QSC CXD4.3Q-NA Power Amplifiers	1	CFCI	
QSC CXD4.5Q-NA Power Amplifiers	2	CFCI	
QSC CX168 Power Amplifiers	2	CFCI	
QSC CX254 Power Amplifiers	4	OFCI	
QSC PS1600G Paging Station	1	OFOI	
JBL Control 29 AV speakers	12	OFOI	
JBL AW526-LS	20	CFCI	
JBL Control 328C speakers + backbox	25	CFCI	

Equipment Schedule			
JBL Control 14C/T speakers + mounting	11	CFCI	
JBL Control 23 speakers	3	CFCI	
JBL ASB6125 speakers	3	CFCI	
JBL Control 67 P/T speakers	41	CFCI	
JBL AM7200/64 speakers	1	CFCI	
JBL CBT 50LA-1 speakers	2	CFCI	

FIGURE 54.1
Sample Equipment Schedule

Source Location:	Source Device:	Source J-Box #:	Destination Location:	Destination Device:	Destination J-Box #:	Cable Type:	QTY:	Cable Type Length Tolerance (Max Feet):
EER AVC Rack #2	Power Amp #1	J2A	Scene A	JBL CBT 50LA-1	J2B	2-wire speaker cable	1	TBD depending on wire gauge
EER AVC Rack #2	Power Amp #1	J2A	Scene A	JBL CBT 50LA-1	J4A	2-wire speaker cable	1	TBD depending on wire gauge
EER AVC Rack #2	Power Amp #1	J2A	Scene B	JBL Control 67 P/T	J2C	2-wire speaker cable	2	TBD depending on wire gauge
EER AVC Rack #2	Power Amp #2	J2A	Scene 2	JBL Control 328C	J2E	2-wire speaker cable	4	TBD depending on wire gauge
EER AVC Rack #2	Power Amp #2	J2A	Scene 2	JBL Control 328C	J2E	2-wire speaker cable	1	TBD depending on wire gauge
EER AVC Rack #2	Power Amp #3	J2A	Scene 6	JBL Control 328C	J2I	2-wire speaker cable	3	TBD depending on wire gauge
EER AVC Rack #2	Power Amp #4	J2A	Scene 4	JBL Control 328C	J2G	2-wire speaker cable	3	TBD depending on wire gauge
EER AVC Rack #2	Power Amp #5	J2A	Scene 6	JBL Control 328C	J2I	2-wire speaker cable	3	TBD depending on wire gauge
EER AVC Rack #2	Power Amp #5	J2A	Scene 7	JBL Control 328C	J2J	2-wire speaker cable	1	TBD depending on wire gauge

FIGURE 54.2
Sample Cable Schedule (A Small Part of a Very Long Document)

Source Location:	Source Device:	Source J-Box #:	Destination Location:	Destination Device:	Destination J-Box #:	Cable Type:	QTY:	Cable Type Length Tolerance (Max Feet):
EER AVC Rack #2	Power Amp #6	J2A	Scene 7	JBL Control 328C	J2J	2-wire speaker cable	1	TBD depending on wire gauge
EER AVC Rack #2	Power Amp #6	J2A	Scene 8	JBL Control 328C	J2K	2-wire speaker cable	2	TBD depending on wire gauge
EER AVC Rack #2	Power Amp #7	J2A	Scene 5	JBL Control 328C	J2H	2-wire speaker cable	3	TBD depending on wire gauge
EER AVC Rack #2	Power Amp #7	J2A	Scene 6	JBL Control 328C	J2I	2-wire speaker cable	1	TBD depending on wire gauge

FIGURE 54.2

Continued

A — RF Schedule

Mic #	Transmission Frequency	Character	Actor	Element	Cap/Color	Transmitter	Receiver	Fit/Rigging:	Notes:
1	554.025	Hero	Bob	DPA 4061	Beige/Beige	ULXD1 - 1	ULXD4Q - 1	Crown/halo	
2	555.025	Villain	Steve	DPA 4061	Beige/Brown	ULXD1 - 2	ULXD4Q - 1	L ear, tape, blonde wig clip	
3	557.025	Guard #1	Harry	DPA 4061	Black/Beige	ULXD1 - 3	ULXD4Q - 1	Changes (see bible)	
4	558.025	Guard #2	Lucy	DPA 4061	none/Black	ULXD1 - 4	ULXD4Q - 1	R ear, floral/blk Hellerman	
5	560	Best Friend	Angela	DPA 4061	Beige/Beige	ULXD1 - 5	ULXD4Q - 2	Crown/halo	
6	561.025	Sidekick	Garth	DPA 4061	Beige/Beige	ULXD1 - 6	ULXD4Q - 2	Hatbrim, elastic web	
7	563.025	Dancer	Aaron	DPA 4061	Beige/Brown	ULXD1 - 7	ULXD4Q - 2	Crown/halo	
8	564.025	Bartender	Liam	DPA 4061	Black/Beige	ULXD1 - 8	ULXD4Q - 2	Crown/halo	
9	565.025	Ghost	Mark	DPA 4061	none/Black	ULXD1 - 9	ULXD4D - 1	L ear, visible boom	
10	567.025	Ghoul	Joe	DPA 4061	Beige/Beige	ULXD1 - 10	ULXD4D - 1	L ear, visible boom	

B — Mic Mounting Guide for One Actor

ACTOR PHOTO HERE w. MIC RIGGING SHOWN

Character:		Actor:	
Transmitter #:			
Transmitter Model/#:			
Transmitter Gain:			
Transmitter Squelch:			
Transmitter Frequency:			
Element Model:			
Element Color:			
Element Cap:			
Rigging Description:			
Rigging Accessories:			
Headpiece/Hat/Wig:			

C — RF Tracking Sheet

Act	Scene	Bob	Steve	Harry	Lucy	Angela	Garth	Aaron	Liam	Mark	Joe	Notes:
1	1											
1	2											
1	3											
1	4											
1	5											
1	6											
1	7											
1	8											
1	9											
1	10											
2	1											
2	2											
2	3											
2	4											
2	5											
2	6											
2	7											
2	8											
3	1											
3	2											
3	3											
4	1											
5	1											
5	2											
5	3											
5	4											
5	5											
5	6											
5	7											
5	8											

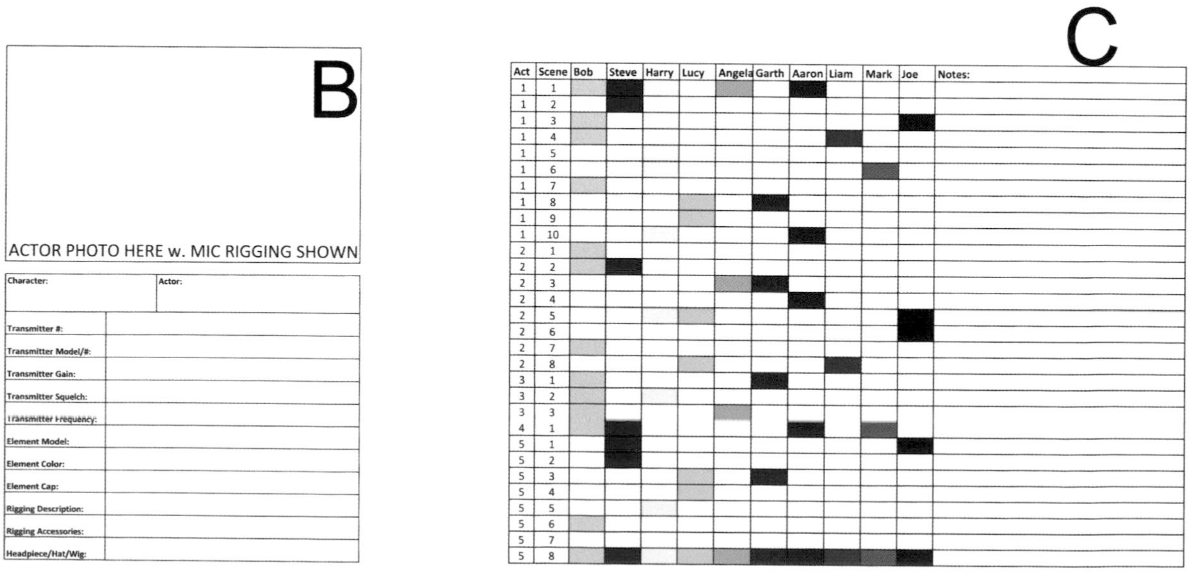

FIGURE 54.3

Sample (A) RF Schedule; (B) Mic Mounting Guide for One Actor; (C) RF Tracking Sheet

- Cable Schedule—a list of every cable to be used in the system (sample in Figure 54.2). The use of this document varies, but the general purpose is twofold: 1) make sure the proper cables are ordered, in the proper lengths, and with the proper terminations,[3] and 2) to list important facts about the cables that may be relevant to the project—like maximum length of cable type before signal degradation, and the like. When specifying, say, mic cables for a concert tour, one simply lists quantities of cables of varying lengths as desired, and the shop prepares a package with those cables at the ready. However, when installing permanent network or fiber-optic runs through a theme park, more info is needed: which junction box does the cable terminate at? What type of fiber (multi-mode or single-mode, what termination type)? As such, this document will vary in complexity quite a bit from project to project. For the designer, it is mainly a way to think through all the signal runs and make sure you know how the signal will get from place to place, and how much cable you need in order to accomplish those runs.
- RF Schedule—aka Wireless Schedule (Figure 54.3). This is a document listing every wireless mic and IEM, with unit numbers, user names (actor/character/musician, etc.), transmission frequencies, unit types for mic elements or in-ear monitor elements, unit types for transmitters/receivers, and any other relevant info (element color, cap type if relevant to mic model, etc.). This is first and foremost used for the setup of wireless world, so mics are assigned to their designated frequencies, transmitters and receivers are labeled with the unit number designated in the SBD, elements are associated with the proper packs, etc. As the crew works on the show, they will either expand this schedule or create a parallel set of documents with further RF info—pack settings (gain, squelch, etc.), mounting details for mic elements (including photos), and so on. In addition to these documents, for theatrical work, an RF run sheet is created, that shows which mics are on which performers in which scenes, and if necessary, where mics switch from one actor to another, or where there is a costume quick change that means the mic needs to be re-rigged, or the like.[4]
- IP Schedule (Figure 54.4)—a list of all network-connected devices in a system, and their IP addresses, and (if needed) subnet masks and other relevant network settings (e.g., DHCP for many Dante devices that don't allow direct IP setting).

IP Schedule		
Computer	**IP Address**	**Subnet Mask**
Yamaha CL5	192.168.0.5	
Yamaha RIO-32	192.168.0.12	
Qlab Main	192.168.5.17	
Qlab Backup	192.168.5.100	
LX Board—ETC Ion	192.168.10.14	
Mainstage A	192.168.5.43	
Mainstage B	192.168.5.44	
DSP Control Computer	192.168.5.11	
DSP 1	192.168.7.70	
DSP 2	192.168.7.94	
Designer Laptop	TBD	
Associate Des. Laptop	192.168.7.23	
RF Station SL	192.168.7.111	

FIGURE 54.4
Sample IP Schedule

As with anything, the larger the project, the more complex these schedules get. The Super Bowl has a much larger and more complex RF and IP schedule set than a 4-person straight play performed in a 99-seat house. On large projects like the Super Bowl, often the RF schedule will be the job of a dedicated RF supervisor, and the IP schedule likewise will fall to a dedicated IT team—rather than forcing the sound system designer(s) to do all of that work by themselves. However, for the vast majority of projects, this documentation is still generated by the sound system designer.

Anything that can be listed can be the subject of a schedule document if the team desires. There could be an amplifier schedule on a project featuring a ton of amps—especially if they are also network connected—for example, some amps in the Crown line of products feature both the ability to connect for network control over Ethernet, but also to send and receive audio data via the proprietary BluLink protocol. As such, each amp will have both an Ethernet control address and a BLULink address. These could be listed as part of the general IP schedule, or it could form the core of a dedicated amplifier schedule. An amp schedule isn't a typical document, but if you or your team find it useful to subdivide the information that way—more *power* to you.[5]

On large installation projects, we might find J-Box schedules (a list of all installed Junction Boxes and their connections), a trigger schedule (all sensors/triggers in a design, their connections, and their data), and so on and so forth into the horizon. Any type of gear that can be coherently subdivided out of the system for examination may well be the subject of its own schedule.

FACILITIES IMPACT REPORT

On large-scale installation projects, it is frequently the case that either a facility is being built new for the project, or an existing facility is being updated and retrofitted for the project. As such, the team needs to know how much power is needed (and in what form—how many circuits, what amperage delivered, what types of plugs), what the HVAC requirements are to keep gear operating smoothly, how heavy a given rack will be, and so on. A facilities[6] impact report (Figure 54.5) is a document designed to produce that information in a clear and easy to read format.

The core of this document is whatever space or spaces will hold the most rack-mounted power and control equipment. It is rare, for example, to find an FIR detailing the loads of speakers flown around a theater—though that info will be provided somewhere, either on the plot, or in elevation, or wherever is easiest for the TD to analyze it and approve weight loads—but it is common to find some form of this document for FOH or a machine room. Again, as I said at the beginning of this chapter, the true FIR isn't a standard document for theatrical or concert work but is a staple of installation design.

What is included in a FIR?

- Equipment broken down first by room/venue and second by rack (if there are multiple racks in one space).
- Unit counts for each piece of equipment.
- Power use in watts per piece of equipment.
- Current draw in amperes per piece of equipment.
- Heat generation in BTU[7]/hr per piece of equipment (some gear will actually list this in specs, for others you must convert from watts—multiply watts by 3.41 and you have a decent estimate of BTU/hr). This figure is important for the HVAC engineers to ensure your gear won't overheat.
- Rack spaces per piece of equipment.
- Weight per piece of equipment.

Using these figures, we also then tally total weight per rack (not forgetting to add the weight of the rack itself), the circuits needed (as a factor of the overall amperage needed in the space), total heat given off in the space, and so on. Different projects will demand different information, as well—for some installation projects, the FIR is the appropriate place to situate information about speaker weight so that the structural engineers can see what weight load you are demanding for your in-ceiling speakers and whatnot.

EER item:	Quantity:	Power/unit (watts):	Weight/unit (lbs):	Heat/unit (BTU/hr/unit):	Current draw (amperes):	Rack spaces/unit:	Notes:
Q-SYS Core 510 Audio DSP Unit	1	370	23	600	3.7	2	CORE 510 serves as show controller and audio playback server as well as DSP
Q-SYS Core 510 (as I/O)	1	370	23	600	3.7	2	
HP 2530–48G-PoE+ Gigabit Switch- managed 48 port	2	300	10.4	1,024	2.5	1	
QSC TSC-80w-G2 Touch Screen	1	180	6	615	1.5	6	requires Middle Atlantic 6 ru blank panel and punch out to mount as if blank panel were "wall"; weight includes panel estimate
Dell R7610 Rack PC (for configuring control system and devices)	1	1,100	45	3,754	9.17	2	
Dell DKMMLED185-G01 Keyboard Mouse Monitor Interface	1	20	10	68.5	0.17	11	~11 RU when open, just over 1 RU when closed
Tripp-Lite SU2200RTXLCD2U Uninterruptible Power Supply	2	200	48.5	683	16	2	
Tripp-Lite BP48V60RT-3U External Battery Pack	2	0	0	0	0	3	
Gefen EXT-DVI-1CAT6 DVI over Cat6 transmitters	2	10	3.5	34	0.083	0.5	receivers are powered via POE from transmitter, transmitters are small, half- rack style units, to be zip tied or otherwise connected to the rear of the rack
Gefen Pro HDFST-MOD-16416-HDELR HDMI-CAT6	1	175	25	598	1.45	3	
QSC CXD4.2Q-NA Power Amps	11	1,200	18.5	4,095	10	2	
QSC CXD4.3Q-NA Power Amps	1	2,250	21	7,677	18.75	2	
QSC CXD4.5Q-NA Power Amps	2	3,500	22	11,943	37.5	2	
QSC CX254 Power Amps	4	696	21	2,375	5.8	2	
QSC CX168 Power Amps	2	600	21	2,048	5	2	

FIGURE 54.5
Sample EER FIR

EER item:	Quantity:	Power/unit (watts):	Weight/unit (lbs):	Heat/unit (BTU/hr/unit):	Current draw (amperes):	Rack spaces/unit:	Notes:
Calculate Circuits:		**Total EER power (watts):**	**Total EER rack gear weight (lbs):**	**Total EER heat load (BTU/hr):**	**Total EER current (amperes):**	**Total EER rack spaces:**	
		29,474	619.3	99,255.5	293.806	75	
40 amp x 2 circuits (5Q amplifiers)							
20 amp x 22 circuits (amplifier per circuit for all non-subwoofer amps),							
20 amp x 2 circuits (?C, processing, switches, etc. EER Rack 1)							

Now that we've prepared schedules and impact reports, we're done with spreadsheets, right?

Oooo, so close. No, we've got one more major spreadsheet to address: budgets![8]

Notes

1 Thrilling!

2 Even though there is equipment—cable—that is not counted in the equipment schedule, this name still sticks on some projects.

3 If the cables are arriving pre-terminated and not being field-terminated. For rental shop orders being used in a live production system, generally, cables arrive pre-terminated. For installation projects, cables are often terminated to exact lengths on-site to avoid coiled slack in dropped ceilings and the like.

4 As a rule, when you can afford enough mics, don't make actors share mics—moving mics during production isn't fun, and can cause problems if someone misses a cue or a change takes longer than expected. That said, in smaller regional theater productions, this sharing of mics still happens more often than we'd like.

5 I will NEVER tire of Dad jokes!

6 I am using the plural form here on purpose, as most FIRs are done for projects with multiple sub-spaces within the project, like an EER or two, an operator station, and different scenes for a theme park ride. As such, the document represents multiple *facilities* in one workbook. That said, if you want to make this singular (especially if you only have one space for which this data is relevant), feel free.

7 British Thermal Units, a standard measurement of heat radiated from equipment.

8 Dum dum *duuuuummmmmm!*

CHAPTER 55

Budget Documents

At last, the spreadsheet to end all spreadsheets, the budget!

So, you've figured out how much your system is going to cost to buy or rent. You've identified areas that are critical and those that are less so, armoring yourself for VE. You now have to present your work to people with actual money who can say yea or nay to your wild ideas. How best to do this?

The form should be different in different circumstances. A performance project (theater, concert, etc.) with a rental system is likely to feature a less detailed budget than one for a large installation project. There are a couple of reasons for this, chiefly that in many circumstances, a rental house will not give you a line-item budget for renting each individual item but will rather quote you a package price for the entire complement of gear being rented. This has some advantages and disadvantages: on the plus side, it makes your paperwork easier. While you still generally want to list all the major pieces of gear in a budget estimate of this type, you don't generally need to show the client every little cable by type but can (for example) just list "cable" as a line. In the pricing column, you may simply have one number (the rental price as quoted) or at most three numbers—the rental price for gear, the cartage rate (for pickup and delivery of the gear) and the installation rate (if the rental house crew is employed to do this work).

For installation projects, there is often considerably more information included in the budget estimate. First, since we assume that we are budgeting at MSRP, we have the purchase price column per unit of gear, multiplied by the unit count, for a type of gear subtotal column. We may subdivide the design by scene or location (as in the case of theme park rides and other major installations, where we want to know how much we are spending in each subspace of the project—see Figure 55.1 for an example). If a project has been budgeted, and then there have been change orders placed (added gear for added functionality, after the initial budget estimate, associated specifically with official change orders in the design process), we may need to subtotal the initial design and the change orders separately, so a client can see what those change orders will cost in and of themselves. Sometimes we need to list whether a piece of equipment is "contractor furnished, contractor installed" or "contractor furnished, owner

Project X Budget Estimate					SCENE HARDWARE TOTAL:	SCENE INSTALLATION ESTIMATE (30% of equipment costs--very rough estimate to include cable, programming, etc):	SCENE TOTAL COST:	
SCENE NAME:	ITEM:	QTY:	COST/UNIT:	ITEM TOTAL:				NOTES:
EER	HP 2530-48G-PoE+ Switch	2	$3,070.17	$6,140.34	$52,310.55	$15,693.17	$68,003.72	
	Dell R7610 Rack PC	1	$2,985.33	$2,985.33				
	Dell DKMMLED185-G01 KMM	1	$1,449.99	$1,449.99				
	VisionTek Connect PCIe card 4x USB 3.0, Dell Part # A6246530	1	$29.99	$29.99				
	QSC Q-Sys Core 1100 DSP	1	$15,860.00	$15,860.00				
	QSC Q-Sys MTP-128 128-channel audio playback package	1	$7,000.00	$7,000.00				
	QSC Q-Sys CIML4 Mic/Line Input Card for Core 1100 (connection from Parkwide Audio System)	1	$300.00	$300.00				
	QSC MD-S "130 Hour" audio playback storage drive	1	$1,670.00	$1,670.00				
	QSC Q-Sys IO8 frame--Original Scope	1	4170	4170				
	QSC Q-Sys CODP4 Dataport/Line Output Card (to CX168 and CX702 amps)	8	$460.00	$3,680.00				
	QSC DPC-10 Dataport 10' cables	16	$30.00	$480.00				
	QSC Q-Sys TSC-8-BK Touch Screen Panel	1	$3,089.00	$3,089.00				
	QSC Q-Sys TSC-8-BX Back Box for touch screen	1	$249.00	$249.00				
	Middle Atlantic HBL6 6 RU blank panel (machineable) for TSC-8 mounting	1	$50.00	$50.00				
	Middle Atlantic EB1 1 RU blank panel for spacing	19	$11.28	$214.32				
	Middle Atlantic EB6 6 RU blank panel for future development spaces	2	$31.98	$63.96				
	Middle Atlantic DGR 45SA 27 45ru equipment rack	3	$1,552.16	$4,656.48				
	Middle Atlantic D3 3RU Rack Drawer	1	$222.14	$222.14				

FIGURE 55.1
Purchase Budget Estimate for an EER

installed" or "owner furnished, contractor installed"—meaning that if we purchase gear from the contractor and they are doing the install work, it is listed as CFCI, but if we purchase from them and we install it (as is common on large theatrical projects that still might buy gear but will often put it in place themselves), it would be listed as CFOI, and so on.[1]

Beyond these subdivisions, we sometimes further itemize by rack (so we know what each rack full of gear costs), by purpose (so we can know what our FOH system costs versus our monitor system), and so on.

The general goals of any budget paperwork should be threefold:

1. Present the budget in a clear, easy-to-understand format.
2. Demonstrate what value is added to the system in each sub-area by showing what each sub-area costs.
3. Never undercut costs.

It is tempting, especially when starting out, to produce a document that shows the absolute lowest price you can get. For some enterprises (regional theater comes to mind) this may, in fact, be requested—so the team can assess real numbers. However, it is always wise to leave a little flexibility in your budget so that when push comes to shove, you can still accomplish what you set out to accomplish, even after some cuts.

Clear headers, color-coding, frequent and logical subtotals, and the like, make for good clear budget estimates. If data needs to spread across multiple pages, make sure your column headers reappear at the top of each subsequent page so that it is easy to understand what you are looking at without scrolling back to the top.[2] There isn't much more to say about budget paperwork that hasn't been said in our earlier chapters on preparing the budget itself—just make sure your paperwork is clean and easy to read, and that your math is correct.

Now, let's examine a document that is very common for architectural and, especially, themed entertainment design, but not common at all in theatrical or concert-style work: the technical narrative.

Notes

1. Sometimes this info is not listed in the budget but instead in the equipment schedule. Note that OFOI might appear on a schedule to indicate gear the owner is counting on using in the design that will come from their stock. This would not be seen on a budget worksheet, as already-owned gear that will be installed by the owners and thus will not cost extra in any way is not relevant to estimating costs.
2. This is good advice for any spreadsheet that takes multiple pages.

CHAPTER 56

Technical Narratives

Technical writing comes in two general forms—writing that is highly technical and geared toward an audience of specialists (e.g., most research on sound technology, and many of the advanced engineering and design books on the market), or writing that is meant to communicate technical information to a *non*-technical audience (e.g., user guides for home audio equipment). Technical narratives fall into the latter category.

The purpose of a technical narrative in our business is generally to communicate the operating goals and major functions of a system to an audience of clients—an audience that needs to know what we're delivering, but who won't understand anything about equipment models, network protocols, or the like. These documents are found most often in the production processes for themed entertainment, and they are often paired with the *creative narrative*, a document walking the client through the story beats of a theme park ride, scene by scene. Technical narratives are sometimes also used in non-themed projects that feature some measure of interactivity, as clients often want a description of how the interactivity works.

Writing such a document can be surprisingly difficult. How do we walk the line between providing enough detail that the client understands what we're going to do with our sound system, without filling a document with specs? Let's walk through an exercise in creating such a document, as an example of the process.

Message in a Bottle

I've learned that despite having all these great technical words, when we write a narrative, we should try to distill it to plain English as much as possible. *Plain English*. It's like—we're providing a loudspeaker system so the loudspeaker goes to a decibel level. The dBs is about *this*—to give you an idea, a rock concert is about *this* in comparison. Really, literally plain English. I think that's really important to convey to your clients.—David Kotch

We are in a conference call.[1] We have been brought aboard to design an audio system for a new stunt show, based on a famous Korean action movie for a theme park in Seoul. The movie is unavailable in the states, so we can't watch it for reference, but that doesn't particularly matter since we're not in charge of the story or content design. The creative director walks us through the show as it is conceived so far—it is early in the process, pre-CD phase, so all of this could change, but we still need to come up with a CD design and a technical narrative of that design, so we must listen closely:

> The basic story is about our hero, Min-jun. He is a crime-fighting detective with a "wild streak." At the start of our story, we're inside a bank. Customers are lined up to do transactions and stuff. Suddenly, BANG! A jeep or a Humvee or something crashes through the wall of the bank, upstage right. Bricks explode onstage, dust everywhere, and a bunch of masked bandits jump out. They stick everyone up and steal a ton of money and everyone's jewelry. A teller trips an alarm but then gets shot. The bandits escape, leaving their jeep (which we realize they stole) halfway in the building.

> Min-jun arrives with the police. He surveys the scene, and after finding a few clues finds out that this is the work of the "Death Gang." They are serious baddies, so he's going to have to rely on his ACTION TEAM (this is the working name of the stunt show) to find and capture the culprits.

> So, the stage rotates, and we are now at the baddies' hideout, which is sort of like a bamboo treehouse over a lagoon, but with one of the walls removed, so we can look inside. There are a few snipers posted in trees nearby, and as the baddies talk about what they are going to do with their haul, one of them mentions their rival Min-jun. "I hope he doesn't find us here." The leader of the baddies calls the one who mentioned Min-jun a coward and throws him out of the treehouse into the lagoon, where angry crocodiles eat him voraciously (lots of water jet fx). The baddies laugh.

> The stage rotates again, and we see Min-jun with his action team, who all enter in ways that set an example of their special skills—like, the taekwondo expert does a flip kick into the room, the explosives expert enters in a cloud of flash powder, and so on—details TBD. The team is five people in all. They talk and realize that through their work, they have narrowed down the DEATH GANG's possible hiding place to one spot—the lagoon. They plan a raid on the lagoon.

> The rest of the show rotates back to the lagoon and consists of the ACTION TEAM invading the DEATH GANG's hideout. There are many explosions, including water explosions, the snipers are felled from their trees and splash down into the lagoon (angry crocodiles feed), Min-jun arrives on a jet-ski, dodging crocodiles. The battle rages in and around the treehouse until finally Min-jun and the leader of the Death Gang have it out, with Min-jun winning, the treehouse being exploded into the air, and the stash of cash and jewels safely rescued (before the treehouse explodes, of course).

> The show is about 20 min and has 30 min to reset between each showing.[2]

The content designer on the phone call says, "Yeah, so, we need a lot of explosions, lots of power, mics for everyone, some music playback, crocodile effects . . . you get the gist." That is the sum total of the creative designer's statement about sound effects at this stage. You have a rough groundplan, and that's it. Oh, and like most stunt shows, this takes place in a small outdoor arena, seating about 700 people.

So! First, we must rough in a concept design for our system. We need vocal replacement—no one cares if we localize to speakers for a stunt show. We need waterproof wireless vocal mics—if we recall from earlier in the book, there's really only one model (made by Lectrosonics) that will do the trick here. We need an FOH mixer and engineer. We need a playback server of some kind for effects and music, but we know we probably need both a live operator (because actor timing isn't always consistent) and the ability to receive show control triggers (so the explosion sounds sync with the scenic automation). We need a lot of subwoofers hidden in and around the set for booming explosions, as well as some hidden top speakers for localized crashes (the jeep through the wall, for example). We need a very waterproof speaker or speakers around the lagoon for crocodile sounds. We need music speakers, but at this stage, there's no reason to think that we'll need them to be separate from the vocal system, as, again, we are not creating any illusions that there are real musicians in the space, just playing BGM under dramatic action scenes. We might *want* it to be separate, as it will still sound clearer, but that's a secondary concern.

Our primary system concerns: vocal intelligibility, effects localization for key moments, and overall high SPL.

So, first, we'll draw a quick CD plan view. We won't bother to separate it into ground versus RCP, as that is unnecessary at this stage. Then, we'll write our narrative.

It is tempting, in this situation, to begin thinking about gear. I know personally, this is how my mind works. Immediately I'm thinking, *well, we need those Lectrosonics mics because they're the only real waterproof models; it's a wide arena audience so probably main line arrays, do I want QLab playback or a hardware server with a touch-panel front end . . . ?* And so on. I can't help it, I am a technically oriented designer, and these are often the first things in my mind when considering a new system. However, at this stage (and in a technical narrative in general), clients don't want or need that kind of information. So, what goes into a technical narrative?

- A location by location—and/or scene by scene—breakdown. This starts with the entry to the attraction or building, includes the queue, the lobby, the retail and snack bar areas—anywhere that AV content might be played.
- Each location is broken down into the following:
  - Description—what happens in that location from a guest perspective
  - Audio (and/or video/control) equipment—a list of types of equipment and their basic functions, without any specifics.
  - Audio (and/or AVC) content—a basic breakdown of what we are communicating with our equipment in that scene.

If we were writing a real narrative, we'd gather information about all of the scenes, including the pre-show. Is there music playing in the arena before the show? Is there AV content in the lobby/queue/snack bar? Are there interactive elements anywhere in the space? And so on. For now, we will just write a technical narrative for the first scene, our bank robbery:

Scene 1:

Description:
Guests are seated in the arena, with BGM playing at a low level. Music swells, a playback announcement is made to please turn off cell phones and so on, and then announcing the title of the show. While the announcement is made, actors get in place as bankers and customers. We hear exaggerated SFX of bank machines from behind the counter. A few seconds go by, and then outside the upstage wall we hear the approach of the speeding jeep. It crashes into the upstage wall. Huge explosion sound—bricks flying everywhere, actors leaping for safety. The Death Gang holds up the bank, stealing everything they can get their hands on, then escapes through the busted wall. Min-jun and the Police arrive, sirens blaring (lights visible through the busted wall). They talk to a few people and discover it is the work of the Death Gang. Min-jun turns to the audience and says that this is a job for the ACTION TEAM. ACTION TEAM theme music plays, and set begins to rotate to the LAGOON HIDEOUT for Scene 2.

Audio equipment:
There is a computer-based playback system, triggered by a live operator, controlling music and sound effect cues. Each speaking actor wears a waterproof wireless mic (Min-jun and other key players wear double mics, for failure-proofing redundancy). Mics are mixed by a live front of house mix engineer with a digital mixing console. Vocals are reinforced through full-range loudspeakers mounted downstage of the performance area, covering the full audience evenly and loudly. Sound effects play through a range of speakers—many hidden in the scenery so that effects can be localized to the spot onstage where the actions happen. The jeep exploding through the wall is played through speakers hidden in the wall around the breakaway section, and subwoofers both onstage and under the stage at the downstage edge. Music is played through the same full-range system as the vocals (or, if space allows, a parallel full-range system for more sound clarity). All sound is routed through a master digital signal processor (DSP). Mixer and playback controls are housed in FOH BOOTH in the main audience section. Computers for control, playback media, and DSP are all housed in EER, underneath the FOH position in a sealed climate-controlled room. Intercoms are provided for all run crew. Monitor speakers are hidden onstage so actors can hear music and effects cues and time their work to the sounds.

Audio content:
BGM for entry music as well as cues for Death Gang and Action Team, effects for bank sounds, jeep/wall explosion, and police sirens and cars, pre-show announcements, microphone-reinforced actors.

That would be a sample narrative for this scene. As we moved through subsequent scenes, anything that repeats can be simplified. So, BGM doesn't have to repeatedly be described as coming through the speakers that it comes through—that would only be mentioned again if the location of those sounds changes in a given scene.

A technical narrative is a bizarre exercise in reducing very technological work to elemental steps of description. I will admit, the first time I wrote a technical narrative, it was still *way too technological* and focused on actual gear. I sent it to the design firm I was working for, and they sent it back, immediately demanding I removed all the detail. It's a weird practice, but besides being useful for communicating basic ideas to clients, it can also be useful for you as a designer. If you can explain your system clearly without referring to specific gear, you truly understand the system's intent and functionality. If you can't, you probably need to think it through a bit more clearly. Sometimes, this can be a good exercise for young designers—even for systems that wouldn't generally demand this document. If you can describe your theatrical or concert system in simple and generic terms, then you really understand what you're trying to do with that system. If you can't explain it in these terms—think through your design again.

Alright! We've now examined most of the standard documents for sound system packages. In the final chapter of this section, we examine some basic practices for presenting these documents for your clients, your bidders, and your integrators.

Notes

1 Everything happens in a conference call. *So many conference calls.*
2 This narrative is a synthesis of several projects I've worked on over the years but describes no single project.

CHAPTER 57

Presentation Packaging

To Clients, Bidders, and Integrators

You've designed a killer system. You've met all your performance needs and come in under budget. So, how do you show off your work so that it looks as professional as possible? There are three basic groups of people to whom you will be showing your work:

- Clients—whether a design firm with whom you are subcontracting, or a producer/director whose show you are designing freelance, or an architecture firm whose new office tower you've detailed, client presentation documents require a bit more gloss (and sometimes less technical detail) than presentations to other people.
- Bidders—companies from whom you are soliciting prices for equipment (and possibly services).
- Integrators—companies that will install your work, and sometimes also provide the equipment. While there is some overlap between bidders and integrators, and on certain projects you might even request bids from a series of integrators for their integration services, for the purposes of this chapter's discussion, we identify bidders as those providing gear (e.g., a rental house), where integrators will be those providing full installation and testing services (typically on larger perm-install projects).

CLIENTS

When presenting work to clients, we want our package to be as slick as possible. It needs to feature every document the client requested, and generally nothing else.[1] As discussed in Section 12, what documents are due to a client will vary quite a bit from project to project, but there are some good general rules we can observe.

- Uniform look—all documents presented to clients should be formatted consistently across all plates of a package. On many projects, this means that everything gets plated in the standard 11″ × 17″ CAD package, with title block in an identical position on each page. This includes PDFs and Excel documents, which would be added to the CAD template as images and plated at the same size as technical drawings.

On some projects, Excel and other such documents will be presented in their native formats, as parallel submissions to the CAD files, but they should still include the same company/project logos as the CAD package, and the same creator, date, and revision info that would be found in a title block. Title blocks, by the way, should be absolutely identical in every way (save for the data listed within each that identifies the plate) everywhere they appear. This means identical look, size, placement, etc., on the page. Even a slight variation (like a title block moving a couple of millimeters from one plate to another) looks very unprofessional.

- Everything is a PDF—clients generally don't want to open AutoCAD files, and in many cases, they are not even capable of doing so (they don't have the software or knowledge). So, every document you submit should be exported as a PDF (Adobe Portable Document Format). PDF has a number of advantages that make it the standard submission format for project documents: it is a format everyone can read, on just about every machine; it is a format that locks your document in place, so—for example—a Word document will preserve its formatting even if the client doesn't have the same font you used; every program you will use for document generation will have an easy export-to-PDF mode; they do not generally corrupt when being emailed or uploaded.

- Make sure notes are readable, and items are identified. While this is a good *general* note, I mention this here specifically because of speaker prediction output graphics. If you've designed a system with 25 speakers, but the model you are presenting to a client is the 4-speaker vocal subsystem, you either need to remove the other speakers from the model before exporting prediction results or do some post-production on the image to identify the speakers in question. Speaker prediction software, as a rule, doesn't identify speakers visually in the modeling images, so this is your job (Figure 57.1).

Beyond these general notes, each project will have different demands. A theater director or producer who knows something about tech may ask to see more detail than one that doesn't. Don't inundate a client with documents unless that is what you're being paid to do. Keep everything as sharp and clear and easy to read as possible, and by all means, *read your PDFs after they have been exported!* Sometimes, from some software, PDFs get blurry, even though the original document was clear. You need to check every PDF you export to verify that it is still as clear and readable as the original document you produced (this is especially relevant to PDF exports of technical drawings from CAD programs).

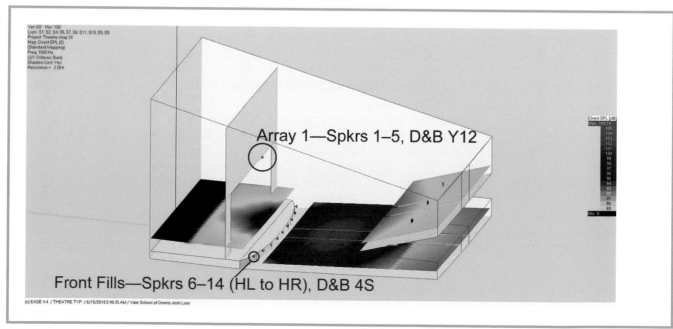

FIGURE 57.1

EASE® Model With Speakers Clearly Labeled and Notes Added for Clarity

> **Message in a Bottle**
>
> You have to speak art, and you have to speak science. You have to be able to then interpret the vision of truly artistic people. They speak fluent art and they are—God love them—they are people that think of stuff I would never think of, including crazy stuff, and also amazingly brilliant stuff. One of the big things is to be able to land those concepts and hear them and interpret them into something that you can implement, some kind of vision—you can take that vision and say, "Well, to make that vision happen, I'm going to need 45 surround speakers and space mapping to do that. But yes, we can do that." But on the other hand, what you're going to implement is scientific, so you have to then be able to be fluent on that other side in order to implement that and to be able to talk to that person who doesn't even, not only doesn't speak, like, artistic, but they speak ones and zeros, wow! The network expert and the digital technology expert that you need in order to make that sound go down and inside and out and also be broadcasting to everybody's telephones or whatever is the newest thing that's going to happen on Broadway.—Bob McCarthy

BIDDERS

If you are sending a system out to bid, particularly for a rental system, the main documents you will present are equipment and cable schedules. Often, all a bidder needs to see is a list of the gear, a location to which it will be delivered, and a rental period (a week? 6 weeks? 2 years?) in order to produce a quote. The more elaborate the system, and the longer term the rental, the more likely that the bidder might want to see a few basic technical drawings, as they might feel that the gear you've selected doesn't 100% match their stock, but if they can see the gear in its intended use context, they can recommend good substitutions for you. Some houses will recommend substitutions without any additional technical documentation, though, so this is a case-by-case situation.

Documents going to bidders should always include the following at the bare minimum:

- All gear needed
- Acceptable substitutions, if any have been identified in advance
- Location(s) of gear use
- Delivery date and return date (and cartage info—do they cart, do you?)
- Client name (who will be paying the bills)
- Contact name (generally you) and info
- Date quote is needed—a rental house will work faster on a bid that you need next week then on one for a project a year away

INTEGRATORS

Integrators get a full document set, minus things like speaker predictions and technical narratives. Predictions and narratives and the like are documents specifically created for clients (and you) to better understand the work being done. Integrators don't need this at all. They don't need working documents, like your magic sheets, intended for use by the production team. And of course, since you are submitting your documents to them either for bidding or because you've already selected them as your integrator, they don't need your budget estimate. What they do need is *literally every other document*. Every technical drawing, schedule, and detail should go to the integrator, so they can best prepare for the work. Integrators will often take the work as designed and then suggest their own refinements or alternate plans (particularly for things like rigging and mounting—which is good, we pay integrators for their expertise in this realm and others of practical import).[2]

The package for an integrator should combine the rules for packages to client and packages to bidders. They should, of course, be as clean and easy to read as possible but should also include all the contact info and project details that are often assumed on client documents. All documents going to an integrator (and in general) should include a default stamp that reads something like

Documents are to demonstrate design intent only and are not for construction. All designs in this package need to be properly engineered by appropriate parties. Others will need to check for code compliance for the references and designs in this package.

This wording is, of course, specific to integrators, as we are expecting them to install the projects, so they need to have their own people sign off on the work. If we're bidding a rental package for a theater show, of course, we assume the theatrical crew will be doing all installation, and the theater production team needs to sign off on the work—so this specific type of disclaimer is not relevant.

LOGOS

Any project for a production company or architecture firm or design firm will use those firm(s) respective logos in document headers, in title blocks, and so on. However, if you find yourself doing a lot of freelance work, it behooves you to create a simple logo that you can use to stamp all of your documents with (digitally, of course, no one needs rubber and ink here). Logos not only make it easy to recognize who has created any given document but make it easy to identify who the "author" of the work is when questions arise. If I am designing for a firm, it is assumed that even if I am off for a day, if there's an urgent question, a client, bidder, or integrator can call my firm and get answers (or at least pass the message along), as they are the "author" of the work. If the logo on the drawing is my own for my freelance work—I am the person who gets the call. Don't make your logo too complex, or it won't print clearly at different scales.

DOCUMENTS IN PROGRESS

All of the aforementioned is describing the presentation of documents that are finalized (for whatever design stage they represent). However, there is one part of many design processes we haven't discussed—*redlining*.

A *redline* document is one that has been marked up with red notes (originally red ink, now red text, and other notes from software) in order to indicate questions, changes, or problems that need to be addressed in the next formal revision. Such documents are used as working documents between submission stages.

Figure 57.2 is a drawing that has been redlined.

It shows two areas of concern, which are outlined with a cloud-shaped red border. These are called "revision clouds," and are the standard tool for identifying an area of a document that needs work. These revision clouds have callout notes showing what needs to be addressed

* * *

FINAL POINTS

All drawing sets should be numbered sequentially and laid out in such a way that the design is easy to understand. Don't start with a mounting detail and then move into elevations, then a technical narrative, then predictions, then a plot. This makes a confusing read. Remember that even in this most technical of processes, we are telling a story, and that story must be logical. Start at the beginning, lay out what you are intending, then detail and refine as the package goes on.[3] Provide a drawing table of contents or index, unless you are submitting for an architecture project, in which case whoever packages your work with the rest of the architectural documentation will provide a master index for the whole set.

Make sure all documents are properly labeled/headed and feature appropriate title blocks and legends.

Make sure it is clear who created each document, when they were created, and what revision the current plate is.

If you follow these major points on your presentation packages, you will be able to produce documents that are clean and clear and easy to read, that provide the proper information to the proper parties, and that communicate (above all) that you and your work are professional.

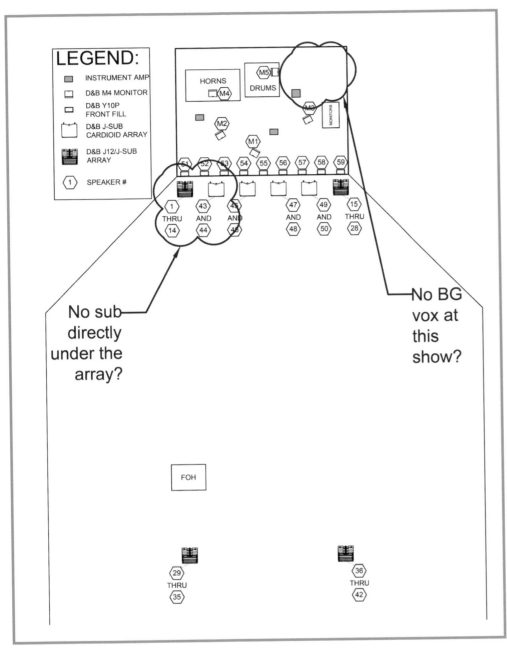

FIGURE 57.2
Redlined Drawing With Revision Clouds

Speaking of professionalism, in our final section of the book, we look into some guidelines for navigating the *process* of presenting your work.

Notes

1 In particular, if you have not been paid to create speaker predictions, you will generally not include them, even if you've done them. Speaker predictions—on some projects—are still seen as an "extra." On one level, this is unfortunate since they should be standard . . . they help everyone and make the work stronger. On another level, this is fortunate, since this means we can still charge extra on many projects for speaker prediction work (which, if done with EASE®, is done with expensive software that you generally have had to buy yourself or for your firm). So, even though I recommend running speaker predictions whenever possible (even when not being paid for them, as they will improve the work created)—don't give clients documents they haven't paid for or asked for, in general. There are exceptions to every rule, of course, and if your company (or you as a freelancer) were to be working on a test basis for a new and large client, and if you impress that client you stand to get a lot more work from them, it can sometimes be worth considering how far above and beyond to go with a project. That said, it is generally frowned upon to give away services that others charge for, as it undermines your colleagues across the industry. If you were, for example, to include speaker prediction work without being paid for it, you would want to mention to the client that you threw this in at no charge, even though it would *normally* cost extra.

2   Though it should always be made clear to any integrator that any change to the design as delivered must be approved by you/the client before proceeding with the alternate plan. The worst thing is to have a design undermined because an integrator changed something whose import they misunderstood without checking with the designer and client first.

3   Some project managers will specify the desired drawing order. If their order doesn't tell a good story, you can inquire as to why they set the order the way they did, but if they are adamant, go with what they request. Pleasing clients and bosses is a very important part of this job and there's no need to argue over trivialities like document order.

CHAPTER 58

First Presentations

Presenting your work to clients can be stressful. You've labored to create something you think will be effective and excellent, and—like in any creative process—there is always the chance that the client will not like your work, will not accept your rationale for certain choices, or will otherwise reject your creation. When we go to present design work, there are a few major guidelines we can keep in mind that will help keep the process from being too stressful. Additionally, we'll briefly discuss some of the differences in the presentation process between different parts of the industry.

First, always remember that if you've done your due diligence and created a design that will work as well as possible given the strictures of your project (budget, timeframe, goals, etc.), you've done the job you are being paid to do. Having a design (or even elements of a design) rejected can hurt, but remember that even in that instance, it is the work itself being rejected—not you as a person. If you've slacked off or not thought through all your choices, then a rejection of your work actually *should* have more personal weight, as it directly impugns your process. But again, so long as you do the work in good conscience and to the best of your ability, you *personally* need to remember to divest your person from your product. This can be hard when you are just starting out, but, eventually, you'll get just as calloused and blasé as the rest of us, and criticisms will roll easily off your back.[1]

Message in a Bottle

Get better contracts. Because as an early sound designer, we were just doing things, we weren't trying to go beyond. When you get a huge hit, that's gone worldwide, you suddenly realize you have no royalties, no further agreements. If I'm doing 14 productions of *Les Misérables* and still living off the per diem, there's something wrong. Starting now—make sure your contract is in place with your producers, make sure you wake up every morning with a smile on your face because you're happy to go to work, you're happy to work with those individuals. Don't work with someone you're not going to get on with, who won't allow you to be who you are work-wise or as an individual. If there would be an order, that would be way more ahead of contracts. Working with people with whom you're having a terrible time is way more detrimental to your health than the other one. It's terribly demoralizing and it'll just kill you.—Jonathan Deans

TYPES OF FIRST PRESENTATION

How you literally present your work will vary on different projects, and some of that will depend on your personal situation, while some of that will depend on the type of project. In an ideal world, we'd all be able to fly to wherever design meetings are happening, so we could present our work in person. While this is less convenient than presenting virtually via Internet/phone, the easiest way to ensure that you are communicating your ideas clearly is to be in the room with those you are presenting to, so you can see their faces as they react to your work. However, a lot of times, this isn't possible. Let's examine a bit of the difference between an in-person presentation versus a virtual presentation.

IN-PERSON

An in-person presentation of design work requires some preparation that a virtual presentation does not. First, and perhaps most obviously, dress well. What "dress well" means for different industries will vary. In the world of architecture and theme park design, this generally means the sharper end of business casual attire, where in the world of concert and theatrical design, less formality is expected but no less attention to how "put together" you are. Make sure your hair is set however you like it, your breath is fresh, and so on. An in-person presentation usually benefits from some kind of visual aid beyond the actual design drawings you've completed. This might be as straightforward as a PowerPoint presentation that summarizes the key points of the sound system design, with some simple text identifying key data that a client needs to understand. Clients generally want to know about speakers and placement, for example, far more than they want to know about your detailed junction box plan. Keep your presentation focused on the kinds of points that you would mention in a technical narrative, even if you're adding some model information to the mix because you are presenting a design that is beyond CD stage.

Of the varying industries we may work for as sound system designers, theater in particular has standards and expectations for design presentations that other disciplines do not. In theater, it is fairly common that all departments' designers will present together in succession in a single day, in a single meeting. The goal of this meeting is to make sure that all departments get a basic picture of what the other departments are doing. In such meetings, it is important that you include in your presentation any parts of the design that might impact other departments—speakers that need to be mounted in a set, or that might be mounted in typical lighting positions, or the like. If conflicts are identified, it is not generally going to be expected that they are solved there and then in that meeting (though if they can be, all the better), but it is expected that you will mention them so that the production manager can note them and make sure they get solved in a timely fashion.

Presenting in person for large installation projects often is more compartmentalized. It is rare that all of the department designers are in one place at one time for first presentations, as they are often working as subcontractors for a variety of firms. As such, your presentation may be a bit longer than in a theatrical context (where you may have only 5–10 min to show your work). I've been involved in architectural projects where my first design presentation ranged from 10 min to 1 hr. It depends on the scale of the project, of course.

As the saying goes: you only get one chance to make a first impression. Even if you know everyone on your project team already, a first presentation is your chance to introduce your design for the first time, and the first impression the design itself makes is important to the success of the project. No one should turn in sloppy work, on any part of a project, in any phase—but if there are two points in the process where sharp work is the most important, they are the first presentation and the final package submission.

If you are a comfortable public speaker, you can make a simple plan for your presentation—either via your PowerPoint, or a series of cards with bullet points on them for your major points, so you don't miss anything—and then speak extemporaneously. This industry favors a seemingly casual attitude, while also prizing precision in the work itself. However, not everyone is a comfortable public speaker. If you are not comfortable, it can help to write a bit of a script, so you know how one point leads into another as you present. If you have any doubt that you can easily speak to a group and explain your design, rehearse your spiel. It can feel weird to stand in your living room talking about your sound system to your dog/roommate/wall, but it is far better to practice a presentation several times than it is to stumble or sound uncertain about your work when presenting. Even a glorious, amazing design can be rejected if the designer seems unconfident and unclear when presenting it.

VIRTUAL

A virtual presentation, on the other hand, is becoming ever more common (particularly for first presentations). In the digital age, designers can be (and are) scattered all over the world, and the idea that they would come physically together to present designs is increasingly old-fashioned for a lot of projects. I've designed projects all over the world, and most of the time I've designed them from my living room in whatever state I happened to be living in at the time.

Some projects will still have a formal-ish first presentation, even if virtual. In these cases, what typically happens is that the designer will submit an initial document package to a shared drive (or email to the relevant team members). This allows them a chance to look over the work before the first presentation (even if, as is so often the case, they fail to do so). Then, making use of online meeting tools,[2] the team will join together to share. It is fairly common, during such a meeting, for either the project manager or the designer to share his or her screen with the team and to walk through the document sets submitted as they talk. Depending on who this presentation is for (what kind of client), this may be a very informal meeting or may retain some formality. If you are designing for an architectural firm, it may still behoove you to make a PowerPoint or a Prezi or what have you and use that as the substance of your virtual presentation. If you are presenting as a subcontractor to an AV design firm, this step is usually not necessary, and you can present directly from the document set (however, at an in-person presentation, even to an AV firm, it may be a good plan to make a presentation document to work from that is different from and summarizes the drawing set).

Virtual presentations can be both easier and more stressful than in-person meetings. On the one hand, unless you are required to join with your camera live, you don't have to worry about what you look like or how you're dressed.[3] You don't have to make sure your breath is fresh. On the other hand, you can't see the room react, so you need to try to be attentive to what people say and to any sonic but nonverbal cues you may hear (sighs, groans, etc.).

For some projects, the first "presentation" will literally be a document submission. These cases are usually where you are working as a subcontractor for a company that understands your discipline (e.g., an AV design firm), and they can be trusted to understand your document set. While this can be nice (to not have to explain our work), it can also be stressful (what if they misunderstand our work?).

For any first presentation, whether the first presentation of a basically finished system design for concert or theater, or the presentation of CD-phase plans for a longer installation project, the goals are the same:

- Identify key points of your system's functionality, and how they will be accomplished
- Include enough detail that it is clear that you have thought through how the design will work, without focusing on detail (unless asked to explain more in follow-up questions)
- Speak clearly, succinctly, and without agitation but with energy

- Tell a story—remember the rule of packaging a document set. Start with the big picture, then drill down as needed.
- Keep it as short as possible, without leaving out important info.

We must always assume (with rare exception) that work presented at a first design presentation is a draft. Even if it's a really *sharp* draft, intended to represent the full design, it is very common that *something* you present in the first round will have to change before the design is finalized. Be prepared for this eventuality and remember that the critiques offered or problems identified are—again—not critiques of you but attempts to make the design work as well as possible.

Now—we've presented our first round. We have some notes from the team, maybe some questions that need resolution before revisions can be made, and so on. Next, let's briefly examine the process of revision presentations, which tend to be much less formal than either first or final presentations.

Notes

1 I'm kidding. Sort of. I am not blasé about my work, nor should you be, but I have designed enough systems to know that if someone disapproves of one of my choices, it has *nothing* to do with me personally—so I am blasé about any personal feeling relating to changes or critiques.
2 Like GoToMeeting, Zoom, or the like.
3 I have joined a ton of virtual meetings where all parties involved have webcams and no one uses them. Webcams are awkward, technical people can be awkward, and we often just avoid them.

CHAPTER 59

Revisions and Value Engineering

We talked previously[1] about the process of value engineering—which is to say, the picking apart of a design in order to save money. As always, our goal during VE and revisions is to preserve needed system functionality while cutting costs as much as possible. The process of VE can be undertaken in all manner of fashions, and in this chapter, we'll examine some of the processes and practices common to the revising of work.

We've submitted our first designs (whether in person or virtually). We've had the chance to explain ourselves and our reasoning, and now we're expecting notes. Sometimes, the first notes and revisions come at the first design presentation, and we leave that meeting with work to do right away. Sometimes, a design will be presented, and days or weeks go by while the team examines your documents, shares them with other parts of the team, or (as is often the case when working with large firms) works on other projects before having time to return to your design to assess it. Regardless of when the first notes or revisions arrive, there are a few things we must do upon their arrival:

1. Carefully ingest the notes. Sometimes they will be delivered in the form of email, other times as redlined drawings, still other times in person or over the phone. However they arrive, be sure you understand them completely before responding to them—and if they are unclear, ask any questions needed in order that you may properly understand them.
2. Once you are certain you understand what the notes are, assess them so that you understand *why* they have been given to you. Sometimes, a revision comes from a change to the underlying project itself, but if you don't know that, you may feel a requested change is arbitrary. Once you know why the notes are what they are, you can proceed.
3. For any notes or changes that you disagree with, express your disagreement . . . within reason. If your initial design would cost $200,000, and the producer comes back and says that you only have half that money available, there's not much that disagreeing will do except make you out to be a jackass. However, if a client requests a change from one-speaker position to another, and you think that the value added by virtue of this

change (e.g., the speaker is less visible) is not worth the trade (e.g., much worse localization), now is the time to express that doubt. When challenging notes, be clear about your reasoning. Acknowledge that the other party's reasoning is valid,[2] but explain why you see your concerns as being more crucial than those driving the change request. In this process—challenging notes—it is imperative that we always remember to keep things from becoming personal. Stick to facts. If the facts don't support you, probably you should just go ahead and change whatever is being requested.

4. Prepare your revised document(s). Revisions are always labeled as such, sometimes with numbers, sometimes with letters, sometimes with a complex series of identifiers specified by the project manager. Whatever the revision protocol:

   a. Always SAVE AS when making changes. It is very common for changes to be requested and then, upon seeing them, to want to revert to the original plan. Make sure you have saved each change as a new draft so that it is easy to revert.

   b. Identify any areas of uncertainty. If the new changes create a situation where something is unclear (e.g., how a new speaker will be mounted in a given location, or the like), identify with a callout, revision cloud, or both.

Handling a revision process is key to the success of a working designer. As the saying goes, people don't remember what you did, they remember how you made them feel.[3] Revisions may be difficult, but if you don't make the client *feel* that revision is difficult, they will love you for your easy working practice.

That said, there are limits. If you've designed a 200-speaker system for a giant installation, and the client comes back and says something awful and vague like "I want to see a totally different version," it is sometimes necessary to make clear to them that this isn't a change so much as a whole new design. Depending on your contract arrangement, such changes may be included in the project so that—while you may want to gently identify that this request means "going back to the drawing board"—it's totally fine and will be ready as soon as possible. If your contract does not include this kind of massive revision, this may present what is called a "change order" and necessitate extra payment. This situation is tricky and will be navigated differently by different designers with different clients. How much to charge will vary depending on your stature as a designer, your contract, the timeframe demanded for these changes, and so on. The faster the changes need to be made, the more it will cost (assuming this is a change order and you aren't on staff).

There is a difference between standing your ground on a design choice because you've made the best plan and the client's notes undermine the success of your design and standing your ground because you don't want to undo difficult work. No one wants to undo work they've already done but remember who is buttering your bread. If the client is requesting changes that make sense, or that are workable, you need to deliver them.

At any point in the design process, it is important to comport oneself professionally and kindly. Give people the benefit of the doubt, while always keeping one eye behind your back to make sure you're not being taken advantage of. Expect that as a young designer just starting out you may need to do more for less money than you will as an

Message in a Bottle

Be willing to ask questions and own every mistake you make. I would say that was the biggest thing—having to learn to own my mistakes, by *not* owning them when I was, like, 24 was really painful. So I would say definitely, absolutely, straight-up letting people know when you don't know something, and totally owning the responsibility when you mess something up. That really frees you—learning those lessons the hard way once freed me to experiment, which is what led to many of the more interesting relationships and designs I created. I learned how to say I don't know, and then I learned how to make mistakes, and then I learned how to take risks—because if you take the risk and you make the mistake and you own it, that's OK.—Robert Kaplowitz

established professional. Don't nickel and dime a client—include general revision guidelines (either how many passes, or what timeframe revisions must take place within) in your scope of work and contract so that you are not being railroaded, but don't charge per revision (unless they are well outside the count or timeframe agreed upon).

Revision documents are not expected to be as pristine as first presentation docs or as final design docs. They still need to be clean and clear, but often contain unresolved notes and questions, callouts, revision clouds, and so on. Identify revisions *as* revisions, make sure alternate versions (for A/B comparison on the part of the clients) are clearly marked as alternates, and that revised documents don't replace originals (at least on your hard drive, and ideally not in the shared project archive).

Revision conversations are often the key place where designers make their names. Many people can deliver initial designs, but the designers who can work through the mud and weeds to get from the first idea to the successful final design while remaining cool, calm, collected, and pleasant to work with are the designers who keep getting hired. Be realistic about timeframes. If someone asks for design changes in a day and you know it'll take a week, let them know right away. One of the cardinal sins a designer can commit is to miss deadlines.[4]

Revision presentations are seldom given full in-person presentations, unless they are part of a project that has regular project team meetings week to week. If you're a staff designer at a firm, you may well be presenting your revisions in person at a check-in meeting, but unless those changes are massive, no one is generally expecting a new PowerPoint on the revisions. In many cases, revisions are handled virtually, even when a team is meeting regularly. The latest revisions will be uploaded/shared, everyone will note the changes, and any responses or further changes will be addressed either in the next in-person meeting or via email.[5] The revision process is never as formal as first and final presentations, and while we still always strive to create accurate, excellent work, the level of pressure is a bit lower in the revision stages as a result.

Message in a Bottle

You have to look at the budget you have, the equipment that's available, the companies you're working with, the personalities that may get involved. Because if you're doing consulting you know the architect's going to be like hey this thing has to be *this way*. You think, "That's going to be really bad," but if the client is like "no, I really like that" you have to be like "well, here's the thing we can do to mitigate the suck." Nothing's ever perfect.—Christopher Plummer

So! We've revised our design, we've made our arguments, we're ready to put the final package together. How do we present that work?

Notes

1 Chiefly in Chapter 33.
2 Assuming it is . . . sometimes you will be asked to change something based on a misunderstanding of how that something will work, and this is the time to clarify and disabuse the client of their incorrect notion—kindly, of course.
3 Let's be clear, they *do* remember what you did, in a general way, it's just that the impressions we leave on people are more felt than thought, and we want to leave our colleagues and clients feeling good about us and our work whenever possible.
4 It does happen, but it leaves a bad taste in everyone's mouth.
5 A note on email—many firms now are implementing a rule that each email thread be focused on one topic alone so that subjects are easy to track. Personally, I kind of hate this rule. If I made ten changes, I don't want to send ten emails, I want to send one email with all the info, but I do understand the logic behind wanting to keep subjects separate for easy searchability. Make sure you know your project's protocols before emailing people about updates and changes.

CHAPTER 60

Finalizing and Signing Off

We've made it through the whole process. We've assessed our design needs, created a system, and specified gear that we believe will meet those needs; worked through revisions and V.E.; and are ready to present our final design. How do we submit our work? Is there a final formal presentation round? What do we need to do so that we're sure we're ready to "sign off" on the design?[1]

SUBMISSION FORMAT

How your work gets physically submitted will vary project by project. For some projects, like theater, your document set will be submitted, likely to the production manager, as a coherent set of audio documents, without being integrated into a show package. For other projects, like most architectural projects, your documents will be submitted to your project manager, who will then integrate them into the overall project document set. If you are working for a firm with good organizational structure, the latter type of project will have someone assigned as the CAD manager, who can take all of your documents and format them with appropriate title blocks, in the proper sizes, and get everything inserted into the overall package. On less well-organized projects, you may be stuck doing this work yourself. Regardless of who does the document finishing, it is expected, on installation projects, that you will be incorporating your documents into the overall set, including architectural drawings, lighting design drawings, scenery, and anything else so that there is one master "final document package" that can be turned to for answers to any questions that develop in the production process.

For some projects, submitting the PDFs will be the final. For others, you will be required to submit the raw source documents (CAD, Excel, whatever) as well as the exported PDFs. For still others, you will submit the raw source documents, and the CAD/document manager will create the PDFs for you, per the standards set internally to the company. Always check with your team to ensure you are delivering whatever documents make up the final set in the proper format(s). For most projects, this final submission will be digital. Even if there will be a final in-person presentation, the work itself is transmitted as files, to a shared drive or via email to the appropriate parties.

FINAL PRESENTATIONS

On any project that has gone through many revisions, there is the possibility of a final presentation round, either in person or virtually. Sometimes, the purpose of this final presentation is to once again get the client up to speed, feeling comfortable, and like they are in on all the latest changes. Sometimes, the purpose of this presentation is once again to get the other design teams to be aware of your work and how it might impact them (again, most common in the theater). Whenever there is a final presentation round, the same rules apply from first presentations (be clean, smell good, speak clearly, be organized, rehearse in advance if you need to, make a PowerPoint or something like it if this is an in-person submission). In addition, a final presentation round typically will go into more detail than a first presentation round, particularly about anything that has changed from the beginning of the project. As always, remember to stay focused on the critical infrastructure, major details, and big picture. Don't get bogged down by model names and manufacturer info unless it is relevant to the others in the meeting. Be professional, don't go over your allotted time, and so on.

Now, for many projects, there won't be a formal final presentation round. You will simply submit your final documents and move on. In this case, you need to make trebly sure that your design is fully documented and that every conceivable issue or question is addressed in the package (at least those that have come up during the process, and that you foresee being relevant to bidders, integrators, or whoever gets these packages next).

SIGNING OFF

Before submitting a final design package, it is imperative that you re-read the full package. This requires mental effort, not just because reading elaborate technical packages is challenging, but because, as the designer, it is virtually guaranteed that you are so close to the work that it is hard to step back and see it clearly. Whenever possible, I recommend taking a "critical break,"[2] which ideally is a day or two (a weekend! Imagine that!) away from the design, before re-reading. This break allows your mind to drift away from the details, such that when you re-read your work, ideally you are doing so with as fresh a perspective as possible. Admittedly, if you've been working on a design for 2 years straight, taking a day or two off probably isn't enough for you to actually see the documents as an outsider might, but it's better than just submitting the moment you've finished your final revision.[3]

The goal of this re-reading is to see the design and documentation with the clearest set of eyes, to identify errors that you may have missed along the way, and to ensure that everything is represented as clearly as possible. Any project step can benefit from this critical break and re-reading, if time allows, but it is especially important before the final submission, as the final submission is what will be bid, ordered, integrated, and built.[4]

Once we've re-read and submitted our final drafts, we will be considered to have "signed-off" on the design, which means that for all intents and purposes, it's ready to move to the bid/execution phase. Once we've signed-off, any changes we need or want to make to our designs will be "change orders" that our equipment providers, integrators, or even clients can bill us for. We want to avoid being charged for changes after we've submitted, so we need to avoid making those changes by solving the problems in the design in advance.

Now, once designs are finalized and submitted, there are still two more rounds of documentation that *sometimes* become necessary. On occasion, when selecting an integrator (for example), the integrator proposes a change—e.g., swapping one speaker for another because they have a better deal with the manufacturer of the latter—and we will then update the drawings as a "bid set" to reflect the changes.

Finally, once a project is installed, we may need to generate "as-built" drawings. As-builts are documents that reflect the entire design as physically realized. These are very common documents in installation work, because, for example, a junction box might be specified in a final design as 2' in from a corner on a wall, but the actual box got moved to 18" from the wall due to an on-site change in wall framing, or whatever. A permanent (or even semi-permanent) installation project will require a comprehensive as-built set delivered to the client, so they can use that set as the permanent in-house reference for what actually exists in the space. Large performance productions sometimes require as-builts as well, particularly if the production is, say, a Broadway show that will later tour. Any changes from the final design package to the realized design need to be updated in as-builts so that the operating document set is 100% accurate.

Message in a Bottle

To me, there's no difference between sound design and sound system design. You can't be a sound designer and not know your system and what it's capable of doing. Just walking in there and someone else has designed a system and I'm gonna go in there and use it . . . yes, you can have a big dog bark and a small dog bark and a loud dog bark and a dog bark to the left—that may be, though you don't know because you didn't design the system. But actually having a dog that, besides all those things, that is actually coming, and you know how it's being covered, where it can come from, what you can do, how you can move it around, how you cannot move it around, and all those different options—that's part of the sound design. To me, system sound design and sound design fall under the same roof. Know what your system is capable of doing, what *you* are capable of doing, and just feel absolutely confident about yourself and your team going in. Have as much fun as possible. Something I say to my team—and there's always a team, even if you designed it yourself, you're probably not going to be sitting there running the show every night, you've probably got someone backstage, whatever—when, as a sound designer who is working with a team, I often will say to them, "Look, all of our issues, all of our good stuff and bad stuff that we have as individuals, that we find as part of this process, we keep to ourselves. We do not go out and share it. We also do not go out and look for a victim in another department. We do not throw stones because we live in a glass house. We respect and give people everything they want." Quite often, I will give time that's been allocated to me as a sound designer to another department, because I can see they need it way more than I do—they need it; otherwise, they're gonna be in trouble, where I've got it worked out, and I can do this. I think that care for collaboration and respect for the final product [are important]. Of course, in order to be able to do that, you have to have experience. You can't go in there as an inexperienced designer without knowing where that line is drawn, where you're going to fall into the quagmire of crap and can't get out, because you've made too many errors by ignoring certain red flags. That's happened to me; every now and then, you do a project, and you realize you've just made the mistake that you're telling everyone not to make. I did that on *Spiderman: Turn Off the Dark*. I was teaching the Broadway Sound Master Class, and I went through with them and counted all the red flags I should've observed, and there were over 40 red flags. Generally, after three red flags, you quit. I did 40-odd, and I was still there, and I was like, "Oh my God, this is terrible."—Jonathan Deans

* * *

Alright!

We've completed our system design! It's been installed, our as-built drawings are done, and we feel pretty impressed with ourselves.[5] What's next? Well, we need to find our next project (or, if we work for a firm, simply move onto our next project). But also, how do we learn from what we've done? How do we keep learning and growing as designers in general? This is the subject of our brief afterword.

Notes

1 To sign-off on a design means to state that to the best of your knowledge and ability, the design is complete, will work as drawn, and is ready to be executed (or bid, as the case may be).

2 I have borrowed this term from Roey Izhaki's excellent book *Mixing Music*, where he uses the term to describe a mix engineer taking a break between "finishing" a mix and listening back to it to ensure that the engineer still feels it's finished, a day or two later with a clearer head.

3 Of course, better to submit without this critical break and reevaluation than to miss a critical production deadline.

4 On every project I've done where I didn't have time for a critical break and re-read, there is almost always an error or two that I spot if and when I look back at those drawings a year later. The goal is to avoid these problems. Humans make mistakes, and it is very rare that anyone can create a huge multi-page design package with no errors, but we want to get as close as possible.

5 As we should, this isn't an easy business!

WHAT COMES NEXT—A PATH FORWARD FOR THE SYSTEM DESIGNER

So, you've read this book. You've (hopefully) understood this book. You've maybe even designed a system or two with the help of this book. What next?

As a sound system designer, your job is complex, and demands near-constant learning if you are to remain employable in this field. The most obvious reason for this is just that technology evolves quickly. The other reason is that techniques are constantly changing—in response to the growing and changing demands of our audiences. One hundred years ago, the very fact of amplifying sound at all was miraculous, and as such, people didn't demand very much from sound systems in terms of fidelity, clarity, power, and so on. Today, however, while the average consumer can't discuss details of how Meyer's Constellation works—or even what it is—they can hear the difference between a room with a Constellation in place versus one without it. As audiences demand more richness and depth of sonic experience, we must continually update our practices to help meet those demands.

The first major thing to make a part of your ongoing practice is to take advantage of whatever situations you find yourself in that allow you to listen to different equipment. Are you working at a production rental shop? They have a ton of gear! Take some downtime and listen to every speaker and microphone available. Are you designing for a theater company? If they own speakers, see if you can take some time and listen to every single one. Listen to, hook up, and test gear wherever you find yourself working so that you build your mental database of what things sound like. As someone who has designed systems for years, I can tell you what certain speakers sound like versus certain others just from memory and experience. Until you build up a large mental database of equipment that you understand and can use intelligently, it's hard to design a wide range of systems.

If you don't find yourself in a situation where you can listen to a ton of gear just for the heck of it, the next best things are to find out when speaker companies (and other sound companies, but especially speaker companies) are doing demos or shootouts of their equipment. Many major manufacturers will periodically arrange shootouts, particularly of new product lines, via regional integrators and vendors. Sign up for manufacturer mailing lists, contact their education divisions (if they have them), talk with local vendors. If you are anywhere near a major metropolitan area, there are bound to be companies that rent sound equipment. If you have a show you're about to spec, but are unsure of a piece of gear, contact the rental providers in your area, explain your situation, and ask if you could come in and listen to the gear sometime soon, so you could decide whether to specify it for your system. Many vendors will accommodate such requests, with the idea that if you like the gear, you may rent it from them.

Go to trade shows. Every year, the Audio Engineering Society hosts national (US), international, and regional events that show off the latest in audio technology. There are mics to listen to, speaker listening rooms, software demos, and the like. AES is a very powerful tool for the designer in learning about new equipment, and if you attend the full conference (instead of just the expo floor) there are countless presentations on new techniques and technology. Find gear you like, make contacts with manufacturers, heck, even find a job!

Subscribe to good trade publications. Tape Op, Lighting and Sound America, Sound on Sound, and many others provide reviews and listings of new gear, descriptions of current shows/tours and the systems they used, and so on. Some of these are free (Tape Op just requires online signup), some are free to those verifiably working in the business (Lighting and Sound America), and some make you pay (Sound on Sound), but these are very valuable tools for understanding the industry in the present tense.

Go to shows, see installations, attend theme parks, and so on. Everywhere you go, pay attention to the sound system. What gear is being used, and how? I pay attention to speakers everywhere I go, from the supermarket to Tomorrowland. If you are at a concert and want to know more about the system but can't see what kind of speakers are being used, linger by the FOH console, and see if you can catch the engineer at a moment of pause. Strike up a conversation. So long as you are not bothering them during a critical moment, most engineers are happy to talk about gear with you—the love of gear is often a big part of why they became engineers in the first place! Even if you can't talk to anyone and can't see exactly what speakers are being used, you can at least get a general sense. How many arrays or point source boxes can you see? How many subs? How are they set up in the space? Even this generic kind of information is useful (especially if you associate the system with the size of venue in your mind).

Talk to other designers. There isn't one central coffee shop where all the sound system designers hang out (and if there were such a spot, it would probably be a bar), but as you work through the industry you will meet other designers. Ask them about their work. Ask them about their approach to specific challenges or types of projects. Ask them whatever questions you can think of that they will answer. Remember, of course, that different designers have different opinions about what sounds good and why, but store that info away for the rainy day where you have to design a system you aren't confident about, and none of your typical solutions are working.

Read good books about audio. Focal Press/Routledge (the publisher of this humble tome) has published a wealth of very useful books on audio, systems, engineering, design, and so on—several of which are listed in the "Bibliography and Further Reading" section of this book. As your skills expand, so should your library. I own a ton of audio-related books, some of which I still haven't read because the specific need has yet to arise, and some of which I have returned to over and over again until the pages are dog-eared. Books are an invaluable part of the process of becoming a better designer, and while they are no substitute for experience, they can provide things that mere experience will not (such as the theory behind the practice, or theory that *refutes* some of the practice—because sometimes people in the field are lazy and do things in ways that could be improved upon with better understanding of the theoretical dimensions of their work).

Now, in your practice as a designer, you provide yourself with good learning opportunities, as you can select gear and plans of operation outside your normal comfort zone. I know a lot of professional designers who say that their general rule is to always test "one new" piece of gear, or system configuration, or whatever, on each show. This is great if you're already an experienced designer, as that way you are just adding to your repertoire, building knowledge, but also limiting risk (if one part of a very complex system doesn't work out, you can always swap back to a piece you know once you've identified the failure). However, I've been at countless student events where seasoned designers will give this advice, and students throw up their hands. *Of course, this advice doesn't work for student designers!* You may be designing your first-ever system! Everything is new to you! So, for a while, you need to just work from best practices that you can identify in readings, coursework, and from designers around you. Don't build systems that are overly complex or ambitious right out of the gate, create simple, effective systems, and *master them*. Once you are rock-solid with basic analog and digital signal flow and setup, add a layer of complexity to the next show, and so on. Each show or project provides learning opportunities.

That said, sometimes you are forced to take a leap way out of your comfort zone in order to accept a given job. I've certainly found myself on projects where instead of adding one new element, I was adding 20—and this is as a professional working designer, not as a student. But I did my research, I read my manuals, I checked with colleagues, and when I had a plan I believed would work, I submitted it (and most of the time it worked)! I've met young designers who feel that they would rather pass on a big opportunity career-wise then risk being in a situation where they didn't know how to solve every problem in advance. So long as you are ready to do the hard work to learn what you don't know in the interim, I tend to advise going for it. My career has expanded because I've always taken the next good opportunity, even if that meant lots of sleepless nights learning new software, or about new equipment.

Most importantly of all, keep your ears and mind open. Listen to the world around you—think about what a system would need to be in order to replicate the sounds you hear. A truck passes on a nearby highway—how loud was that? Where did the sound localize to? How much bass and treble were present? Ask these questions of the world around you, and when it comes time to create a system to replicate these sounds, you'll already have some idea of how it might be done. Think sonically, listen actively, and remember that there is always more than one way to amplify a dragon!

Now, let's go make the world sound better, one room at a time!

Happy designing!

—Josh Loar
January 2019

The following is a guide both to works that have proved useful to me in my career as a designer and in the writing of this book, and to other books that are simply good resources for sound system designers. This is by no means a comprehensive list but aims to at least serve as an introduction to the field of literature. Works are listed alphabetically by author and each includes a brief statement as to why I find it valuable.

Books:

American Radio Relay League (95th edition, 2017). *The ARRL Handbook for Radio Communications.* American Radio Relay League, Springer, Connecticut, USA

If you want to get hardcore into RF engineering, this is a good resource. It is dense, long, and filled with real engineering data. If you suddenly find yourself working on a giant RF project, this is a good resource to have on your shelf (and even if you're only working on small projects, it's still filled with great info, even if some of it focuses literally on radios and not on RF mics).

Ballou, Glen (editor, 5th edition, 2015). *Handbook for Sound Engineers.* Routledge, Abingdon, UK

One of the pillars of practical engineering knowledge. While design and engineering are not the same discipline, the best designers usually have a wealth of engineering experience and knowledge, and this book is a great resource.

Ballou, Glen (2017). *A Sound Engineer's Guide to Audio Test and Measurement.* Routledge, Abingdon, UK

A handy resource on measuring audio. A shorter book, with lots of good charts and info.

Baxter, Dennis (2007). *A Practical Guide to Television Sound Engineering.* Focal Press, Waltham, MA, USA

If you find yourself needing to design a system for TV or other such video/broadcast media, this is a great resource.

Brixen, Eddy B. (2nd edition, 2014). *Audio Metering: Measurements, Standards, and Practices.* Focal Press, Waltham, MA, USA

Another good resource on measuring sound. Lots of info on standard terminology, different types of tests, and how to read those charts you see on devices from old-school oscilloscopes to contemporary software measurement tools.

Carter, Paul, & Chiang, George (3rd edition, 1994). *The Backstage Handbook: An Illustrated Almanac of Technical Information.* Broadway Press, Louisville, KY, USA

Not technically a sound book, but indispensable all the same. This book features information on so many parts of the stage production process, including all kinds of construction and rigging info that you may either need to understand to converse with the team responsible for that on your project, or understand in order to make use of it when designing your own custom mounts and the like. I've had this book on my shelf for decades, and I'm never sorry I have it.

Davis, Don, Patronis, Eugene, & Brown, Pat (4th edition, 2013). *Sound System Engineering.* Focal Press, Waltham, MA, USA

This is a serious engineering book. It has a lot of math. It is very valuable. If you want to get much more mathematical about how sound, electricity, and systems work, this is one of the best books you can read. If you are math-phobic, it's a challenging read, but if you are ready to put in the number-crunching, you'll have a much stronger understanding of the principles underlying our craft.

Davis, Gary, & Jones, Ralph (2nd edition, 1988). *Sound Reinforcement Handbook.* Hal Leonard Corporation, Milwaukee, WI, USA

This book is arguably out of date (it really pre-dates the digital audio revolution in many ways), but in terms of analog signal flow and basic reference, it's still great.

Eargle, John (1989). *Handbook of Sound System Design*. Elar Publishing Company, N/A (defunct)

Another out of date work, but one by John Eargle, who was a genius. Worth examining if you want some background, or a sense of the state of system design in the late 80s.

Eargle, John, & Foreman, Chris (2002). *JBL Audio Engineering for Sound Reinforcement*. Hal Leonard Corporation, Milwaukee, WI, USA

Useful, practical.

Eargle, Jon, & Rayburn, Ray A. (3rd edition, 2011). *Eargle's Microphone Book*. Focal Press, Waltham, MA, USA

This is THE work for anyone seeking a clearer understanding of the design and functionality of microphones. If you'd like to design a microphone, you can't go wrong starting here, and if you just want a deeper grasp of how microphones work, it's a great resource.

Everest, F. Alton, & Pohlmann, Ken (5th edition, 2009). *Master Handbook of Acoustics*. McGraw-Hill/TAB Electronics, New York City, NY, USA

The classic tome on acoustics. There are a ton of other useful acoustics books out there, but this one set the standard for a reason—it is comprehensive, clearly written, and totally useful.

Fischer, Walter (3rd edition, 2010). *Digital Video and Audio Broadcasting Technology: A Practical Engineering Guide*. Springer, Berlin, Heidelberg, Germany

More broadcast fun. Includes the video side of things for those who are interested.

Hechtman, John (2008). *Audio Wiring Guide*. Focal Press, Waltham, MA, USA

Most designers aren't wiring their own systems, but every designer needs to understand how wiring works. This is a great encyclopedia of wiring types, terminations, specs, and so on.

Holman, Tomlinson (2nd edition, 2007). *Surround Sound: Up and Running*. Focal Press, Waltham, MA, USA

THX (the Tomlinson Holman eXperiment) was invented by this guy. If you have any desire to know more about systems for cinema surround, this is the place to go.

Horowitz, Paul, & Hill, Winfield (3rd edition, 2015). *The Art of Electronics*. Cambridge University Press, Cambridge, UK

If you find yourself needing to understand fundamental electronics better (if, for example, you find yourself designing a theme park system and don't know what PLCs are—or what they do) this is a good place to start. A demystification of electronic engineering for those in need.

Huntington, John (2nd edition, 2017). *Show Networks & Control Systems*. Zircon Designs Press, Brooklyn, NY, USA

The bible of show control systems. Huntington is a master explainer of complex systems and principles. This book in earlier editions was called "Control Systems for Live Entertainment" and went through several editions under that name, so while this is the second edition under the current title, this is something like the 5th edition over all.

Izhaki, Roey (3rd edition, 2017). *Mixing Audio: Concepts, Practices & Tools*. Routledge, Abingdon, UK

Not technically a system book, but Izhaki's explanations of various effects (from EQ and dynamics through the most modern modeling effects) are the clearest I've ever read. And, if you have any interest in mixing music, this is the best book on the subject that I've found.

Jaffe, J. Christopher (2010). *The Acoustics of Performance Halls: Spaces for Music from Carnegie Hall to the Hollywood Bowl*. W.W. Norton & Company, New York City, NY, USA

A wonderful assessment of the sonic qualities of top-flight performance halls, from one of the world's most acclaimed acousticians and concert hall designers.

Katz, Bob (2nd edition, 2007). *Mastering Audio: The Art and Science*. Focal Press, Waltham, MA, USA

Again, not technically a system book, but Katz talks a lot about how a mastering studio should be set up, so it's valuable to system designers in that sense. Also, a great resource for those mixing and mastering music.

Kurose, James, & Ross, Keith (7th edition, 2016). *Computer Networking: A Top-Down Approach*. Pearson, London, UK

A very good book that goes into more detail about computer networking than we had room for in this one. This is not an entertainment-focused book, it is hardcore IT knowledge, but any system designer just getting into the field today needs not only a strong background in networking but understanding of subtopics we used to think of as marginal (QoS, network security) that are increasingly vital. This book is a great place to start.

McCarthy, Bob (3rd edition, 2016). *Sound Systems: Design and Optimization*. Routledge, Abingdon, UK

The master. Bob McCarthy has tuned more sound systems (and more COMPLEX sound systems) around the world than anyone else. As you expand your systems from basic to complex, as you work extensively with arrayed speakers, this book is the best resource for understanding how sound interacts in a space, and how we as system designers can manipulate it for the most clarity and control. I return to this book over and over again when I am designing projects, and Bob is also a super nice guy! Any system designer needs this on their shelf.

Moulton, David (2000). *Total Recording*. KIQ Productions, Sherman Oaks, CA, USA

Moulton's tome is a great resource for studio engineers, but also talks extensively about setting up studio spaces, and the use of speakers in such spaces, and as such is a great resource for anyone designing a recording, mixing, or other production facility. Plus, it's just fascinating from start to finish.

Newell, Philip (4th edition, 2017). *Recording Studio Design*. Routledge, Abingdon, UK

One of the towering works on studio design. Great information for anyone traveling down that road, and a fascinating read.

Newell, Philip, & Holland, Keith (2006). *Loudspeakers: For Music Recording and Reproduction*. Routledge, Abingdon, UK

If loudspeakers are the most important part of any system, it behooves us as system designers to understand speakers as well as we can! This work cites a ton of research and presents a really detailed examination of the design of loudspeakers.

Porter, Tom (2004). *Archispeak: An Illustrated Guide to Architectural Terms*. Routledge, Abingdon, UK

Anyone getting into the architectural design world would be well served to read this book. Architecture has a lexicon that is very rich and deep, and a good visual guide is an indispensable aid when working in this area.

Roginska, Agnieszka & Geluso, Paul (editors, 2018). *Immersive Sound: The Art and Science of Binaural and Multi-Channel Audio*. Routledge, Abingdon, UK

A wonderful survey of the world of immersive audio. While some of the book deals with binaural in headphones, which may not be terribly relevant to sound system designers, it then goes into binaural production via speakers, and further multi-channel developments. Relevant for theatrical design, art installations, cinema, and any project that needs to immerse the audience.

Strong, Judith (editor, 2010). *Theatre Buildings: A Design Guide*. Routledge, Abingdon, UK

A great resource for those seeking to better understand the design of performing arts spaces and theaters in particular.

Talbot-Smith, Michael (editor, 2nd edition, 2001). *Audio Engineer's Reference Book*. Focal Press, Waltham, MA, USA

I've never met anyone who has read this cover to cover, but that's not what it's designed for (have you ever read a dictionary start to finish?). A true audio reference manual, and a great tool to have in the arsenal.

Toole, Floyd E. (3rd edition, 2017). *Sound Reproduction: The Acoustics and Psychoacoustics of Loudspeakers and Rooms*. Focal Press, Waltham, MA, USA

Again, a very valuable resource on loudspeakers, and particularly on how we perceive them in acoustic environments. Well written, clear, and useful.

Other Publications

Bob McCarthy's Blog: www.bobmccarthy.com

Bob writes frequently about new work and ideas he's contemplating. Worth a read.

National Association of the Deaf: www.nad.org/resources/technology/assistive-listening/assistive-listening-systems-and-devices/

Good information about Assistive Listening Systems and Devices.

Rational Acoustics (2018). *SMAART® v8 User Guide*. Rational Acoustics

Yes, this is a user manual for a software, but it's also a fine guide to the process of system measurement and optimization itself. I'd recommend starting here before diving into Bob McCarthy's book, especially if you have a SMAART® license and can work along.

United States Access Board: www.nps.gov/dscw/upload/AssistiveListeningSystemsTechBulletin_Aug2003.pdf

Good white paper on Assistive Listening, ADA compliance, etc.

White Papers

Check out any technical articles by the following companies. They produce a ton of them, and they are all fascinating.

Ahnert-Feistel Media Group

Makers of EASE®, acoustics and speaker experts.

d&b audiotechnik

Speaker designers/manufacturers who have a particular interest in many topics, including array theory and practice.

Meyer Sound

Another fine designer and manufacturer of speakers. Lots of interesting research.

Walters-Storyk Design Group

Preeminent studio designers, acousticians, and venue gurus. They conduct cutting-edge research when they are between jobs. Always worth a read.

Note: Page numbers in *italics* indicate a figure on the corresponding page.

579